Protein Tyrosine Phosphatases in Cancer

Benjamin G. Neel • Nicholas K. Tonks
Editors

Protein Tyrosine Phosphatases in Cancer

Springer

Editors
Benjamin G. Neel
Laura and Isaac Perlmutter Cancer Center
New York, NY, USA

Nicholas K. Tonks
Cold Spring Harbor Laboratory
Cold Spring Harbor, NY, USA

ISBN 978-1-4939-3647-2 ISBN 978-1-4939-3649-6 (eBook)
DOI 10.1007/978-1-4939-3649-6

Library of Congress Control Number: 2016948122

Printed on acid-free paper

This Springer imprint is published by Springer Nature
The registered company is Springer Science+Business Media New York

Preface

A crucial aspect of signal transduction, the mechanism by which cells respond to environmental cues, is the organization of coordinated networks of protein–protein interactions, a process mediated by protein phosphorylation. Disruption of the normal patterns of phosphorylation results in aberrant regulation of signal transduction and has been implicated in the etiology of a variety of major human diseases, including cancer. The ability to modulate signaling pathways selectively holds enormous therapeutic potential. Although inhibitors of protein tyrosine kinases (PTKs) have yielded some spectacular success stories, such as Gleevec/Imatinib and Herceptin/Trastuzumab, challenges remain. In particular, although initially there may be a dramatic response, patients soon develop resistance to the therapy. Therefore, despite the obvious potential of targeting PTKs, particularly in the context of developments in personalized medicine, it is anticipated that alternative therapies, to target simultaneously different signaling enzymes and events, would be more effective than targeting PTKs alone.

The focus on PTKs for drug development ignores the other major component of phosphorylation-dependent regulation of signaling. Protein phosphorylation is a reversible process, in which the coordinated and competing activities of kinases and phosphatases are important for determining signaling outcome. One of the challenges facing those working on the protein tyrosine phosphatases (PTPs), which function in combination with the PTKs, has been to overcome the prejudice in the field that sees these enzymes dismissed as a family of constitutively active, nonspecific housekeeping enzymes. As will become evident upon reading the reviews assembled in this book, this view of the PTPs is totally without foundation. By contrast, PTPs function as specific regulators of tyrosine phosphorylation-dependent signaling pathways. Furthermore, direct links have been established between the disruption of PTP function and the etiology of major diseases, including metabolic diseases and cancer. As such, the PTPs have been validated as potential therapeutic targets and offer a complementary perspective to the protein kinases through which to develop novel strategies to treat major diseases, including cancer.

In this book, we present a collection of reviews that describe recent developments in our understanding of the function of members of the PTP family, with a particular focus on their roles in cancer and their potential to form the basis for therapeutic strategies to address this and other major diseases. Our goal is to illustrate that the regulation of signal transduction depends upon integrating the function of PTKs and PTPs, with *both* classes of enzymes serving crucial roles. Contrary to the view of the PTPs as housekeeping enzymes, their structural diversity, their specificity, their ability to function both positively and negatively to regulate signaling, and their links to the etiology of major human diseases, including cancer, all attest to a fundamental role for the PTPs. As further examples of the crucial role of PTPs in the regulation of signaling under normal and pathophysiological conditions are established in the coming years, we hope that these enzymes will be exploited fully for therapeutic development.

New York, NY, USA B.G. Neel
Cold Spring Harbor, NY, USA N.K. Tonks Toronto

Contents

Time to Shine the Spotlight on the Protein Tyrosine Phosphatase Family of Signal Transducing Enzymes

Nicholas K. Tonks

Abstract Although these are exciting times to be working on the protein tyrosine phosphatase (PTP) family of enzymes, they are also challenging times. The detailed characterization of many of the PTPs has revealed fundamentally important roles for these enzymes in the control of cell function. Furthermore, direct links have been established between disruption of PTP function and the etiology of major diseases, including metabolic diseases and cancer. As such, the PTPs have been validated as therapeutic targets and offer a complementary perspective to the protein kinases through which to develop novel strategies to treat major diseases, including cancer. This book presents a collection of reviews that describe recent developments in our understanding of the function of members of the PTP family, with a particular focus on their roles in cancer and their potential to form the basis for therapeutic strategies to address this and other major diseases.

Keywords Signal transduction • Protein tyrosine phosphatase • Protein tyrosine kinase • Drug discovery • Oncogene • Tumor suppressor • Cancer • Diabetes and obesity • Cell adhesion • Redox

The reversible addition and removal of phosphate to proteins, which is termed protein phosphorylation, is the central feature of the mechanism of signal transduction—the process by which cells respond to stimuli in their environment. The activities of the enzymes that mediate the addition (kinases) and removal (phosphatases) of phosphate groups are coordinated in signal transduction pathways to mediate the cellular response to environmental stimuli and the function of these enzymes is frequently disrupted in major human diseases, including cancer. The ability to modulate such signal transduction pathways selectively with drugs holds enormous

N.K. Tonks (✉)
Cold Spring Harbor Laboratory, Cold Spring Harbor, New York, NY 11724, USA
e-mail: tonks@cshl.edu

B.G. Neel, N. Tonks (eds.), *Protein Tyrosine Phosphatases in Cancer*,
DOI 10.1007/978-1-4939-3649-6_1

therapeutic potential. Drugs that target the protein kinases represent breakthroughs in cancer therapy. For example, the HER2 protein tyrosine kinase is amplified and/or overexpressed in several cancers, in particular in ~25 % of breast cancer, where it is associated with poor prognosis. The humanized HER2-directed antibody, Herceptin (Trastuzumab), is an example of a "rational cancer therapy" for treatment of HER2-positive metastatic breast tumors. It targets HER2 as a unique marker of the cancer cell. Although Herceptin is used frequently and is presented as a frontline treatment of choice, the overall success rate is low and patients develop resistance to the therapy. This problem of resistance, both de novo and acquired, has become an obstacle to the successful application of kinase-directed therapies in general. Therefore, despite the obvious potential, it is anticipated that new alternative therapies, administered alone or in combination with kinase-directed drugs, will represent the way forward. The challenge is to identify such alternative therapies. In light of the intimate cooperation between kinases and phosphatases in the regulation of signal transduction under normal and pathophysiological conditions, it would seem prudent also to consider the protein phosphatases in this context. The protein tyrosine phosphatases (PTPs), in particular, have been garnering attention as potential therapeutic targets; nevertheless, they remain a largely untapped resource for drug development. The purpose of this book is to present a collection of reviews that describe recent developments in our understanding of the function of members of the PTP family, with a particular focus on their roles in cancer and their potential to form the basis for therapeutic strategies to address this and other major diseases.

A misconception that still pervades the field is the view of PTPs as a family of constitutively active, broad specificity, housekeeping enzymes. Furthermore, the dismissal of phosphatases as "erasers" in systems biology models of signal transduction only serves to reinforce such prejudice. In contrast, myriad studies have established that structural and functional diversity is a hallmark of the PTP family that matches such diversity within the protein tyrosine kinases. In fact, the function of PTKs and PTPs are integrated such that both classes of enzymes are critical regulators of signal transduction. Furthermore, disruption of both classes of enzymes has been shown to underlie major human diseases, including cancer.

The PTP family comprises 107 genes in humans [1]; this structural diversity alone points to functional specificity beyond a general housekeeping role. Of these, 37 encode "classical" pTyr-specific enzymes (Fig. 1). Within these classical PTP genes, 21 encode transmembrane, receptor-like proteins (RPTPs), the structures of which suggest the potential for modulation of signaling events directly through ligand-regulated protein tyrosine *dephosphorylation*. Again this is indicative of a direct role in signaling, beyond a housekeeping function. Similar to the PTKs, the diversity in the extracellular segments of the RPTPs presumably reflects an equivalent diversity in the nature of the ligands to which they respond. As discussed in the chapter by Andrew Stoker, however, the identity and function of such ligands remain a largely unresolved issue in the field.

Initial studies of the prototypic receptor PTP, CD45, revealed another important facet of the function of members of the PTP family—their ability to function positively as well as negatively in the regulation of cell signaling. This reinforces further

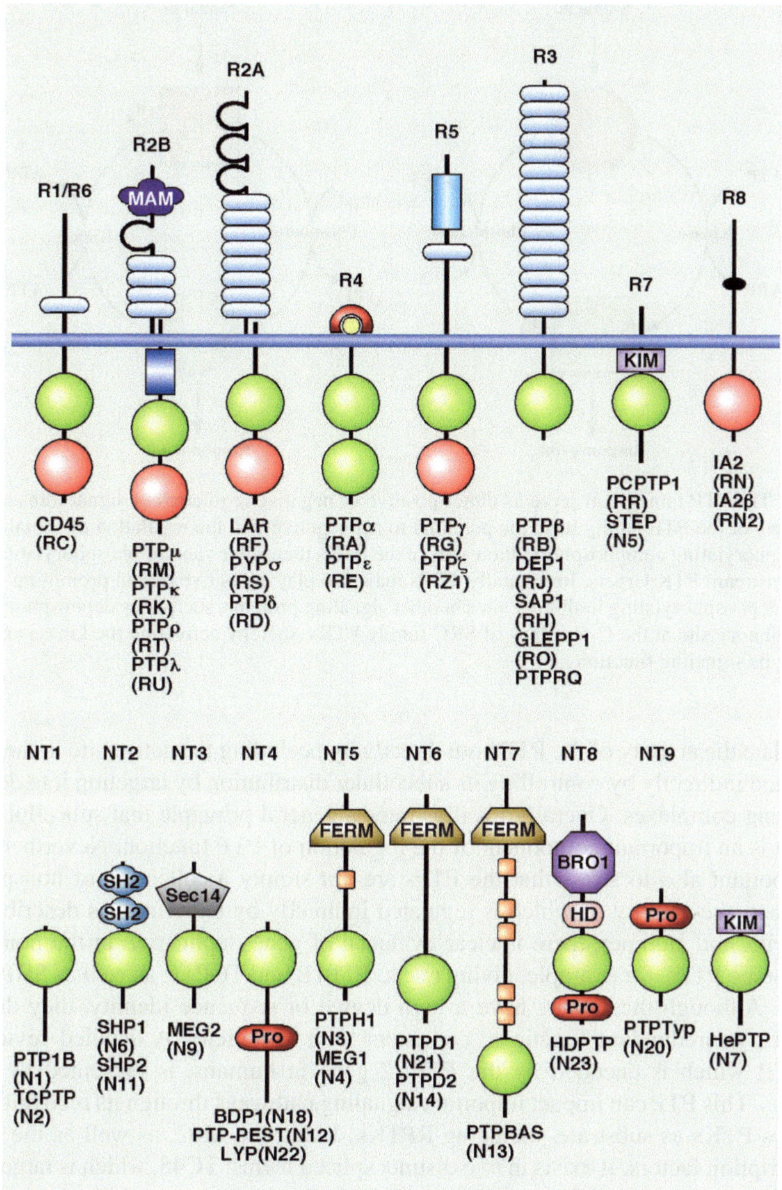

Fig. 1 The classical, pTyr-specific members of the PTP family

the concept that PTPs may play an important direct role in switching on signaling pathways, rather than simply acting as passive antagonists of PTK function (Fig. 2).

The remaining 16 classical PTPs are nontransmembrane, cytoplasmic enzymes, in which the catalytic domain is flanked by sequences that may serve a regulatory function (Fig. 1). This includes SH2 domains, which, as described by Chan and Neel, can

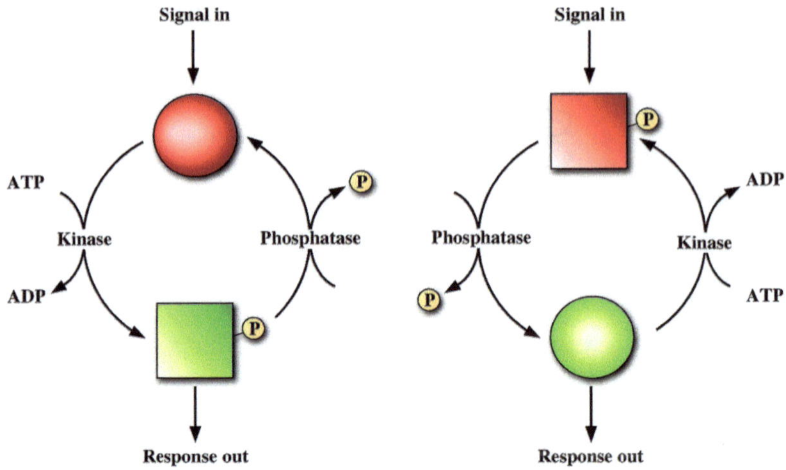

Fig. 2 The PTP family can serve as either positive or negative regulators of signal transduction. Members of the PTP family have the potential to act negatively in the regulation of signaling, by dephosphorylating autophosphorylation sites in the PTKs themselves and/or phosphorylation sites in downstream PTK targets. Importantly, PTPs may also play a positive role in promoting signaling by dephosphorylating inhibitory sites in other signaling proteins, such as by dephosphorylating the inhibitory site at the C-terminus of SRC family PTKs, thereby activating the kinase and promoting its signaling function

modulate the activity of the PTP both directly by occluding the active site in the basal state and indirectly by controlling its subcellular distribution by targeting it to defined signaling complexes. Overall, this illustrates a general principle that subcellular targeting is an important component of the regulation of PTP function. Nevertheless, it is important also to stress that the PTPs are *not* simply a collection of nonspecific enzymes, the activity of which is regulated indirectly by tethering. As described by Hendriks and Bohmer, there is clear evidence of gene duplication in the nontransmembrane PTPs, for example, giving rise to PTP1B and TCPTP, as well as SHP1 and SHP2. Although these pairs have a high degree of sequence identity, they display distinct, nonredundant functions, consistent with specificity. A detailed review of TCPTP, which is encoded by the *PTPN2* gene in humans, is presented by Tony Tiganis. This PTP can impact important signaling pathways through its recognition of various PTKs as substrate, including RPTKs, JAKs and SRC, as well as the STAT transcription factors. It exists in two distinct spliced forms: TC48, which is targeted to the cytoplasmic face of ER membranes, and TC45, which has the ability to shuttle in and out of the nucleus. These isoforms, which share the same catalytic domain, nonetheless have the ability to act on distinct substrates, or to act on common substrates in a spatially and temporally distinct manner. This review focuses on the identification of TCPTP as a tumor suppressor in T cell acute lymphoblastic leukemias, as well as its potential tumor suppressive role in breast cancer, and discusses the intriguing possibility that alterations in the distribution of TCPTP between its two spliced isoforms may have an impact in cancer.

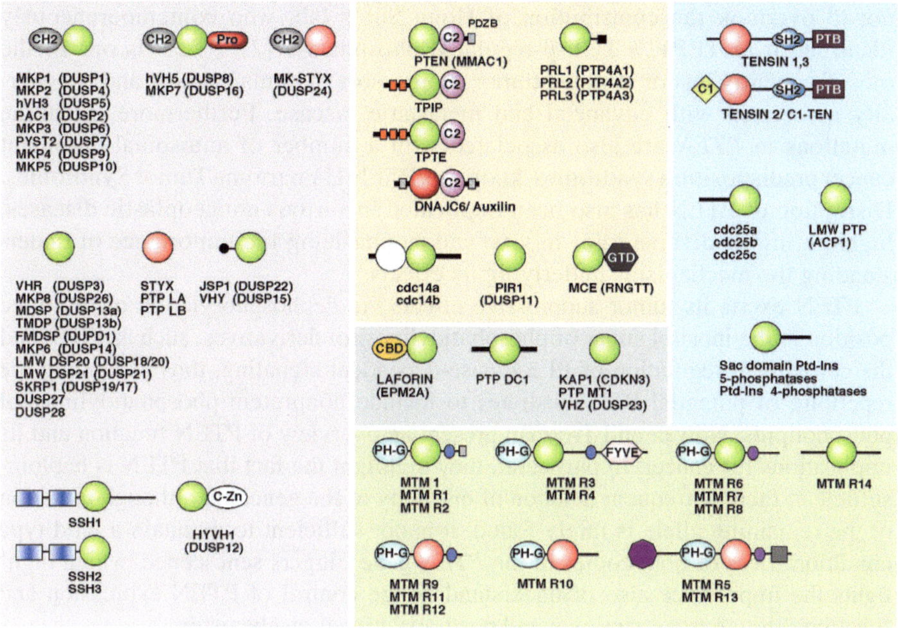

Fig. 3 The dual specificity phosphatases (DUSPs)

The largest portion of the PTP family is classified under the global heading of "dual specificity phosphatases," also known as DUSPs (Fig. 3). These enzymes have the capacity to dephosphorylate Ser/Thr residues in proteins, as well as pTyr proteins; in addition, as discussed below, they can control signal transduction via the dephosphorylation of nonprotein substrates, such as inositol phospholipids. Many of the DUSPs have been implicated in the control of fundamentally important physiological processes, including the cell cycle and cytoskeletal function. Kidger and Keyse review the role of MAP kinases (ERK, JNK, and p38) in cancer, and focus on those DUSPs that serve as MAP Kinase Phosphatases (MKPs). They review how the function of the MKPs is controlled at the level of expression, the stimuli that lead to their induction, their subcellular location (particularly nuclear vs cytoplasmic distribution), and their intrinsic specificity for particular MAP kinases. In addition, they describe how abnormal regulation of MKP function is encountered in a wide variety of tumors. In short, the complex roles of the MAP kinases in cancer are matched by the complexity in function of the DUSP MKPs that control them.

Considering the established role of PTKs as drivers of tumorigenesis, it was thought originally that the PTPs would serve predominantly as tumor suppressors. Nevertheless, the identification and characterization of the first PTP tumor suppressor took a long time, and when it was published it included some surprises. In 1997, three groups identified independently the tumor suppressor from the chromosome 10q23 locus. Although this discovery is primarily associated with the work of Wigler/Parsons (PTEN) and Steck (MMAC1), it is important not to overlook the contribution of Hong Sun's lab, who contemporaneously identified it as TEP1, a TGF-β-regulated phosphatase [2]. PTEN is one of the most frequently lost or mutated tumor suppressors in human cancer and is generally associated with advanced and metastatic disease. Furthermore, germline mutations in *PTEN* are also associated with a number of autosomal dominant cancer predisposition syndromes, known as PTEN Hamartoma Tumor Syndromes. Disruption of PTEN has also been implicated in various nonneoplastic diseases, highlighting its diverse roles in vivo and emphasizing the importance of understanding the mechanisms underlying its effects.

PTEN exerts its tumor suppressive effects via dephosphorylation of the three position in the inositol sugar of phosphatidylinositol derivatives, such as PIP3, and the consequent regulation of PI 3-kinase-dependent signaling, thus expanding the repertoire of potential PTP substrates to include nonprotein phosphatidylinositol phospholipids. Nowak and Trotman present an overview of PTEN function and its implications for cancer. In particular, they highlight the fact that PTEN is haploinsufficient; there is frequent deletion of one copy of the gene, but although mutation of the remaining allele is rarely found, it is not sufficient to maintain a wild type condition. Loss of both copies of the *PTEN* gene triggers senescence, which highlights the importance also of understanding the control of PTEN expression and function through transcriptional and posttranslational mechanisms.

PTEN is a complex protein that features the PTP domain in its N-terminal segment; this forms an extensive interface with a C2 domain, which serves to bind phospholipid membranes. The C-terminus of the protein is characterized by the presence of multiple phosphorylation sites and a binding motif for PDZ domain-containing regulatory proteins. PTEN has an unusual architecture to its active site, which is sufficiently large to accommodate the sugar head-group of inositol phospholipids as a substrate; however, it displays intrinsic phosphatase activity towards both protein and lipid substrates. Furthermore, tumor-derived mutations are not restricted to the catalytic core of PTEN, but have been identified throughout the coding sequence. This suggests that there may also be phosphatase-independent functions of PTEN, a topic that is reviewed by Papa and Pandolfi. In particular, they highlight mechanisms for nuclear import of PTEN and discuss a scaffolding role of the protein in the control of chromosome stability and DNA repair. The authors also highlight examples of the covalent modification of PTEN, which includes not only phosphorylation, but also ubiquitylation and sumoylation. As described also by others, such covalent modifications are turning about to be an important feature of the mechanisms by which PTP function is regulated in vivo.

Subsequent to the identification of PTEN, a large body of evidence now supports tumor suppressive roles for other members of the PTP family, including the classical pTyr-specific enzymes. Andrew Stoker focuses on the receptor-like PTPs in this regard. Interestingly, RPTPs are subject to a range of mutations in a wide variety of cancers, including in their extracellular segments, which highlights the potential importance of ligand binding for the normal regulation of their function. Furthermore, there are RPTPs that serve potential oncogenic functions. For example, expression of RPTPα alone was sufficient to transform fibroblasts, with the mechanism involving activation of SRC family kinases through dephosphorylation of an inhibitory C-terminal site of phosphorylation. This activation of SRC family kinases turns out to be a recurring theme as a mechanism by which PTPs may exert a positive effect on signaling. Similarly, as reviewed by Hendriks and Bohmer, roles for the nontransmembrane PTPs are also being revealed in various cancers; however, many of these enzymes remain to be extensively characterized, illustrating the need for further study in the future.

Although the study of tumor suppressor PTPs may suggest new therapeutic targets from the signaling pathways that these enzymes regulate, there are now a growing number of oncogenic PTPs that themselves may prove to be therapeutic targets. Chan and Neel present a detailed examination of SHP2, which serves as a positive regulator for full activation of RAS/MAP kinase signaling downstream of multiple growth factors and cytokines. In its basal state, SHP2 exists in a low activity conformation in which the active site is occluded by an intramolecular interaction with the N-terminal SH2 domain. Under normal circumstances, following stimulus the SH2 domains are engaged by appropriate pTyr ligands as SHP2 is recruited into a signaling complex, which induces a conformational change that releases the autoinhibitory interaction and generates an active form of the phosphatase with an open catalytic site. Now, gain-of-function mutations in SHP2 have been identified in human disease, initially in Noonan syndrome. This includes mutations in residues in and around the N-SH2 domain, which may facilitate activation by pTyr ligands, and in key residues at the interface between the N-SH2 and catalytic domains, which would induce the active conformation in the absence of a stimulus. Activating mutations in SHP2 are also associated with increased risk of malignancies, such as juvenile myelomonocytic leukemia and acute myeloid leukemia. As a result, *PTPN11*, which encodes SHP2, was recognized as the first PTP oncogene. More recently, a tumor suppressor function for *PTPN11* in metachondromatosis, which is associated with the development of cartilage and bone tumors, has been identified and inactivating mutations in the enzyme have been characterized. This illustrates how a single member of thc PTP family can display major differences in function depending upon signaling context—a situation far removed from a housekeeping role. Although SHP2 plays a fundamentally important role in the regulation of growth factor and cytokine signaling, further studies will be required to define the critical substrates of the enzyme and thereby the mechanism by which it exerts its effects.

Members of the PTP family that play a positive, oncogenic role in signaling are also found in the DUSPs. Gabrielli and Burgess review the example of cdc25, which is a critical regulator of cell cycle progression and checkpoint function. Cdc25 catalyzes the dephosphorylation and activation of the kinase CDK in cyclin-CDK complexes. In higher eukaryotes, there are three forms of the enzyme, cdc25A, B, and C, which act at different stages of the cell cycle. Cdc25A is upregulated in various cancers, and cdc25A and B play important roles in checkpoints, such as in response to DNA damage. Interestingly, cdc25 is another member of the PTP family for which posttranslational modifications play a key regulatory role. Phosphorylation of cdc25, such as in response to Chk1, generates binding sites for 14-3-3, which controls the activity and subcellular distribution of the phosphatase. More recently, cdc25 has also been implicated in functions other than control of cell cycle, such as regulation of apoptosis, suggesting that future studies, particularly aimed at the generation of isoform-specific inhibitors of the enzyme, may make important contributions to exploiting multiple aspects of the cancer phenotype.

The PRLs (Phosphatase of Regenerating Liver) are another interesting example from the DUSP category. Rogers and Pallen review this intriguing group of PTPs, which are the only farnesylated members of the PTP family, with prenylation being important for membrane localization. The PRLs are also unusual in that the Ser/Thr residue following the invariant Arg in the signature motif (H-C-$(X)_5$-R-[S/T]) is replaced by Ala. This hydroxyl residue plays an important role in the second step of catalysis for the active classical PTPs, suggesting that the intrinsic activity of PRLs may be low. PRL-1 and PRL-2 have been shown to behave as oncogenes when overexpressed, and expression of PRL-3 is dramatically upregulated in metastatic colorectal cancer. As discussed by Rogers and Pallen, the lack of identification of the physiological substrates for the PRLs remains a major obstacle to understanding their function. Current studies have focused on the characterization of PRL-interactomes and the development of mouse models to define more precisely their role in normal and pathophysiological growth and development.

PTP1B, the prototypic member of the family, is a further example of a PTP that can be either a positive or a negative regulator of signaling, depending upon context. PTP1B plays a well-established role in downregulating signaling in response to insulin and leptin and is a highly validated therapeutic target for diabetes and obesity. PTP1B-null mice are healthy, display enhanced insulin sensitivity, do not develop type 2 diabetes, and are resistant to obesity when fed a high fat diet. In fact, excitement about this function of PTP1B initially fuelled the interest in PTPs in general as therapeutic targets; however, it is important also to note that PTP1B function is not restricted to metabolic regulation. The *PTPN1* gene, which encodes PTP1B, is located at chromosome 20q13, which is a region that is frequently amplified in breast cancer and associated with poor prognosis. It has also been reported that PTP1B is overexpressed in breast tumors together with the HER2 oncoprotein tyrosine kinase. Mice expressing activated alleles of HER2 in mammary glands develop multiple mammary tumors and frequent metastases to the lung; however, when such mice were crossed with *PTP1B*-knockout mice, tumor development was delayed and the incidence of lung metastases was decreased. Conversely, targeted overexpression of PTP1B alone was sufficient to drive mammary tumorigenesis in mice. These observations illustrate that PTP1B not only plays a role in attenuating insulin and leptin signaling, but also plays a positive role in promoting signaling events associated with breast tumorigenesis, positioning the enzyme with the potential to play an important role in the interplay between obesity and cancer. In separate reviews, Bence, and Labbe and Tremblay, address this issue of metabolic disease and its possible role in predisposition to cancer, together with the potential importance of PTPs, such as PTP1B, in this process.

Finally, as we obtain a more detailed picture of the physiological functions of members of the PTP family, and how those functions are disrupted in disease, the importance of exploiting these enzymes therapeutically becomes more apparent.

Although this is a hugely exciting development for the field, it is not without its challenges. He and Zhang review the current status of PTP-based therapeutics, focusing on PTP1B and SHP2 from the classical PTPs, as well as the DUSPs cdc25 and PRL. In following established approaches to developing small molecule drugs that bind to the active site of an enzyme, industry has found PTPs to be challenging targets for therapeutic development. For example, although it was possible to generate potent, specific, and reversible inhibitors of PTP1B, such molecules were highly charged, due to the chemistry underlying PTP-mediated catalysis, and lacked oral bioavailability. Consequently, it was decided that they were of limited drug development potential; in fact, a prevailing view in industry is that PTPs may be "undruggable." Nevertheless, this may prove to be a huge opportunity for scientists in the academic sector to develop creative new approaches to inhibiting PTP function with small molecule drug candidates and to reinvigorate drug discovery efforts. With this in mind, are there any lessons that can be learned from the experience gained in the generation of protein kinase-directed therapeutics?

Initially, there were also reservations about whether or not protein kinases could be targeted therapeutically. Despite that, the global market for kinase inhibitors reached $28 billion in 2010, with a projected market of $40 billion in 2016! Perhaps more importantly, kinase inhibitors have opened up new approaches to treating major diseases, including cancer. Nevertheless, the issue of de novo and acquired resistance to such kinase inhibitors remains a challenge. The opportunity to address this challenge through targeting the phosphatases as the complementary family of enzymes to the kinases in the regulation of signal transduction has obvious attractions. A major obstacle to the acceptance of PTPs as therapeutic targets has been the issue of specificity—both the specificity of the enzymes themselves and the specificity of the inhibitors that target them. But to what extent is this really a problem? On the one hand, it is becoming clear that individual protein kinases may also recognize a broad array of substrates—for example, ATM, which regulates the response to genotoxic stress, recognizes hundreds of targets [3]. On the other hand, the extent to which PTPs are actually promiscuous is far from clear. In fact, recent studies illustrate that the suppression of individual PTPs by RNA-interference is accompanied by specific changes in signaling and by extremely subtle changes in tyrosine phosphorylation patterns [4].

One concern about targeting PTP1B therapeutically has always been the extent to which inhibitors of the enzyme would also inhibit TCPTP, its closest relative in the PTP family. Although targeted ablation of the *PTPN1* gene suggested beneficial effects of inhibiting PTP1B in diabetes and obesity, homozygous deletion of *PTPN2* illustrated that the global loss of TCPTP from birth produced lethal defects in the immune system. In contrast, mice that were subjected to heterozygous loss of TCPTP appeared normal [5]. Furthermore, as discussed by Tony Tiganis, it has been shown that partial inhibition of TCPTP, such as may be achieved with a small molecule inhibitor, may actually enhance further the beneficial effects of inhibiting PTP1B on insulin and leptin signaling [6, 7]. Similar objections were raised initially to the pursuit of kinase inhibitors, with the view that the sequence similarity between kinases would be an obstacle to generating specific inhibitors. Of course, now we know that not only is it possible to generate effective kinase inhibitors, but also the vast majority also target the ATP-binding site, the feature that all kinases share in common! Furthermore, the extent to which such absolute specificity is required is also open to question. Some of the kinase inhibitors that are the most effective clinical candidates actually target multiple kinases. As an extreme example, Sunitinib, which is an FDA-approved drug for renal cell carcinoma, targets >30 % of the kinome [8, 9]! In fact, a new term, polypharmacology, has been introduced to describe the study of single drugs that act on multiple targets in a disease pathway. Such precedents, together with the extensive validation of PTPs as therapeutic targets, must surely suggest that it would be imprudent simply to dismiss the PTP family as "undruggable" without further investigation.

In recognizing concerns about specificity, one strategy that would avoid targeting the PTP active site would be to look for allosteric inhibitors, which bind at a site remote from the catalytic center, but induce conformational changes in the enzyme that result in inhibition. In my lab, we have characterized a small molecule natural product, TRODUSQUEMINE, as such an allosteric inhibitor of PTP1B [10]. This is a unique mechanism of inhibition that will have been missed in the efforts to target PTP1B that have been conducted to date in industry. We have demonstrated that this inhibitor antagonizes the function of HER2 in cell and animal models of breast cancer. In particular, it essentially abrogates metastasis of HER2-positive tumor cells in the NDL2 mouse model of breast cancer. Now, in a collaboration with the Monter Cancer Center at Northwell Health, TRODUSQUEMINE/MSI-1436 is being taken into a Phase 1 clinical trial for HER2-positive cancer. Similar efforts are being made in other labs to target SHP2 with allosteric inhibitors also for cancer. This approach targets the least conserved portions of individual PTPs and so offers an excellent opportunity to build specificity into any resulting inhibitors.

In several of the following chapters, the tight control over the activity of the PTPs in vivo by posttranslational modifications is discussed. As highlighted by Kendra Bence, several labs have established the importance of reversible oxidation and inactivation of PTPs as a regulatory mechanism. The architecture of the PTP active site is such that the essential cysteinyl residue, which is a defining feature of members of the PTP family, displays unique properties that favor its role as a nucleophile in catalysis, but also render it prone to oxidation. This has now been identified as a normal physiological regulatory mechanism that represents a new tier of control over tyrosine phosphorylation-dependent signaling pathways. For example, insulin stimulation of mammalian cells leads to enhanced production of intracellular H_2O_2, which causes reversible oxidation of PTP1B and inhibition of its enzymatic activity, which in turn concomitantly promotes the signaling response to insulin. Therefore, if PTP1B could be sequestered in this insulin-induced, oxidized, inactive state by a small molecule drug, the result would be an enhanced insulin signaling response. Furthermore, this would have the advantage of focusing attention specifically on that pool of PTP1B that is responsible for antagonizing insulin signaling, the pool that the cell itself targets for oxidation so as to fine-tune the signaling response. Consequently, one would anticipate that this strategy would also reduce the potential for side effects that may accompany more system-wide inhibition of PTP1B function.

The conformational changes that are the structural consequences of mild oxidation of PTP1B, such as occurs in response to insulin, have been defined by X-ray crystallography [11]. We hypothesized that a conformation-sensor antibody that recognizes the structural features of the reversibly oxidized form of PTP1B selectively may stabilize the inactive state, inhibit its reactivation by reducing agent and thereby inhibit phosphatase activity. Using antibody phage display we have generated such conformation-sensor antibodies and discovered that expression of these antibodies in cells led to enhanced and sustained insulin-induced signal transduction [12]. Furthermore, these antibodies are able to inhibit PTP1B function specifically without an apparent effect on TCPTP function. These data provide proof-of-concept that stabilization of the oxidized, inactive form of PTP1B with appropriate therapeutic molecules may offer a novel paradigm for phosphatase drug development. They also illustrate how an understanding of PTP biology at a basic science level in an academic setting may lead to new approaches to drug discovery in this exciting area.

Overall, the deep insights into the structure, regulation, and function that have developed over the last 25 years and are highlighted in this book have validated the PTPs as potential therapeutic targets and as the basis for new approaches to drug development. New examples of the importance of PTPs in the etiology of disease, as well as in potential treatments, continue to be revealed. Recent studies of the importance of receptor PTPs in axonal regrowth after injury [13] and the potential involvement of PTPs in autism spectrum disorders [14] are a case in point. In addition, PTP inhibitors may offer approaches alone, or in combination with kinase inhibitors, to address the problem of resistance. Hopefully, continued efforts to exploit these important, but challenging, targets will ultimately benefit patients through discovery of new therapies for major human diseases.

Acknowledgments Research in the author's lab was supported by NIH grants CA53840 and GM55989, and the CSHL Cancer Centre Support Grant CA45508. The author is also very grateful for support from the following foundations: The Gladowsky Breast Cancer Foundation, The Don Monti Memorial Research Foundation, Hansen Memorial Foundation, West Islip Breast Cancer Coalition for Long Island, Glen Cove CARES, Find a Cure Today (FACT), Constance Silveri, Robertson Research Fund, and the Masthead Cove Yacht Club Carol Marcincuk Fund.

References

1. Tonks NK. Protein tyrosine phosphatases: from housekeeping enzymes to master-regulators of signal transduction. FEBS J. 2013;280:346–78.
2. Li D, Sun H. TEP1, encoded by a candidate tumor suppressor locus, is a novel protein tyrosine phosphatase regulated by transforming growth factor β. Cancer Res. 1997;57:212402129.
3. Shiloh Y, Ziv Y. The ATM protein kinase: regulating the cellular response to genotoxic stress, and more. Nat Rev Mol Cell Biol. 2013;14:197–210.
4. St. Germain JR, Taylor P, Zhang W, Li Z, Ketela T, Moffat J, Neel BG, Trudel S, Moran MF. Differential regulation of FGFR3 by PTPN1 and PTPN2. Proteomics. 2015;15:419–33.
5. You-Ten KE, Muise ES, Itie A, Michaliszyn E, Wagner J, Jothy S, Lapp WS, Tremblay ML. Impaired bone marrow microenvironment and immune function in T cell protein tyrosine phosphatase-deficient mice. J Exp Med. 1997;186:683–93.

6. Galic S, Klingler-Hoffmann M, Fodero-Tavoletti MT, Puryer MA, Tonks NK, Tiganis T. Regulation of insulin receptor signaling by the protein tyrosine phosphatase TCPTP. Mol Cell Biol. 2003;23:2096–108.
7. Dodd GT, Decherf S, Loh K, Simonds SE, Wiede F, Balland E, Merry TL, Münzberg H, Zhang ZY, Kahn BB, Neel BG, Bence KK, Andrews ZB, Cowley MA, Tiganis T. Leptin and insulin act on POMC neurons to promote the browning of white fat. Cell. 2015;160:88–104.
8. Karaman MW, Herrgard S, Treiber DK, Gallant P, Atteridge CE, Campbell BT, Chan KW, Ciceri P, Davis MI, Edeen PT, Faraoni R, Floyd M, Hunt JP, Lockhart DJ, Milanov ZV, Morrison MJ, Pallares G, Patel HK, Pritchard S, Wodicka LM, Zarrinkar PP. A quantitative analysis of kinase inhibitor selectivity. Nat Biotechnol. 2008;26:127–32.
9. Davis MI, Hunt JP, Herrgard S, Ciceri P, Wodicka LM, Pallares G, Hocker M, Treiber DK, Zarrinkar PP. Comprehensive analysis of kinase inhibitor selectivity. Nat Biotechnol. 2011;29:1046–51.
10. Krishnan N, Koveal D, Miller DH, Xue B, Akshinthala SD, Kragelj J, Jensen MR, Gauss C-M, Page R, Blackledge M, Muthuswamy SK, Peti W, Tonks NK. Targeting the disordered C-terminus of PTP1B with an allosteric inhibitor. Nat Chem Biol. 2014;10:558–66.
11. Salmeen A, Andersen JN, Myers MP, Meng TC, Hinks JA, Tonks NK, Barford D. Redox regulation of protein tyrosine phosphatase 1B involves a novel sulfenyl-amide intermediate. Nature. 2003;423:769–73.
12. Haque A, Andersen JA, Salmeen A, Barford D, Tonks NK. Conformation-sensing antibodies stabilize the oxidized form of PTP1B and inhibit its phosphatase activity. Cell. 2011;147: 185–98.
13. Lang BT, Cregg JM, DePaul MA, Tran AP, Xu K, Dyck SM, Madalena KM, Brown BP, Weng YL, Li S, Karimi-Abdolrezaee S, Busch SA, Shen Y, Silver J. Modulation of the proteoglycan receptor PTPσ promotes recovery after spinal cord injury. Nature. 2015;518:404–8.
14. Krishnan N, Krishnan K, Connors CR, Choy MS, Page R, Peti W, Van Aelst L, Shea SD, Tonks NK. PTP1B inhibition suggests a therapeutic strategy for Rett syndrome. J Clin Invest. 2015;125:3163–77.

RPTPs and Cancer

Andrew W. Stoker

Abstract In the human phosphatome, there are 21 tyrosine-specific receptor-like PTPs (RPTPs), falling into subtypes with distinct ectodomains. In this chapter, we will present a broad overview from an RPTP-centric standpoint, describing what is known about the implied or actual role of each RPTP in cancers and cell transformation, and their roles as putative tumour suppressor or promoter of tumour growth and survival. It must be made clear that many of the studies discussed have included functional studies of the RPTPs, so for most RPTPs their roles in cancer remain unproven. The gene family is evolutionarily conserved and the cell adhesion molecule-like form of many RPTPs has historically focussed attention of their roles in cell adhesion and motility [1, 2]. The initial excitement about RPTPs as being the key foils to the oncogene kinases did not bear much fruit initially, but 20 years on they are now firmly part of the jigsaw of cancer cell signalling. In 2005, MacKeigan and co-workers completed an ambitious project to test kinases and phosphatases and define "cell survival enzymes" and "cell death enzymes". Surprisingly, half of the RPTPs were identified as one or the other [3]. Table 1 provides an overview of these and other RPTPs, although we must apologize that not all the documented evidence can be discussed in this fast moving field. Readers are encouraged to see further discussions in recent reviews {Julien 2011 #10086; Ostman 2006 #7679; Navis 2010 #8562}. PTPRR will not be discussed as it is covered in the chapter *N*.

In the human phosphatome there are 21 tyrosine-specific receptor-like PTPs (RPTPs), falling into subtypes with distinct ectodomains. In this chapter, we will present a broad overview from an RPTP-centric standpoint, describing what is known about the implied or actual role of each RPTP in cancers and cell transformation, and their roles as putative suppressors or promoters of tumour growth and survival. It must be made clear that many of the studies discussed have not included functional studies of the RPTPs, so for most RPTPs their roles in cancer remain unproven. The gene family is evolutionarily conserved and the cell adhesion

A.W. Stoker (✉)
Cancer Section, Developmental Biology & Cancer Programme, UCL Institute of Child Health, 30 Guilford Street, London WC1N 1EH, UK
e-mail: a.stoker@ucl.ac.uk

B.G. Neel, N. Tonks (eds.), *Protein Tyrosine Phosphatases in Cancer*,
DOI 10.1007/978-1-4939-3649-6_2

molecule-like form of many RPTPs has historically focussed attention on their roles in cell adhesion and motility [1, 2]. The initial excitement about RPTPs as being the key foils to the oncogene kinases did not bear much fruit initially, but 20 years on they are now firmly part of the jigsaw of cancer cell signalling. In 2005, MacKeigan and co-workers completed an ambitious project to test kinases and phosphatases and define "cell survival enzymes" and "cell death enzymes". Surprisingly, half of the RPTPs were identified as one or the other [3]. Table 1 provides an overview of these and other RPTPs, although we must apologize that not all the documented evidence can be discussed in this fast moving field. Readers are encouraged to see further discussions in recent reviews [4–6]. PTPRR will not be discussed as it is covered in the chapter ***N***.

PTPRA (PTPα)

PTPα as a potential oncoprotein. It has been known for many years that cells can be transformed by chronic tyrosine kinase activation. The archetypal oncogene is $pp60^{v\text{-}src}$, a mutated and constitutively active form of $pp60^{c\text{-}src}$ (SRC). The cellular kinase is normally tightly regulated and suppressed by carboxy-terminal tyrosine phosphorylation of Y527. In 1992, one of the pivotal PTPs to target Y527 was discovered to be PTPα, a type R4 RPTP [7–9]. Expression of PTPα in rodent fibroblasts could activate SRC, transform the cells and even induce tumours in animals. This insight into PTPs as potential oncogenes exposed them as potential therapeutic targets. Through its indirect regulation of FAK, PTPα can control cell adhesion and spreading, which may have an influence over cancer cell motility and invasiveness [10–12].

In a challenging study, an isoform of PTPα with a nine amino acid ectodomain insertion could activate SRC, but could not transform NIH3T3 cells. The ectodomain of PTPα, whose role has been rather shrouded in mystery, may thus influence its oncogenic capacity [13]. A study by Kapp [14] showed that certain catalytic mutants of PTPα could transform NIH3T3 cells, not wt forms, suggesting that non-SRC effectors are critical for PTPα-driven transformation. The transforming potential of PTPα is thus still somewhat controversial. There are nevertheless studies that do link this RPTP to cancer. In oral squamous cell carcinomas (OSCCs) cancers with significant differentiation, *PTPRA*/PTPα was expressed predominantly in the stromal compartment, whereas in higher grade tumours they were expressed in carcinoma cells themselves [15]. PTPα could thus be linked to either stromal activation, or poor differentiation of the cancer cells. Expression of PTPα is elevated in colon carcinomas when compared to normal mucosa [16], and in gastric tumours PTPα protein expression correlated positively with poor prognosis, metastases and lymphovascular invasion [17]. Experimental suppression of PTPα induced apoptosis in oestrogen receptor (ER)-negative breast cancer and colon cancer cells, but not in immortalized ER-positive breast cancer cells. This may be significant, since ER status is a key prognostic indicator [18]. Recent research also indicates that PTPα activity maintains HER2/neu-mediated mammary tumours in mice, and promotes initiation (Mohamed Bentires-Alj, pers.comm; in press).

Table 1 Documented relationships between RPTPs and cancer types

Gene	Suppressor potential		Oncogenic potential	
	Cancer type		Cancer type	
PTPRA	Breast	[21]	CRC, head and neck,	[15, 16]
			Gastric, breast	[17, 18]
			Cell transformation in culture	[8, 9, 11, 19]
PTPRD	Glioma, neuroblastoma, melanoma, skin, CRC, liver, head and neck, lung, pancreatic	[54, 55, 58, 62, 166, 167]		
	Hepatocellular carcinomas	[70, 71]		
	siRNA screen	[3]		
PTPRE			Mammary	[22–24, 26, 168].
PTPRF	CRC, breast, lung	[27]	Thyroid and breast	[73, 74] [75]
PTPRG	Nasopharyngeal carcinoma	[38]	Glioma, gastric and lymphoma	[42]
	Breast, ovarian, renal and lung	[27, 41–44, 46]		
	Lymphoma/leukaemia	[34, 35]		
	Gastric and CRC	[27, 29, 30]		
	Merkel cell carcinoma	[36, 37]		
PTPRH	Liver	[81]	Gastrointestinal and CRC	[80, 82, 83]
PTPRJ	Colon, CRC, breast and lung	[87, 90, 91, 93, 156, 169]		
	Thyroid, pancreatic	[94, 95, 97, 161]		
	Renal	[100]		
	Meningioma	[157]		
PTPRK	Melanoma and Lymphoma	[105–107]		
PTPRM	Glioma and melanoma	[113, 114]	Glioma	[113, 114]
	CRC	[141]		
	Prostate	[120]		
PTPRO	Breast, lung, leukaemia and CRC	[124–126]		
	Liver (rat)	[123]		
PTPRS	CRC	[139]		
	Head and neck	[140]		
PTPRT	CRC, gastric and lung	[27, 141]		
PTPRZ			Glioblastoma	[53, 148–153]
			Gastric, melanoma, lung	[17, 151, 165, 170]

Mutations in *PTPRA* provided insight into how PTPα may promote neoplasia. Around 30 % of breast, colon and liver tumours contain *PTPRA* mutations (amino acids 245, 445 and 652) [19]. PTPα 245, which lacks the catalytic domains, still activates SRC, but endogenous PTPα is necessary for cell transformation. Rat fibroblasts expressing PTPα245 also formed tumours. One explanation for PTPα245's actions is that it dominantly activates endogenous PTPα, possibly by driving it into heterodimers with the endogenous molecule. Homodimers of wild type PTPα are catalytically inhibited by a mutual "wedge" domain [20]. Heterodimers with PTPα245, however, might leave the full-length partner exposed and active. PTPα thus represents a potential therapeutic target in a range of cancers.

PTPα as a suppressor. Although PTPα does have oncogenic potential, there is one study study that demonstrates a potential suppressive influence. Ardini et al. showed that overexpression of *PTPRA* in MCF7 or N202.1A mammary tumour lines could reduce their growth rates and in the case of N202.1A, reduce tumour growth and metastases in vivo [21].

PTPRE (PTPε)

PTPε as an oncogene. PTPε is a type R4 RPTP closely related to PTPα. It is strongly expressed in mammary tumours driven by RAS or NEU [22]. MMTV-driven PTPε in the mammary epithelium of transgenic mice leads to a low frequency of long latency mammary tumours [23]. Cells from NEU-induced tumours from PTPε-deficient animals were no longer transformed [24]. PTPε may thus be a necessary component of mammary tumourigenesis. PTPε, like PTPα, potently activates SRC [24], FYN and YES [25], although SRC provides the critical link between PTPε and NEU-induced mammary cell transformation [25]. NEU seemingly phosphorylates Y695 in PTPε, inducing docking and activation of SRC [26]. PTPε thus presents an opportunity as a potential drug target in mammary tumourigenesis.

PTPRG (PTPγ)

PTPγ as a suppressor. PTPγ is a type R5 RPTP, related to PTPζ. Human *PTPRG* maps to chromosome 3p14-21, a region commonly deleted or translocated in tumours. *PTPRG* is subject to mutation, LOH and promoter methylation in a remarkable range of cancers (Table 1). The collective data provide a convincing picture of a candidate tumour suppressor gene.

Gastric and CRC. *PTPRG* was mutated in 5 % of CRC analysed by Wang [27], with a mixture of intracellular and extracellular mutations found (Fig. 1). The extracellular mutations are outside the carbonic anhydrase (CA) domain and unlikely to directly affect binding of PTPγ to contactin, a PTPγ ligand [28]. CpG island microarray analysis of right-sided colorectal cancers has pinpointed a locus in the first intron of *PTPRG* as being subject to methylation in 17/18 carcinomas examined [29]. Methylation was also seen in precancerous adenomas and hyperplastic polyps.

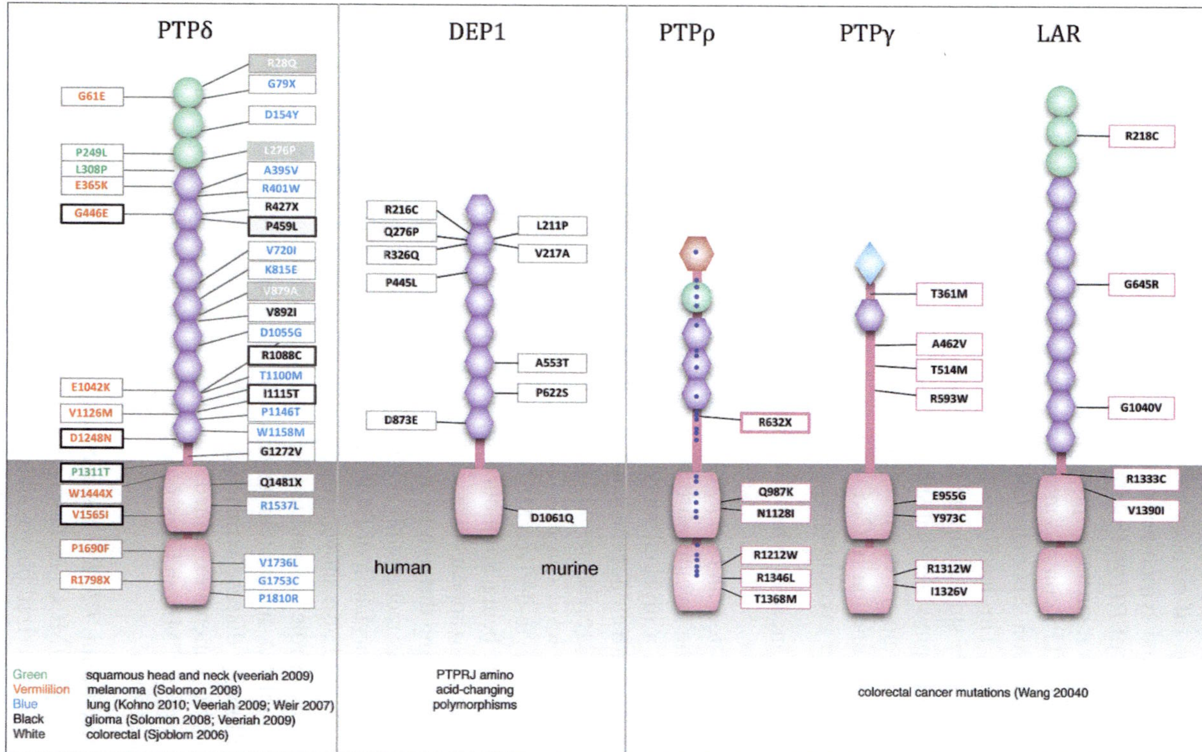

Fig. 1 Sample of mutations and polymorphisms found in RPTP genes linked with cancer. For PTPδ, the mutations are from a range of cancers and these are colour coded. Boxes in bold are mutations that have been functionally tested and shown to knock out PTPδ signalling. For DEP-1, polymorphisms linked to potential promotion of cancer are shown, with both human and murine examples {Ruivenkamp [90] #10547; Mita [91] #10562}. For PTPρ, PTPγ and LAR, points where mutations occur in colorectal cancers are indicated either in blue dots (PTPρ) or in boxes. For PTPρ, the named boxes indicate mutations that have been shown to knock out enzyme activity and R632X has been shown to have no tumour suppressive function {Wang [27] #5039}. Green circles indicate immunoglobulin-like domains, purple hexagons indicate fibronectin type III domains, the red hexagon is a MAM domain and the blue diamond is a carbonic anhydrase-like domain

Interestingly, this intron 1 methylation did not affect *PTPRG* mRNA expression, but did affect binding of the zinc finger protein CTCF, which could have far reaching influences over surrounding chromatin. This specific methylation pattern may lend itself to being a novel biomarker for colon cancers.

Promoter methylation differences were also observed in *PTPRG*, between primary gastric tumours and lymph node metastases [30]. Metastases had a significantly higher frequency of *PTPRG* methylation and gene silencing, suggesting that the critical step towards metastatic spread may require suppression of PTPγ function. It is not known if PTPγ controls motility, invasiveness or some other parameter of the metastatic transition. Sporadic CRC is most commonly associated with loss of function in APC [31]. This leads to activation of β-catenin and the WNT pathway [32]. Interestingly, PTPγ studies in the developing chick spinal cord indicate that β-catenin is a possible, physiological substrate of PTPγ [33] and that neurogenesis is perturbed after either up- or down-regulation of PTPγ. Thus a potential linkage between PTPγ loss of function in CRC and the β-catenin/WNT axis of signalling may be worthy of further investigation.

Lymphoma. Another cancer in which promoter methylation of *PTPRG* has been implicated is lymphoma. Sporadic hypermethylation occurred in 27 % of T-cell lymphomas (CTCL) tested and *PTPRG* could be reactivated by demethylating agents in vitro [34]. In the myeloproliferative disease chronic myelogenous leukaemia (CML), *PTPRG* is a myeloid marker on CD34(+) cells. *PTPRG* expression correlated negatively with clonogenic capacity of four tumour cell lines in methocel culture, while its overexpression could suppress colony survival [35]. Experimental suppression of *PTPRG* level can also increase colony formation in the *PTPRG*-expressing, megakaryocyte-derived MEG-01 and LAMA-84 cell lines. These data all point towards PTPγ being a tumour suppressor protein in these cell lineages. Here PTPγ signalling also appears to associate with myeloid differentiation and also the inhibition of signalling through BCR/ABL. BCR/ABL and associated substrate CRKL were shown to interact directly with PTPγ in pull-down assays and PTPγ can reduce tyrosine phosphorylation of BCR/ABL and CRKL; STAT5 is a downstream target of BCR/ABL and also shows reduced phosphorylation [35]. More evidence of a tumour suppressor role is found in CML patients themselves, where *PTPRG* is downregulated at mRNA and protein levels in diseased leukocytes, and *PTPRG* is re-expressed following disease remission. Here, a loss of promoter methylation occurs in 55 % of the patients examined. Thus the *PTPRG* gene is a good candidate tumour suppressor in CML and has potential as a direct or indirect therapeutic target, or diagnostic tool.

Merkel Cell Carcinoma. *PTPRG* has been implicated recently in the aetiology of Merkel cell carcinoma (MCC) an aggressive, but rare form of human skin cancer. A previously uncharacterized polyomavirus was found integrated into the genome in 80 % of MCC tumours with viral DNA integrated clonally in most tumours [36]. The *PTPRG* locus was targeted in one of eight virus-integrated tumours, implicating the gene potentially in clonal growth of the tumour. A second study analysed potential suppression of *PTPRG* expression by hypermethylation in MCC [37].

Nasopharyngeal Carcinoma. *PTPRG* is a candidate tumour suppressor gene in nasopharyngeal carcinoma (NPC). When compared to normal control samples, 41 % of tumours tested by Cheung and co-workers showed down-regulation of *PTPRG* expression [38]. This occurred through both promoter hypermethylation and loss of heterozygosity. When *PTPRG* was experimentally overexpressed in NPC cell lines, growth suppression occurred along with reduced anchorage-independent growth. Moreover, significant tumour growth suppression occurred in nude mice when *PTPRG* was induced using the Tet-induction system in vivo. Although all of these findings point to *PTPRG* being a good tumour suppressor candidate, the authors also found there was no significant association between *PTPRG* expression in tumour samples and either tumour stage or histology. Also, both low and high levels of *PTPRG* can be found in these, and other, NPC tissues, so the relationship between *PTPRG* loss of function and tumour aetiology remains somewhat muddied. Chueng also studied the potential downstream effector pathway from PTPγ and concluded that PTPγ indirectly suppresses cyclinD1, E2F1 and Rb phosphorylation, the combined effect of which is to block G1-S and shift cells out of the cell cycle. Interestingly, these effects appear to be substrate dependent, in that cell cycle exit only occurred when the tumour cells expressing PTPγ were placed on Matrigel, an ECM extract from Engelbreth-Holm-Swarm (EHS) mouse sarcoma cells [39]. It is possible that some of these Matrigel components are promoting PTPγ signalling in a similar way that Matrigel can stimulate DEP-1 activity [40]. Recent evidence from other labs has pinpointed contactins as PTPγ-interacting molecules in the nervous system [28], although contactins are GPI anchored and are probably unlikely to be present in Matrigel.

Ovary, Lung, renal. PTPγ is apparently underexpressed in a set of tumour samples from lung and ovary [41] and similar conclusions were made from subsets of ovarian and lung cancers [42]. One allele of the gene was also found to be deleted in a 3/5 renal rumour cell lines and 5/10 lung tumour samples [43]. *PTPRG* is close to the 3p14.2 breakpoint of the t(3;8) familial renal cell carcinoma (RCC)-associated constitutional chromosome translocation, although it is not clear if disruption of the gene is a critical component of this cancer [44]. No direct, experimental evidence of suppressor function for PTPγ was demonstrated in these cancers.

Breast. Liu has also demonstrated that in MCF-7 cells, re-expression of PTPγ can suppress proliferation, quench the cell proliferation response of MCF-7 to estradiol-17beta, and suppress soft agar growth [45]. In a follow-up study, Shu and co-workers went on to confirm that re-expression of PTPγ in MCF-7 cells could block tumour formation effectively in immunocompromised mice [46]. Overexpression of PTPγ also caused cells to reinitiate the cell cycle more slowly after serum starvation, and elevate their levels of p21(cip) and p27 (kip), key, negative cell cycle regulators that act in tumour suppressor pathways. This may occur because of the ability of PTPγ to also reduce ERK1/2 phosphorylation. In a similar vein to the studies on NPC (above), *PTPRG* promoters in breast cancer cell lines SK-Br-3 and MCF-7 appeared to suffer increased methylation and demethylation agents reactivated *PTPRG* expression. Vezzalini has also shown that PTPγ expression is low in 56 %

of breast cancer samples studied, indicating a potential link to this disease [42]. These studies therefore further support the idea that *PTPRG* may suffer promoter silencing during the aetiology of some breast cancers, removing a tumour suppressor protein and releasing control over the cell cycle.

PTPγ summary. Although there is no direct evidence that PTPγ has a tumour-supportive role, the door is left open by the admittedly rather meagre data that some lymphomas and astrocytomas have significantly elevated levels of PTPγ protein compared to normal tissues [42]. Also, the morphological differentiation of phaeochromocytoma PC12 cells is suppressed by PTPγ [47]. Thus PTPγ may yet have some tissue-specific, oncogenic potential in cells of neuronal origin. Nevertheless, the bulk of the current data clearly support the fact that PTPγ is a likely tumour suppressor, whose loss of function can play a significant part in the aetiology of several key human cancers.

PTPRD (Encoding PTPδ)

The type R2A RPTP subgroup comprises three enzymes, PTPδ, LAR (leukocyte common antigen-related) and PTPσ [48, 49], which are expressed in overlapping patterns in many tissues (reviewed in [50]). The first R2A enzyme to discussed here is PTP. Gene deficiency in mice results in poor growth and deficits in hippocampal LTP [51, 52]. Given its strong, neural expression, it is unsurprising that *PTPRD* has surfaced as a potential tumour suppressor in neural and glial tumours. However, the gene is also implicated in several other cancers.

Glioma. The most aggressive form of brain tumour is glioblastoma multiforme (GBM), a tumour that is locally invasive and also relatively resistant to chemotherapy. The involvement of PTPs in glioma aetiology has been reviewed recently [53]. A high frequency of focal deletions is found in *PTPRD* (15 % of samples) and loss of *PTPRD* in larger deletions (33 % of cases) in gliomas, with somatic and germline mutations also found [54]. Functional assessment of tumour suppressor activity confirmed that *PTPRD* re-expression in GBM and melanoma cells could suppress their proliferation [54]. These data all point to *PTPRD* being a significant suppressor candidate in this aggressive brain tumour.

A further study by Veeriah et al., also demonstrated that *PTPRD* is frequently subject to microdeletions and larger deletions, as well as point mutations in 6 % of GBM studied. Moreover, in those tumours that have not suffered *PTPRD* mutations, promoter methylation occurs in more than a third of these tumours, leading to loss of *PTPRD* expression [55]. *PTPRD* loss or inactivation also correlated well with poor prognosis. To back up the hypothesis that PTPδ has suppressive potential, experimental knock-down of the gene in immortalized astrocytes induced greater growth in culture and increased xenograft growth [55]. Significantly, PTPδ proteins harbouring mutations found in GBM also lost their suppressive potential. These data again all strongly implicate *PTPRD* and its encoded PTPδ protein as being growth suppressive in glioma tumours.

Neuroblastoma. Neuroblastoma is the most common extracranial solid tumour in children under the age of 5 [56] and prognosis remains poor for the more aggressive forms of the disease. As a paediatric tumour of neural crest origin, one of its properties is the presence of gene amplifications, in particular MYCN, and there are also increased frequencies of 17q trisomy and deletion of 1p or 11q. There are also imbalances in other chromosomal locations including 9p, which harbours the *PTPRD* gene. Indeed, a number of genetic and functional studies have now highlighted *PTPRD* as a potential tumour suppressor locus in this disease.

With the advent of high-resolution oligonucleotide array comparative genomic hybridization (CGH) at the genome level, studies of metastatic NB have revealed common microdeletions in the 5' UTR of *PTPRD* [57], as well as abnormal mRNA splicing in over half of NBL samples and cell lines [58]. The most frequent microdeletion involved the *PTPRD* locus, in 12 % of samples studied, indicating a possible tumour suppressor function for this gene [57]. Supporting this, the different NB tumour stages studied by Nair and co-workers suggest that more aggressive stages have relatively lower *PTPRD* expression, as well as lower expression than in fetal neuroblasts of the adrenals, the origin of many NB tumours [59]. Another study more recently, looking at a form of chromosome "shredding" called chromothripsis, indicated that this kind of chromosomal damage occurs in nearly 20 % of high-grade neuroblastomas. Moreover, the *PTPRD* locus itself was found to be subject to this chromothripsis, as were other genes involved in growth cone motility and neuritogenesis [60].

Although genetic associations are useful, direct functional studies can provide more convincing evidence for suppressive potential. Two such functional studies have been reported on PTPδ in NB cells. The first showed that PTPδ interacts with and dephosphorylates aurora kinase A (AURKA), a major player in NB tumour promotion. This PTPδ interaction causes AURKA down-regulation and consequent destabilization of MYCN. This novel finding was matched by data showing that after transfection of PTPδ expression vectors into KELLY NB cells (MYCN-amplified), it could suppress growth largely through apoptosis [61]. Another study has also analysed the potential of PTPδ to reduce colony formation in a distinct set of NB cell lines. In this case, however, no evidence was found for a growth suppressive role, even though PTPδ protein was strongly expressed in stably derived sublines [62]. This apparent conflict could have technical explanations, including the fact that different PTPδ protein isoforms were used in the studies, or there may be distinct, tumour cell line differences in response to PTPδ signalling. The ability of PTPδ to act as a growth suppressor in NB cells is therefore somewhat unresolved and it may be that its suppressive function is highly dependent on the individual tumour genetic makeup, or the specific PTPδ isoform that is expressed.

Melanoma. Like neuroblastoma, melanoma is of neural crest origin and once again *PTPRD* has been pinpointed here as a potential tumour suppressor. About 10 % of melanomas studied by Stark et al. showed homozygous deletions (HD) in *PTPRD* [63]. Somatic mutations have also been found in around 12 % of tumours examined by Solomon [54] and re-expression of wild type *PTPRD* in melanoma cells with defective *PTPRD* genes did inhibit cell proliferation and also induced cell death. Re-expression of mutated forms of *PTPRD*, however, could not do this. Not all forms of melanoma may be so influenced by *PTPRD* though, since there was no evidence of a link between *PTPRD* mutations and familial "melanoma-astrocytoma syndrome", indicating that *PTPRD* may not be functionally important in this inherited form of the disease [64]. In non-familial melanoma, as well as in neuroblastoma, it does seem that the PTPδ has potential as a suppressor of neuroectodermal tumours.

Skin and Lung Carcinoma. Cutaneous squamous cell carcinoma (SCC) is an increasingly common form of skin cancer, whose genetic causation both at onset and progression is still fairly unclear. A SNP microarray study by Purdie et al. pinpoints *PTPRD* yet again, with microdeletions in the majority of tumours analysed [166]. Furthermore, the Veeriah study identified *PTPRD* point mutations in 13 % of head and neck squamous cell carcinomas [55]. A more recent SCC study by Lambert et al. identified a very high rate (37 %) of mutations in *PTPRD* in SCC, correlating with metastatic disease, although there was not much sign of promoter methylation in these tumour samples. *PTPRD* thus seems to be one of the most highly mutated genes in SCC and may offer a useful marker of metastatic SCC disease [65].

PTPRD was the second most common locus undergoing homozygous deletion in more than 10 % of lung cancer cell lines studied [66, 67]. This was broadly corroborated by other studies where point mutations were found in 9 % of lung cancers [55], or HD were again found [68]. Deletions in the 5' untranslated region of *PTPRD* were found in 4 % of primary human lung adenocarcinomas [69]. In these skin and lung cancers, there has yet to be convincing functional analysis carried out on PTPδ to support the genetic data.

Liver and Clear Cell Renal Cell Carcinoma. In a large copy number variation (CNV) study by Nalesnik et al. on 98 hepatocellular carcinomas (HCCs) [70], CNV was found in *PTPRD* in 21 cases (4 amplifications/17 deletions), with the deletions strangely all clustering within the intron between exons 24 and 25. The potential significance of this very selective *PTPRD* disruption may relate to the fact that PTPδ acts on STAT3 as a substrate in glioma cells [55]: STAT3 is also an important component of signalling that controls hepatocyte regeneration after liver damage, controlling entry into S phase. *PTPRD* was also previously implicated in hepatocarcinogenesis in a gene expression study by Urushibara [71]. Here *PTPRD* mRNA was reduced selectively in chemically induced hepatoma in rats and also in HepG2 cells. *PTPRA*, *PTPRG* and *PTPRF* were not affected, suggesting that specific loss of *PTPRD* mRNA was linked with this cancer [71]. In the kidney, clear cell renal cell carcinoma (cCRCC) has been screened for microsatellite alterations and satellite D9S168 (9p23-22), which resides in the 5'-untranslated region of PTPRD, was altered at high frequency [72]. Immunohistochemistry of normal kidney proximal tubules revealed high PTPδ expression, whereas cancerous epithelia contained little if any of the protein. This D9S168 microsatellite in *PTPRD* thus might be valuable as a predictor of poor prognosis in cCRCC, and PTPδ may yet be proven to be growth suppressive in the kidney tubule epithelia.

***PTPRD*/PTPδ summary**. PTPδ has thus been flagged up numerous times as a potential tumour suppressor, in a wide range of cancers. The siRNA screen of MacKeigen et al. also revealed *PTPRD* as having the potential to enhance cancer drug sensitivity [3].

These collective data strongly suggest that this phosphatase has tumour suppressor potential in a wide range of tissues, although its influence may be protein isoform-specific. Whether *PTPRD* mutations or epigenetic silencing promote tumourigenesis, or are required for tumour maintenance, has yet to be clarified in vivo.

PTPRF (LAR)

***PTPRF* as an oncogene**. Konishi and co-workers examined a range of tumour tissues for the expression of LAR (glioma, lung, breast, prostate, gastric), but found high expression only in thyroid carcinomas [73]. This strong expression was seen in 95 % of thyroid carcinomas, but in only 12 % of adenomas; LAR was not seen in benign thyroid disease. They hypothesized that this heightened carcinoma expression was a potential indication of a functional role for LAR in the aetiology of this cancer. However, functional evidence of this is yet to be forthcoming.

Another expression study, this time in breast cancers, also supported a potential role of LAR as an oncogene in these cancers. Yang et al., showed that LAR protein and mRNA were both significantly elevated in breast cancer tissue compared to normal breast tissue and that aberrant patterns of alternatively spliced LAR mRNAs were present in the tumours. Again, no functional evidence for LAR as an oncogene was provided at the time, but the authors suggested that there might be prognostic potential in the high expression and splicing changes [74]. A later study by Levea et al. examined this further, comparing LAR expression and parameters such as metastatic state and ER-positivity [75]. They concluded that there was significant association of high LAR expression with metastatic potential of rat mammary adenocarcinoma cell lines and with 50 % of metastatic human tumours (as opposed to 21 % of non-metastatic tumours). LAR expression also correlated strongly with oestrogen receptor positivity, providing some preliminary support for LAR being a potential prognostic tool in breast cancer [75].

It is perhaps significant that LAR can not only inactivate PTKs, but also cause direct and indirect activation of others, suggestive of oncogenic potential. In neurons, LAR indirectly activates TrkB by first directly activating SRC ($pp60^{c\text{-}src}$). Experimental suppression of LAR induces loss of activation of TrkB, Shc and AKT [76]. This ability to activate SRC is similar to that of PTPα, a type R4 RPTP with oncogenic capacity (see above). In a related vein, Zheng showed that although LAR can inactivate the PTK ABL, this leads counter-intuitively to PDGF-R activation in mouse fibroblasts [77].

PTPRF as a suppressor. There are clear examples in which LAR has been shown to suppress RTK signalling. Suppression of LAR in rat hepatoma cells (McA-RH7777) causes activation of IRS-1, Shc and MAP-K, and enhanced autophosphorylation of IGFR, EGFR and HGFR upon appropriate RTK ligand stimulation [78]. This would be more consistent with a tumour suppressive potential for LAR. This is consistent with the data found by Wang, wherein 9 % of lung and breast cancers had sustained *PTPRF* mutations [27] (Fig. 1), although the sample numbers were low ($n = 11$) and the significance of these mutations has yet to be functionally tested. Overexpression of LAR in several cell types leads to apoptosis without affecting cell adhesion, acting through p130cas [79]. Arguably, LAR could thus be important in controlling cell survival and might act against uncontrolled proliferation and survival of cancer cells. In conclusion, there is evidence for both an oncogenic and suppressive potential with LAR, but which activity is dominant may be highly context and cell type dependent.

PTPRH (SAP-1)

***PTPRH* as a suppressor**. *PTPRH* encodes SAP-1 (stomach cancer-associated protein tyrosine phosphatase-1), an R3 RPTP expressed most strongly in liver and brain and to a lesser degree in stomach and heart [80]. In hepatic carcinomas, SAP-1 expression was much reduced or undetectable in more aggressive tumours, compared to low-grade tumours and normal tissue [81]. In support of SAP-1 having a tumour suppressive role, re-expression of *PTPRH* in hepatic carcinoma cell lines led to a significant decrease in growth rate and in migration in culture. SAP-1 thus may influence the differentiation state of hepatic cells and play a role in the onset of some liver cancers.

***PTPRH* as an oncogene**. SAP-1 protein was not detected in normal colon, whereas in pancreatic and colorectal cancer cell lines SAP-1 was abundant [80]. Around 40 % of human colon adenocarcinomas also strongly express SAP-1, but expression is lower in adenomas [82]. Overexpression of SAP-1 may therefore be functionally important in later stages of colorectal cancer evolution. Curiously, the murine expression of SAP-1 is different from that reported in humans in that SAP-1 is found most strongly in the gastrointestinal tract [83]. Moreover, SAP-1 localizes to gastrointestinal epithelial cells, on brush border microvilli. Although SAP-1-deficient animals have apparently normal guts, when these mice were crossed to mice with the adenomatous polyposis coli (APC) gene, SAP-1 deficiency reduced tumour growth [83]. Although the human and mouse expression of *PTPRH* and SAP-1 may be different, in both species the gene behaves apparently with tumour-promoting properties in gut tissues.

PTPRJ (DEP1)

PTPRJ encodes the R3 RPTP DEP-1, characterized by extracellular FNIII domains and a single catalytic domain. Studies suggest that this phosphatase may be a negative regulator of cell adhesion and motility [84] and an effector of contact inhibition of cell proliferation [85], making it a logical candidate for having tumour suppressive function. Indeed, DEP-1 was one of the first RPTPs to be implicated as a potential tumour suppressor and has subsequently become a major focus of research in diverse cancers. Although germline ablation of *PTPRJ* is compatible with normal development and is not sufficient to lead to tumourigenesis itself [86], the data below strongly suggest that DEP-1 signalling is a necessary component in cellular growth control and that its loss is at least a step in the process of carcinogenesis.

Breast Cancer. The first significant study on DEP-1 as a TS was done in breast cancer cell lines. Here it was found that DEP-1 expression correlated positively with cell differentiation. Furthermore, experimental expression of DEP-1 in cancer cells (both estrogen receptor (ER)-positive and ER-negative) could dramatically hinder their proliferation [87]. A recent study by Smart has shown that PTPRJ expression is reduced in breast cancer compared to normal breast epithelia and that this correlates with poorer patient survival [88]. DEP-1 localization inside breast epithelial cells also appears to shift from the normal, apical localization in differentiated cells, to a cytosolic location in many tumour cells that retain DEP-1 expression. This is likely to affect DEP-1 signalling, since DEP-1 is believed to form complexes with adhesion molecules in the plasma membrane [85], and is also implicated in regulating tight junctions in epithelial cells [1]. It is not yet clear whether the change in location is a consequence or a cause of the cancer. DEP-1 has also been implicated in directly regulating EGFR actions, by modulating EGFR endocytosis and signalling [89]. The disruption of DEP-1 localization in breast cancer cells could therefore affect both its junctional role and EGFR regulatory role, disrupting two potential lines of suppressive actions in epithelia.

Colon. It has been possible to take advantage of cancer susceptibility models in mice to implicate DEP-1 in colon cancer. The Scc1 (Susceptibility to colon cancer 1) locus in mice is a quantitative trait locus (QTL). In 2002, it was found through haplotype mapping that *PTPRJ* was the underlying, causative suppressor gene [90]. Examination of human colon cancers, as well as breast and lung cancers, revealed 50 % or more had LOH in *PTPRJ*. Of the amino acid polymorphisms found, A1176C(Gln276Pro) showed preferential loss of the allele A1176 versus C1176, suggesting that the presence of Proline in this FNIII domain was somehow reducing the suppressive potential of DEP-1 [90]. Several further amino acid polymorphisms were found in human tumours as well as the original Scc1 locus in mice. All but one of these polymorphisms affects extracellular FNIII domains and it is predicted that several are very likely to affect domain structure or interactions with potential ligands [90]. Interestingly, *PTPRJ* haploinsufficiency appears to be enough to drive tumour susceptibility, since the second alleles were always intact at the sequence level, although epigenetic silencing was not ruled out. Similar *PTPRJ* polymorphisms have also been implicated in susceptibility to a range of other cancers [91].

Thyroid and Pancreatic. Rat *Ptprj* is a differentiation marker in thyroid cells. Significantly, *Ptprj* was suppressed after oncogenic transformation of thyroid cells [92]. Later studies by the same group showed major reductions in DEP-1 expression in human thyroid carcinomas. Moreover, *Ptprj* expression in malignant, rat thyroid cells could neutralize their cancerous behaviour [93]. This was also found by Iuliano after viral transduction of *Ptprj* in thyroid carcinoma cells, inhibiting cell proliferation and blocking xenograft growth [94]. This coincided with reduced pERK levels, increased p27kip1 and loss of PLCγ1 phosphorylation. This involvement of p27kip was also highlighted in the Trapasso study, where the stabilization of p27kip1 by DEP-1 was clearly implicated as being a major effector pathway of DEP-1 in its suppressive capacity [93].

PTPRJ suffers LOH in about a seventh of human thyroid tumours (adenomas and carcinomas), and as with colon cancers, amino acid polymorphisms evidently influence tumour incidence. Homozygosity at the Gln276Pro and Arg326Gln polymorphisms occurred more frequently in cancer patient genomes, and LOH was more frequent in tumours of patients heterozygous for the polymorphism. These data point to DEP-1 being linked albeit at low penetrance, with thyroid carcinogenesis [95]. A more recent study by the same group has looked at a large sample of papillary thyroid carcinomas (PTC) and the non-synonymous Gln276Pro and Asp872Glu polymorphisms. A significant overabundance of homozygotes in Asp872 was seen in the cancer group, but no increased frequency of homozygotes of the Gln276Pro polymorphism. Thus, here homozygosity in Asp872 may be the significant factor in risk of PTC [96].

DEP-1 expression is low in cell lines from pancreatic carcinomas compared to normal tissue. Expression of rat *Ptprj* in human pancreatic tumour-derived cell lines suppresses their growth and induces apoptosis. Similarly, adeno-associated virus delivery of *Ptprj* to PSN1 cell-derived xenograft tumours reduced tumour growth by two thirds [97].

Renal: DEP-1 expression has been found to correlate positively with that of the tumour suppressor protein von-Hippel Lindau (pVHL) [98, 99], raising interest in the potential suppressive role played by DEP-1 in clear cell renal cell carcinoma (cCRCC). In pVHL-negative tumour cell lines, *PTPRJ* expression was lower than after reconstitution of the cells with pVHL. *PTPRJ* expression is low in 80 % of cCRCCs, and in cCRCC patients low *PTPRJ* mRNA had poorer prognosis. pVHL is a regulator of Hypoxia Inducible Factor α (HIF-α), and pVHL inactivation can seemingly suppress *PTPRJ* expression through a mechanism determined at least in part by HIF-2α stabilization [100].

Meningioma. Meningiomas make up 20–25 % of all brain tumours in adults, although their genetic origins are fairly unclear. Petermann and co-workers have found in fact that unlike its suppressor roles, DEP-1 appears to play a tumour-supporting role in meningioma [101]. Mapping of these tumours for heterozygosity loss shows that *PTPRJ* is a target in a number of these tumours, backed up by a similar reduction in DEP-1 protein levels. Moreover, experimental knock-down of DEP-1 in DEP-1-expressing meningioma cell lines causes these cells to become more motile, to grow better in suspension cultures and to become more invasive in a xenotransplant model in vivo. In these cells, PDGF signalling was also enhanced, paxillin phosphorylation reduced and cell adhesion to the ECM was perturbed. In this brain tumour, therefore, DEP-1 may again play a suppressive role, acting through the control of PDGF signalling and cell adhesion.

***PTPRJ* Summary**. Similar to *PTPRD*, *PTPRJ* is clearly a good candidate tumour suppressor in a range of cancers. The gene was similarly flagged up in a recent siRNA screen, again identifying it as a tumour suppressor able to somehow effect the cytotoxicity of well-known cancer drugs [3]. How the mutations and polymorphisms found in *PTPRJ* affect the molecular signalling of DEP1 is still unclear. The identity of the ligand for this molecule is also shrouded in mystery currently.

In addition to its suppressive role in cancer cells themselves, DEP-1 may also exert an influence over tumour vascularization. DEP-1 can negatively regulate VEGFR in endothelial cells, blocking cell proliferation once they reach confluence [85]. DEP-1 may operate by preventing internalization of the VEGR complex and onward transmission of its signals [102]. One might conjecture that in a tumour environment, cancer cells may have paracrine influences over DEP-1 activity in local endothelia, block DEP-1's activity. This might alleviate growth control in the vasculature. Furthermore, the *PTPRJ* gene expression is subject to negative regulation by HIF-1α under hypoxia conditions in vasculature smooth muscle cells, again hinting that local, hypoxic tumour environments may locally knock out DEP-1 in vasculature-associated SMCs, enhancing PDGFR action and cell proliferation [103]. As a potential therapeutic target, the DEP-1 ectodomain has been shown to be targetable with a specific bivalent antibody, which is proposed to activate DEP-1 and suppress endothelial cell proliferation in culture [104].

PTPRK (PTPκ)

Encoding the type R2B RPTP PTPκ, the human *PTPRK* gene was located to 6q22.2–q22.3, a region thought to contain a tumour suppressor locus, as evidenced by its common deletion in melanomas, ovarian and haematological cancers and others [105]. Closer examination of melanomas later showed that *PTPRK* was downregulated in around one fifth of cell lines derived from these cancers, compared to normal melanocytes [106]. Furthermore, in primary central nervous system lymphomas (PCNSLs), mostly thought to be non-Hodgkin's lymphoma, LOH mapping identified *PTPRK* as a site with frequent deletions and this correlated with poorer survival. PTPκ expression was reduced in most tumours analysed immunohistochemically [107]. Beyond this, there had been no direct evidence yet that *PTPRK* represents a genuine suppressor gene. Recently, however, in a study of the LOX-PP tumour suppressor in lung cancer cells, Sanchez-Morgan et al. demonstrated a novel interaction of LOX-PP with PTPκ [108]. PTPκ is known to undergo proteolytic cleavage to release a nuclear, catalytic domain fragment that binds β-catenin and triggers transcription [109]. They argued that this interaction of LOC-PP with nuclear PTPκ drove the PTPκ fragment to destruction by proteolysis, releasing β-catenin back into the cytoplasm where it reassociates with cadherins and increases cell–cell adhesion. This leaves little β-catenin to now signal transcriptionally in the nucleus, one possible way in which LOX-PP acts as a suppressor. In melanoma cells, it has also been shown that full-length PTPκ can also drive β-catenin into cadherin complexes and block transcriptional activation by β-catenin. This blocks cyclin D1 and c-myc expression and reduces the tumour cells' motility in culture [110].

PTPRM (PTPμ)

Glioblastoma. One of the most clinically challenging aspects of GBM is the diffuse nature of the tumour within the brain. The ability of the cells to disperse has raised interest in the possibility of a role here for PTPμ, another R2B RPTP related to PTPκ. PTPμ is known to regulate cell–cell adhesion and cell motility and is expressed on neurons and glia [111]. It has homophilic binding properties and interacts directly with cadherins [1], forming a bridge of defined gap width between cells [112]. PTPμ, like PTPk, undergoes controlled proteolytic cleavage in cells, ultimately releasing a cytoplasmic, catalytic fragment [113]. Unusually, PTPμ proteins are excessively cleaved and fragmented in many GBM, with the normal, full-length PTPM molecule being difficult to detect. Full-length protein is retained in less aggressive astrocytomas [114]. Tellingly, when full-length PTPμ is re-expressed in LN-299 glioma cells (deficient in full-length PTPμ), it dampens cell migration and reduces growth factor-independent survival. In contrast, in U87 MG GBM cells, which do express full-length PTPμ, the knock-down of PTPμ induces greater motility in culture and greater invasion of brain tissue in vivo [114]. The excessive proteolytic cleavage of PTPμ in LN-299 cells, however, also plays a critical role, since shRNA treatment of LN-299 cells to remove cleaved PTPμ actually blocks their motility and survival. Significantly, a peptide inhibitor of the nuclear form of PTPμ can also block motility and survival. A more recent study by the same group confirmed these findings and also showed that shRNA-induced suppression of *PTPRM* in xenografted GBM cells, that contain proteolyzed PTPμ, also leads to reduced tumour growth [115]. Thus the soluble catalytic domain of PTPμ found in the nucleus may act to promote tumour cell survival and invasiveness, whereas the full-length protein may suppress them. As this aberrant proteolysis appears to be key, it raises the possibility of using proteolysis as a biomarker of tumour grade and also its potential as a target for therapeutics development [113]. The cleaved ectodomain of PTPμ also represents a potentially useful biomarker in GBM [116]. As proteolysis is also found in PTPκ and other RPTPs [109, 117], it will be interesting in future to see whether similar dual, but opposing, roles of full-length and cleaved proteins are found. As a further route to perhaps a similar goal, PTPε is found in alternatively translated cytoplasmic forms that have altered biochemical activity [118].

Interest in microRNAs in cancer aetiology has been growing in recent years and a study in 2012 by Quintavalle has shown that micro RNAs (miR) miR-221 and 222 are more highly expressed in tumourigenic glioma cell lines and can target the 3' non-translated region of PTPμ mRNA [119]. Significantly, overexpression of these miRs reduced PTPμ expression and promoted greater motility in the tumour cells, as well as increasing their proliferation in suspension cultures. Moreover, re-expression of miR-resistant *PTPRM* mRNA in the cells could dampen motility. These findings do suggest that these miRs might govern the aggressiveness of glioma cells through specific targeting of PTPμ.

Prostate Cancer. Although there is no genetic evidence that *PTPRM* is linked to prostate cancer, one study that examined the adhesion of LNCaP prostate cancer cells did note that reintroduction of *PTPRM* restored adhesion of the cells both to cadherin and to PTPμ, even if the PTP was catalytically inactive [120]. This is the only such study in prostate cancer cells to date, however, and although normal prostate does express PTPμ, it does not yet imply that loss of PTPμ is directly relevant in this cancer's aetiology.

PTPRO (PTPro)

PTPRO encodes a type R3 RPTP [49] PTPRO (GLEPP1, CRYP-2). Implicated in neuronal development in particular [121, 122], this protein also falls into the tumour suppressor bracket once again. Dietary-induced hepatocellular carcinomas can be induced in mice, wherein the tumours show global changes in DNA methylation. *PTPRO* was identified in a screen of genes differentially methylated during this hepatocarcinogenesis. *PTPRO* was methylated and expression reduced in the cancers [123]. Later studies also showed that this gene is hypermethylated in lung tumours and cell lines [124], chronic lymphocytic leukaemia [125] and breast cancer cells lines [126]. More convincing evidence that the gene could be a suppressor came from cell culture studies showing that re-expression of *PTPRO* in A549 cells suppressed suspension growth and triggered apoptosis [124]. Moreover, *PTPRO* expression in the CLL line WaC3CD5 promoted the inhibitory influence of the therapeutic fludarabine [125], and reintroduction of PTPRO into MCF-7 cells made them more prone to tamoxifen suppression [126]. In the last study, it was suggested that oestrogen-induced suppression of PTPRO expression is a facet of early tumour induction. These PTPRO studies provide an exciting prospect of using demethylation as an epigenetic route for driving potent tumour suppressor therapies in several tumour types.

PTPRS (Encoding PTPσ)

The PTPσ protein has been studied extensively in the nervous system, where it has roles in axon growth and guidance [127, 128] and in nerve regeneration [129, 130]. The protein is expressed in two major isoforms [131], the longer of which is fairly widely expressed in tissues. Its function outside the nervous system is less clear, but like other R2A RPTPs, it can target RTKs [127, 132] and its extracellular domain plays roles in cell–cell and cell–matrix adhesion [133–135]. PTPσ has also been implicated recently as a suppressor of autophagy in U2OS cells [136], a potential pointer to a role in cancer cell evolution [137, 138].

Microsatellite instability studies of colorectal cancers and adenomas had hinted that several PTP genes, including *PTPRS*, might be selectively mutated in these diseases [139]. Looking at long mononucleotide repeat mutation rates, it was concluded that although PTPRS was subject to mutation, the frequencies were too low to present a significant association with these cancers and could not be used as a "Positive Selective Target Gene". In contrast to this study, more recent research into head and neck squamous cell carcinomas has more firmly shone the spotlight on

PTPRS as a candidate suppressor. Morris and co-workers examined the EGFR signalling pathway in these tumours, being aware that PTPσ had previously been implicated in EGFR regulation [132, 140]. They found significant gene copy number alterations in the majority of head and neck tumours examined, and *PTPRS* microdeletions were seen in 26 % of samples. *PTPRS* microdeletions resulted in reduced mRNA expression in all tumours examined. Of particular interest, experimental suppression of *PTPRS* using siRNA in HNSCC cancer cells promoted prominent EGFR activation, and also desensitized HCC27 and H3255 cells to the EGFR inhibitory drug erlotinib. Lung cancer patient survival studies moreover showed that when EGFR is activated and *PTPRS* expression is low, then survival is significantly worsened. These data suggest that the PTPσ protein has tumour suppressive properties by actively suppressing the EGR signalling pathway in head and neck tissues. There is therefore potential prognostic value of determining the *PTPRS* gene state for this malignancy and for further understanding the downstream targets of PTPσ [140]. *PTPRS* has also fallen out of an siRNA screen in HELA cells, where it was identified as a putative tumour suppressor that could maintain cell sensitivity to apoptosis-inducing drugs [3].

PTPRT (PTPρ)

In the seminal study of colorectal and other cancers by Wang and co-workers, the *PTPRT* gene was revealed as the most commonly mutated gene in CRC, with mutations also in two of eleven lung cancers and two of twelve gastric cancers [27]. The mutations were spread fairly evenly along the whole length of the encoded protein (Fig. 1) and five catalytic region (D1 and D2) mutations in either D1 or D2 were confirmed as catalytically defective. Wild type *PTPRT* and one form with an extracellular nonsense mutation were re-expressed in HCT116 CRC cells, where the wild type dramatically suppressed colony formation and the mutant did not [27]. *PTPRT*, along with several other RPTP genes, has also recently been found to undergo increased incidence of promoter methylation in CRC [141]. At a molecular level, PTPρ has recently been shown to target paxillin directly in CRC-derived cells [142]. By generating cells that had wild type paxillin replaced with a Y88F mutant, the authors showed that cells had reduced motility and growth in suspension, as well as poor xenograft growth. Mice lacking the PTPRT gene were also more susceptible to the effects of azoxymethane, a carcinogen. This study reinforces the likelihood that PTPRT is indeed a key suppressor in CRC in man.

PTPRZ (PTPζ)

As discussed above, the type 5 PTPγ is implicated in a range of cancer scenarios. Its sister protein, PTPζ, has been of equal interest over the years and has added a significant amount to our understanding of how RPTPs can act to promote cancer behaviours. PTPζ (originally also named PTPβ; not to be confused now with the accepted nomenclature of the type R3 PTPβ), like PTPγ, has a carbonic anhydrase domain and two catalytic domains. It was originally identified as an RPTP with strong expression in the nervous system [143, 144] and it occurs in several protein isoforms, encoded by alternatively spliced mRNAs. As well as transmembrane forms, one of the most well-known isoforms is phosphacan, a highly glycosylated, secreted form of the PTPζ ectodomain [145]. In the early days of ligand searching, PTPζ was one of the first to be matched up to binding patterns, including pleiotrophin (Ptn) [146] and contactin [147]. Over the years these ligands have become central to our understanding of PTPζ regulation and tissue function, both in normal tissues and in cancer. In the cancer area, the evidence overwhelmingly points to PTPζ as being supportive of cancer growth.

Glioblastoma. The relationship of PTPζ to the biology of glioma has been elegantly reviewed recently [53]. Because of this, we will more briefly review the historical background and provide the most recent information available at this time. Early studies of gliomas showed that there was a bias towards different PTPζ isoforms being expressed in a range of tumour grades. The receptor forms were found in all lower grade tumours (WHO grade 1-3 gliomas), but around half of high-grade GBM did not express them. In these cases, the GBMs expressed the soluble form phosphacan [148]. In lower grade II gliomas, it has also been shown that PTPζ expression may be able to distinguish between histological subtypes such as astrocytomas, and oligodendrogliomas, with the latter expressing high PTPζ protein level and the former not [149]. PTPζ expression patterns might therefore be of some use in distinguishing between tumour grades in glioma.

A study by Muller in 2003 showed that *PTPRZ* mRNA was elevated in glioblastomas compared to normal human brain tissue, as was its ligand Ptn. RNAi was used to suppress *PTPRZ* expression in GBM lines G122, resulting in a 20–30 % reduction in migration in Boyden chamber assays [150]. To complement this, G84 cells (lacking PTPζ) and D566 cells (expressing PTPζ) were treated with pleiotrophin. D566 respond by increasing motility several-fold, whereas G84 are unresponsive; RNAi knock-down of PTPζ in D566 abolished this effect. Importantly, under these various treatment conditions, there was no evidence that PTPζ influenced the proliferative capacity of the GBM cells; this was also seen in U87 cells by Foehr [151]. Here, therefore, the principal influence of PTPζ is in positively influencing GBM cell migration. Given that such cell motility is a key facet of GBM disease in the brain, it is suggested that PTPζ expression could present a suitable therapeutic target [150].

Ulbricht and co-workers also showed that PTPζ isoforms and pleiotrophin were overexpressed in various astrocytic glioma grades [152]. PTPζ levels correlated positively with grade, where the PTPζ was localized to the tumour cells themselves. Ptn was elevated in lower grade astrocytomas compared to normal brain and there was no clear correlation with increasing grade. Ptn had chemotactic and strong haptotactic effects on GBM cell migration, but the latter in particular was inhibited when either an antibody to PTPζ was introduced into the media [152] or the *PTPRZ* mRNA was reduced with RNAi [153]. There may therefore be some form of autocrine, positive reinforcement in the motility of cells in some gliomas, based on the co-expression of both PTPζ and Ptn. It is interesting that in this and a later study it is suggested that Ptn is most effective at stimulating migration when it is substrate- or matrix-bound [152, 153]. Whether Ptn is also supplied in a paracrine fashion by tumour stroma in gliomas, as it may be in other cancers [154], could also be a key question to answer. A later study of Ulbricht further showed that PTPζ knockdown in U251-MG glioma cells potently blocked the tumourigenicity of these cells in xenografts and orthotopic grafts [153]. In contrast to the studies by Muller [150] and Foehr [151], the proliferation of PTPζ-depleted cells in vitro was also significantly reduced compared with that of control clones. Foehr and co-workers have developed anti-PTPζ antibodies that showed some antitumour potential when linked to the toxin saporin and delivered to U87 tumour-bearing mice [151]. These studies collectively demonstrate the potential of antagonizing PTPζ function, either genetically or through immunological means, as a therapeutic approach against glioma tumour cells. The delivery of such PTPζ antagonist across the blood brain barrier will be a constant challenge of course, although direct intracranial delivery may be a viable option.

Other cancers. The apparently elevated expression of *PTPRZ* or PTPζ has been described in a range of tumours. *PTPRZ* overexpression has been seen in gastric cancers correlating with tumour progression and metastasis [17], whereas PTPζ has been detected in melanomas but not in normal pigment cells of the skin [170]. In small-cell lung cancer cell lines, Makinoshima very recently showed that PTPζ is highly expressed [165]. PTPζ is also found in other human neuroendocrine tumours. Makinoshima carried out functional assays and found that Ptn could activate calmodulin phosphorylation, suggesting that PTPζ may be upstream of this event. Finally, they show that *PTPRZ1* has a role in promoting tumour growth in a xenograft model. PTPζ could thus emerge as a therapeutic target in another common set of human cancers.

RPTPs: Cancer Diagnostics and Therapeutics

Although PTPs in cancer diagnostics is to be covered in a further chapter of this book, there are a few things that can be said here that relate specifically to RPTPs. From the evidence presented above, it is clear that the subfamily of RPTP genes is subject to a significant rate of genetic and epigenetic damage or dysregulation in a range of common cancers. How many of these changes are "driver" or "passenger" mutations remains to be seen, but the mounting evidence for at least some driver RPTP gene alterations is becoming quite compelling, in combination with other oncogenic changes. The fact that most of these RPTP genes fall into the putative tumour suppressor subclass means that they are certainly of potential use in diagnostics and prognostics. However, as therapeutic targets these suppressors are less obviously tackled, given that the "target" protein is often not there or has lost activity. In some cases the mutations in RPTPs may lead to dominant negative actions, in which case we may need to find ways to either disinhibit the wt enzymes or reintroduce wt type genes to counter the mutants. In practical terms, however, this may be very difficult in vivo and it is more likely that these genetic/epigenetic alterations will be most useful at a diagnostic level. Our future efforts to reinstate the signalling pathways affected by loss of function in suppressor RPTPs may instead be better served by identifying and therapeutically targeting the key downstream substrates or interactors (Table **2**).

Table 2 RPTPs and their signalling effectors in cancer

RPTP protein	Effectors [direct (d) or indirect (id) interactor, where known]	Consequence for cancer cells where known
PTPα	SRC (d) [8]; FYN (d) [7]; FAK (d) [10, 11] [12]	PTPα activates SRC and can transform cells; In A431 cells, PTPα activates SRC but induces FAK activation and greater adhesion; In PTPα-deficient fibroblasts, SRC, FYN and integrin signalling are reduced, leading to poor cell spreading
PTPδ	STAT3 (d?)	Suppresses glioma proliferation [55]
PTPε	SRC (d) [24], FYN (d) and YES (d) [25]	PTPε activates SRC and maintains mammary tumour cell transformation; FYN and YES are also activated by PTPε, but effects in mammary tumour cells are less clear
DEP-1	MET (d); GAB-1 (D); p120 catenin (d) [156]	DEP-1 negatively regulates MET, GAB-1 and p120catenin in breast cancer cells, targeting very specific tyrosine residues
	PDGF-R signalling (d); Paxillin (id) [157, 158]	DEP-1 inactivates PDGFR through selective tyrosine residues, while paxillin phosphorylation and cell adhesion are maintained by DEP-1; meningioma cell motility and invasiveness is increased in absence of DEP-1
	RET (d?) [159]	RET is often mutated in thyroid cancers; DEP-1 suppresses wt and RET(C634R) mutant RET signalling, but not M918T RET
	P27kip1 (id) [93, 94]	P27kip1 stabilization by DEP-1 suppresses transformed cell phenotype in thyroid carcinoma cells
	VEGFR (d?) [85, 102]	In endothelial cells, VEGF binds VEGFR; VEGR moves to the cadherin complexes where DEP-1 can inactivate VEGFR signalling; this process could underpin contact inhibition of cell growth
	pVHL (d) [100]	DEP-1 binds directly to tumour suppressor pVHL; pVHL expression positively regulates the expression level of DEP-1 in renal cancer cells through HIF-dependent process
	SRC (d) and TRKB (d) [76]	LAR activates SRC and this sustains TRKB signalling
LAR	ABL (d) [77]	LAR suppresses basal ABL activity, thereby maintaining PDGFR activity
	IGFR, EGFR and HGFR (d) [78, 160]	LAR suppresses PI3-K activation downstream of several key RTKs
	P130cas (d) [79]	LAR dephosphorylates p130cas, inducing loss of p130cas stability and apoptosis
PTPγ	SRC (d); FAK (id?); paxillin (id?) [161]	SRC activation; increased malignant thyroid cell adhesion
PTPκ, PTPμ, LAR (nuclear forms)	β-catenin (d) [109, 117] [113]	Altered transcription by β-catenin; some evidence that cleaved RPTPs preferentially bind to β-catenin
PTPκ (full-length form)	β-catenin (d) [110]	Negatively regulates β-catenin signalling in the nucleus in melanoma cells; suppresses cyclin D1 and motility
	LOX-PP (d) (Sanchez-Morgan [108])	LOX-PP causes proteolytic destruction of nuclear form of PTPk in lung cancer cells, suppressing β-catenin signalling

(continued)

Table 2 (continued)

RPTP protein	Effectors [direct (d) or indirect (id) interactor, where known]	Consequence for cancer cells where known
PTPμ	RACK1 (d) [120]	PTPμ indirectly suppresses PKC through its RACK1 interaction, promoting cadherin-based adhesion
PTPσ	TRKs (d) [127]; EGFR (d?) [132]	PTPσ suppresses TRK signalling; Soft agar growth of A431 cells and EGFR signalling are reduced by PTPσ re-expression
PTPρ	Paxillin [142]	Paxillin is direct target of PTPρ
		In colorectal cancer cells, regulating migration and anchorage-independent growth
PTPζ	β-catenin (d) [162]	Pleiotrophin binding suppresses PTPζ activity, leading to β-catenin activation
	β-adducin (d); FYN (d) [163, 164]	Pleiotrophin binding to PTPζ induces phosphorylation of these binding partners of PTPζ, possibly inducing the adhesive and other changes seen in Ptn-treated cancer cells.
	Calmodulin (d?) [165]	

Nonetheless, one major plus point of RPTPs as therapeutic targets, at least hypothetically, is the presence of an extracellular domain. There are already examples where bivalent antibodies to RPTP ectodomains clearly influence RPTP signalling [104]. As our knowledge of RPTP ligands also grows there hopefully will be opportunities to modulate RPTP activities with biochemicals that block or mimic these ligands. This may be most practicable in instances where we might wish to reactivate an RPTP by preventing its self-inhibitory dimerization in the membrane. Use of extracellular reagents has the obvious benefit that they need not be designed to crossed cell membranes, thus avoiding a critical problem that has beset the field of catalytic domain inhibitors. In cases where intracellular proteolysis generates oncogenic forms of RPTPs [109, 117], we would need to seek other novel mechanisms of specific proteolytic blockade.

Conclusions

When the first RPTP to be discovered, CD45, was defined in 1988 [155], a new field was born. As the protein family grew, there was much excitement surrounding the potential of these receptor-like molecules as regulators of oncogenic, receptor tyrosine kinases and other oncogenic pathways dependent on phosphotyrosine-based signalling. However, it has taken more than 20 years for us to begin to really see the potential significance of these RPTPs in the context of cancer biology. With technological advances in genomic and proteomic sciences we can now appreciate that RPTPs sit centrally in cell signalling networks and that their perturbation, either genetic, epigenetic or post-translation, can have profound consequences for cell behaviour, and can influence several facets of the tumour cell behaviour (Fig. 2).

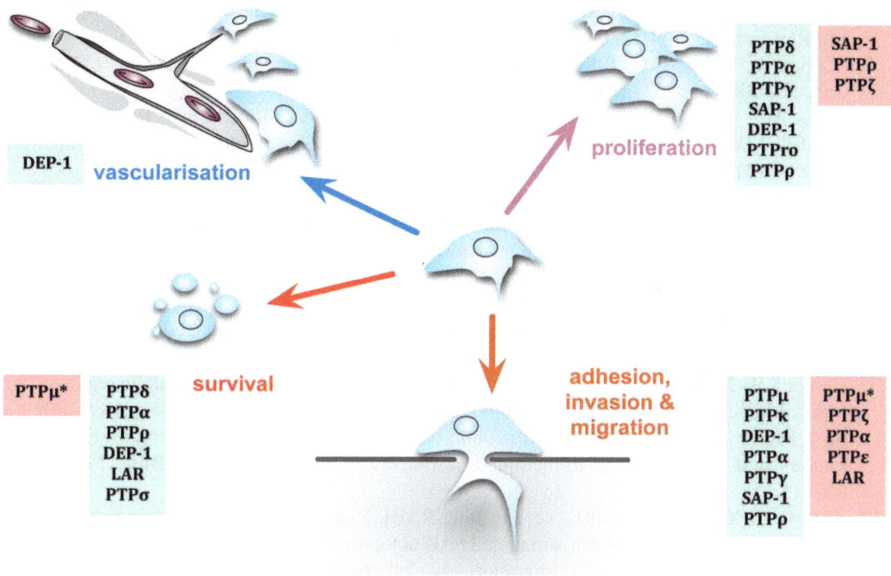

Fig. 2 RPTP enzymes have been implicated in functional assays in a range of cancer cell behaviours, either as suppressors of these behaviours or enhancers of them. This schematic gives examples of RPTPs linked to cell survival, adhesion, invasion and migration, proliferation and potentially vascularization. In each parameter, a green box indicates RPTPs with tumour suppressor behaviour and the pink boxes those with oncogenic-like behaviour. Where there is an asterisk on PTPμ this indicates the proteolytically cleaved, nuclear form of the protein. The role for DEP-1 in tumour vascularization is more putative, based on the role of DEP-1 in endothelial growth regulation. See text for more details

Their actions as promoters of tumour cell proliferation, survival and motility, and their actions as growth suppressors, are bringing them into increasing focus for cancer biologists. We now expect them to teach us much more about cancer cell biology and at the same juncture they, or their immediate signalling networks, hold many opportunities in the area of therapeutics.

References

1. Sallee JL, Wittchen ES, Burridge K. Regulation of cell adhesion by protein-tyrosine phosphatases: II. Cell-cell adhesion. J Biol Chem. 2006;281(24):16189–92. doi:R600003200 [pii] 10.1074/jbc.R600003200.
2. Burridge K, Sastry SK, Sallee JL. Regulation of cell adhesion by protein tyrosine phosphatases 1. Cell-matrix adhesion. J Biol Chem. 2006;281(23):15593–6.
3. MacKeigan JP, Murphy LO, Blenis J. Sensitized RNAi screen of human kinases and phosphatases identifies new regulators of apoptosis and chemoresistance. Nat Cell Biol. 2005;7(6):591–600. doi:ncb1258 [pii] 10.1038/ncb1258.

4. Julien SG, Dubé N, Hardy S, Tremblay ML Inside the human cancer tyrosine phosphatome. Nat Rev Cancer. 2011;11(1):35–49. doi:10.1038/nrc2980.
5. Ostman A, Hellberg C, Böhmer FD. Protein-tyrosine phosphatases and cancer. Nat Rev Cancer. 2006;6(4):307–320. doi:10.1038/nrc1837.
6. Navis AC, van den Eijnden M, Schepens JTG, Hooft van Huijsduijnen R, Wesseling P, Hendriks WJAJ. Protein tyrosine phosphatases in glioma biology. Acta Neuropathol. 2010;119(2):157–175. doi:10.1007/s00401-009-0614-0.
7. Bhandari V, Lim KL, Pallen CJ. Physical and functional interactions between receptor-like protein- tyrosine phosphatase alpha and p59fyn. J Biol Chem. 1998;273(15):8691–8.
8. Zheng XM, Wang Y, Pallen CJ. Cell transformation and activation of pp60$^{c\text{-}src}$ by overexpression of a protein tyrosine phosphatase. Nature. 1992;359:336–9.
9. Ponniah S, Wang DZ, Lim KL, Pallen CJ. Targeted disruption of the tyrosine phosphatase PTPalpha leads to constitutive downregulation of the kinases Src and Fyn. Curr Biol. 1999;9(10):535–8.
10. Harder KW, Moller NP, Peacock JW, Jirik FR. Protein-tyrosine phosphatase alpha regulates Src family kinases and alters cell-substratum adhesion. J Biol Chem. 1998;273(48):31890–900.
11. Su J, Muranjan M, Sap J. Receptor protein tyrosine phosphatase alpha activates Src-family kinases and controls integrin-mediated responses in fibroblasts. Curr Biol. 1999;9(10): 505–11.
12. Zeng L, Si X, Yu W-P, Le HT, Ng KP, Teng RMH, Ryan K, Wang DZ-M, Ponniah S, Pallen CJ. PTP alpha regulates integrin-stimulated FAK autophosphorylation and cytoskeletal rearrangement in cell spreading and migration. J Cell Biol. 2003;160(1):137–46. doi:10.1083/jcb.200206049.
13. Tremper-Wells B, Resnick RJ, Zheng X, Holsinger LJ, Shalloway D. Extracellular domain dependence of PTPalpha transforming activity. Genes Cells. 2010;15(7):711–24. doi:10.1111/j.1365-2443.2010.01410.x.
14. Kapp K, Siemens J, Weyrich P, Schulz JB, Haring HU, Lammers R. Extracellular domain splice variants of a transforming protein tyrosine phosphatase alpha mutant differentially activate Src-kinase dependent focus formation. Genes Cells. 2007;12(1):63–73.
15. Berndt A, Luo X, Bohmer FD, Kosmehl H. Expression of the transmembrane protein tyrosine phosphatase RPTPalpha in human oral squamous cell carcinoma. Histochem Cell Biol. 1999;111(5):399–403.
16. Tabiti K, Smith DR, Goh HS, Pallen CJ. Increased mRNA expression of the receptor-like protein tyrosine phosphatase alpha in late stage colon carcinomas. Cancer Lett. 1995;93(2): 239–48.
17. Wu CW, Kao HL, Li AF, Chi CW, Lin WC. Protein tyrosine-phosphatase expression profiling in gastric cancer tissues. Cancer Lett. 2006;242(1):95–103.
18. Zheng X, Resnick RJ, Shalloway D. Apoptosis of estrogen-receptor negative breast cancer and colon cancer cell lines by PTP alpha and src RNAi. Int J Cancer. 2008;122(9):1999–2007. doi:10.1002/ijc.23321.
19. Huang J, Yao L, Xu R, Wu H, Wang M, White BS, Shalloway D, Zheng X. Activation of Src and transformation by an RPTPalpha splice mutant found in human tumours. EMBO J. 2011;30(15):3200–11. doi:10.1038/emboj.2011.212.
20. Majeti R, Bilwes AM, Noel JP, Hunter T, Weiss A. Dimerization-induced inhibition of receptor protein tyrosine phosphatase function through an inhibitory wedge. Science. 1998;279(5347): 88–91.
21. Ardini E, Agresti R, Tagliabue E, Greco M, Aiello P, Yang LT, Menard S, Sap J. Expression of protein tyrosine phosphatase alpha (RPTPalpha) in human breast cancer correlates with low tumor grade, and inhibits tumor cell growth in vitro and in vivo. Oncogene. 2000;19(43):4979–87.
22. Elson A, Leder P. Protein-tyrosine phosphatase epsilon: an isoform specifically expressed in mouse mammary tumors initiated by v-Ha-ras or neu. J Biol Chem. 1995;270(44):26116–22.
23. Elson A. Protein tyrosine phosphatase epsilon increases the risk of mammary hyperplasia and mammary tumors in transgenic mice. Oncogene. 1999;18(52):7535–42.

24. Gil-Henn H, Elson A. Tyrosine phosphatase epsilon activates Src and supports the transformed phenotype of Neu-induced mammary tumor cells. J Biol Chem. 2003;278(18):15579–86.
25. Granot-Attas S, Elson A. Protein tyrosine phosphatase epsilon activates Yes and Fyn in Neu-induced mammary tumor cells. Exp Cell Res. 2004;294(1):236–43. doi:10.1016/j.yexcr.2003.11.003.
26. Berman-Golan D, Granot-Attas S, Elson A. Protein tyrosine phosphatase epsilon and Neu-induced mammary tumorigenesis. Cancer Metastasis Rev. 2008;27(2):193–203. doi:10.1007/s10555-008-9124-0.
27. Wang Z, Shen D, Parsons DW, Bardelli A, Sager J, Szabo S, Ptak J, Silliman N, Peters BA, van der Heijden MS, Parmigiani G, Yan H, Wang TL, Riggins G, Powell SM, Willson JK, Markowitz S, Kinzler KW, Vogelstein B, Velculescu VE. Mutational analysis of the tyrosine phosphatome in colorectal cancers. Science. 2004;304(5674):1164–6.
28. Bouyain S, Watkins DJ. The protein tyrosine phosphatases PTPRZ and PTPRG bind to distinct members of the contactin family of neural recognition molecules. Proc Natl Acad Sci U S A. 2010;107(6):2443–8. doi:0911235107 [pii] 10.1073/pnas.0911235107.
29. van Roon EH, de Miranda NF, van Nieuwenhuizen MP, de Meijer EJ, van Puijenbroek M, Yan PS, Huang TH, van Wezel T, Morreau H, Boer JM. Tumour-specific methylation of PTPRG intron 1 locus in sporadic and Lynch syndrome colorectal cancer. Eur J Hum Genet. 2011;19(3):307–12. doi:10.1038/ejhg.2010.187.
30. Wang JF, Dai DQ. Metastatic suppressor genes inactivated by aberrant methylation in gastric cancer. World J Gastroenterol. 2007;13(43):5692–8.
31. Samowitz WS. Genetic and epigenetic changes in colon cancer. Exp Mol Pathol. 2008;85(1):64–7. doi:10.1016/j.yexmp.2008.03.008.
32. Scholer-Dahirel A, Schlabach MR, Loo A, Bagdasarian L, Meyer R, Guo R, Woolfenden S, Yu KK, Markovits J, Killary K, Sonkin D, Yao YM, Warmuth M, Sellers WR, Schlegel R, Stegmeier F, Mosher RE, McLaughlin ME. Maintenance of adenomatous polyposis coli (APC)-mutant colorectal cancer is dependent on Wnt/beta-catenin signaling. Proc Natl Acad Sci U S A. 2011;108(41):17135–40. doi:10.1073/pnas.1104182108.
33. Hashemi H, Hurley M, Gibson A, Panova V, Tchetchelnitski V, Barr A, Stoker AW. Receptor tyrosine phosphatase PTPgamma is a regulator of spinal cord neurogenesis. Mol Cell Neurosci. 2011;46(2):469–82. doi:10.1016/j.mcn.2010.11.012.
34. van Doorn R, Zoutman WH, Dijkman R, de Menezes RX, Commandeur S, Mulder AA, van der Velden PA, Vermeer MH, Willemze R, Yan PS, Huang TH, Tensen CP. Epigenetic profiling of cutaneous T-cell lymphoma: promoter hypermethylation of multiple tumor suppressor genes including BCL7a, PTPRG, and p73. J Clin Oncol. 2005;23(17):3886–96. doi:10.1200/JCO.2005.11.353.
35. Della Peruta M, Martinelli G, Moratti E, Pintani D, Vezzalini M, Mafficini A, Grafone T, Iacobucci I, Soverini S, Murineddu M, Vinante F, Tecchio C, Piras G, Gabbas A, Monne M, Sorio C. Protein tyrosine phosphatase receptor type {gamma} is a functional tumor suppressor gene specifically downregulated in chronic myeloid leukemia. Cancer Res. 2010;70(21):8896–906. doi:10.1158/0008-5472.CAN-10-0258.
36. Feng H, Shuda M, Chang Y, Moore PS. Clonal integration of a polyomavirus in human Merkel cell carcinoma. Science. 2008;319(5866):1096–100. doi:10.1126/science.1152586.
37. Helmbold P, Lahtz C, Enk A, Herrmann-Trost P, Marsch W, Kutzner H, Dammann RH. Frequent occurrence of RASSF1A promoter hypermethylation and Merkel cell polyomavirus in Merkel cell carcinoma. Mol Carcinog. 2009;48(10):903–9. doi:10.1002/mc.20540.
38. Cheung AK, Lung HL, Hung SC, Law EW, Cheng Y, Yau WL, Bangarusamy DK, Miller LD, Liu ET, Shao JY, Kou CW, Chua D, Zabarovsky ER, Tsao SW, Stanbridge EJ, Lung ML. Functional analysis of a cell cycle-associated, tumor-suppressive gene, protein tyrosine phosphatase receptor type G, in nasopharyngeal carcinoma. Cancer Res. 2008;68(19):8137–45.

39. Benton G, Kleinman HK, George J, Arnaoutova I. Multiple uses of basement membrane-like matrix (BME/Matrigel) in vitro and in vivo with cancer cells. Int J Cancer. 2010;128(8): 1751–7.
40. Sorby M, Sandstrom J, Ostman A. An extracellular ligand increases the specific activity of the receptor-like protein tyrosine phosphatase DEP-1. Oncogene. 2001;20(37):5219–24.
41. van Niekerk CC, Poels LG. Reduced expression of protein tyrosine phosphatase gamma in lung and ovarian tumors. Cancer Lett. 1999;137(1):61–73.
42. Vezzalini M, Mombello A, Menestrina F, Mafficini A, Della Peruta M, van Niekerk C, Barbareschi M, Scarpa A, Sorio C. Expression of transmembrane protein tyrosine phosphatase gamma (PTPgamma) in normal and neoplastic human tissues. Histopathology. 2007;50(5):615–28. doi:10.1111/j.1365-2559.2007.02661.x.
43. LaForgia S, Morse B, Levy J, Barnea G, Cannizzaro LA, Li F, Nowell PC, Boghosian-Sell L, Glick J, Weston A, Harris CC, Drabkin H, Patterson D, Croce CM, Schlessinger J, Heubner K. Receptor protein-tyrosine phosphatase γ is a candidate tumor suppressor gene at human chromosome region 3p21. Proc Natl Acad Sci U S A. 1991;88:5036–40.
44. Kastury K, Ohta M, Lasota J, Moir D, Dorman T, LaForgia S, Druck T, Huebner K. Structure of the human receptor tyrosine phosphatase gamma gene (PTPRG) and relation to the familial RCC t(3;8) chromosome translocation. Genomics. 1996;32(2):225–35.
45. Liu S, Sugimoto Y, Sorio C, Tecchio C, Lin YC. Function analysis of estrogenically regulated protein tyrosine phosphatase gamma (PTPgamma) in human breast cancer cell line MCF-7. Oncogene. 2004;23(6):1256–62.
46. Shu ST, Sugimoto Y, Liu S, Chang HL, Ye W, Wang LS, Huang YW, Yan P, Lin YC. Function and regulatory mechanisms of the candidate tumor suppressor receptor protein tyrosine phosphatase gamma (PTPRG) in breast cancer cells. Anticancer Res. 2010;30(6):1937–46.
47. Shintani T, Noda M. Protein tyrosine phosphatase receptor type Z dephosphorylates TrkA receptors and attenuates NGF-dependent neurite outgrowth of PC12 cells. J Biochem. 2008;144(2):259–66. doi:10.1093/jb/mvn064.
48. Barr AJ, Ugochukwu E, Lee WH, King ONF, Filippakopoulos P, Alfano I, Savitsky P, Burgess-Brown NA, Müller S, Knapp S. Large-scale structural analysis of the classical human protein tyrosine phosphatome. Cell. 2009;136(2):352–63. doi:10.1016/j.cell.2008.11.038.
49. Andersen JN, Mortensen OH, Peters GH, Drake PG, Iversen LF, Olsen OH, Jansen PG, Andersen HS, Tonks NK, Moller NP. Structural and evolutionary relationships among protein tyrosine phosphatase domains. Mol Cell Biol. 2001;21(21):7117–36.
50. Johnson KG, Van Vactor D. Receptor protein tyrosine phosphatases in nervous system development. Physiol Rev. 2003;83(1):1–24.
51. Uetani N, Kato K, Ogura H, Mizuno K, Kawano K, Mikoshiba K, Yakura H, Asano M, Iwakura Y. Impaired learning with enhanced hippocampal long-term potentiation in PTPdelta-deficient mice. EMBO J. 2000;19(12):2775–85.
52. Uetani N, Asano M, Mizuno K, Yakura H, Iwakura Y. Targeted disruption of the murine protein tyrosine phosphatase delta (MPTPdelta) results in growth retardation and behavioural abnormalities. In: Yakura H, editor. Kinases and phosphatases in lymphocyte and neuronal signalling. Tokyo: Springer; 1997. p. 313–4.
53. Navis AC, van den Eijnden M, Schepens JTG, Hooft van Huijsduijnen R, Wesseling P, Hendriks WJAJ. Protein tyrosine phosphatases in glioma biology. Acta Neuropathol. 2010;119(2):157–75. doi:10.1007/s00401-009-0614-0.
54. Solomon DA, Kim JS, Cronin JC, Sibenaller Z, Ryken T, Rosenberg SA, Ressom H, Jean W, Bigner D, Yan H, Samuels Y, Waldman T. Mutational inactivation of PTPRD in glioblastoma multiforme and malignant melanoma. Cancer Res. 2008;68(24):10300–6.
55. Veeriah S, Brennan C, Meng S, Singh B, Fagin JA, Solit DB, Paty PB, Rohle D, Vivanco I, Chmielecki J, Pao W, Ladanyi M, Gerald WL, Liau L, Cloughesy TC, Mischel PS, Sander C, Taylor B, Schultz N, Major J, Heguy A, Fang F, Mellinghoff IK, Chan TA. The tyrosine phosphatase PTPRD is a tumor suppressor that is frequently inactivated and mutated in glioblastoma and other human cancers. Proc Natl Acad Sci U S A. 2009;106(23):9435–40. doi:0900571106 [pii] 10.1073/pnas.0900571106.

56. Brodeur GM. Neuroblastoma: biological insights into a clinical enigma. Nat Rev Cancer. 2003;3(3):203–16.
57. Stallings RL, Nair P, Maris JM, Catchpoole D, McDermott M, O'Meara A, Breatnach F. High-resolution analysis of chromosomal breakpoints and genomic instability identifies PTPRD as a candidate tumor suppressor gene in neuroblastoma. Cancer Res. 2006;66(7):3673–80.
58. Nair P, De Preter K, Vandesompele J, Speleman F, Stallings RL. Aberrant splicing of the PTPRD gene mimics microdeletions identified at this locus in neuroblastomas. Genes Chromosomes Cancer. 2008;47(3):197–202. doi:10.1002/gcc.20521.
59. De Preter K, Vandesompele J, Heimann P, Yigit N, Beckman S, Schramm A, Eggert A, Stallings RL, Benoit Y, Renard M, De Paepe A, Laureys G, Pahlman S, Speleman F. Human fetal neuroblast and neuroblastoma transcriptome analysis confirms neuroblast origin and highlights neuroblastoma candidate genes. Genome Biol. 2006;7(9):R84.
60. Molenaar JJ, Koster J, Zwijnenburg DA, van Sluis P, Valentijn LJ, van der Ploeg I, Hamdi M, van Nes J, Westerman BA, van Arkel J, Ebus ME, Haneveld F, Lakeman A, Schild L, Molenaar P, Stroeken P, van Noesel MM, Ora I, Santo EE, Caron HN, Westerhout EM, Versteeg R. Sequencing of neuroblastoma identifies chromothripsis and defects in neuritogenesis genes. Nature. 2012;483:589–93. doi:10.1038/nature10910.
61. Meehan M, Parthasarathi L, Moran N, Jefferies CA, Foley N, Lazzari E, Murphy D, Ryan J, Ortiz B, Fabius AW, Chan TA, Stallings RL. Protein tyrosine phosphatase receptor delta acts as a neuroblastoma tumor suppressor by destabilizing the aurora kinase a oncogene. Mol Cancer. 2012;11(1):6. doi:10.1186/1476-4598-11-6.
62. Clark O, Schmidt F, Coles CH, Tchetchelnitski V, Stoker AW. Functional analysis of the putative tumor suppressor PTPRD in neuroblastoma cells. Cancer Invest. 2012;30(5):422–32. doi:10.3109/07357907.2012.675383.
63. Stark M, Hayward N. Genome-wide loss of heterozygosity and copy number analysis in melanoma using high-density single-nucleotide polymorphism arrays. Cancer Res. 2007;67(6):2632–42. doi:10.1158/0008-5472.CAN-06-4152.
64. Solomon DA, Kim JS, Yang XR, Tucker MA, Goldstein AM, Samuels Y, Waldman T. Lack of inherited mutations of PTPRD in familial melanoma and melanoma-astrocytoma syndrome. Pigment Cell Melanoma Res. 2009;22(4):489–91. doi:10.1111/j.1755-148X.2009.00587.x.
65. Lambert SR, Harwood CA, Purdie KJ, Gulati A, Matin RN, Romanowska M, Cerio R, Kelsell DP, Leigh IM, Proby CM. Metastatic cutaneous squamous cell carcinoma shows frequent deletion in the protein tyrosine phosphatase receptor type D gene. Int J Cancer. 2012;131(3):E216–26. doi:10.1002/ijc.27333.
66. Nagayama K, Kohno T, Sato M, Arai Y, Minna JD, Yokota J. Homozygous deletion scanning of the lung cancer genome at a 100-kb resolution. Genes Chromosomes Cancer. 2007;46(11):1000–10. doi:10.1002/gcc.20485.
67. Sato M, Takahashi K, Nagayama K, Arai Y, Ito N, Okada M, Minna JD, Yokota J, Kohno T. Identification of chromosome arm 9p as the most frequent target of homozygous deletions in lung cancer. Genes Chromosomes Cancer. 2005;44(4):405–14. doi:10.1002/gcc.20253.
68. Zhao X, Weir BA, LaFramboise T, Lin M, Beroukhim R, Garraway L, Beheshti J, Lee JC, Naoki K, Richards WG, Sugarbaker D, Chen F, Rubin MA, Janne PA, Girard L, Minna J, Christiani D, Li C, Sellers WR, Meyerson M. Homozygous deletions and chromosome amplifications in human lung carcinomas revealed by single nucleotide polymorphism array analysis. Cancer Res. 2005;65(13):5561–70.
69. Weir BA, Woo MS, Getz G, Perner S, Ding L, Beroukhim R, Lin WM, Province MA, Kraja A, Johnson LA, Shah K, Sato M, Thomas RK, Barletta JA, Borecki IB, Broderick S, Chang AC, Chiang DY, Chirieac LR, Cho J, Fujii Y, Gazdar AF, Giordano T, Greulich H, Hanna M, Johnson BE, Kris MG, Lash A, Lin L, Lindeman N, Mardis ER, McPherson JD, Minna JD, Morgan MB, Nadel M, Orringer MB, Osborne JR, Ozenberger B, Ramos AH, Robinson J, Roth JA, Rusch V, Sasaki H, Shepherd F, Sougnez C, Spitz MR, Tsao MS, Twomey D, Verhaak RG, Weinstock GM, Wheeler DA, Winckler W, Yoshizawa A, Yu S, Zakowski MF, Zhang Q, Beer DG, Wistuba II,

Watson MA, Garraway LA, Ladanyi M, Travis WD, Pao W, Rubin MA, Gabriel SB, Gibbs RA, Varmus HE, Wilson RK, Lander ES, Meyerson M. Characterizing the cancer genome in lung adenocarcinoma. Nature. 2007;450(7171):893–8. doi:10.1038/nature06358.

70. Nalesnik MA, Tseng G, Ding Y, Xiang GS, Zheng ZL, Yu Y, Marsh JW, Michalopoulos GK, Luo JH. Gene deletions and amplifications in human hepatocellular carcinomas: correlation with hepatocyte growth regulation. Am J Pathol. 2012;180(4):1495–508. doi:10.1016/j.ajpath.2011.12.021.
71. Urushibara N, Karasaki H, Nakamura K, Mizuno Y, Ogawa K, Kikuchi K. The selective reduction in PTPdelta expression in hepatomas. Int J Oncol. 1998;12(3):603–7.
72. Li X, Tan X, Yu Y, Chen H, Chang W, Hou J, Xu D, Ma L, Cao G. D9S168 microsatellite alteration predicts a poor prognosis in patients with clear cell renal cell carcinoma and correlates with the down-regulation of protein tyrosine phosphatase receptor delta. Cancer. 2011;117(18):4201–11. doi:10.1002/cncr.26028.
73. Konishi N, Tsujikawa K, Yamamoto H, Ishida E, Nakamura M, Shimada K, Yane K, Yamashita H, Noguchi S. Overexpression of leucocyte common antigen (LAR) P-subunit in thyroid carcinomas. Br J Cancer. 2003;88(8):1223–8. doi:10.1038/sj.bjc.6600876.
74. Yang T, Zhang JS, Massa SM, Han X, Longo FM. Leukocyte common antigen-related tyrosine phosphatase receptor: increased expression and neuronal-type splicing in breast cancer cells and tissue. Mol Carcinog. 1999;25(2):139–49.
75. Levea CM, McGary CT, Symons JR, Mooney RA. PTP LAR expression compared to prognostic indices in metastatic and non-metastatic breast cancer. Breast Cancer Res Treat. 2000;64(2):221–8.
76. Yang T, Massa SM, Longo FM. LAR protein tyrosine phosphatase receptor associates with TrkB and modulates neurotrophic signaling pathways. J Neurobiol. 2006;66(13):1420–36.
77. Zheng W, Lennartsson J, Hendriks W, Heldin CH, Hellberg C. The LAR protein tyrosine phosphatase enables PDGF beta-receptor activation through attenuation of the c-Abl kinase activity. Cell Signal. 2011;23(6):1050–6. doi:10.1016/j.cellsig.2011.01.024.
78. Kulas DT, Goldstein BJ, Mooney RA. The transmembrane protein-tyrosine phosphatase LAR modulates signaling by multiple receptor tyrosine kinases. J Biol Chem. 1996;271(2):748–54.
79. Weng LP, Wang X, Yu Q. Transmembrane tyrosine phosphatase LAR induces apoptosis by dephosphorylating and destabilizing p130Cas. Genes Cells. 1999;4(3):185–96.
80. Matozaki T, Suzuki T, Uchida T, Inazawa J, Ariyama T, Matsuda K, Horita K, Noguchi H, Mizuno H, Sakamoto C, et al. Molecular cloning of a human transmembrane-type protein tyrosine phosphatase and its expression in gastrointestinal cancers. J Biol Chem. 1994;269(3):2075–81.
81. Nagano H, Noguchi T, Inagaki K, Yoon S, Matozaki T, Itoh H, Kasuga M, Hayashi Y. Downregulation of stomach cancer-associated protein tyrosine phosphatase-1 (SAP-1) in advanced human hepatocellular carcinoma. Oncogene. 2003;22(30):4656–63. doi:10.1038/sj.onc.1206588.
82. Seo Y, Matozaki T, Tsuda M, Hayashi Y, Itoh H, Kasuga M. Overexpression of SAP-1, a transmembrane-type protein tyrosine phosphatase, in human colorectal cancers. Biochem Biophys Res Commun. 1997;231(3):705–11.
83. Sadakata H, Okazawa H, Sato T, Supriatna Y, Ohnishi H, Kusakari S, Murata Y, Ito T, Nishiyama U, Minegishi T, Harada A, Matozaki T. SAP-1 is a microvillus-specific protein tyrosine phosphatase that modulates intestinal tumorigenesis. Genes Cells. 2009;14(3):295–308.
84. Jandt E, Denner K, Kovalenko M, Ostman A, Bohmer FD. The protein-tyrosine phosphatase DEP-1 modulates growth factor-stimulated cell migration and cell-matrix adhesion. Oncogene. 2003;22(27):4175–85.
85. Grazia Lampugnani M, Zanetti A, Corada M, Takahashi T, Balconi G, Breviario F, Orsenigo F, Cattelino A, Kemler R, Daniel TO, Dejana E. Contact inhibition of VEGF-induced proliferation requires vascular endothelial cadherin, beta-catenin, and the phosphatase DEP-1/CD148. J Cell Biol. 2003;161(4):793–804.
86. Trapasso F, Drusco A, Costinean S, Alder H, Aqeilan RI, Iuliano R, Gaudio E, Raso C, Zanesi N, Croce CM, Fusco A. Genetic ablation of PTPRJ, a mouse cancer susceptibility gene, results in normal growth and development and does not predispose to spontaneous tumorigenesis. DNA Cell Biol. 2006;25(6):376–82.

87. Keane MM, Lowrey GA, Ettenberg SA, Dayton MA, Lipkowitz S. The protein tyrosine phosphatase DEP-1 is induced during differentiation and inhibits growth of breast cancer cells. Cancer Res. 1996;56(18):4236–43.
88. Smart CE, Askarian Amiri ME, Wronski A, Dinger ME, Crawford J, Ovchinnikov DA, Vargas AC, Reid L, Simpson PT, Song S, Wiesner C, French JD, Dave RK, da Silva L, Purdon A, Andrew M, Mattick JS, Lakhani SR, Brown MA, Kellie S. Expression and function of the protein tyrosine phosphatase receptor J (PTPRJ) in normal mammary epithelial cells and breast tumors. PLoS One. 2012;7(7), e40742. doi:10.1371/journal.pone.0040742.
89. Tarcic G, Boguslavsky SK, Wakim J, Kiuchi T, Liu A, Reinitz F, Nathanson D, Takahashi T, Mischel PS, Ng T, Yarden Y. An unbiased screen identifies DEP-1 tumor suppressor as a phosphatase controlling EGFR endocytosis. Curr Biol. 2009;19(21):1788–98.
90. Ruivenkamp CA, van Wezel T, Zanon C, Stassen AP, Vlcek C, Csikos T, Klous AM, Tripodis N, Perrakis A, Boerrigter L, Groot PC, Lindeman J, Mooi WJ, Meijjer GA, Scholten G, Dauwerse H, Paces V, van Zandwijk N, van Ommen GJ, Demant P. PTPRJ is a candidate for the mouse colon-cancer susceptibility locus Scc1 and is frequently deleted in human cancers. Nat Genet. 2002;31(3):295–300. doi:10.1038/ng903.
91. Mita Y, Yasuda Y, Sakai A, Yamamoto H, Toyooka S, Gunduz M, Tanabe S, Naomoto Y, Ouchida M, Shimizu K. Missense polymorphisms of PTPRJ and PTPN13 genes affect susceptibility to a variety of human cancers. J Cancer Res Clin Oncol. 2010;136(2):249–59. doi:10.1007/s00432-009-0656-7.
92. Zhang L, Martelli ML, Battaglia C, Trapasso F, Tramontano D, Viglietto G, Porcellini A, Santoro M, Fusco A. Thyroid cell transformation inhibits the expression of a novel rat protein tyrosine phosphatase. Exp Cell Res. 1997;235(1):62–70. doi:10.1006/excr.1997.3659.
93. Trapasso F, Iuliano R, Boccia A, Stella A, Visconti R, Bruni P, Baldassarre G, Santoro M, Viglietto G, Fusco A. Rat protein tyrosine phosphatase eta suppresses the neoplastic phenotype of retrovirally transformed thyroid cells through the stabilization of p27(Kip1). Mol Cell Biol. 2000;20(24):9236–46.
94. Iuliano R, Trapasso F, Le Pera I, Schepis F, Sama I, Clodomiro A, Dumon KR, Santoro M, Chiariotti L, Viglietto G, Fusco A. An adenovirus carrying the rat protein tyrosine phosphatase eta suppresses the growth of human thyroid carcinoma cell lines in vitro and in vivo. Cancer Res. 2003;63(4):882–6.
95. Iuliano R, Le Pera I, Cristofaro C, Baudi F, Arturi F, Pallante P, Martelli ML, Trapasso F, Chiariotti L, Fusco A. The tyrosine phosphatase PTPRJ/DEP-1 genotype affects thyroid carcinogenesis. Oncogene. 2004;23(52):8432–8.
96. Iuliano R, Palmieri D, He H, Iervolino A, Borbone E, Pallante P, Cianflone A, Nagy R, Alder H, Calin GA, Trapasso F, Giordano C, Croce CM, de la Chapelle A, Fusco A. Role of PTPRJ genotype in papillary thyroid carcinoma risk. Endocr Relat Cancer. 2010;17(4):1001–6. doi:10.1677/erc-10-0143.
97. Trapasso F, Yendamuri S, Dumon KR, Iuliano R, Cesari R, Feig B, Seto R, Infante L, Ishii H, Vecchione A, During MJ, Croce CM, Fusco A. Restoration of receptor-type protein tyrosine phosphatase {eta} function inhibits human pancreatic carcinoma cell growth in vitro and in vivo. Carcinogenesis. 2004;25(11):2107–14.
98. Frew IJ, Krek W. Multitasking by pVHL in tumour suppression. Curr Opin Cell Biol. 2007;19(6):685–90. doi:10.1016/j.ceb.2007.10.001.
99. Boysen G, Bausch-Fluck D, Thoma CR, Nowicka AM, Stiehl DP, Cima I, Luu VD, von Teichman A, Hermanns T, Sulser T, Ingold-Heppner B, Fankhauser N, Wenger RH, Krek W, Schraml P, Wollscheid B, Moch H. Identification and functional characterization of pVHL-dependent cell surface proteins in renal cell carcinoma. Neoplasia. 2012;14(6):535–46.
100. Casagrande S, Ruf M, Rechsteiner M, Morra L, Brun-Schmid S, von Teichman A, Krek W, Schraml P, Moch H. The protein-tyrosine phosphatase receptor type J is regulated by the pVHL-HIF axis in clear cell renal cell carcinoma. J Pathol. 2012;229(4):525–34. doi:10.1002/path.4107.
101. Petermann A, Haase D, Wetzel A, Balavenkatraman KK, Tenev T, Guhrs KH, Friedrich S, Nakamura M, Mawrin C, Bohmer FD. Loss of the protein-tyrosine phosphatase DEP-1/PTPRJ drives meningioma cell motility. Brain Pathol. 2011;21(4):405–18. doi:10.1111/j.1750-3639.2010.00464.x.

102. Lampugnani MG, Orsenigo F, Gagliani MC, Tacchetti C, Dejana E. Vascular endothelial cadherin controls VEGFR-2 internalization and signaling from intracellular compartments. J Cell Biol. 2006;174(4):593–604. doi:10.1083/jcb.200602080.
103. ten Freyhaus H, Dagnell M, Leuchs M, Vantler M, Berghausen EM, Caglayan E, Weissmann N, Dahal BK, Schermuly RT, Ostman A, Kappert K, Rosenkranz S. Hypoxia enhances platelet-derived growth factor signaling in the pulmonary vasculature by down-regulation of protein tyrosine phosphatases. Am J Respir Crit Care Med. 2011;183(8):1092–102. doi:10.1164/rccm.200911-1663OC.
104. Takahashi T, Takahashi K, Mernaugh RL, Tsuboi N, Liu H, Daniel TO. A monoclonal antibody against CD148, a receptor-like tyrosine phosphatase, inhibits endothelial-cell growth and angiogenesis. Blood. 2006;108(4):1234–42.
105. Zhang Y, Siebert R, Matthiesen P, Yang Y, Ha H, Schlegelberger B. Cytogenetical assignment and physical mapping of the human R-PTP-kappa gene (PTPRK) to the putative tumor suppressor gene region 6q22.2-q22.3. Genomics. 1998;51(2):309–11. doi:10.1006/geno.1998.5323.
106. McArdle L, Rafferty M, Maelandsmo GM, Bergin O, Farr CJ, Dervan PA, O'Loughlin S, Herlyn M, Easty DJ. Protein tyrosine phosphatase genes downregulated in melanoma. J Invest Dermatol. 2001;117(5):1255–60.
107. Nakamura M, Kishi M, Sakaki T, Hashimoto H, Nakase H, Shimada K, Ishida E, Konishi N. Novel tumor suppressor loci on 6q22-23 in primary central nervous system lymphomas. Cancer Res. 2003;63(4):737–41.
108. Sanchez-Morgan N, Kirsch KH, Trackman PC, Sonenshein GE. The lysyl oxidase propeptide interacts with the receptor-type protein tyrosine phosphatase kappa and inhibits beta-catenin transcriptional activity in lung cancer cells. Mol Cell Biol. 2011;31(16):3286–97. doi:10.1128/mcb.01426-10.
109. Anders L, Mertins P, Lammich S, Murgia M, Hartmann D, Saftig P, Haass C, Ullrich A. Furin-, ADAM 10-, and {gamma}-secretase-mediated cleavage of a receptor tyrosine phosphatase and regulation of {beta}-Catenin's transcriptional activity. Mol Cell Biol. 2006;26(10):3917–34.
110. Novellino L, De Filippo A, Deho P, Perrone F, Pilotti S, Parmiani G, Castelli C. PTPRK negatively regulates transcriptional activity of wild type and mutated oncogenic beta-catenin and affects membrane distribution of beta-catenin/E-cadherin complexes in cancer cells. Cell Signal. 2008;20(5):872–83.
111. Ensslen-Craig SE, Brady-Kalnay SM. Receptor protein tyrosine phosphatases regulate neural development and axon guidance. Dev Biol. 2004;275(1):12–22.
112. Aricescu A, Jones E. Immunoglobulin superfamily cell adhesion molecules: zippers and signals. 2007. Curr Opin Cell Biol.
113. Burgoyne AM, Phillips-Mason PJ, Burden-Gulley SM, Robinson S, Sloan AE, Miller RH, Brady-Kalnay SM. Proteolytic cleavage of protein tyrosine phosphatase mu regulates glioblastoma cell migration. Cancer Res. 2009;69(17):6960–8. doi:10.1158/0008-5472.can-09-0863.
114. Burgoyne AM, Palomo JM, Phillips-Mason PJ, Burden-Gulley SM, Major DL, Zaremba A, Robinson S, Sloan AE, Vogelbaum MA, Miller RH, Brady-Kalnay SM. PTPmu suppresses glioma cell migration and dispersal. Neuro Oncol. 2009;11(6):767–78.
115. Kaur H, Burden-Gulley SM, Phillips-Mason PJ, Basilion JP, Sloan AE, Brady-Kalnay SM. Protein tyrosine phosphatase mu regulates glioblastoma cell growth and survival in vivo. Neuro Oncol. 2012;14(5):561–73. doi:10.1093/neuonc/nos066.
116. Burden-Gulley SM, Gates TJ, Burgoyne AM, Cutter JL, Lodowski DT, Robinson S, Sloan AE, Miller RH, Basilion JP, Brady-Kalnay SM. A novel molecular diagnostic of glioblastomas: detection of an extracellular fragment of protein tyrosine phosphatase mu. Neoplasia. 2010;12(4):305–16.
117. Haapasalo A, Kim DY, Carey BW, Turunen MK, Pettingell WH, Kovacs DM. Presenilin/gamma-secretase-mediated cleavage regulates association of LAR receptor tyrosine phosphatase with beta-catenin. J Biol Chem. 2007;11(3):897–907.
118. Gil-Henn H, Volohonsky G, Toledano-Katchalski H, Gandre S, Elson A. Generation of novel cytoplasmic forms of protein tyrosine phosphatase epsilon by proteolytic processing and translational control. Oncogene. 2000;19(38):4375–84.

119. Quintavalle C, Garofalo M, Zanca C, Romano G, Iaboni M, del Basso De Caro M, Martinez-Montero JC, Incoronato M, Nuovo G, Croce CM, Condorelli G. miR-221/222 overexpression in human glioblastoma increases invasiveness by targeting the protein phosphate PTPmu. Oncogene. 2012;31(7):858–68. doi:10.1038/onc.2011.280.
120. Hellberg CB, Burden-Gulley SM, Pietz GE, Brady-Kalnay SM. Expression of the receptor protein tyrosine phosphatase, PTPmu, restores E-cadherin-dependent adhesion in human prostate carcinoma cells. J Biol Chem. 2002;277(13):11165–73.
121. Stepanek L, Sun QL, Wang J, Wang C, Bixby JL. CRYP-2/cPTPRO is a neurite inhibitory repulsive guidance cue for retinal neurons in vitro. J Cell Biol. 2001;154(4):867–78.
122. Stepanek L, Stoker AW, Stoeckli E, Bixby JL. Receptor tyrosine phosphatases guide vertebrate motor axons during development. J Neurosci. 2005;25(15):3813–23.
123. Motiwala T, Ghoshal K, Das A, Majumder S, Weichenhan D, Wu YZ, Holman K, James SJ, Jacob ST, Plass C. Suppression of the protein tyrosine phosphatase receptor type O gene (PTPRO) by methylation in hepatocellular carcinomas. Oncogene. 2003;22(41): 6319–31.
124. Motiwala T, Kutay H, Ghoshal K, Bai S, Seimiya H, Tsuruo T, Suster S, Morrison C, Jacob ST. Protein tyrosine phosphatase receptor-type O (PTPRO) exhibits characteristics of a candidate tumor suppressor in human lung cancer. Proc Natl Acad Sci U S A. 2004;101(38):13844–9.
125. Motiwala T, Majumder S, Kutay H, Smith DS, Neuberg DS, Lucas DM, Byrd JC, Grever M, Jacob ST. Methylation and silencing of protein tyrosine phosphatase receptor type O in chronic lymphocytic leukemia. Clin Cancer Res. 2007;13(11):3174–81.
126. Ramaswamy B, Majumder S, Roy S, Ghoshal K, Kutay H, Datta J, Younes M, Shapiro CL, Motiwala T, Jacob ST. Estrogen-mediated suppression of the gene encoding protein tyrosine phosphatase PTPRO in human breast cancer: mechanism and role in tamoxifen sensitivity. Mol Endocrinol. 2009;23(2):176–87. doi:10.1210/me.2008-0211.
127. Faux C, Hawadle M, Nixon J, Wallace A, Lee S, Murray S, Stoker A. PTPsigma binds and dephosphorylates neurotrophin receptors and can suppress NGF-dependent neurite outgrowth from sensory neurons. Biochim Biophys Acta. 2007;1773(11):1689–700.
128. Rashid-Doubell F, McKinnell I, Aricescu AR, Sajnani G, Stoker A. Chick PTPσ regulates the targeting of retinal axons within the optic tectum. J Neurosci. 2002;22(12):5024–33.
129. Shen Y, Tenney AP, Busch SA, Horn KP, Cuascut FX, Liu K, He Z, Silver J, Flanagan JG. PTPsigma is a receptor for chondroitin sulfate proteoglycan, an inhibitor of neural regeneration. Science. 2009;326(5952):592–6. doi:1178310 [pii] 10.1126/science.1178310.
130. Fry EJ, Chagnon MJ, Lopez-Vales R, Tremblay ML, David S. Corticospinal tract regeneration after spinal cord injury in receptor protein tyrosine phosphatase sigma deficient mice. Glia. 2010;58(4):423–33. doi:10.1002/glia.20934.
131. Stoker AW, Gehrig B, Haj F, Bay BH. Axonal localisation of the CAM-like tyrosine phosphatase CRYP alpha: a signalling molecule of embryonic growth cones. Development. 1995;121(6):1833–44.
132. Suarez Pestana E, Tenev T, Gross S, Stoyanov B, Ogata M, Bohmer FD. The transmembrane protein tyrosine phosphatase RPTPsigma modulates signaling of the epidermal growth factor receptor in A431 cells. Oncogene. 1999;18(28):4069–79.
133. Coles CH, Shen Y, Tenney AP, Siebold C, Sutton GC, Lu W, Gallagher JT, Jones EY, Flanagan JG, Aricescu AR. Proteoglycan-specific molecular switch for RPTPsigma clustering and neuronal extension. Science. 2011;332(6028):484–8. doi:10.1126/science.1200840.
134. Aricescu AR, McKinnell IW, Halfter W, Stoker AW. Heparan sulfate proteoglycans are ligands for receptor protein tyrosine phosphatase sigma. Mol Cell Biol. 2002;22(6):1881–92.
135. Takahashi H, Arstikaitis P, Prasad T, Bartlett TE, Wang YT, Murphy TH, Craig AM. Postsynaptic TrkC and presynaptic PTPσ; function as a bidirectional excitatory synaptic organizing complex. Neuron. 2011;69(2):287–303. doi:10.1016/j.neuron.2010.12.024.
136. Martin KR, Xu Y, Looyenga BD, Davis RJ, Wu CL, Tremblay ML, Xu HE, MacKeigan JP. Identification of PTPsigma as an autophagic phosphatase. J Cell Sci. 2011;124(Pt 5):812–9. doi:10.1242/jcs.080341.

137. Chen N, Karantza V. Autophagy as a therapeutic target in cancer. Cancer Biol Ther. 2011;11(2):157–68.
138. Chen S, Rehman SK, Zhang W, Wen A, Yao L, Zhang J. Autophagy is a therapeutic target in anticancer drug resistance. Biochim Biophys Acta. 2010;1806(2):220–9. doi:10.1016/j.bbcan.2010.07.003.
139. Korff S, Woerner SM, Yuan YP, Bork P, von Knebel DM, Gebert J. Frameshift mutations in coding repeats of protein tyrosine phosphatase genes in colorectal tumors with microsatellite instability. BMC Cancer. 2008;8(1):329.
140. Morris LG, Taylor BS, Bivona TG, Gong Y, Eng S, Brennan CW, Kaufman A, Kastenhuber ER, Banuchi VE, Singh B, Heguy A, Viale A, Mellinghoff IK, Huse J, Ganly I, Chan TA. Genomic dissection of the epidermal growth factor receptor (EGFR)/PI3K pathway reveals frequent deletion of the EGFR phosphatase PTPRS in head and neck cancers. Proc Natl Acad Sci U S A. 2011;108(47):19024–9. doi:10.1073/pnas.1111963108.
141. Laczmanska I, Karpinski P, Bebenek M, Sedziak T, Ramsey D, Szmida E, Sasiadek MM. Protein tyrosine phosphatase receptor-like genes are frequently hypermethylated in sporadic colorectal cancer. J Hum Genet. 2013;58(1):11–5.
142. Zhao Y, Zhang X, Guda K, Lawrence E, Sun Q, Watanabe T, Iwakura Y, Asano M, Wei L, Yang Z, Zheng W, Dawson D, Willis J, Markowitz SD, Satake M, Wang Z. Identification and functional characterization of paxillin as a target of protein tyrosine phosphatase receptor T. Proc Natl Acad Sci U S A. 2010;107(6):2592–7. doi:10.1073/pnas.0914884107.
143. Levy JB, Canoll PD, Silvennoinen O, Barnea G, Morse B, Honegger AM, Huang JT, Cannizzaro LA, Park SH, Druck T, Huebner K, Sap J, Ehrlich M, Musacchio JM, Schlessinger J. The cloning of a receptor-tpye protein tyrosine phosphatase expressed in the central nervous system. J Biol Chem. 1993;268(14):10573–81.
144. Krueger NX, Saito H. A human transmembrane protein-tyrosine-phosphatase, PTPζ, is expressed in brain and has an N-terminal receptor domain homologous to carbonic anhydrases. Proc Natl Acad Sci U S A. 1992;89:7417–21.
145. Maurel P, Rauch U, Flad M, Margolis RK, Margolis RU. Phosphacan, a chondroitin sulfate proteoglycan of brain that interacts with neurons and neural cell-adhesion molecules, is an extracellular variant of a receptor-type protein tyrosine phosphatase. Proc Natl Acad Sci U S A. 1994;91(7):2512–6.
146. Maeda N, Nishiwaki T, Shintani T, Hamanaka H, Noda M. 6B4 proteoglycan/phosphacan, an extracellular variant of receptor-like protein-tyrosine phosphatase zeta/RPTPbeta, binds pleiotrophin/heparin- binding growth-associated molecule (HB-GAM). J Biol Chem. 1996;271(35):21446–52.
147. Peles E, Nativ M, Campbell PL, Sakurai T, Martinez R, Lev S, Clary DO, Schilling J, Barnea G, Plowman GD, et al. The carbonic anhydrase domain of receptor tyrosine phosphatase beta is a functional ligand for the axonal cell recognition molecule contactin. Cell. 1995;82(2):251–60.
148. Norman SA, Golfinos JG, Scheck AC. Expression of a receptor protein tyrosine phosphatase in human glial tumors [In Process Citation]. J Neurooncol. 1998;36(3):209–17.
149. Hagerstrand D, Smits A, Eriksson A, Sigurdardottir S, Olofsson T, Hartman M, Nister M, Kalimo H, Ostman A. Gene expression analyses of grade II gliomas and identification of rPTPbeta/zeta as a candidate oligodendroglioma marker. Neuro Oncol. 2008;10(1):2–9. doi:10.1215/15228517-2007-041.
150. Muller S, Kunkel P, Lamszus K, Ulbricht U, Lorente GA, Nelson AM, von Schack D, Chin DJ, Lohr SC, Westphal M, Melcher T. A role for receptor tyrosine phosphatase zeta in glioma cell migration. Oncogene. 2003;22(43):6661–8.
151. Foehr ED, Lorente G, Kuo J, Ram R, Nikolich K, Urfer R. Targeting of the receptor protein tyrosine phosphatase {beta} with a monoclonal antibody delays tumor growth in a glioblastoma model. Cancer Res. 2006;66(4):2271–8.
152. Ulbricht U, Brockmann MA, Aigner A, Eckerich C, Muller S, Fillbrandt R, Westphal M, Lamszus K. Expression and function of the receptor protein tyrosine phosphatase zeta and its ligand pleiotrophin in human astrocytomas. J Neuropathol Exp Neurol. 2003;62(12):1265–75.
153. Ulbricht U, Eckerich C, Fillbrandt R, Westphal M, Lamszus K. RNA interference targeting protein tyrosine phosphatase zeta/receptor-type protein tyrosine phosphatase beta suppresses glioblastoma growth in vitro and in vivo. J Neurochem. 2006;98(5):1497–506.

154. Orr B, Vanpoucke G, Grace OC, Smith L, Anderson RA, Riddick AC, Franco OE, Hayward SW, Thomson AA. Expression of pleiotrophin in the prostate is androgen regulated and it functions as an autocrine regulator of mesenchyme and cancer associated fibroblasts and as a paracrine regulator of epithelia. Prostate. 2011;71(3):305–17. doi:10.1002/pros.21244.
155. Charbonneau H, Tonks NK, Walsh KA, Fischer EH. The leukocyte common antigen (CD45): a putative receptor-linked protein tyrosine phosphatase. Proc Natl Acad Sci U S A. 1988;85(19):7182–6.
156. Palka HL, Park M, Tonks NK. Hepatocyte growth factor receptor tyrosine kinase met is a substrate of the receptor protein-tyrosine phosphatase DEP-1. J Biol Chem. 2003;278(8):5728–35.
157. Petermann A, Haase D, Wetzel A, Balavenkatraman KK, Tenev T, Guhrs KH, Friedrich S, Nakamura M, Mawrin C, Bohmer FD. Loss of the protein-tyrosine phosphatase DEP-1/PTPRJ drives meningioma cell motility. Brain Pathol. 2010. doi:10.1111/j.1750-3639.2010.00464.x.
158. Kovalenko M, Denner K, Sandstrom J, Persson C, Gross S, Jandt E, Vilella R, Bohmer F, Ostman A. Site-selective dephosphorylation of the platelet-derived growth factor beta-receptor by the receptor-like protein-tyrosine phosphatase DEP-1. J Biol Chem. 2000;275(21):16219–26.
159. Iervolino A, Iuliano R, Trapasso F, Viglietto G, Melillo RM, Carlomagno F, Santoro M, Fusco A. The receptor-type protein tyrosine phosphatase J antagonizes the biochemical and biological effects of RET-derived oncoproteins. Cancer Res. 2006;66(12):6280–7.
160. Kulas DT, Zhang WR, Goldstein BJ, Furlanetto RW, Mooney RA. Insulin receptor signaling is augmented by antisense inhibition of the protein tyrosine phosphatase LAR. J Biol Chem. 1995;270(6):2435–8.
161. Le Pera I, Iuliano R, Florio T, Susini C, Trapasso F, Santoro M, Chiariotti L, Schettini G, Viglietto G, Fusco A. The rat tyrosine phosphatase eta increases cell adhesion by activating c-Src through dephosphorylation of its inhibitory phosphotyrosine residue. Oncogene. 2005;24(19):3187–95.
162. Meng K, Rodriguez-Pena A, Dimitrov T, Chen W, Yamin M, Noda M, Deuel TF. Pleiotrophin signals increased tyrosine phosphorylation of beta beta-catenin through inactivation of the intrinsic catalytic activity of the receptor-type protein tyrosine phosphatase beta/zeta. Proc Natl Acad Sci U S A. 2000;97(6):2603–8.
163. Pariser H, Perez-Pinera P, Ezquerra L, Herradon G, Deuel TF. Pleiotrophin stimulates tyrosine phosphorylation of beta-adducin through inactivation of the transmembrane receptor protein tyrosine phosphatase beta/zeta. Biochem Biophys Res Commun. 2005;335(1): 232–9.
164. Pariser H, Ezquerra L, Herradon G, Perez-Pinera P, Deuel TF. Fyn is a downstream target of the pleiotrophin/receptor protein tyrosine phosphatase beta/zeta-signaling pathway: Regulation of tyrosine phosphorylation of Fyn by pleiotrophin. Biochem Biophys Res Commun. 2005;332(3):664–9.
165. Makinoshima H, Ishii G, Kojima M, Fujii S, Higuchi Y, Kuwata T, Ochiai A. PTPRZ1 regulates calmodulin phosphorylation and tumor progression in small-cell lung carcinoma. BMC Cancer. 2012;12(1):537. doi:10.1186/1471-2407-12-537.
166. Purdie KJ, Lambert SR, Teh M-T, Chaplin T, Molloy G, Raghavan M, Kelsell DP, Leigh IM, Harwood CA, Proby CM, Young BD. Allelic imbalances and microdeletions affecting the PTPRD gene in cutaneous squamous cell carcinomas detected using single nucleotide polymorphism microarray analysis. Genes Chromosomes Cancer. 2007;46(7):661–9. doi:10.1002/gcc.20447.
167. Calhoun ES, Hucl T, Gallmeier E, West KM, Arking DE, Maitra A, Iacobuzio-Donahue CA, Chakravarti A, Hruban RH, Kern SE. Identifying allelic loss and homozygous deletions in pancreatic cancer without matched normals using high-density single-nucleotide polymorphism arrays. Cancer Res. 2006;66(16):7920–8. doi:10.1158/0008-5472.CAN-06-0721.
168. Chiusaroli R, Knobler H, Luxenburg C, Sanjay A, Granot-Attas S, Tiran Z, Miyazaki T, Harmelin A, Baron R, Elson A. Tyrosine phosphatase epsilon is a positive regulator of osteoclast function in vitro and in vivo. Mol Biol Cell. 2004;15(1):234–44.
169. Ruivenkamp C, Hermsen M, Postma C, Klous A, Baak J, Meijer G, Demant P. LOH of PTPRJ occurs early in colorectal cancer and is associated with chromosomal loss of 18q12-21. Oncogene. 2003;22(22):3472–4.
170. Goldmann T, Otto F, Vollmer E. A receptor-type protein tyrosine phosphatase PTP zeta is expressed in human cutaneous melanomas. Folia Histochem Cytobiol. 2000;38(1):19–20.

Non-transmembrane PTPs in Cancer

Wiljan J.A.J. Hendriks and Frank-D. Böhmer

Abstract In this chapter, we will discuss the role of 14 non-transmembrane (NT) type protein-tyrosine phosphatases (PTPs, encoded by the genes *PTPN3, PTPN4, PTPN5, PTPN6, PTPN7, PTPN9, PTPN13, PTPN14, PTPN18, PTPN20, PTPN21, PTPN22, PTPN23*, and *PTPRR*) in cancer. The presentation of NTPTPs includes a brief description of general features of the individual molecules in structure and activity regulation, as well as key facts about their physiological functions. Genetic or epigenetic alterations of NTPTP genes in cancer cells and, if available, the molecular consequences for the specific phenotypic effects are subsequently discussed. A role for defects of PTPN6, PTPN13, or PTPN23 function in certain tumor types has been established best. For other NTPTPs, indications for putative roles in cancer rest on gene mutations in cancer tissues, phenotypes of cell lines with altered NTPTP status, and known physiological functions, but need confirmation in future studies. Clearly, NTPTP effects on cancer phenotypes are dependent on the specific cell context. Roles of NTPTPs in the cancer microenvironment and for therapeutic responses to anti-cancer drugs are emerging. Functions of NTPTPs in immune cells, notably of PTPN6, might also allow therapeutic exploitation in the future.

Keywords Non-receptor PTP • Non-transmembrane PTP • NTPTP • Cancer • PTPN3 • PTPN4 • PTPN6 • PTPN7 • PTPN9 • PTPN13 • PTPN14 • PTPN18 • PTPN20 • PTPN21 • PTPN22 • PTPN23 • PTPRR

Wiljan J.A.J. Hendriks (✉)
Department of Cell Biology, Radboud Institute for Molecular Life Sciences, Radboud University Medical Centre, Geert Grooteplein 28, 6525 GA Nijmegen, The Netherlands
e-mail: WiljanJAJ.Hendriks@radboudumc.nl

F.-D. Böhmer
Institute of Molecular Cell Biology, Center for Molecular Biomedicine, Jena University Hospital, Jena, Germany
e-mail: Boehmer@med.uni-jena.de

B.G. Neel, N. Tonks (eds.), *Protein Tyrosine Phosphatases in Cancer*,
DOI 10.1007/978-1-4939-3649-6_3

General Introduction

The classical non-receptor type (NT) PTPs comprise protein products of 18 genes, of which many encode multiple variants with in part proven differential functions (Table 1 and Fig. 1). Roles in cancer have been established for the members of this subfamily to a variable extent. NTPTPs for which an implication in cancer has been most intensely studied and firmly validated—PTPN1 (PTP1B), PTPN2 (TCPTP), PTPN11 (SHP-2), and PTPN12 (PTP-PEST)—will be covered in the specific Chaps. 4–7, respectively. Here, we will discuss the cancer relevance of the remaining NTPTP-encoding genes. We included *PTPRR* because its encoded protein products are highly related to the NTPTPs PTPN5 (STEP) and PTPN7 (HePTP), and as such, complete the R7 subtype of classical PTPs [1]. While it is already becoming clear that some of these PTPs are important players in the context of cancer, such as PTPN6 (SHP-1), PTPN13 (PTP-BAS, PTP1E, PTPL1, FAP1), and PTPN23 (HD-PTP), only few data exist for others (Table 2). This does not mean, however, that they are not important, but rather might reflect a need for more attention. For most of the genes discussed here, indeed some indications for an involvement in cancer do exist, such as relatively frequent mutations in certain cancer types (Table 2). However, in many cases, more work is needed to causally link such observations with alterations of the biological functions of the affected PTPs in cancer cells.

Multiple genetic lesions are required before a cell becomes malignant. These are directly or indirectly related to the multiple phenotypic aberrations, which have been termed cancer "hallmarks" [2]. It is therefore not surprising that loss- or gain-of-function of PTP genes usually only affects a subset of features of cancer cells or has only subtle effects in certain contexts (Table 2). Obviously, functional redundancies of different PTPs, provided they are co-expressed in a particular cell type, will also play a role for the magnitude of effects of PTP gene alterations. Depending on the context of cell type and other genetic lesions, cancer-associated changes of the PTP status may even cause opposite phenotypes. Although many of the PTPs we consider here appear to affect cell features mainly in a tumor-suppressive direction, they could promote cancer phenotypes in a few other contexts. For example, PTPN6 is likely to be tumor-suppressive in hematopoietic cells, but might be oncogenic when over-expressed in breast cancer. Similarly, PTPN13 promotes apoptosis in the context of hepatocellular carcinoma, but over-expression of PTPN13 has been associated with apoptosis resistance in Ewing's sarcoma (Table 2).

In the following sections, we will discuss in detail the roles of individual NTPTPs in cancer. We chose to group the PTPs according to their structural similarities, and thus use the order of NT-types as defined a decade ago [1]. This will facilitate highlighting common structural features and possible functional redundancies, as well as functional differences despite structural relatedness. We will first briefly present general features in structure and regulation of the respective NT type, and then summarize the current knowledge on physiological functions for the corresponding NTPTPs. This will provide an essential basis for the subsequent discussion of their specific roles in certain cancer types.

Table 1 General features of non-transmembrane PTPs

	Synonyms	Subfamily[a]	Chromosomal localization	Exons	Protein, alternate gene products	Expression	X-ray/NMR[b] structure	References
PTPN1	PTP1B	NT1	20q13.13	10	435 aa	See further details in Chap. 6		
PTPN2	TCPTP	NT1	18p11.21	9	415 aa	See further details in Chap. 5		
PTPN3	PTPH1	NT5	9q31	26	6 isoforms: 913, 868, 782, 737, 626 and 581 aa	Broadly expressed	2B49 (PTP domain)	[136, 137, 178]
PTPN4	PTP MEG-1	NT5	2q14.2	27	926 aa	Ubiquitously expressed	2CS5 (PDZ domain) 2I75 (PTP domain)	[139, 178]
PTPN5	STEP	R7	11p15.1	15	565, 533, 393, 232 aa	Central nervous system neurons, esp. in the striatal area	2BIJ, 2BV5, 2CJZ (all PTP domain only)	[178, 323, 352, 370]
PTPN6	SHP-1, SH-PTP1, HCP, PTP1C	NT2	12p13.31	17	Hematopoietic (595aa), epithelial (597aa), long form (SHP-1 L, 624aa)	Hematopoietic, epithelial, endothelial, neuronal, skeletal muscle	2B3O (full-length protein) 3PS5 (full-length, open conformation) 2YU7 (C-terminal SH2 domain)	[9, 49, 52, 53, 371–375]
PTPN7	PTPN7, HePTP, BPTP-4, LPTP, LC-PTP, PTPNI	R7	1q32.1	10	3 isoforms: 465, 434 and 399 residues	Primarily hematopoietic	2A3K, 2GP0, 2HVL (PTP domain) 2QDP, 3D42, 3D44, 3O4S, 3O4T, 3O4U (PTP with ERK peptide)	[338, 339, 352, 376, 377]
PTPN9	MEG-2	NT3	15q24.2	13	593aa	Widely expressed, including brain, lung, heart, liver, kidneys, spleen, and bone marrow	2PA5 (catalytic domain)	[98, 100, 101, 178]

(continued)

Table 1 (continued)

	Synonyms	Subfamily[a]	Chromosomal localization	Exons	Protein, alternate gene products	Expression	X-ray/NMR[b] structure	References
PTPN11	SHP-2	NT2	12q24.13	16	593 aa	See further details in Chap. 4		
PTPN12	PTP-PEST	NT4	7q11.23	18	780 aa	See further details in Chap. 7		
PTPN13	PTPL1, FAP1, PTP1E, PTP-BAS, PNP1; PTP-BL (mouse)	NT7/8	4q21.3	48	4 variants: 2490, 2485, 2466 and 2294 residues	Widely expressed, most notably in epithelia and brain	3LNX, 3LNY (2nd PDZ domain); 1WCH (PTP domain)	[222, 378–384]
PTPN14	PTP Pez, PTPD2, PTP36	NT6	1q32.2	19	1187 aa	Broadly expressed, including hematopoietic lineages	2BZL (PTP domain)	[178, 181, 182]
PTPN18	BDP1; PTP20 (rat); PTP-HSCF, PTP-K1 (mouse)	NT4	2q21.1	15	460 aa (alternate transcripts were described but not yet characterized on protein level)	High in murine hematopoietic stem cells, decreasing with differentiation; also in lung, kidney, brain, colon, ovarian granulosa cells, several human tumor cell lines	2OC3 (catalytic domain) 1OZU, 1GM1, 1VJ6 (NMR structure 2nd PDZ domain) 4GFU, 4GFV, 4NND (catalytic domain in complex with substrate phosphopeptides from ERBB2)	[118, 120, 178, 385–387]
PTPN20	Typ (mouse); cancer/testis antigen 126	NT7/8	10q11.22	Exceeding 35 (8 +27)	Many reported variants, in fact spanning two previously annotated genes; PTPN20A and FMRPD2	High in testis, at low levels in many other tissues	–	[210, 297]

PTPN21	PTPD1, PTP-RL10	NT6	14q31.3	18	Two variants; 1174 and 336 aa residues. The latter essentially consists of the PTP domain only	Expressed throughout development in multiple tissues	–	[200, 201]
PTPN22	LYP; PEP (mouse)	NT4	1p13.2	21	807aa; shorter alternate transcripts reported, e.g., LYP2; PTPN22.6 (no PTP domain, dominant negative) characterized at protein level	High in lymphoid cells, at low levels in myeloid cells, negligible in other tissues	3H2X, 2P6X, 3OLR, 2QCT, 3BRH (PTP domain alone and in different complexes)	[121, 122, 178, 388–390]
PTPN23	HD-PTP; PTP-TD14 (rat)	NT9	3p21.3	25	1636 aa	Multiple tissues, enriched in epithelia, low expression in heart, spleen, and muscle	3RAU (Bro1-like domain) PTP domain [178] (no PDB entry)	[301–304, 308, 391]
PTPRR	PTPRQ, EC-PTP1, NC-PTPCOM1, Ch-1PTPase, PTPPBSα, PTPPBSβ and PTPPBSγ; PTPBR7, PTP-SL (mouse); PCPTP1, PC12 PTP, (rat)	R7	12q15	14	Four isoforms: a 657 aa receptor-type, a 545 aa vesicle-associated transmembrane isoform, and 451 and 412 aa residues-long cytosolic proteins. All contain the KIM and PTP domain	Expressed in neurons throughout development. Very low expression levels in other tissues including cartilaginous cells	2A8B, 1JLN (PTP domain)	[352, 354, 357, 380, 392–395]

[a]Classification according to [1]

[b]Selected entries in the Research Collaboratory for Structural Bioinformatics (RCSB) protein data bank (PDB) http://www.rcsb.org

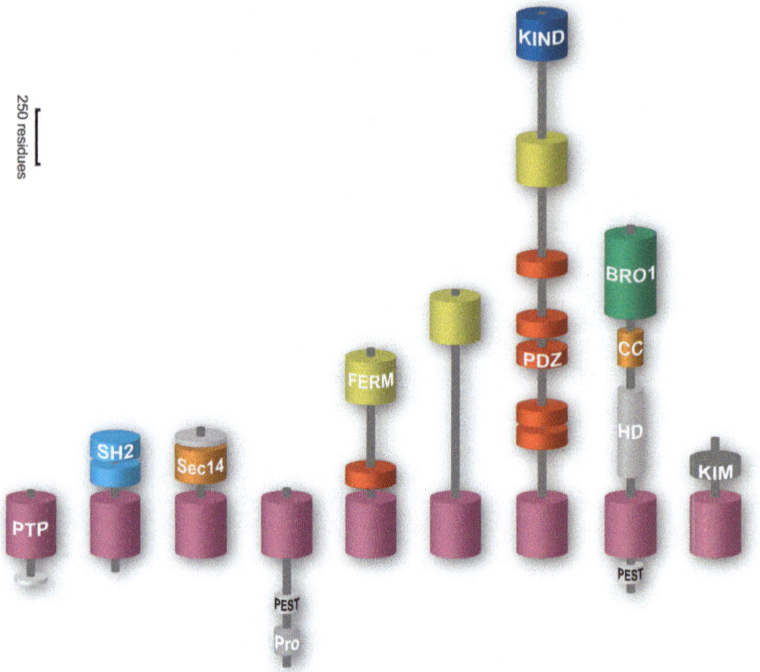

Fig. 1 Domain composition for subfamilies of non-transmembrane, classical phosphotyrosine-specific PTPs. Schematics show a representative, common isoform for each subtype, with N-terminus on top and C-terminus pointing down. Other isoforms may differ in their content of protein domains or subcellular targeting sequences. Below, the subtype classification, according to Andersen et al. [1] and adjusted for the most recent insight in PTPN20 annotation [210], and the respective members are listed. PTPN1, PTPN2, PTPN11, and PTPN12 will be discussed later on, in Chaps. 4–7, respectively. Domain abbreviations used: BRO1, yeast Bro1 homologous domain 1; CC, V-domain with Coiled Coil motifs; FERM, band 4.1/Ezrin/Radixin/Moesin homology; HD, His Domain; KIM, Kinase Interaction Motif; KIND, Kinase N-lobe-like Domain; PDZ, PSD-95/Dlg/ZO-1 homology; PEST, proline-, glutamic acid-(E), serine-, and threonine-rich sequence; Pro, Proline-rich sequence; Sec14, Sec14 homology; SH2, SRC homology 2 domain. The light grey sequence at the end of the NT1-type protein symbolizes the ER membrane targeting tail in PTPN1. The grey disc in PTPN9 represents a CRAL–TRIO N-terminal homology domain. Proteins and domains are drawn to scale and a protein size indication for 250 amino acid residues is depicted

NT2 Type, SH2 Domain-Containing PTPs

This PTP class comprises PTPN6 and PTPN11. Both molecules are structurally related, harboring tandem SH2 domains in front of the PTP domain (see Fig. 1). In the absence of activating signals, they exist in a "closed" conformation, with the

Table 2 Role of non-transmembrane PTPs in cancer[a]

	Alterations in cancer cells		Affected aspects of cancer cell biology ("hallmarks"[c])	
	Genetic mutations[b]/Expression changes	Occurrence, cancer type	Cell process	Description
PTPN3	Mutation	22 % of liver cancer[2]	n.i.[e]	
	Mutation	Translocation in NSCLC line	Survival and proliferation	Loss of function increases colony formation in cell line
	Expression	HPV E6-mediated breakdown in cervical cancer	Survival and proliferation	Loss of function increases colony formation in cell lines
	Expression	Increased in esophageal and gastric cancer	n.i.	
	Expression	Increased in colon cancer	Proliferation	Cooperates with p38γ in facilitating RAS-mediated oncogenic signaling
	Expression	Increased in half of breast cancer specimen, correlating with metastatic potential	Proliferation	Interferes with growth-suppressive role of Vitamin D receptor
PTPN4	Mutation	35 % of liver cancer[2]	n.i.	
	Mutation	31 % of large intestinal cancer[1]	n.i.	
			Survival and proliferation	Over-expression inhibits growth and colony formation in cell line
			Apoptosis	Ligand binding to PDZ domain induces apoptosis in glioblastoma cells
PTPN5	Mutation	20 % of liver cancer[2]	n.i.	

(continued)

Table 2 (continued)

	Alterations in cancer cells		Affected aspects of cancer cell biology ("hallmarks"[c])	
	Genetic mutations[b]/Expression changes	Occurrence, cancer type	Cell process	Description
PTPN6	Mutation Expression	12 % of large intestinal cancer[1] Promoter silencing in different hematopoietic cancers Over-expressed in subset of breast cancer	Proliferation	Negatively regulating in hematopoietic and possibly further cell types
			Proliferation	Possibly promoting in mammary epithelial cancer
	Expression		Apoptosis	Promoting in response to different drugs
			Immune surveillance	Negative regulator of anti-tumor immune responses
			Angiogenesis[d]	Negative regulator of VEGFR2-mediated angiogenesis
PTPN7	Mutation	LOH in non-Hodgkin lymphomas	n.i.	
	Mutation	Amplified in MDS	n.i.	
	Expression	Lower levels in lymphomas	n.i.	
	Expression	Increased levels in acute Leukemias	n.i.	
PTPN9	Mutation Mutation	14 % of large intestinal cancer[1] 16–18 % of liver cancer[2]	n.i. n.i. Motility and invasion Immune surveillance[d] Angiogenesis[d]	Negatively regulating. Knockdown enhances HER2 and EGFR signaling. Over-expression inhibits colony formation and matrigel invasion
				Positive effector of immune cell cytokine secretion
				Negative regulator of VEGFR2 in endothelial cells

PTPN13	Mutation	LOH in hepatocellular carcinoma and lung cancer	Proliferation and survival	Inhibitory effect of PTPN13 on (anchorage-independent) tumor growth
	Mutation	Germ line mutations associated with multiplex familial hepatocellular carcinoma	Apoptosis and motility	Two mutations impaired PDZ domain binding to FasR and TRIP6
	Mutation	9 % of colon cancer, and in HPV-negative squamous cell carcinomas.	Proliferation and survival	Two mutations abrogated PTP activity
	Expression	HPV E6-mediated breakdown in head and neck squamous carcinomas	Proliferation and survival	Knock-down increases colony formation and synergizes with oncogenic ERBB2-RAS signaling
	Expression	Hypermethylation of promoter in tumor cells	n.i.	
	Expression	Increased expression due to EWS-FLI1 translocation in Ewing's Sarcoma	Apoptosis	Increased colony forming potential and apoptosis resistance, possibly via interactions with FAS receptor and VCP
	Expression	Increased levels in multiple tumor types	Apoptosis	Levels correlate with relative (FAP-mediated) apoptosis resistance
	Expression	Increased levels in TEL translocation-positive myeloid leukemias	Apoptosis	TEL fusion protein up-regulates PTPN13 and boosts (FAP-mediated) apoptosis resistance
	Expression	Anti-estrogen-induced upregulation in breast cancer cases	Apoptosis	Increased PTPN13 counteracts PI3K/AKT survival pathway via IRS1 dephosphorylation, hence pro-apoptotic

(continued)

Table 2 (continued)

	Alterations in cancer cells		Affected aspects of cancer cell biology ("hallmarks"[c])	
	Genetic mutations[b]/Expression changes	Occurrence, cancer type	Cell process	Description
PTPN14	Mutation	Colorectal, breast, head and neck, kidney, liver, ovarian and skin cancer	n.i.	
	Expression	Lower levels in liver metastasis as compared to primary pancreatic tumor	Metastasis	Counteracts SRC by dephosphorylating p130Cas in mouse tumor model
			Survival	Downstream effector of AKT pathway, SRC counteractor by binding and. dephoshorylating YAP/TAZ that mediate Hippo signaling
PTPN18	-		Proliferation	Potentially negatively regulating. Inhibits HER2 signaling in breast cancer cell lines
PTPN20	–		–	
PTPN21	Mutation	10 % of liver cancer[2]	n.i	
	Mutation	17 % of intestinal cancer[1]	n.i.	
	Expression	Levels correlate with grade and invasiveness in bladder cancer	Proliferation and motility	Knock-down attenuates cell growth and motility
			Chemoresistance	Knock-down sensitizes cell lines to cisplatin treatment
PTPN22	Mutation	44 % of skin cancer[1]	n.i.	
	Mutation	20 % of gastric cancer[2]	n.i.	
	Mutation	14–27 % of liver cancer[2]	n.i.	
			Apoptosis	Preventing apoptosis in CLL cells by interfering with pro-apoptotic BCR signaling
			Immune surveillance[d]	Negative regulator of T-cell receptor signaling

PTPN23		Region containing *PTPN23* frequently deleted in human solid cancers.	Proliferation	Negatively regulating. Over-expression inhibits colony formation of renal cancer cells
		Mutations in 14 % of large intestinal cancer[1]	Motility and invasion	Negatively regulating. Knockdown enhances motility and invasion of mammary cancer cells
				Positive effector of Hippo-tumor suppressor pathway in *Drosophila melanogaster*
			Angiogenesis	Negative regulator of endothelial cell migration; knockdown enhances motility
PTPRR	Mutation	Translocation event in leukemia involving TEL	–	Dominant-negative effect of TEL fusion protein rather than PTPRR loss-of-function explains the tumorigenic effect of the translocation
	Expression	Lower levels in cervical cancers due to promoter hypermethylation	n.i.	

[a]For details, see discussion in text and references therein
[b]Data derived from COSMIC[1], the catalogue of somatic mutations in cancer [173] http://cancer.sanger.ac.uk/cosmic, the International Cancer Genome Consortium[2] [152] http://dcc.icgc.org, and references quoted in the text. Only alterations occurring in >10 % of at least >5 analyzed cancer samples are listed
[c]For definition of "hallmarks" see [2, 94, 396]
[d]Cancer relevance not yet analyzed
[e]n.i., not investigated

N-SH2 domain binding to, and preventing access to, the catalytic domain. A major common mechanism of activation involves binding of pY-peptide ligands to the N-terminal SH2 domain, leading to conformational changes, opening of the structure, and activation of the enzymes. The cellular functions of PTPN6 and PTPN11 are, however, largely divergent (for reviews [3–5]; see also below and Chap. 4). The selectivity of cellular actions is determined by a combination of the selectivity of SH2 domain binding and catalytic specificity (e.g. [6–8]). In this chapter, we will discuss PTPN6; PTPN11 (SHP-2) is covered in Chapter 4.

PTPN6

Regulation: Two different promoters, P1 and P2, drive expression of the *PTPN6* gene in a tissue-selective manner [9–12]. Importantly, promoter P2 harbors CpG-rich sequences, which make it prone to silencing by methylation. Apart from SH2 domain ligand binding, PTPN6 activity is also modulated by tyrosine phosphorylation in the C-terminus, i.e., Y536 and Y564 in the major hematopoietic isoform [13, 14]. The C-terminus can also mediate nuclear translocation [15] and binding of acidic phospholipids [16, 17], affecting activity and substrate access [18]. Finally, PTPN6 can be inhibited by reversible oxidation, leading to formation of an intramolecular disulfide bridge [19, 20]. Chemical inhibitors of PTPN6 comprise sodium stibogluconate and the recently developed, selective, and potent inhibitor TPI-1 and its orally administratable analogs [21].

Physiological Functions: Two naturally occurring mouse strains with defective *Ptpn6* gene, designated "*motheaten* (*me*)" and "*motheaten viable* (*me v*)" [22, 23], respectively, have contributed importantly to our insights into the biological functions of PTPN6. Mice homozygous for the *Ptpn6 me* allele do not express functional PTPN6 and die at an age of 3 weeks. *Ptpn6 me-v* mice express a PTPN6 version with reduction of activity to 20 % and live only 6–9 weeks. *Ptpn6 me/me* and *Ptpn6 mev/mev* mice suffer from chronic inflammation and autoimmunity, and heterozygous animals develop lymphoma and leukemia at higher age [22]. The reduced lifetime of these animals is caused by invasion of macrophages and neutrophils into different organs, notably the lung. Inflammatory infiltrations in the skin cause a patchy absence of hair and pigment, the *motheaten* phenotype. More recently, mouse strains with further hypomorphic *Ptpn6* alleles have been described. The *spin* ("spontanous inflammation") mutation of the *Ptpn6* locus was identified among mutant strains generated by *N*-ethyl-*N*-nitrosourea treatment of C57BL/6J mice [24]. This mutation affects the C-terminal SH2 domain of PTPN6 (Y208N) and has been predicted to cause a mild reduction in PTPN6 function by compromised recruitment of binding partners. The spontaneous mutation *Ptpn6meB2* consists of the insertion of a B2 repeat in exon 6 of the *Ptpr6* gene, causing small sequence alterations as well as reduced PTPN6 expression and activity [25]. The *Ptpn6spine* and *Ptpn6meB2* mice exhibit similar but milder inflammatory phenotypes as observed in the *motheaten* strains.

Several hematopoietic compartments are enlarged in *Ptpn6 me/me* and *Ptpn6 mev/mev* mice, indicating that PTPN6 functions as a negative regulator of the proliferation of their progenitors. A well-studied mechanism is the negative regulation of erythropoietin-receptor (EpoR) signaling by PTPN6 [26], causing strongly elevated numbers of erythroid progenitors (CFU-E) in *Ptpn6 me/mev* mice [23]. Further, abrogation of PTPN6-mediated negative regulation of the receptor tyrosine kinases (RTK) c-KIT [27–29] and CSF-1 receptor [30, 31], as well as that of the interleukin-3 receptor [32, 33], presumably contribute to alterations in hematopoiesis in the *Ptpn6* mutant animals. Further hematological abnormalities in the *Ptpn6* mutant may relate to PTPN6-mediated negative regulation of immune cell functions, including B- and T-cell receptor (BCR, TCR) signaling [34–37], natural killer cell activity [38], migratory responses of multiple immune cells to chemokines [39], oxidant production by neutrophils [40], and integrin-mediated adhesion of neutrophils [40] and macrophages [41]. Proper control of these processes often requires interactions of PTPN6 with members of different inhibitory immunoreceptor families (for reviews see [4, 42, 43]).

Loss of SHP-1 in all hematopoietic cells and their multiple interactions in immune responses have, however, limited and partially compromised the conclusions obtained from the analysis of *motheaten* mice, notably with respect to the cell-autonomous character of observed signaling alterations. Recent analyses using cell-type-specific *Ptptn6* gene inactivation have uncovered more detailed roles of PTPN6 in specific immune cells. For example, conditional mutagenesis of *Ptpn6* specifically in mouse B-cells revealed profound cell-autonomous effects of PTPN6 on B-cell differentiation. PTPN6 deficiency in B-cells caused elevated serum levels of immunoglobulins and severe autoimmunity at older age with features similar to systemic lupus erythematosus [44]. Specific *Ptpn6* knockout in T-cells revealed an important role of PTPN6 in regulation of peripheral T cell homeostasis [45]. PTPN6 deficiency caused enhanced IL-4 signaling via STAT6, associated with increased frequencies of memory T-cells. Surprisingly, and in contrast to earlier observations in mice with systemic PTPN6 deficiency, the T-cell-specific *Ptpn6* gene inactivation did not affect T-cell selection, responses to TCR stimulation, or the frequency of regulatory T-cells [45]. By generating mice with specific inactivation of the *Ptpn6* gene in neutrophils and dendritic cells, the contribution of PTPN6 loss in these cells to the inflammatory *motheaten* phenotype has recently also been further dissected [46]. The neutrophil-specific *Ptpn6* knockout caused cutanous inflammation. Enhanced integrin signaling in PTPN6-deficient neutrophils via enhanced activities of SRC-family kinases (SFKs) and SYK kinase was identified as underlying molecular mechanism. The dendritic cell-specific *Ptpn6* knockout caused severe autoimmunity, which was related to elevated MyD88-dependent signaling. The direct substrates of PTPN6 modulating the latter pathway remain yet to be identified. *Ptpn6* gene inactivation in neither of these cell types caused a lethal inflammatory lung disease as seen in *motheaten* mice [46].

The functions of PTPN6 in non-hematopoietic cells are less well-characterized. PTPN6 has the capacity to regulate multiple receptor-tyrosine kinases, including the EGF (epidermal growth factor) receptor (EGFR, also designated ERBB1) in epithelial cell lines [47, 48], the VEGF (vascular endothelial growth factor) receptor 2 in endothelial cells [49], and possibly the PDGF receptor [50]. The receptor tyrosine kinase ROS, which is an important regulator of epididymal differentiation in mice, has been identified as an epithelial target of PTPN6 [51]. The ROS intracellular domain contains two high-affinity binding sites for PTPN6 SH2 domains, which mediate efficient association and negative regulation of ROS signaling. ROS is hyperphosphorylated in the epididymis of *Ptpn6 mev/mev* mice, indicating the in vivo relevance of the interaction. In neuronal cells, PTPN6 can directly down-regulate the activity of the NGF receptor TRKA, modulating the development of sympathetic neurons. PTPN6 deficiency resulted in elevated neuron numbers with improved survival ex vivo [52]. PTPN6 also has emerged as a negative regulator of insulin signaling in muscle and liver [53]. *Ptpn6 mev/mev* mice exhibited increased glucose tolerance and insulin sensitivity, associated with elevated signal transduction. These findings were further supported by experiments using adenoviral gene transfer of a dominant-negative PTPN6 variant and by down-regulation of PTPN6 levels using shRNA. Recently, liver-specific inactivation of the *Ptpn6* gene was shown to protect mice from high-fat diet (HFD)-induced hepatic insulin resistance. In wild-type mice, HFD induces *Ptpn6* expression, which causally contributes to hepatic insulin resistance [54]. A common PTPN6 target downstream of different receptors may be the Gα-interactive vesicle-associated protein (GIV, also designated Girdin), an enhancer of AKT signaling. GIV becomes tyrosine phosphorylated downstream of several RTK and G-protein-coupled receptors (GPCR) and serves to bind the p85 regulatory subunit of phosphoinositide-3 kinases (PI3K). PTPN6 associates with GIV and dephosphorylates it [55].

Several reports have provided evidence for a positive role of PTPN6 in activation of the RAS-MEK (mitogen-activated protein kinase (MAPK)/extracellular signal regulated kinase (ERK) kinase)-ERK1/2 pathway [56, 57], but the underlying mechanism is still not clear. Potentially, GRB2 (growth factor receptor-bound protein 2, an upstream activator of RAS-MEK-ERK1/2 signaling) binding sites, generated by tyrosine phosphorylation in the C-terminus of PTPN6, may play a role [58]. Moreover, in CaCo2 colon epithelial cells, PTPN6 and PTPN11 can form a ternary complex with the signaling adaptor GAB1 (GRB2-associated binder-1) and then cooperate to mediate EGF-induced ERK1/2 activation [59]. In a cell-type specific manner, PTPN6 may possibly have some redundant or cooperative functions with its relative PTPN11 (see Chap. 4).

Role in Cancer: Phenotypes of the *me/me v* mouse strains suggested early on that loss of *PTPN6* might contribute to the development of hematological malignancies [22]. Enhanced proliferation of hematopoietic progenitors, but also enhanced apoptosis resistance of PTPN6-deficient cells [60], could play an oncogenic role in this context. Interestingly, a cell population enriched in hematopoietic stem cells from *Ptpn6 mev/mev* mice could initiate a myeloproliferative disease (MPD) when transplanted into lethally irradiated recipient mice [61]. Restoration of *PTPN6* expression or expression of dominant-negative STAT5 (signal transducer and activator of transcription 5) inhibited MPD development. The authors proposed that PTPN6 mediates an anti-proliferative effect by blocking the oncogenic action of STAT5.

Loss of PTPN6 expression has been detected in different human hematopoietic cancers (Table 2). DNA methylation of the relevant *PTPN6* P2 promoter is the underlying mechanism in the majority of 60 cases of T-cell leukemia, NK/T-lymphoma, ALL (acute lymphocytic leukemia), AML (acute myeloid leukemia), and CML (chronic myeloid leukemia) cases analyzed in an early study. Allele loss (loss of heterozygosity, LOH) has additionally been observed in a number of these cases [62]. *PTPN6* promoter methylation was also found in the vast majority of 90 different leukemia and lymphoma cases [63], and in 27 of 34 multiple myloma cases [64]. The methylation status of *PTPN6* may be associated with a worsened prognosis in patients with myelodysplastic syndrome [65]. Downregulation of PTPN6 levels in anaplastic large cell lymphoma could also be correlated with methylation [66]. Signaling of the anaplastic lymphoma kinase-nucleophosmin (ALK-NPM) oncoprotein was later found to be negatively regulated by PTPN6 [67, 68]. Similarly, the BCR-ABL oncoprotein in CML cells may be negatively regulated by PTPN6 [69, 70]. These findings suggest that PTPN6 down-regulation may exacerbate the signaling of these oncoproteins. In diffuse large B-cell lymphoma, *PTPN6* promoter methylation was observed at a CpG island within promoter P2 distinct from that found methylated in other hematological cancers (denoted CpG2). Methylation and, in addition, silencing H3K27me3 histone marks at *PTPN6* P2 were associated with decreased PTPN6 levels, and treatment with a demethylating agent or a histone methyltransferase inhibitor restored PTPN6 expression [71].

Alternative mechanisms of PTPN6 loss in hematological cancers were also observed. In AML cells, alternative splicing of the *PTPN6* gene based on RNA hyper-editing leads to mRNAs predictably encoding non-functional proteins [72]. Disruption of PTPN6 expression occurs also as consequence of infection with the human T cell leukemia virus-1 (HTLV-1), the causative agent for adult T cell leukemia [73]. The HTLV-1-encoded Tax oncoprotein represses the P2 promoter activity of *PTPN6* by displacing NFκB and recruiting histone deacetylase-1 [10, 12]. In diffuse large B-cell lymphoma, in addition to DNA-methylation, silencing histone-methylation marks were found associated with the *PTPN6* promoter, and pharmacological inhibition of histone methylation resulted in re-expression of *PTPN6* [71]. Although these findings are suggestive for a role of PTPN6 loss in human hematopoietic cancers, possibly by synergizing with other genetic lesions in driving proliferation, they are largely correlative. Moreover, promoter silencing by methylation also affected genes of further antiproliferative signaling molecules, such as *SOCS1*, making it difficult to judge the contribution of PTPN6 loss to tumorigenesis. Notably, *Ptpn6* deficiency alone was insufficient to cause neoplasias in mice with T- or B-cell-specific knockout of the gene ([44, 45] and B.G. Neel, personal communication). Further investigations in mouse models, e.g., the combination of oncoprotein expression with *Ptpn6* loss in specific hematopoietic compartments, appear highly warranted to firmly establish the contribution of PTPN6 down-regulation to hematological malignancies.

Relatively little is known about alterations of PTPN6 in cancers of non-hematopoietic tissues. The capacity of PTPN6 to negatively regulate several RTKs predicts similar roles as in hematopoietic malignancies. Indeed, in fibroblast models

PTPN6 can suppress ligand-stimulated transforming activity of ROS and of an activated RET (rearranged during transfection) RTK version that was identified in multiple endocrine neoplasia (MEN) type 2A [51, 74]. Similar observations were reported for TRK-T3, an oncogenic fusion protein of the TRKA RTK, which has been found in a papillary thyroid carcinoma [75]. However, *PTPN6* was found to be over-expressed, rather than being down-regulated, in some epithelial cancers and in ovarian cancer cell lines [76]. *PTPN6* mRNA over-expression was also detected in a subset of mammary carcinoma and matched over-expression of mRNA encoding *GRB2* [77]. In a recent tissue microarray study involving 2,081 breast cancer samples, expression of PTPN6 protein was not detectable in normal breast tissue. However, PTPN6 immunoreactivity was observed for 7.2 % of tumors and correlated with negative parameters of prognosis including high histological grade, over-expression of the ERBB2 (also designated HER2) oncoprotein, estrogen receptor negativity, and reduced survival [78]. These correlative findings suggest a contributing role of PTPN6 to the malignancy of mammary cancer. A possible underlying mechanism may relate to the paradoxical inhibitory effect of the TRKA RTK, a PTPN6 substrate, on proliferation of breast cancer cells. A recent study showed negative regulation of *PTPN6* expression by p53, leading to enhanced activity of TRKA and in turn attenuation of cell growth [79]. In a mouse model of MMTV-neu-driven mammary tumorigenesis, *Ptpn6* deficiency in mammary tissue had, however, neither an attenuating nor a promoting effect on tumor formation (B.G. Neel, personal communication), arguing against a causal contribution of PTPN6 for tumorigenesis in this cancer entity. In a set of high-grade gliomas, PTPN6 expression detected by IHC has recently been correlated to reduced patient survival and increased chemotherapy resistance [80]. Conversely, hypermethylation of the *PTPN6* promoter has been found positively correlated with grade and overall survival in another set of glioma patients. PTPN6 expression was, however, not assessed in the latter study [81]. Further investigation of the PTPN6 status in human tumors and the analysis of mouse models with selective *Ptpn6* gene inactivation or forced PTPN6 over-expression in non-hematopoietic tissues are desirable to further explore its possible role for development of respective tumor entities.

Interestingly, both in non-hematopoietic and hematopoietic cells, PTPN6 appears mediating effects of cytostatic and cytotoxic agents (Fig. 2). A well-studied example is the role of PTPN6 in signaling of somatostatin (SST) and its therapeutic analogs. PTPN6 associates with the subtype 2 somatostatin receptor (SSTR2) [82], whose activation is essential for the antiproliferative effect of SST on different cell lines, such as pancreatic cancer cells [83], medullary thyroid carcinoma [84], and pituitary tumor cells. Generally, the mechanisms downstream of PTPN6 in this context are still enigmatic. However, induction of the candidate tumor suppressor gene *ZAC1* depends on PTPN6 activity and is essential for the antiproliferative effect of the SST analog Octreotide [85]. PTPN6 is also important downstream of CD37/Tetraspanin, a candidate therapeutic target in B cell malignancies. Anti-CD37 antibodies and CD37-binding peptides such as SMUP-016 are currently being explored in clinical trials. Ligation of CD37 leads to tyrosine phosphorylation and recruitment of signaling molecules including PTPN6, which is essential for apoptosis induction [86].

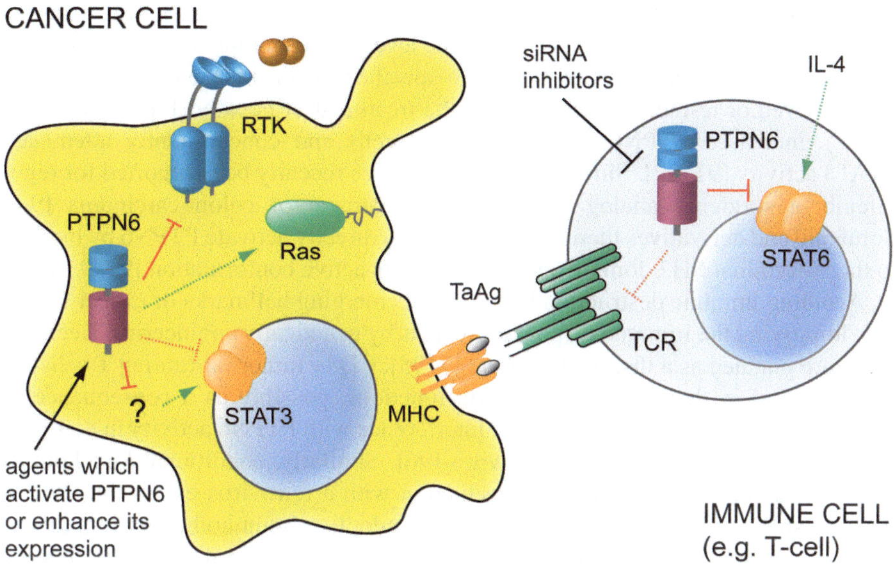

Fig. 2 Potential relevance of PTPN6 for cancer therapy. In cancer cells, PTPN6 can interfere with proliferation through negative regulation of receptor tyrosine kinases (RTK), and the transcription factor signal transducer and activator of transcription 3 (STAT3). Although STAT3 has been proposed as a direct substrate of PTPN6 (*dotted line*), its inactivation presumably occurs more likely by inhibition of upstream activators. Activation of the RAS pathway by PTPN6 in epithelial cells has also been reported. A number of anti-proliferative agents, such as somatostatin, have been shown to elevate PTPN6 activity and/or *PTPN6* expression, thereby inhibiting cancer cell proliferation and/or promoting apoptosis. Immune cell responses against cancer cells, e.g., the cytotoxic activity of T-cells (triggered through recognition of major histocompatibility complex (MHC)-presented tumor-associated antigens (TaAg) via specific T-cell receptors (TCR)), are compromised in tumors. Experimental studies indicated that enhancement of immunological anti-tumor responses may be achieved by suppression of PTPN6 activity or expression selectively in immune cells, for example, in the context of adoptive immunotherapy. Recent studies have shown that enhanced T-cell responses in absence of PTPN6 may not be related to enhanced TCR signaling (as initially believed (*dotted line*)), but rather to an elevated frequency of memory T-cells through more sustained IL-4/STAT6 signaling

Induction or re-expression of PTPN6 by several naturally occurring or synthetic antiproliferative agents has also been described. For example, guggulsterone, a farnesoid X receptor antagonist, induces PTPN6 expression in human multiple myeloma cells, in turn causing STAT3 down-regulation, inhibition of proliferation, and induction of apoptosis [87]. Inhibition of metastasis and angiogenesis in the context of gastric cancer by honokiol, a component of traditional Asian medicine, has recently also been related to induction of PTPN6 expression [88]. The direct targets of PTPN6 in these systems have not been identified and may represent upstream kinases rather than STAT3 itself, which has hitherto not been firmly established as PTPN6 substrate [89]. PTPN6 appears also important for the efficacy of cancer therapy by some tyrosine kinase inhibitors. Low levels of PTPN6 expression were associated with resistance to the therapeutic drug imatinib for CML, both in a cell line and when comparing expression levels and therapy response in CML patients [90]. Sorafenib and dovitinib, pan-specific tyrosine kinase inhibitors, which are approved or tested, respectively, for the treatment of hepatocellular carcinoma (HCC), increased PTPN6 activity in HCC cells and concomitantly attenuated STAT3 activity [91, 92]. Similar observations have recently been reported for regorafenib, a sorafenib analog approved for treatment of colon carcinoma [93]. Sorafenib and derivatives thereof can apparently directly activate PTPN6 by binding to the N-terminal SH2 domain and promoting its active conformation [92, 93].

Avoiding immune destruction is one of the emerging hallmarks of cancer [2, 21, 94]. Improving the immune responses, such as by genetically engineering T-cells, is therefore pursued as a therapeutic strategy [95]. CD8+ tumor infiltrating T cells are compromised in their activity and lytic functions, possibly as a consequence of tumor-mediated activation of PTPN6. Interference with PTPN6 activity in non-lytic cells restored their lytic activity in vitro [96]. Similarly, conditional knockout of *Ptpn6* in mature CD8 T cells or transduction with a retrovirus expressing *Ptpn6*-directed shRNA greatly improved efficacy of adoptive immunotherapy in a mouse leukemia model [97]. Similar mechanisms may play a role in the recently demonstrated anti-tumor activity of novel PTPN6 inhibitors. These compounds were active against B16 mouse melanoma cells in the immunocompetent C57BL/6 mouse background, but not in nude mice [21]. Thus, an exciting therapeutic relevance of PTPN6 may rest on the inhibition of its activity in immune cells, either systemically to enhance endogenous anti-tumor immune responses or in cells used for adoptive immunotherapy of cancer (Fig. 2).

NT3 Type, Sec14-Homology Domain Containing PTPN9

PTPN9 (MEG2) is the only member of the NT3 class, and hence is unique in its structure: a lipid-binding domain with homology to cellular retinaldehyde binding protein (CRALBP, also designated CRAL_TRIO domain) and the yeast protein SEC14p is followed by the PTP domain and a short C-terminal segment (Fig. 1) [98].

Regulation: *PTPN9* expression can be induced by phorbol ester in U937 cells [99] and phytohemagglutinin or IL-2 in T lymphocytes [100]. The CRALBP/SEC14p domain binds phospholipids with high affinity [101, 102], a feature that is important for PTPN9's biological function [103] (see below). This domain could play a negative regulatory role, as its truncation is activating [101]. A crystal structure of the full-length molecule remains to be determined and might shed light on putative intramolecular interactions.

Physiological Function: A well-established function of PTPN9 is the positive regulation of homotypic fusion of secretory vesicles, e.g., in mast cells and T-cells, which depends on PTP activity [103]. The localization of PTPN9 on secretory vesicles was unique among 15 PTPs analyzed [104] and is mediated by interaction of the CRALBP domain with vesicle-resident proteins such as TIP47 and Arfaptin2 [105]. Vesicle association of PTPN9 is independent of lipid binding, but its fusion-promoting activity depends on interaction with vesicle proteins and lipid binding [102, 105]. A critical PTPN9 substrate in this context is *N*-ethylmaleimide-sensitive factor (NSF), a key regulator of vesicle fusion. Tyrosine phosphorylation of NSF at Y83 increases ATPase activity and reduces αSNAP binding. This site is dephosphorylated by PTPN9, thereby promoting vesicle fusion and secretion, e.g., of IL-2 in T-lymphocytes [103].

Ptpn9-deficient mice display multiple neurodevelopmental defects, craniofacial abnormalities, hemorrhages, defects in long bone formation and dental development, and late embryonic lethality [106]. These severe phenotypes were at least partially attributed to functions in vesicle fusion and secretion. For example, secretory vesicles were absent in lymphocytes and agonist-induced IL-2 secretion was impaired. Aggregation of PTPN9-deficient platelets in response to thrombin was strongly compromised, related to defective release of platelet granules. The mechanisms underlying defects in other tissues are not yet clear, but could be related to defective migration and differentiation of cranial neural crest cells and endothelial cells, caused by defective secretion of growth factors [106]. Further functions of PTPN9 were found in regulation of insulin signaling in hepatocytes and VEGF signaling in endothelial cells. Negative regulation of both pathways may be mediated, at least in part, by PTPN9-mediated dephosphorylation of the upstream receptor tyrosine kinases, insulin receptor (IR), and VEGF receptor 2, respectively [107, 108]. *PTPN9* is a target of miRNAs-126/126* and has been proposed as positive regulator of erythroid differentiation [109]. An essential role for erythropoiesis has recently also been described for the zebrafish *PTPN9* ortholog *ptpn9a* [110].

Role in Cancer: At this point, knowledge about roles of PTPN9 in cancer is rather limited. Mutations with yet unknown functional relevance have been observed in some epithelial cancers (see Table 2). In Polycythemia vera (PV), a high membrane-bound PTP activity in erythroid progenitors could be assigned to PTPN9. The activity decreased with erythroid differentiation, but remained higher in PV patient cells. Dominant-negative PTPN9 inhibited erythroid differentiation of PV and normal cells in vitro. Although the data suggest that PTPN9 may play a contributing role in erythroid differentiation, its relevance for the development of PV remains to be explored [111]. More recently, data were obtained suggesting ERBB2 and EGFR down-regulation by PTPN9 in breast cancer cell lines over-expressing these RTKs [112]. By a PTP-targeted siRNA screen for effects on tyrosine- hyperphosphorylation of ERBB2 in SKBR3 cells, PTPN9 was identified as a negative regulator. Experiments with PTPN9 trapping mutants and PTPN9 over-expression supported the relevance of direct PTPN9-ERBB2/EGFR interaction and indicated negative regulation of downstream signaling by PTPN9, notably of the activation of STAT3 and STAT5, but not of ERK1/2 or AKT. Over-expression of PTPN9 inhibited colony formation, matrigel invasion, and MMP-2 (matrix metalloproteinase-2) secretion of breast cancer cell lines. The authors proposed that these effects are mediated by a small pool of PTPN9 in the plasma membrane, as opposed to the main PTPN9 pool on secretory vesicles [112]. PTPN9 is subject to down-regulation by miRNA miR-24, which is over-expressed in breast cancer tissue. In keeping with a role of PTPN9 for regulation of EGFR, over-expression of miR-24 enhanced EGFR phosphorylation and signaling as well as growth, invasion, and metastasis of tumors from engineered cells in mice [113]. In another recent study, PTPN9 over-expression in human mammary tumor cells was shown to inhibit growth in a mouse xenograft model. PTPN9 expression and STAT3 phosphorylation were found inversely correlated in human breast cancer samples, and the direct dephosphorylation of cytoplasmic STAT3 was proposed as mechanism of the anti-transforming effects of PTPN9 over-expression [114]. It will be important to further test the relevance of these interactions using in vivo models and to study also other cancer entities for possible alterations in the PTPN9 status.

NT4 Type PTPs with PEST and Proline-Rich Motifs

The members of this family, PTPN12, PTPN18, and PTPN22, are structurally distinguished by long C-terminal domains harboring proline-rich motifs and sequences with similarity to PEST sequences (Fig. 1). The former motifs can interact with SH3 (src homology 3) domains and potentially WW domains. The PEST motifs are indicative of short half-lives in other proteins; however, the members of the NT4 PTP subfamily are stable. The three related PTPs may have partially redundant functions, notably in hematopoietic cells. PTPN12 (PTP-PEST) will be covered extensively in Chapter 7; here, we will discuss PTPN18 and 22.

PTPN18

Physiological Function: Current knowledge regarding PTPN18 function is very limited because, in the absence of a mouse knockout strain or genetic link to a human disease, it rests entirely on experiments in cell lines and in vitro. Availability of constructs for targeted gene inactivation in mice (International Knock-out Mouse Consortium (www.knockoutmouse.org)) could change this situation in the near future. PTPN18 can associate with cellular SRC kinase (CSK). Association is mediated by binding of tyrosine phosphorylated sites in the C-terminal part of PTPN18 with the CSK SH2 domain. PTPN18 was shown to selectively dephosphorylate the SRC-family kinase LYN at the activating pY417, as opposed to the inhibitory pY528, but, in another study, had little activity towards pY418 of c-SRC [115, 116]. In ovarian granulosa cells, PTPN18 appears to be involved in regulation of actin reorganization upon stimulation with follicle-stimulating hormone (FSH), and p190RhoGAP has been proposed as its substrate in this process [116]. PTPN18 associates effectively with PSTPIP1 (proline, serine, threonine phosphatase interacting protein 1) via its proline-rich motifs in the C-terminus and can also dephosphorylate PSTPIP1. As PSTPIP1 is an actin binding protein and a putative regulator of cytokinesis, these findings suggest that PTPN18 may also play a role in cytokinesis [117]. Finally, dephosphorylation of several receptor tyrosine kinases by PTPN18 was shown in co-expressing HEK293 cells [118].

Cancer Relevance: PTPN18 is robustly expressed in breast cancer cell lines and mouse mammary cancer tissues that over-express the oncoprotein ERBB2 [119]. This observation prompted analysis of a possible regulatory interaction. Indeed, in HEK293 cells PTPN18 was capable of dephosphorylating endogenous ERBB2endogenous ERBB2, thereby attenuating ERBB2 downstream signaling, but not EGF or IGF1 (insulin-like growth factor-1) receptors. Moreover, *PTPN18* knockdown enhanced ERBB2 autophosphorylation in the T47D breast cancer cell line [119]. The mechanism of interaction of ERBB2 with PTPN18 has recently been studied in some detail. Interestingly, PTPN18 has not only the capacity to dephosphorylate ERBB2 at several sites, it can also promote its ubiquitination and degradation. The latter function is mediated by the PTPN18 PEST domain [120]. Clearly, the possibility that PTPN18 regulates oncogenic signaling of ERBB2 in breast cancer cells warrants further investigation.

PTPN22

Regulation: The crystal structure of the PTPN22 catalytic domain revealed a unique motif in the putative substrate-binding region, which harbors a PKC phosphorylation site potentially associated with regulation [121]. In T-cells, *PTPN22* expression can be induced by TCR activation and treatment with the phorbol ester TPA [122]. Relatively selective PTPN22 inhibitors with low micromolar potency that are active in intact cells have been developed. Examples are I-C11 and LTV-1, whose characteristics of PTP-domain interaction have been elucidated by crystallography or molecular modeling, respectively [121, 123].

Physiological Function: PTPN22 has been firmly identified as a negative regulator of the proximal events of TCR signaling. Interaction with CSK, another negative regulator of TCR signaling, through one of the proline-rich motifs (designated P4) in the PTPN22 C-terminus is important for this function [124, 125]. PTPN22 can selectively dephosphorylate the activating site pY394 in the SFK LCK, a critical signaling molecule for the TCR, and further potential substrates in T cells have been identified. Inactivation of *Ptpn22* in mice augmented TCR-induced tyrosine phosphorylation, e.g., of LCK pY394, and led to accumulation of effector-memory T-cells. Older mice developed lymphadenopathy and splenomegaly [126]. The relatively mild phenotype may relate to partial functional redundancy with the related PTPN12, which is also expressed in hematopoietic cells [4, 127]. Recently, preferred down-regulation by PTPN22 of TCR-signals induced by weak agonists and self-ligands as opposed to high-affinity ligands was shown. These findings suggest a role of PTPN22 in preventing inflammatory reactions to weaker ligands including autoantigens [128]. A frequent polymorphism in *PTPN22* (C1858T, encoding a R620W PTPN22 variant) was earlier identified as a risk factor for autoimmune diseases, including type I diabetes, rheumatoid arthritis, systemic lupus erythematosus, and myasthenia gravis, but is protective with respect to Crohn's disease [129, 130]. In carriers of the R620W version, a reduced removal of autoreactive B-cells [131] and reduced apoptosis upon BCR-engagement [132] were also observed. There is a controversy in the literature whether the R620W version confers gain or loss of function [127, 129]. The recent analysis of a mouse knock-in strain indicated reduced stability of the mouse protein (PepR619W) caused by increased calpain- and proteasome-mediated degradation, which was associated with reduced protein expression and a phenotype similar to the *Ptpn22* knockout strain [133].

Cancer Relevance: An early report indicated a role of PTPN22 in negative regulation of oncogenic BCR-ABL signaling in CML cells [134]. PTPN22 was found expressed in a number of CML cell lines and in small set of primary CML samples. Inducible over-expression of PTPN22 in a BCR-ABL expressing CML cell line caused dephosphorylation and degradation of BCR-ABL, associated with down-regulation of signaling, and inhibition of colony formation [134]. However, loss-of function analyses will be important to further establish PTPN22 as a negative regulator of BCR-ABL.

Recently, a role of PTPN22 in CLL has been uncovered. This disease is characterized by an expansion and progressive accumulation of mature B lymphocytes. BCR signaling is important for growth and survival of the leukemic cells. BCR activation elicits, however, anti- and pro-apoptotic signals, and the latter are attenuated in CLL cells. Interestingly, this may at least be partially caused by PTPN22. The analysis of PTPN22 levels in 91 cases of CLL revealed an over-expression of the protein in >70 % of cases [135]. Down-regulation of PTPN22 with siRNA enhanced the pro-apoptotic effect of BCR activation by soluble IgM. It also enhanced a number of signaling events, including activation of p38 MAPK, LYN, SYK, ERK, N-terminal c-Jun kinase (JNK), BLNK, and SHIP, which are at least in part pro-apoptotic. In contrast, the activation of AKT, which is anti-apoptotic, was attenuated. Over-expression of PTPN22 had opposite effects. PTPN22 expression was dependent on PKC signaling, and PKC inhibitors reduced PTPN22 expression and enhanced apoptosis induction by soluble IgM. This study identified PTPN22 as both negative and positive regulator of BCR signaling and indicated a role of PTPN22 over-expression in CLL cells by preventing apoptosis through chronic BCR stimulation. Thus, PTPN22 and PKC inhibitors may be therapeutically relevant in CLL [135].

NT5 Type, FERM and PDZ Domain-Containing PTPs

The human genes *PTPN3* and *PTPN4* comprise the NT5 type of non-transmembrane PTPs [1]. NT5, NT6, NT7, and NT8 type PTPs have in common a so-called FERM (acronym for Band 4.1, Ezrin, Radixin, Moesin) domain (Fig. 1) that is responsible for the targeting of these enzymes to submembranous, often cell cortical, sites in the cell.

In addition, NT5-type PTPs contain a single PDZ (acronym for PSD95, Dlg, ZO-1) protein domain that allows specific binding to C-terminal recognition sequences in target proteins. By contrast, NT6 type PTPs lack such a domain, and NT7/8 types contain no less than five PDZ domains. *PTPN3* and *PTPN4* are broadly expressed in many tissues throughout development (Table 1) [136–139], and the presence of FERM and PDZ domains suggests that these proteins may act at junctions between the membrane and the cytoskeleton.

PTPN3

Physiological Functions: The first interaction partner for PTPN3 was not related to FERM or PDZ-mediated binding by this PTP. It rather pointed to phosphoserine residues in PTPN3 as a preferred binding spot for protein 14-3-3β, raising the possibility that this regulatory protein may function as an adaptor molecule for PTPN3 [140]. PTPN3 also interacts with and dephosphorylates vasolin-containing protein (VCP), which is implicated in the control of membrane fusions via the t-SNARE syntaxin 5 and in regulating cell cycle events [141]. The interaction involves the (phospho)tyrosine residues-containing C-terminal tail of VCP, but it is unclear whether the PTPN3 PDZ domain also contributes. It is of note that, recently, VCP was also identified as a substrate for the NT7 type PTPN13 protein [142]. Additional PTPN3 substrates were suggested following its over-expression in Jurkat T cells. PTPN3 not only reduced the TCR-induced activation of reporter genes, but in addition, the MAP kinase ERK2, its upstream activator MEK, and the JNK kinases were less active [143]. In fact, PTPN3's capacity to dephosphorylate TCRζ is based on its PDZ domain-mediated interaction with the complex [144]. Also, the reduction in phorbol ester-stimulated shedding of tumor necrosis factor-α in cells expressing catalytically active, compared with catalytically inactive, PTPN3 results from PTPN3 PDZ domain binding, in this case to the C-terminus of the tumor necrosis factor α-convertase TACE [145]. The PTPN3 PDZ-domain also binds to the C-terminal tail of the sodium channel protein Nav1.5, which facilitates its subsequent dephosphorylation by PTPN3, thereby destabilizing its inactivated state [146]. Additional interacting proteins are reminiscent of what was learned from PTPN13 and include the RTKs ERBB2, growth hormone receptor, and hepatocyte growth factor receptor. In mouse knockout models, normal hematopoietic function but enhanced growth due to augmented growth hormone signaling were observed [147, 148]. PTPN3 is also expressed in the central nervous system and behavioral tests with knock-out animals revealed subtle, sex-specific effects on working memory and motor learning [149]. Additional studies hinted at potential roles in pain sensitivity and, in contrast to the suggested negative regulatory effect on the immune response, in facilitating LPS (lipopolysaccharide)-induced cytokine release [150].

Role in Cancer: Genetic analysis of human colon cancer showed that PTPN3 is mutated in 3 % of the tumor specimens, but the consequence of these mutations was not investigated [151]. Also, somatic mutations have been noted in 22 % of liver cancer cases [152]. Next-generation sequencing of the transcriptome of a non-small cell lung carcinoma (NSCLC) cell line disclosed a translocation event between *ALK* and *PTPN3* that effectively resulted in a *PTPN3* null allele. Subsequent re-expression of PTPN3 in NSCLC lines inhibited colony formation [153], consistent with a tumor-suppressive role. Furthermore, it has been reported that also PTPN3 is targeted for degradation by the oncoviral E6 protein via a PDZ-mediated interaction [154, 155]. In a comparative study of reported E6 PDZ substrate levels following E6 ablation in high-risk human papilloma virus (HPV)-positive cells, however, the tight-junction protein MAGI-1 represented a major degradation target of E6, and PTPN3 levels in fact hardly changed [156].

Instead, several lines of evidence point to a tumor-promoting role for PTPN3. In esophageal carcinogenesis, expression of PTPN3 mRNA was significantly increased in cancer lesions compared with that in normal mucosa [157]. In gastric cancer, a significant negative correlation between PTPN3 levels and the differentiation status of the tumor tissue was observed, although PTPN3 expression could not be linked to patient survival data [158]. A recent search for mutations in the second-most common hepatic cancer-type, intrahepatic cholangiocarcinoma (ICC), revealed that about half of the specimen carried mutations in *PTPN3* [159]. Intriguingly, the mutations reside in or close to the FERM domain, leaving the PDZ and PTP domains in PTPN3 unaltered, yet demonstrate gain-of-function effects in proliferation, colony formation, and migration assays with cholangiocarcinoma cells in vitro. Notably, the rate of disease recurrence was found to be higher for patients with increased PTPN3 expression levels or presence of activating PTPN3 mutations in ICC tumors [159], underscoring the oncogenic nature of *PTPN3*. Cardinale and Alvaro pointed out that this genetic pattern may well be associated with hepatitis B virus (HBV) infection [160]. PTPN3 activity reduces HBV gene expression and replication and binds to the viral core via its PDZ domain [161]. The PDZ-mediated interaction, however, is not required for the PTPN3 effect on HBV replication; rather PTPN3's catalytic activity and its FERM domain are essential for viral suppression. PTPN3 FERM domain mutations may thus be enriched in HBV-infected individuals because of the resulting increased viral replication and consequent evolution of hepatitis towards oncogenesis.

PTPN3 is also over-expressed in half of primary breast cancer cases and levels positively correlate with metastatic potential [162]. Interestingly, this growth stimulatory effect of PTPN3 turned out independent of its PTP activity and rather involved its binding to the vitamin D receptor (VDR). It has not been investigated whether this binding requires the PTPN3 PDZ domain, but, judging from the sequence of VDR C-terminal residues, this seems not very likely. PTPN3–VDR complex formation leads to a mutual stabilization and cytoplasmic accumulation, thus preventing VDR from traveling to the nucleus and exerting its transcriptional, growth inhibitory effect [162].

It has become apparent that KRAS-mediated oncogenic signaling requires the MAP kinase family member p38γ, which has a unique C-terminal sequence that represents a PDZ domain docking site, preferably in the non-phosphorylated form [163]. A search for protein phosphatases that may facilitate this p38γ regulation of RAS transformation resulted in the identification of PTPN3 [164]. PTPN3 and p38γ cooperate in K-RAS oncogenesis through direct PDZ-mediated binding and p38γ

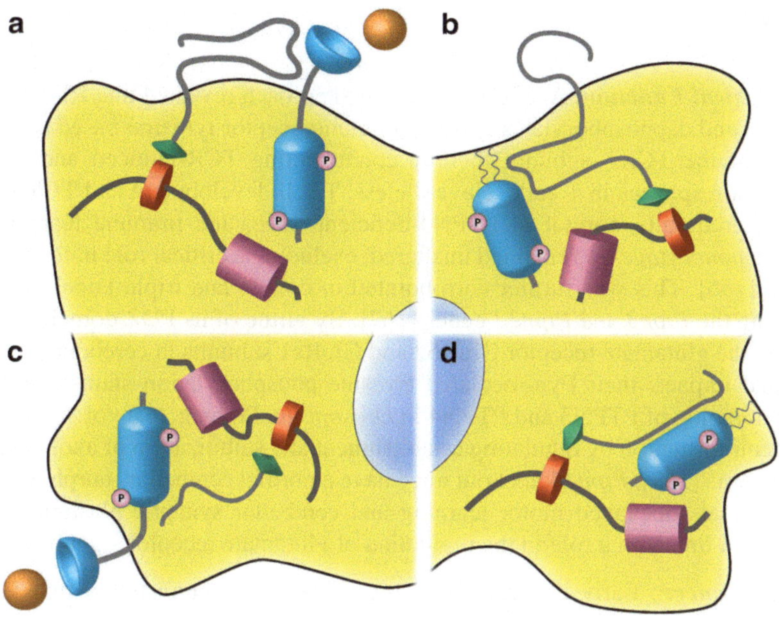

Fig. 3 PDZ-based scaffolds as important regulators of oncogenic signaling by tyrosine kinases and non-transmembrane PTPs. (**a**) By means of their C-terminal PDZ target sequence (*green diamond*) transmembrane molecules like NECL-2 and EFNB recruit PDZ domain-containing PTPs (e.g., PTPN13) to sites where phosphotyrosine-containing growth factor receptors (e.g., the ERBB family) serve as substrates. (**b**) Likewise, cell surface proteins like EFNB are able to recruit PTPN13 to submembranous areas where cytosolic targets like SRC await dephosphorylation. (**c**) The MAP kinase p38γ serves as a cytoplasmic scaffold that enables PTPN3 to use EGFR as a transmembrane substrate. (**d**) Cytosolic scaffolds like RIL also are capable of directing PTPN13 to its substrate SRC. These four types of enzyme/scaffold/substrate PDZ-based signalosomes have been discussed in the text, in the sections dealing with the NT5 and NT7/8 types of non-transmembrane PTPs. PDZ domains are depicted as orange discs and the PTP catalytic segment is presented as a pink cylinder. Cell cytoplasm is in yellow, nucleus in light blue. Blue rods and cups represent the tyrosine kinase and growth factor binding domains in TKs, respectively. Extracellular balls represent receptor ligands and phosphotyrosine residues within the substrates are symbolized by the character P on a pink circle. Thin curly grey lines represent membrane anchoring lipid modifications, as encountered for example in SRC

dephosphorylation by PTPN3 (Fig. 3). In line with a coordinative oncogenic role, p38γ and PTPN3 protein levels are increased by K-RAS, and this is also evidenced in primary colon cancer tissues [164]. Vice versa, the p38γ–PTPN3 interaction also results in p38γ-mediated PTPN3 phosphorylation at Ser-459. Interestingly, this phosphorylation had no effect on PTPN3's catalytic activity toward p38γ, but was required for dephosphorylation of the EGFR by PTPN3 [165]. Furthermore, PTPN3 phosphorylation on Ser-459 aided RAS-transformed cells to withstand sorbitol, but not anysomycin-induced cell death. Thus, PTPN3 participates in RAS oncogenesis and stress signaling by means of phosphoserine 459-dependent interactions with downstream molecules, including dephosphorylation of EGFR receptors and stimulus-specific anti-apoptotic effects [165].

PTPN4

Physiological Function: A substrate-trapping approach revealed that PTPN4 interacts with and dephosphorylates ITAMs (immunoreceptor tyrosine-based activation motifs) of the TCRζ subunit, thereby counteracting TCR-induced and NF-κB-dependent responses in T cells. Nevertheless, T cell development and TCR signaling appeared fully normal in PTPN4-deficient mice, and immune responses to *Listeria monocytogenes* remained unaltered, excluding a critical role in the immune system [166]. This was further corroborated in double and triple knock-out mice involving the *Ptpn3* and *Ptpn13* genes [167]. By virtue of its PDZ domain, PTPN4 binds to the glutamate receptor GluRδ2 and GluRε1 subunits in cerebellar Purkinje cells and impacts their Fyn-mediated tyrosine phosphorylation status [168]. The single homolog of PTPN3 and PTPN4 in *Drosophila*, Ptpmeg, is involved in neuronal circuit formation by regulating establishment and stabilization of axonal projection patterns [169]. *Ptpn4* knockout mice have a normal cerebellar morphology, but these animals displayed motor learning and cerebellar synaptic plasticity abnormalities, in line with a role in the regulation of glutamate receptor signaling [170].

Role in Cancer: Early experiments using COS cells demonstrated that PTPN4 over-expression inhibited proliferation and colony-forming ability independent of its PTP activity [171]. This suggested a growth-suppressive, or perhaps pro-apoptotic effect that is mediated by the FERM and/or PDZ domain part of the PTP. More recent experiments, in which the adapter protein CRK-I was identified as a PTPN4 substrate [172], rather point to an essential role for PTPN4's phosphatase activity in growth regulation. It appeared that the SH3 domain of CRK-I binds to a proline-rich region in PTPN4, between residues 462 and 468, which then induces CRK-I dephosphorylation and subsequent inhibition of CrkI-mediated proliferative and migratory effects in the cell lines studied [172].

Somatic mutations in 35 % of liver cancer cases and copy number variations in 7 % of pancreatic cancer specimens have been noted [152]. Different from PTPN3, a role for PTPN4 in gastric cancer etiology was not proposed [158]. One nonsense mutation and four missense mutations in *PTPN4* have been detected by whole genome sequencing of 16 large intestine tumor samples [173], but no mutations in colon cancer specimens were encountered in the PTP exome study by Wang et al. [151].

A recent serendipitous finding [174] may, however, put PTPN4 biology at center stage in the design of new pro-apoptotic compounds for the treatment of cancers [175]. Viruses are masters in manipulating life and death decisions in the host cell. Virulent rabies strains do so by virtue of the C-terminus of their envelope glycoprotein G that enables them to sequester the PDZ domain-containing host serine-threonine kinases MAST1 and MAST2, thereby promoting neuronal cell survival. A single Q to E amino acid change in the C-terminus of glycoprotein G from an attenuated viral strain, however, turned the fate from survival to apoptotic death for infected neurons [174]. The mutant C-terminus now bound the PDZ domain of PTPN4 and, reasoning that perhaps normally PTPN4 may keep cells from entering apoptosis, the authors tested whether providing the PTPN4 PDZ domain with an appropriate peptide ligand would abrogate this protective function of PTPN4. Indeed, glioblastoma cells were triggered to die following intracellular delivery of the viral peptide or of one derived from an endogenous PTPN4 PDZ ligand like the GluN2A C-terminus [176], opening up novel ways to design pro-death compounds by targeting PDZ-containing proteins such as PTPN4. As to the mechanism involved, Maisonneuve et al. recently reported that PTPN4's catalytic activity is actually inhibited by its PDZ domain, and only upon peptide binding to the PDZ domain this catalytic inhibition is released [177]. Further insight should now come from the identification of the critical PTPN4 substrates that, upon dephosphorylation by the activated PTPN4, gear up the apoptotic machinery of the cell.

NT6 Type, FERM Domain-Containing PTPs

The genes *PTPN14* and *PTPN21* encode NT6 type non-transmembrane PTPs, which are characterized by a C-terminal catalytic PTP domain and an N-terminal FERM domain [1]. As opposed to the NT5 and NT7/NT8 types, however, PTPN14 and PTPN21 proteins do not contain PDZ, KIND (putative kinase non-catalytic C-lobe domain), or other additional protein domains (Fig. 1). It is of note that the NT6 type PTPs contain some atypical residues in conserved loops within the catalytic domain [1] and failed to dephosphorylate any of the 38 phosphopeptides representing a panel of potential PTP substrates [178]. Still, protein phosphatase activity of PTPN14 [179] and PTPN21 [180] has been reported, suggesting that the activation of these PTPs might be highly context-dependent.

PTPN14

Physiological Function: PTPN14 expression patterns in cells of hematopoietic and non-hematopoietic origin (including kidney, skeletal muscle, lung, and placenta [181]) suggested early on a regulatory role in T cell development [182]. Inducible over-expression of PTPN14 in HeLa cells revealed a sub-membranous localization for the protein and resulted in reduced proliferation, cell spreading, and adhesion, paralleled by decreased actin stress fibers and focal adhesions [183]. In proliferating cells cultured at low density, however, PTPN14 resided in the nucleus and nuclear localization was lost upon serum starvation or following addition of TGFβ (transforming growth factor β), an inhibitor of cell proliferation but not migration [184]. In confluent, quiescent cells PTPN14 is found at intercellular junctions where it dephosphorylates adherens junction components, including β-catenin, thereby stabilizing cell–cell interactions and limiting cell motility [179]. TGFβ signaling is instrumental in embryonic epithelial–mesenchymal transition (EMT) and contributes to cancer metastasis. PTPN14 knockdown in zebrafish embryos resulted in defective organ development and over-expression in epithelial MDCK cells induced TGFβ signaling and caused EMT [185].

Further proof for a crucial role of PTPN14 in organogenesis results from studies on Hippo signaling, which underlies midgut homeostasis in flies and intestinal regeneration in mammals. Activation of Hippo is orchestrated by the scaffolding protein Kibra, which recruits FERM domain proteins, such as the tumor suppressors Merlin and Expanded [186]. The *Drosophila* PTPN14 homolog Pez has now been identified as an additional Kibra binding partner, involving the Pez proline-rich segment and Kibra's WW domain [187] (Fig. 4). Intriguingly, Pez promoted Hippo signaling in the fly, thereby restricting intestinal stem cell proliferation, but it was the Pez FERM domain rather than its enzymatic PTP activity that was essential for this growth suppressive effect [187]. Therefore, the mammalian Pez homologs PTPN14 and PTPN21 might be expected to play decisive roles in the maintenance of intestinal tissue homeostasis.

In humans, loss-of-function mutations in *PTPN14* are causative for autosomal-recessive lymphedema-choanal atresia syndrome [188]. The involved frameshift mutation (p.S194fs212X) interrupts the protein's FERM domain and results in an unstable and catalytically inactive protein, demonstrating a crucial role for PTPN14 in lymphatic and nasal development. However, in a mouse *Ptpn14* gene trap mutant, expressing a PTP14 mutant that is truncated at position 170 in the middle of its FERM domain, no signs of any intestinal phenotype were apparent [188]. Instead, the lymphatic hyperplasia with lymphedema phenotype as observed in human patients was recapitulated. Early-onset cancer has not been documented for the p.S194fs212X family, and perhaps expression of the paraloguous gene *PTPN21* in intestine prevents oncogenesis in this tissue. With regard to the mechanism underlying the lymphatic endothelial hyperplasia, VEGF receptor 3 (VEGFR3, a receptor tyrosine kinase essential for lymphangiogenesis) has been put forward as the critical PTPN14 substrate [188, 189]. By analogy with PTPN21, which regulates proliferative signaling pathways through a scaffolding function (as will be described later), one could think of a ligand-mediated signal that unites VEGFR3 and PTPN14, thus leading to proper signal termination.

Role in Cancer: PTPN14 mutations have been identified in colorectal, liver, and intestinal cancer specimens [151, 152, 173], suggesting that PTPN14 may play an important role in tumorigenesis of multiple types of cancers. Recently, Y128 in the p130 CRK-associated substrate (CAS) was identified as a prominent PTPN14 substrate site through phosphoproteomic analyses of colorectal cancer cells [190]. SRC is responsible for the CAS phosphorylation on Y128 and, by doing so, augments the proliferative and migratory potential of colon carcinoma cells. In line with this, cells with high pY128-CAS levels are much more sensitive to the SRC family kinase inhibitor Dasatinib, and PTPN14 mutations may thus serve as predictive markers for Dasatinib response in cancer patients [190]. By comparing gene expression profiles of primary pancreatic cancer cells and their liver metastases in a xenograft model in SCID mice [191], PTPN14 was identified as being consistently down-regulated in metastatic tumor cells, as compared to the primary lesion. This would support a metastasis-suppressive role of PTPN14, possibly through CAS dephosphorylation.

In search for biomarkers that would predict cisplatin and paclitaxel chemosensitivity in non-small cell lung cancer (NSCLC), a signature of 11 genes—among

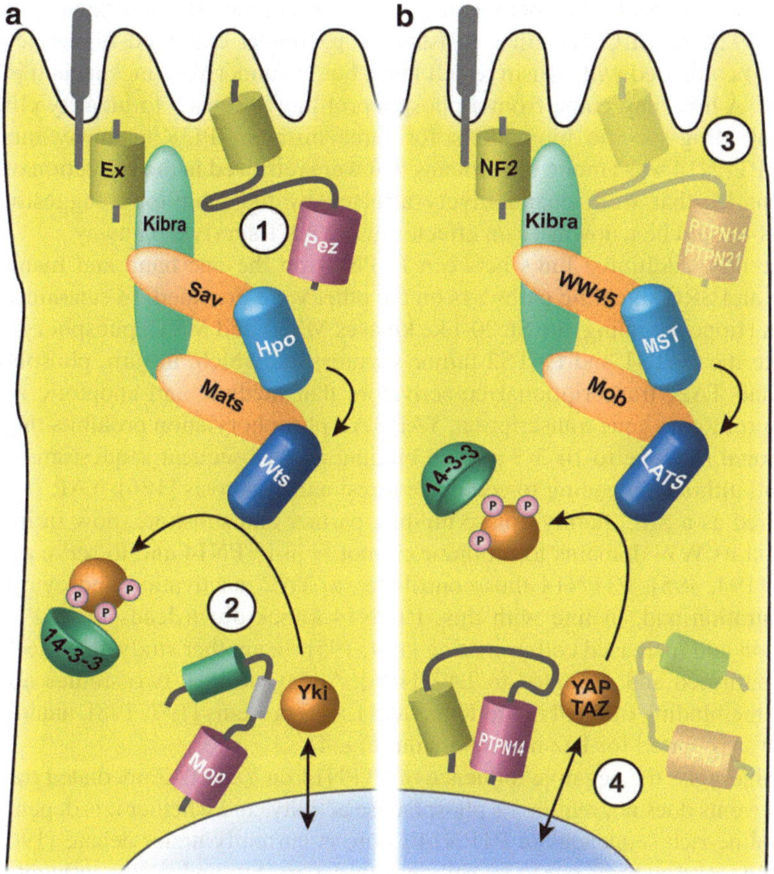

Fig. 4 Non-transmembrane PTPs as major regulators of the SWH signaling pathway. A schematic of the current working model for the Hippo pathway in *Drosophila* (**a**) and mammalian (**b**) cells (in *yellow*, with nucleus in *light blue*) is presented (adapted from [369]). In the fly, cell surface proteins and submembranous proteins such as Ex and Kibra are upstream of the Hippo signalosome, which consists of the scaffold proteins Sav and Mats and the serine/threonine kinases Hpo and Warts (Wts). When activated upon phosphorylation by Hpo, Wts phosphorylates the downstream effector Yorkie (Yki). As a result, Yki is sequestered by 14-3-3 proteins and cannot enter the nucleus to exert its activity as a transcriptional regulator of downstream genes to promote cell proliferation and/or survival. In mammals, the structural components are well-conserved and functional conservation has partly been established. For the downstream effector Yki, two mammalian homologs exist, YAP and TAZ. A *Drosophila* homolog of PTPN14 and PTPN21 called Pez binds to Kibra (interaction 1) and promotes growth-suppressive Hippo signaling independent of its PTP activity. Likewise, the *Drosophila* homolog of PTPN23, Mop, binds Yki, thereby preventing its entry into the nucleus (interaction 2). Again, PTP activity is not required and Mop probably acts as a competitive inhibitor of the binding of other proteins to the Yki WW domains. In mammals, the NT-6 type PTPN14 is part of the Kibra/LATS complex, but it remains to be investigated whether the paralogous PTPN21 is also capable of augmenting Hippo signaling at this level (interaction 3). Furthermore, evidence has been provided that PTPN14 binds to the downstream targets YAP and TAZ (interaction 4), again positively contributing to Hippo growth-suppressive signaling in a manner that is equally well taken care of by a PTPN14 CS mutant. It is tempting to speculate that PTPN23, and perhaps PTPN21, have a similar capacity and thus may carry tumor suppressive potential. For details and references, see the text on the respective PTPs

which was PTPN14—has been extracted that was predictive for paclitaxel resistance [192]. NSCLC cell lines resistant to paclitaxel exhibited higher levels of PTPN14 compared with sensitive cell lines, but the underlying mechanism is as yet unclear. A hint may come from a phosphoprofiling approach to identify clinically relevant, drug-specific biomarkers for three different PI3K pathway inhibitors [193]. PTPN14 was among the proteins that were retrieved in the collection of phosphopeptides that were quantitatively affected by the inhibitors, suggesting that PTPN14 could be a downstream effector of the AKT survival pathway.

Recently, additional links between PTPN14 on the one hand and mammalian Hippo and SRC signaling pathways on the other were provided. In canonical mammalian Hippo signaling, the Ste20-like kinases MST1 and MST2 phosphorylate and activate the LATS1 and LATS2 tumor suppressors, which, in turn, phosphorylate YAP and TAZ, transcriptional co-activators that mediate anti-apoptotic and cell cycle-promoting gene transcription. YAP/TAZ phosphorylation prohibits their transcriptional role due to 14-3-3 protein binding and subsequent sequestration in the cytosol, ultimately leading to cell cycle arrest and apoptosis [186]. YAP, originally identified as a SRC family kinase binding partner and substrate, now turns out to bind via its WW domains to proline-rich motifs in PTPN14 and to serve as a substrate [194, 195]. PTPN14 thus contributes to YAP inactivation and cytoplasmic sequestration and, in line with this, PTPN14 knock-down leads to YAP nuclear retention and increased cell migration [194, 195]. In another study, this mechanism was confirmed and extended to TAZ [196]. Adding to this, two studies now also report the binding of PTPN14 to Kibra and LATS proteins [197, 198], analogous to what was reported for Pez in Drosophila (Fig. 4).

Intriguingly, the negative influence of PTPN14 on YAP/TAZ-mediated transcriptional events does not require its phosphatase activity, and whether it is dependent on the proline-rich sequences in PTPN14's core is currently under debate [196, 199]. PTPN14 stability turns out to be cell density-dependent; under subconfluent culture conditions PTPN14 is targeted for proteasomal degradation by the E3 ligase CRL2 [195]. Thus, PTPN14's inhibitory effect on YAP/TAZ transcriptional activities only kicks in at high cell densities.

Importantly, a PTPN14 fragment that contained the YAP/TAZ binding motifs acted as dominant-negative and sensitized the cancer cells to EGFR inhibitors or drugs like S12 that promote apoptosis [196], opening up novel therapeutic strategies to suppress YAP and TAZ oncogenic activities via interaction blocking agents. It will be interesting to test whether its closest homolog PTPN21 is also able to inhibit YAP/TAZ function in a similar manner but, based on what will be discussed below, this seems rather unlikely.

PTPN21

Physiological Function: The PTPN21 cDNA was first cloned from human skeletal muscle and transient coexpression experiments positioned this PTP as a substrate and binding partner for the proto-oncogenic SRC tyrosine kinase [200].

Later, an alternative splicing variant that essentially consists of the catalytic domain and lacks the SRC-binding part was discovered [201]. Yeast two-hybrid screens identified the kinesin-like protein KIFIC as a tyrosine-phosphorylatable binding protein for the PTPN21 FERM domain and pointed to a potential involvement of the phosphatase in the retrograde transport of Golgi-derived vesicles to the ER membrane [202]. In a similar manner, the PH domain of the cytoplasmic tyrosine-protein kinase BMX was found to specifically recruit PTPN21, which results in enhanced BMX kinase activity, increased phosphotyrosine content on both proteins, and enhanced STAT3 signaling [203].

Via studies addressing the A-kinase anchor protein AKAP121, it was found that PTPN21-mediated activation of SRC and augmentation of the magnitude and duration of EGF signaling is inhibited by AKAP121 through binding and redistribution of PTPN21 to mitochondria [180]. Additionally, PTPN21 is able to bind to focal adhesion kinase (FAK), a scaffold protein enriched at adhesion plaques, thereby promoting cell migration. Functional PTPN21 and SRC were both required for EGF-induced FAK autophosphorylation and downstream signaling, demonstrating that by modulating SRC-FAK signaling at adhesion sites, PTPN21 contributes to cell adhesion and migration [204].

A recent meta-analysis of genome wide association studies revealed that two polymorphisms in the coding segments of *PTPN21* showed significant association with schizophrenia [205]. Mutant mouse models for *Ptpn21* are not yet available, but ES cells in which *Ptpn21* is targeted are on stock at the International Knock-out Mouse Consortium (www.knockoutmouse.org) and such studies could well be on their way.

Role in Cancer: *PTPN21* mutations have been documented in liver cancer cases [152], and sequencing of intestinal cancer specimens has yielded 1 nonsense and 14 missense mutations in the gene [173]. Comparison of gene expression profiles for microsatellite instable (MIN) and chromosomal instable (CIN) types of sporadic colorectal cancers revealed that *PTPN21* is over-expressed in MIN lines [206]. Because the gene contains a long mononucleotide repeat in its coding region, mutation rates for PTPN21 were investigated. Although a high frequency (17 %) of mutations was encountered in MIN samples, this alone does not provide evidence that PTPN21 mutations play a role in MIN etiology [207]. A genome-wide expression analysis in normal and cisplatin-resistant HeLa cells demonstrated that *PTPN21* is among the genes that are consistently up-regulated in resistant cells [208]. Knock-down of PTPN21 sensitized various cell lines to cisplatin and it reversed the acquired chemoresistance in the HeLa derivatives. It may therefore represent a potential target to inhibit and by that way increase chemotherapy efficacy [208].

Because PTPN21 localizes at adhesion sites and the actin cytoskeleton and stimulates SRC, EGF, and FAK signaling, it has some hallmarks of a proto-oncogene. Indeed, EGF stimulation triggers a rapid KIF16B-mediated recruitment of PTPN21 to endosomes that contain EGFR, and PTPN21 knock-down results in EGFR degradation and inhibition of downstream signaling [209]. Moreover, PTPN21 levels increase with increased grade and invasiveness in bladder cancer, and blocking PTPN21 activity attenuates growth and motility of bladder cancer cells [209]. Together with its mysterious contribution to chemoresistance, this renders PTPN21 a tumor-promoting candidate, as opposed to the tumor-suppressive character of its closest homolog PTPN14.

NT7 and NT8 Type PTPs, Containing KIND, FERM, and PDZ Domains

PTPN13 and PTPN20 were originally classified as sole members of NT7 and NT8 type non-transmembrane PTPs, respectively [1]. Quite recently, however, it was shown that PTPN20 represents a true paralog of PTPN13, which also harbors a KIND, a FERM, and five PDZ domains in addition to the single catalytic PTP domain [210]. This finding thus unites the NT7 and NT8 types (Fig. 1). *PTPN13* and *PTPN20* both undergo extensive alternative splicing, generating multiple isoforms that essentially differ in their content of protein interaction domains. They represent by far the largest non-receptor PTPs, and it will not come as a surprise that their extensive anchoring and scaffolding potential contributes significantly to their *modus operandi.*

PTPN13

Physiological Function: The *PTPN13* transcription start site is separated by just a few hundred base pairs from the c-Jun N-terminal Kinase 3 gene *JNK3*, in a head-to-head orientation. These genes thus share a bidirectional promoter region with potential binding sites for transcriptional regulators like Sp1, AP-1, AP-2, GATA-1, NFκB, NF-IL6, and p53 [211, 212]. In addition, the interferon regulatory factor ICSBP/IRF8 and signal transducer and activator of transcription 3 (STAT3) have been identified as *PTPN13* gene repressors [213, 214]. Mainly due to its FERM domain, PTPN13 displays an apical, submembranous localization in polarized epithelia, but the protein also associates with cytoskeletal structures, especially during mitosis, and shuttles into the nucleus [215, 216]. Its five PDZ domains allow many proteins to bind to PTPN13: RIL, BP75, p75NTR, PRK2, IκBα, EphrinB (EFNB), β-catenin, cSRC, APC, STAT4, TAPP1, TRPM2, TRIP6, and PARG1, to name a few. Some of these also appear on the list of PTPN13 substrates: RIL, TRIP6, IκBα, SRC, STAT4, and EFNB. In addition, IRS-1 (insulin-receptor substrate 1) and ERBB2 were shown to be dephosphorylated by the phosphatase. Its best known interaction partners are, however, the apoptosis-regulating membrane receptor FAS and the E6 protein of tumor-associated human papillomaviruses [215, 216].

Both up- and down-regulation of PTPN13 protein levels in HeLa cells lead to a striking increase in the number of multinucleated cells as a result of defective cytokinesis [217]. Deletion of the *Ptpn13* catalytic domain-encoding part [218] or even the complete gene [219] in genetically engineered mice, however, revealed only very subtle phenotypes. A mild reduction in peripheral nerve repair capacity [218] and an improved host defense against intrapulmonary *K. pneumoniae* infection [219] in vivo has been reported. In cultured cells, also growth impairment of retinal glial cells [220], repressed differentiation of adipocytes [221], and increased and prolonged activation of STAT4 and STAT6, and consequently enhanced T helper cell differentiation [219] have been documented. Even cross-breeding with mouse strains that lack the two NT5 type FERM and PDZ-containing PTPs

(PTPN3 and PTPN4) did not reveal any major phenotypic consequences [167]. In retrospect, one can now postulate that remaining presence of PTPN20 may have confounded these studies. Experiments ex vivo, which addressed PTPN13's potential functions based on the nature of its interactome, could be misleading as well because of PTPN20-mediated redundancy.

The PTPN13 PDZ domains enable an interaction with the C-terminus of the human FAS receptor [222], thereby preventing its translocation from intracellular stores to the cell surface and protecting cells from FAS-mediated apoptosis [223–225]. In addition, in some cell types PTPN13 activity results in NFκB activation and Bcl-2 up-regulation, further increasing their resistance to FAS-mediated apoptosis [226]. Because PTPN13 expression is stimulated by NFκB activity and NFκB activity and PTPN13 expression levels both increase in several metastatic tumor types [227], the phosphatase would classify as a strong anti-apoptosis regulator. PTPN13 may also suppress p75(NTR)-mediated pro-apoptotic signals [228] or TRPM2-mediated oxidative stress-induced apoptosis [229] by virtue of its PDZ binding capabilities. Multiple reports point to a functional correlation of PTPN13 levels and resistance to FAS-mediated apoptosis [225, 230–243]. However, the FAS–PTPN13 PDZ-mediated interaction is not conserved in mammals [244] and also reports that do not support an anti-apoptotic role for PTPN13 have emerged [245–249]. For example, in prostate cancer cells PTPN13 facilitates FAS-mediated apoptosis through PKCδ-mediated effects on pro-apoptotic protein expression levels and activation of caspases [250]. Even FAS-independent, pro-apoptotic effects of PTPN13 have been reported [251–253]. Quite recently, it was found that, whereas in two-dimensional cultures of polarized epithelial cells PTPN13 kept the FAS receptor from reaching the plasma membrane, in 3D cultures the PTPN13 scaffold actually provided a FAS receptor reservoir that could rapidly emerge at the cell surface [254]. Taken together, these data emphasize that the impact of PTPN13 on cell survival and apoptosis is highly cell type and context-dependent.

Latest data further add to the complexity of PTPN13 impact on cell and protein homeostasis. While studying how the FBXL2 subunit in ubiquitin ligase complexes dictates substrate-recognition for ubiquitin-proteasomal degradation, Kuchay et al. identified PTPN13 and the PI3K regulatory subunit p85β as interacting proteins [255]. PTPN13 dephosphorylated p85β, enabling its interaction with FBXL2 and subsequent degradation, thus effectively eliminating surplus p85β that could act as a competitor for the recruitment of p110 PI3K catalytic subunit to phosphotyrosine docking sites at the cell membrane. Reduced PTPN13 levels indeed were found to result in attenuation of PI3K-AKT-mTOR signalling and promotion of autophagy [255]. Depending on the stimulus and cell type autophagy can influence cell death both positively and negatively, and in another study, it turned out that the cell's autophagic activity in fact determines how cells respond to apoptotic death receptor activation. It turned out that B-cell lymphoma cells become prone to FAS-induced apoptosis through p62-mediated selective autophagic degradation of PTPN13, but that under a low autophagic flux the PTPN13 levels suffice to efficiently dephosphorylate FAS and lower its cell surface levels [256]. This novel interplay between autophagy and apoptosis regulation may help explain the plethora of effects concerning cell fate regulation that are attributed to PTPN13.

Role in Cancer: PTPN13 was viewed early on as a protein with tumor-suppressive potential based on its homologies to the human neurofibromatosis type 2 (NF2) [257, 258] and *Drosophila* expanded [259] and discs-large [260] tumor suppressors. The identification of PTPN13 as a FAS receptor binding protein [222] and its ability to protect human cells from FAS-induced apoptosis (as described above), however, rather point to an anti-apoptotic role in tumorigenesis. In line with this, increased PTPN13 mRNA or protein levels have been witnessed in multiple cancer types, including prostate, breast, ovarian, renal, oral, and colon carcinomas [261–265]. Furthermore, a positive correlation between PTPN13 levels and apoptosis resistance has been documented for colon carcinoma specimens [239] and many different cancer cell lines [225, 236–238, 240, 241].

An additional line of evidence comes from studies on the chromosomal translocation event that in Ewing's Sarcoma results in the EWS-FLI1 fusion protein. EWS-FLI1 acts as a transcription factor that elevates *PTPN13* expression levels, thereby contributing to the increased apoptosis resistance and increased colony-forming potential of these tumors [266]. In line with this, mouse embryonic fibroblasts that lack PTPN13 catalytic activity have reduced foci-forming ability when transfected with oncogenes, and VCP/p97 was identified as a novel PTPN13 substrate important in this transformation [142]. Tyrosine phosphorylation of VCP enables its midbody localization during cytokinesis [142], a cellular niche where also PTPN13 operates [267]. Similarly, leukemia-associated chromosomal translocations involving the *TEL* gene result in fusion proteins that in myeloid cells hinder the repression of *PTPN13* gene transcription that is normally exerted by the interferon regulatory transcription factor ICSBP [242]. As a result, PTPN13 protein levels and FAS-induced apoptosis resistance were found to be increased, revealing an additional mechanism by which breakpoint fusion proteins can contribute to leukemogenesis. Subsequent studies by this group aimed at implicating PTPN13 in BCR-ABL-driven myeloid tumor formation. They demonstrate that BCR-ABL expression in myeloid leukemia cells leads to an up-regulation of PTPN13 levels and propose that this contributes to the increased FAS-induced apoptosis resistance [268] and β-catenin activity [269] in these cells. This functional implication is solely based on the use of what they call a "PTPN13-blocking SLV peptide" [268, 269] that has been shown to disrupt the PDZ domain-mediated interaction of human FAS and PTPN13 [230]. It is to be expected, however, that such a peptide will hinder the majority of PDZ domain interactions in a cell [270], leaving the alleged role for PTPN13 questionable.

In a multistep mouse tumorigenesis model, the malignant progression of oral carcinoma cells was paralleled by an increase in both PTPN13 levels and apoptosis resistance [271]. Also, for human oral squamous cell carcinoma specimens, a positive correlation of PTPN13 protein levels, chemo-radio resistance, and poorer survival was noted [264]. Surprisingly, though, over-expression of PTPN13 in the much studied colon carcinoma cell line SW480 rather led to an increased sensitivity for the extrinsic apoptosis pathway [253]. Also, many other studies point to tumor-suppressive or even pro-apoptotic roles for PTPN13, as we will discuss now.

Anti-estrogen treatment is widely used in breast cancer adjuvant therapy and, in addition to impairing steroid receptor action, these agents also impede growth factor signaling. Freiss and colleagues noted that anti-oestrogens up-regulated PTPN13 expression levels in breast cancer cells [272]. This had a profound negative effect on the IGF-1 signaling pathway; PTPN13 dephosphorylated IRS-1, leading to reduced PI3K/AKT survival signals and induction of apoptosis [251, 252]. In breast cancer specimens, PTPN13 levels were higher in estrogen receptor-positive samples than in ER-negative ones [273] and, as an independent marker, correlated well with a better survival prognosis [274].

In prostate cancer cell lines, PTPN13 knockdown resulted in enhanced cell migration and the up-regulation of invasion-related genes [275], suggesting that PTPN13 may not only suppress survival but could impair cancer cell invasion also. Inactivation of PTPN13 activity would thus aid the tumorigenic process and indeed studies report *PTPN13* silencing in cancer cell lines as a result of promoter hypermethylation [276, 277]. An alternative silencing mechanism was revealed by a search for anti-correlated expression of conserved mRNA–miRNA pairs in matched tumor and normal kidney samples. The oncogenic miR-185 was found to be up-regulated in renal cell carcinomas, and concomitant reduced expression levels were noted for its targets, which include PTEN and PTPN13 [263].

PTPN13 resides at chromosome 4q21.3, a region displaying allelic loss in hepatocellular carcinomas. Indeed in most HCC cases, either 4q loss or PTPN13 down-regulation through promoter methylation was encountered [277]. Additionally, several gene mutations were identified, and SNP p.Y2081D (in between the last PDZ domain and the catalytic PTP domain) significantly associated with multiplex familial HCC [277]. SNP p.Y2081D also associates with breast cancer risk in a Chinese Han population [278]. Several SNPs in the *PTPN13* gene (p.F1356L, p. I1522M and p.Y2081D) also associated with susceptibility to head and neck squamous cell carcinoma (HNSCC) [279]. Furthermore, homozygosity for SNP p. I1522M increases the risk for lung carcinoma and, in combination with p.Y2081D, for colorectal cancer in the Japanese population [280]. In fact, a growing list of PTPN13 mutations is being extracted from a diverse set of tumor types. It started with a PTP-wide screen in colon cancer specimens, where PTPN13 was among the six PTPs that appeared frequently mutated [151]. Some of the PTPN13 mutants have been investigated at the functional level. For example, p.F1356L and p.L1419P in PDZ domain 2 negatively affect its binding to FAS receptor and TRIP6 C-termini [277]. Mutations p.2408L and p.2423P, encountered in HPV-negative squamous cell carcinoma samples, and p.Y2260* and p.E2455D, detected in colorectal tumors, were all shown to impair PTPN13 catalytic activity [281, 282].

Perhaps the most unexpected way of eliminating PTPN13 activity comes from studies on HPV-positive squamous cell carcinomas. Oncogenic HPV16/18 variants encode an E6 protein that has a C-terminal tail that recruits and targets PDZ domain-containing host proteins for proteasomal degradation. PTPN13 is among the targets destined for degradation when bound to HPV E6 [283]. This is corroborated by the finding that curcumin treatment, which inhibits E6/E7 expression, blocks PTPN13 degradation in HPV16-positive cells [284]. PTPN13 loss in HPV-positive epithelial cells resulted in augmented ERBB2–RAS–ERK signaling, and PTPN13 knock-down synergized with mutants of ERBB2 and RAS in xenograft tumor growth [282]. A similar conclusion was drawn from a knock-down screen in HeLa cells which identified PTPN13 as the single classical PTP that consistently regulated

ERBB2 phosphorylation levels [281]. Intriguingly, ERBB2 activity induces PTPN13 expression in MDA-MB-453 breast carcinoma cells, which places the phosphatase center stage in a negative feedback loop for this pathway [285].

RNAi-mediated PTPN13 knockdown in MCF-7 breast cancer cells confirmed its suppressive effect on growth and invasion and, in addition, pointed to a crucial role for PTPN13 in SRC Y419 dephosphorylation [221]. SRC and PTPN13 had been connected years before, in a study addressing EFNB (Ephrin-B) reverse signaling [286]. Using neuronal cells, it was shown that upon EFNB-triggered tyrosine phosphorylation of EFNB by SRC, PTPN13 joins the complex through PDZ-mediated binding to the EFNB tail. PTPN13 then dephosphorylates EFNB and SRC. Intriguingly, PTPN13 does not target SRC's C-terminal inhibitory phosphotyrosine but rather the p-Y419 autoactivation site and consequently inhibits the kinase [286]. Quite recent experiments [287] now tie these observations together. EFNB1 was shown to interact with ERBB2 in a kinase activity-dependent manner and then recruitment of SRC, to mediate EFNB1 phosphorylation, and subsequent MAP kinase signaling takes place. PTPN13 recruitment then should revert these events and indeed decreased PTPN13 function augments oncogenic signals coming from its partner substrates [287]. A similar interplay exists between EFNB1 and ERBB1 and provides a rationale for disappointing results from ERBB1-targeted therapies [288].

The picture that emerges is that PDZ-mediated interactions recruit PTPN13 to sites where phosphotyrosine-based signals are ongoing and need to be diminished (Fig. 3), reminiscent of ideas that came from studies on NECL-2 in epithelial and neuronal cells [289, 290]. Knock-down of NECL-2 in colon cancer cells leads to enhanced heregulin-induced ERBB2-mediated phosphorylation of ERBB3 and downstream signaling. The growth-suppressive effect of NECL-2 depends not only on its extracellular domain, which facilitates an interaction with ERBB3, but also on the NECL-2 C-terminal PDZ binding site that needs to recruit PTPN13 for ERBB2/ERBB3 dephosphorylation [289, 290]. Recently, a similar scenario was revealed for NECL-4 at hemidesmosomes [291]. Also, the protein RIL serves as an adaptor to recruit PTPN13 and enable SRC dephosphorylation [292]. Inactivating mutations in PTPN13 or transcriptional repression of RIL, as observed in colon cancer specimens, will thus both disrupt SRC inactivation and contribute to colon cancer progression [292].

A recent corroboration of PTPN13's tumor suppressive nature comes from the dedicated screen for mutations in the 38 classical PTP genes in lung cancer specimens [293]. PTPN13 turned out to be the sole non-receptor PTP that was found down-regulated in cell lines and in the majority of primary tumor samples, but methylation changes were not the cause. Eight percent of the samples displayed LOH for the *PTPN13* locus and homozygous deletions occurred frequently within this group. Independent studies underscored the inverse correlation of PTPN13 expression levels with lung carcinoma grade [294] and recurrent gene *PTPN13* mutations in lung adenocarcinomas [295]. Functional studies disclosed a clear inhibitory effect of PTPN13 on anchorage-dependent and anchorage-independent growth in vitro and on tumorigenicity in vivo, which was paralleled by reduced tyrosine phosphorylation levels for EGFR and ERBB2 [293].

PTPN20

Physiological Function: *PTPN20*-encoded proteins initially were thought to consist essentially of a catalytic PTP domain, which displays highest homology to PTPN13 within the human PTP family [1]. Just upstream of *PTPN20*, in a head-to-tail orientation, the gene *FRMPD2* is annotated, which contains the remaining PTPN13-type domains (i.e., KIND, FERM, and PDZ). This triggered the den Hertog group, while establishing the zebrafish PTP gene collection, to investigate whether perhaps *PTPN20* and *FRMPD2* together make up a *PTPN13* look-alike. Indeed, they found ample evidence that the currently annotated *FRMPD2* and *PTPN20* genes in fact represent one single transcription unit, and thus collectively form a true paralogue of *PTPN13*. In zebrafish, PTPN13 and PTPN20 both act as negative regulators of RhoA in convergence and extension cell movements during gastrulation [210]. Interestingly, this PTP activity-dependent effect on RhoA required the GAP (GTPase-activating protein) PARG1, an established PDZ-domain target of PTPN13 [296].

Expression of mammalian PTPN20 was found primarily in testis and, not uncommon for testicular transcripts, in a wide variety of tumor cell lines [297, 298]. *Ptpn20* mouse models are not available, but the International Knock-out Mouse Consortium (www.knockoutmouse.org) has targeted ES cells ready for use. Only very few studies have addressed mammalian PTPN20 (or for that matter FRMPD2) function. Current experiments in mammalian cells [297] should be viewed as studies of the PTPN20 phosphatase domain only, and behavior of more full-length (FRMPD2-including) PTPN20 protein variants is eagerly awaited. The PTPN20 PTP segment was found on membranous structures in the cell and also in the nucleus, and in response to extracellular stimuli, it targeted to sites of actin polymerization [297]. The FRMPD2 part is expressed in many epithelial cell lines, and it localizes at the basolateral membrane in polarized cells [299] in sharp contrast to PTPN13, which is found on the apical side [300]. Recruitment of FRMPD2 to cell–cell contact sites requires the presence of both its FERM and PDZ2 domains and is further dependent on E-cadherin. By virtue of their particular C-terminus (-DSWV*), the catenin family members p0071, ARVCF, and δ-catenin were shown to serve as binding partners for the PDZ2 domain of FRMPD2, further pointing to a role in tight junction formation [299].

Role in Cancer: By analogy to PTPN13, one might expect data implicating PTPN20 as a tumor-suppressor to emerge within the coming years. For the moment, however, no mutations in the "PTP part" of the gene have been deposited [173]. For the "FRMPD2 part," on the contrary, 17 missense and 2 nonsense mutations have been encountered in 264 cancer specimens that span ovary, kidney, breast, skin, and large intestine tumors. In addition, some copy number variations in pancreatic cancers were noted [152] but, in view of a possible re-annotation of these genes, we refrain from further discussion.

NT9 Type, BRO-1 Like Domain-Containing PTPN23

PTPN23 is unique among NTPTPs by virtue of its N-terminal BRO1-like domain, which has homology to the vacuolar sorting protein Bro1 in yeast and its human relative ALIX, a regulator of endosomal sorting. This domain is followed by a "V-domain" with coiled-coiled motifs, the "His-domain (HD)" (14 His and 3 Cys linked by proline-rich segments, containing numerous putative SH3 binding motifs), the PTP domain, and a C-terminal PEST motif [301–304] (Fig. 1).

Regulation: Due to a divergence in the catalytic domain sequence from other members of the PTP family (A for S in the consensus motif VHCSAGXGRTG containing the catalytic site cysteine, and E for D in the conserved WPD motif), the catalytic activity of PTPN23 is impaired, but there is some controversy regarding whether PTPN23 has no [178, 305], or low, possibly very specific, catalytic activity [306, 307].

Physiological Function: Mutagenesis of the *Ptpn23* gene in mouse (leading to expression of only the N-terminal BRO1-like domain of PTPN23 in fusion with beta-galactosidase-neomycin-phosphotransferase II (β-geo)) caused embryonic lethality around E9.5, revealing an important non-redundant function in early embryonal development. Data suggest that fertilization and implantation were not affected, but at day E8.5 the embryos were severely retarded, and in appearance similar to E6.5 or early E7.5 stages. The authors concluded that PTPN23 plays a critical role in early stages of development between E4.5 and E7.5, where gastrulation takes place [308].

Studies in endothelial cells revealed negative regulation of cell motility by PTPN23, which was associated with interactions with SRC and FAK. The phosphorylation of SRC was not affected, whereas the phosphorylation of FAK was negatively regulated by PTPN23 in intact cells. Treatment with fibroblast growth factor (FGF) or Ca^{2+}-elevation destabilized PTPN23 protein through proteasomal and calpain-mediated degradation [306, 309–311]. A recent yeast-2-hybrid screen identified GRB2 and GrpL (GRB2-related protein of the lymphoid system) as novel PTPN23 interaction partners [304]. The functional relevance of this interaction remains yet to be explored.

The existence of a BRO1 homology domain in PTPN23 prompted testing of its involvement in endocytosis and formation of multivesicular bodies (MVB), a pathway involved in down-regulation of surface receptors. Indeed, PTPN23 depletion in HeLa cells led to defects in endosomal cargo sorting and MVB formation. The phenotype could be rescued with full-length PTPN23 and was dependent on the BRO1 domain together with the "V-domain" motif, but did not require the PTP domain [303]. Retroviruses employ the endosomal machinery to promote virus budding. Consistent with a role of PTPN23 in endosomal sorting, its BRO1-domain was also capable of stimulating HIV-1 virus-like particle production [312, 313]. Further, "Myopic" (Mop), the *Drosophila* ortholog of PTPN23, was shown to play a promoting role for EGFR trafficking and signalling in photoreceptor cell differentiation. Rescue experiments revealed the necessity of the BRO1 and presumably the PTP domain for Mop to function, but a PTP variant with the conserved catalytic cysteine replaced by a serine (CS mutant) was as effective as the wildtype in rescuing *mop* deficiency [314]. A central role of PTPN23 for sorting of the EGFR to luminal vesicles of MVB has recently also been established in mammalian cells [315]. Taken together, these data underscore an important role of PTPN23 in regulation of endocytosis.

Cancer Relevance: *PTPN23* has been considered a strong candidate tumor suppressor because it is part of a gene cluster that is frequently deleted in different solid human cancers [302] (Table 2). Early experiments indeed revealed that PTPN23 over-expression inhibits Ha-RAS mediated transformation of NIH3T3 fibroblasts [301]. More recent experiments also showed that PTPN23 inhibits colony formation of human renal cancer cell lines, an activity which was observed both with wildtype PTPN23 as well as with the CS mutant protein. The BRO1-like domain alone was, however, insufficient for suppression of the transformed phenotype [305].

A recent comprehensive analysis of the involvement of PTPs in regulating mammary epithelial cell migration identified PTPN23 as a key player. In a screen of shRNAs targeting 105 PTP gene family members for effects on basal and ERBB2-stimulated mammary epithelial cell migration, knockdown of PTPN23 prominently enhanced basal and ERBB2-activated cell motility and invasion in a 3D matrigel-collagen mixture. Suppression of PTPN23 also stimulated cell scattering, impaired cell–cell adhesion, and enhanced expression of EMT markers, such as the transcription factor SNAIL, and also increased expression of the secreted proteinase MMP-9. Increased phosphorylation of the activating site pY419 in the tyrosine kinase SRC was observed, and a PTPN23 trapping mutant formed a stable complex with SRC. These and other data suggested SRC as important mediator in PTPN23 regulation of mammary cancer cell motility [307]. It is of note that siRNA-mediated silencing of PTPN23 in T24 human bladder carcinoma cells also increased motility, but had no effect on SRC phosphorylation. Rather, in these cells the PTPN23 levels inversely correlated with FAK phosphorylation and its accumulation in focal contacts [316]. Recent studies have suggested a tumor suppressive function of PTPN23 also in the context of testicular germ cell tumors (TGCT). *PTPN23* was identified as a direct target gene of miR-142-3p. Expression of *PTPN23* mRNA in the seminomal tissue was reduced as compared with normal testicular tissue, whereas expression of miR-142-3p was much higher in the tumor samples [317].

Interestingly, a screen for *Drosophila* genes that require a block in apoptosis to promote tissue overgrowth led to the identification of the *PTPN23* ortholog *myopic* (*mop*). Loss of *mop* in the apoptosis-suppressed background enhanced proliferation of epithelial cells and caused similar phenotypes as mutants in the conserved Salvador/Warts/Hippo (SWH) tumor-suppressor pathway. Further analysis revealed that the effect of *mop* loss was at least partially mediated through inhibiting the activity of the transcription factor Yorkie, an oncoprotein, and terminal target of the SWH pathway. Yorkie was identified as binding partner of Mop via a proline-rich motif in Mop and a WW domain in Yorkie, reminiscent of interactions described for *Drosophila* Pez and its mammalian homologs PTPN14 with WW-domain-containing components of the SWH pathway [187, 194–196] (Fig. 4). Yorkie and Mop colocalized in late endosomes, and depletion of *mop* led to change in Yorkie localization and enhanced its protein levels and target gene expression. These experiments suggested that Mop defines a regulatory step in determining outputs of the SWH pathway, perhaps through effects on endosomal sorting. The human Yorkie ortholog YAP promotes proliferation and invasion in NSCLC lines, and its expression correlates with poor prognosis [318].

These findings prompt a number of interesting questions: Does PTPN23 feed into the SWH pathway and affect YAP activity in human cancer cell lines? Is its regulatory role in endocytotic trafficking important for regulation of this pathway in human cells? Is the role of PTPN23 in negative regulation of cell migration and invasion linked to a possible SWH pathway modulation and regulation of endocytosis? Which domains of the PTPN23 multidomain protein are important for these functions? Can alterations of these functions by loss of PTPN23 be found in human tumors and perhaps be modelled in mice? And finally, what is the extent of overlap between PTPN14, PTPN21, and PTPN23 modes of action within the SWH pathway and can this knowledge be translated in tailor-made therapeutic options? Answers to these questions will surely reveal further insights into the role of this exciting molecule in cancer.

R7 Type, KIM Domain-Containing PTPs

The genes *PTPN5*, *PTPN7,* and *PTPRR* encode PTPs that together display a unique domain composition within the phosphotyrosine-specific subfamily of PTPs. Because *PTPRR* also codes for transmembrane, receptor-like PTPs, the three genes have been grouped as R7 type PTP genes [1]. As reflected in Fig. 1, human PTPN5 and PTPN7 isoforms represent cytosolic proteins; however, *PTPN5* does encode some membrane-associated variants. The unique feature of R7-type PTPs is their so-called kinase-interacting motif (KIM) proximal to the catalytic PTP domain [319]. The KIM mediates an interaction with specific representatives of the MAPK family [320], which can be blocked through cyclic-AMP-dependent protein kinase (PKA)-mediated phosphorylation of an essential serine residue within the KIM [321]. An additional stretch in between KIM and PTP domain has been termed the kinase-specificity sequence (KISS) because it directs the MAPK preferences of the R7 type PTPs [322]. It is of note that these three classical PTPs are not the only phosphatases that can bind, dephosphorylate, and inactivate MAP kinases via KIM domain-mediated interactions. Multiple dual-specificity PTPs can do a similar or perhaps even better job, and their involvement in cancer will be discussed extensively in Chap. 8.

PTPN5

Physiological Function: *PTPN5* is expressed primarily in CNS neurons and protein variants include two main isoforms that contain the KIM, KISS, and PTP segments, and two additional variants that lack the PTP domain and might act as dominant-negatives [323]. PTPN5 immunoreactivity reveals a Golgi-like staining pattern [324], implicating the protein functionally at somata, dendritic arbors, and axonal processes, where it is thought to regulate synaptic transmission through pre- and post-synaptic mechanisms that require MAPK signaling. In *Drosophila,* a single homolog of the R7 type PTPs is present, and this cytoplasmic PTP-ER indeed negatively regulates duration, extent, and timing of ERK-type MAPKs [325].

In cultured rat neurons, the glutamate-mediated activation of *N*-methyl-D-aspartate (NMDA) receptors leads to a rapid but transient phosphorylation of ERK. In parallel, PTPN5 is activated by the calcium-dependent phosphatase calcineurin, which dephosphorylates the PTPN5 KIM. This allows the cytosolic sequestration and dephosphorylation of ERK by PTPN5, thus limiting ERK nuclear translocation and signaling [326]. PTPN5 is also able to associate with and dephosphorylate FYN, but not other SRC family members present in the post-synaptic density, by means of its KIM domain and the SH2 domain in FYN [327]. NMDA receptors are regulated by SRC family kinase-mediated tyrosine phosphorylation, and PTPN5 qualifies as an NMDA receptor phosphatase [328]. This puts the protein at a central position in a negative feedback circuitry to spatiotemporally counterbalance NMDA receptor-mediated signal pathways that involve ERK MAP kinases, in order to prevent excitotoxicity and modulate synaptic strengthening [329–331].

In *Ptpn5* knock-out mice, morphological abnormalities in the brain were not encountered, but enhanced ERK phosphorylation in specific brain areas was evident [332], corroborating PTPN5 as a MAPK phosphatase. These PTPN5-deficient mice displayed improved hippocampal-dependent learning and memory and a more dominant behavior, but motivation, motor coordination, visual acuity, and social interactions remained normal [333]. Furthermore, enhanced phosphorylation of NR2B, an NMDA receptor subunit, and proline-rich tyrosine kinase (Pyk2) and increased synaptosomal expression of NMDA and α-amino-3-hydroxy-5-methyl-4-isoxazole-propionic acid (AMPA) receptors were noted, in line with improved cognitive performance. In humans, an association of *PTPN5* with cognitive functioning has become apparent [334], and its relevance for neuropathological disorders, including schizophrenia, Alzheimer's, and Fragile X syndrome, is emerging [335, 336].

Role in Cancer: *PTPN5* expression is rather tissue-specific, so the somatic *PTPN5* mutations that have been detected in liver cancer specimens [152, 173] could represent bystander effects. Also, despite the presence of a long mononucleotide repeat in the *PTPN5* coding region that shows a mutation rate that is higher than the mean in colon tumors displaying microsatellite instability [207], there is no indication for a specific role for PTPN5 in colon cancer etiology. An RNA interference (RNAi) screen in the HT1080 fibrosarcoma cell line yielded PTPN5 as being required for dephosphorylating and activating BAK, a pro-apoptotic BCL-2 family member, in response to chemically induced apoptosis [337]. This rather points to a pro-apoptotic role for PTPN5, in contrast to its anti- excitotoxic nature in nerve cells. As yet, there are no indications that PTPN5 plays a role in brain cancers.

PTPN7

Physiological Function: *PTPN7* expression is, like for *PTPN5*, rather tissue-specific and predominantly involves cells of the hematopoietic system [338, 339]. As *PTPN7* levels were enhanced following interleukin-2 addition, a role in T cell proliferation or differentiation was suspected. PTPN7 was subsequently shown to block TCR-induced signal transduction through dephosphorylation of ERK2 [340, 341]. Also, p38α, but not JNK, represents a PTPN7 target. PKA phosphorylation of a serine residue within the PTPN7 KIM domain effectively blocks the interaction with these MAP kinases, preventing their inactivation by dephosphorylation [342]. In fact, PTPN7 is under continuous control by PKA and a serine-specific phosphatase, and KIM phosphoserine levels can change rapidly in response to external stimuli [343]. In the immune synapse, PTPN7 additionally needs to be phosphorylated by PKCθ, on serine-225 within the PTP domain, to enable PTPN7 targeting to lipid rafts where it then down-regulates TCR signaling [344]. This targeting to cholesterol-rich rafts additionally requires PTPN7 binding to the C-terminal half of oxysterol-binding protein OSBP [345]. In macrophages, PTPN7 apparently acts as a repressor of basal TNF-α production, likely by downtoning ERK and p38 activity. LPS induction of TNF-α production therefore depends on a transient decrease in PTPN7 mRNA and protein levels [346], but how this is exerted remains to be studied.

To uncover its relevance in vivo, *Ptpn7* knock-out mice have been generated. No phenotypic consequences other than an enhanced activation of ERK following TCR stimulation in vitro could be noted, pointing to efficient compensatory mechanisms in the developing mouse [347]. The subtle effects of PTPN7 on the duration and amplitude of MAPK signaling could be recapitulated in the PMA-induced megakaryocytic differentiation of K562 myelogenous leukemia cells [348]. Some years back, a *PTPN7* SNP was reported to associate with height during early childhood in Vietnamese-Korean families [349], which may reflect *PTPN7* polymorphism-related immunomodulatory setbacks during infancy.

Role in Cancer: *PTPN7* resides on chromosome 1q32.1, a region that is found deleted in non-Hodgkin lymphoma cases [339]. By comparing expression levels in pediatric B-cell lymphoma cases and lymphoid tissues, indeed significant lower PTPN7 levels were encountered in the tumor material [350]. Conversely, the chromosomal region is also frequently amplified in preleukemic myeloproliferative diseases. Triplication of gene *PTPN7* in one myelodysplastic syndrome case, and concomitant elevated protein expression in neoplastic myelomonocytic cells, indeed has been reported [351]. Furthermore, also in some acute leukemia cases elevated levels of PTPN7 were found [351], which may contribute in part to the abnormal myeloid cell growth. High-resolution structures for the catalytic domain in all KIM-containing R7 type PTPs are available [352]. This should facilitate development of selective small molecule compounds that modulate PTPN7 functioning and may become valuable in treating hematopoietic malignancies such as T cell acute lymphoblastic leukemia (T-ALL) and acute myelogenous leukemia (AML). A compound that increases MAP kinase activation in T cells through PTPN7 inhibition upon low-dose intraperitoneal injection in mice was recently reported [353], warranting further studies to reveal its value in treatment regimes of acute leukemic conditions.

PTPRR

Physiological Function: Like *PTPN5* and *PTPN7*, *PTPRR* expression is highly regulated, involving tissue-specific promoter use and alternative splicing [354, 355]. High levels of a medium-sized transcript that encodes a transmembrane vesicle-associated isoform are present in adult cerebellar Purkinje cells. A larger transcript results in a receptor-type transmembrane protein on the cell surface of neuronal cells throughout development, most notably in the hippocampus and olfactory bulb. Based on these findings, roles in Purkinje cell maturation and vesicle transport and signaling have been proposed [356]. Much lower amounts of a third transcript have been described for other tissues, most notably cartilage cells, and this results in the synthesis of two cytosolic variants [357].

PTPRR proteins are effective MAP kinase phosphatases, as proposed by the seminal work of Pulido and coworkers [320]. Although effects on magnitude and duration of MAPK signaling can be quite subtle [358, 359], *Ptprr* knock-out mice revealed a significant up-regulation of MAPK phosphorylation levels in relevant brain areas [360]. PTPRR-deficient animals show mild ataxia-like deficits in motor coordination despite normal gross brain morphology [360] and behaviorally demonstrate an increased exploratory activity and reduced novel object recognition ability [361]. Reduced ERK signaling has been noted in the pathogenesis of major depressive disorder (MDD), a mental disease with a genetic component. The rs1513105(C) SNP in the *PTPRR* locus was shown to associate, in female subjects, with an increased risk for MDD [362], but much more data are required before one may conclude that PTPRR plays a (gender-specific) role in MDD.

Role in Cancer: No *PTPRR* mutations were found in 469 brain tumor samples investigated thus far [173]. Several translocations involving *PTPRR*, however, are deposited. A specific type of leukemia is genetically characterized by an inversion event on chromosome 12 that affects the *TEL* (translocation ETS leukemia) gene, which encodes a transcription factor of the ETS (E-26 transforming specific) family. In one case, an inv(12)(p13q13) translocation resulted in a *TEL/PTPRR* fusion gene that expresses multiple different transcripts, one of which produces a fusion protein of the TEL transcriptional repressor and an incomplete PTPRR catalytic domain. Consequently, this TEL-PTPRR fusion protein has no enzymatic activity and its contribution to leukemia development would rely on dominant-negative effects on wild-type TEL functioning [363].

PTPRR has been primarily viewed and studied as a neuronal phosphatase, yet low transcript levels are found in other tissues as well. High-throughput expression studies revealed that PTPRR transcripts were markedly down-regulated, by means of de novo promoter methylation and transcription-repressive histone modification, upon cellular transformation of colon tissue [364]. This hypermethylation was confirmed in an independent study [365] and may represent an early alteration in colorectal tumorigenesis that aids the constitutive activation of the RAS/MAPK pathway. This was recently corroborated by studies in which DNA methylation in cervical cancer metastasis was addressed [366, 367]. *PTPRR* expression was found to be silenced by DNA methyltransferases in cervical cancer specimens and its methylation status correlated well with disease severity [367], but subsequent evaluation of the clinical efficacy rather pointed to ZNF582 instead of PTPRR as a useful biomarker of invasive cervical cancer [368].

Conclusions and Outlook

For some NTPTPs, roles in cancer have been firmly established, and the most appealing examples will be discussed extensively in the coming four chapters. But for many of the remaining 14 NTPTP genes, evidence of a contribution to cancer pathogenesis is emerging, and related knowledge is expanding rapidly. First indications of such roles for most of the PTPs covered in this chapter usually come from mutations or alterations in expression in specific cancer types. Further focused investigations are highly warranted to better establish causal links with cancer biology, as we have outlined in the treatise of the individual NTPTP molecules. For example, for the NPTPs described here, relatively few studies have been performed using cancer models in mice. Such work will be required to more definitively assess causal contributions of PTP gene alterations to cancer phenotypes. As the survey of current knowledge has shown, identified NTPTP functions may be rather specific both with respect to affected cancer cell processes ("hallmarks") and for certain cancer entities. Assessing the status of NTPTPs in the cancer cells themselves will not only uncover mechanisms underlying cancer etiology, but, as first examples show, will certainly help predicting the efficiency of existing and developing novel therapeutic modalities. Finally, some NTPTPs are important regulators in cell types with relevance in the cancer microenvironment. This applies to immune cells, but certainly also to vascular cells and other components of the tumor stroma. NTPTPs in these cell types may be equally valid targets for therapeutic interventions, such as for enhancement of anti-tumor immune responses.

References

1. Andersen JN, Mortensen OH, Peters GH, Drake PG, Iversen LF, Olsen OH, Jansen PG, Andersen HS, Tonks NK, Møller NP. Structural and evolutionary relationships among protein tyrosine phosphatase domains. Mol Cell Biol. 2001;21(21):7117–36.

2. Hanahan D, Weinberg RA. The hallmarks of cancer. Cell. 2000;100(1):57–70.
3. Neel BG, Gu H, Pao L. The 'Shp'ing news: SH2 domain-containing tyrosine phosphatases in cell signaling. Trends Biochem Sci. 2003;28(6):284–93.
4. Pao LI, Badour K, Siminovitch KA, Neel BG. Nonreceptor protein-tyrosine phosphatases in immune cell signaling. Annu Rev Immunol. 2007;25:473–523.
5. Grossmann KS, Rosario M, Birchmeier C, Birchmeier W. The tyrosine phosphatase Shp2 in development and cancer. Adv Cancer Res. 2010;106:53–89.
6. Beebe KD, Wang P, Arabaci G, Pei D. Determination of the binding specificity of the SH2 domains of protein tyrosine phosphatase SHP-1 through the screening of a combinatorial phosphotyrosyl peptide library. Biochemistry. 2000;39(43):13251–60.
7. Imhof D, Wavreille AS, May A, Zacharias M, Tridandapani S, Pei D. Sequence specificity of SHP-1 and SHP-2 Src homology 2 domains. Critical roles of residues beyond the pY+3 position. J Biol Chem. 2006;281(29):20271–82.
8. Ren L, Chen X, Luechapanichkul R, Selner NG, Meyer TM, Wavreille AS, Chan R, Iorio C, Zhou X, Neel BG, Pei D. Substrate specificity of protein tyrosine phosphatases 1B, RPTPalpha, SHP-1, and SHP-2. Biochemistry. 2011;50(12):2339–56.
9. Banville D, Stocco R, Shen SH. Human protein tyrosine phosphatase 1C (PTPN6) gene structure: alternate promoter usage and exon skipping generate multiple transcripts. Genomics. 1995;27(1):165–73.
10. Cheng J, Kydd AR, Nakase K, Noonan KM, Murakami A, Tao H, Dwyer M, Xu C, Zhu Q, Marasco WA. Negative regulation of the SH2-homology containing protein-tyrosine phosphatase-1 (SHP-1) P2 promoter by the HTLV-1 Tax oncoprotein. Blood. 2007;110(6): 2110–20.
11. Wlodarski P, Zhang Q, Liu X, Kasprzycka M, Marzec M, Wasik MA. PU.1 activates transcription of SHP-1 gene in hematopoietic cells. J Biol Chem. 2007;282(9):6316–23.
12. Nakase K, Cheng J, Zhu Q, Marasco WA. Mechanisms of SHP-1 P2 promoter regulation in hematopoietic cells and its silencing in HTLV-1-transformed T cells. J Leukoc Biol. 2009;85(1):165–74.
13. Zhang Z, Shen K, Lu W, Cole PA. The role of C-terminal tyrosine phosphorylation in the regulation of SHP-1 explored via expressed protein ligation. J Biol Chem. 2003;278(7): 4668–74.
14. Frank C, Burkhardt C, Imhof D, Ringel J, Zschornig O, Wieligmann K, Zacharias M, Böhmer FD. Effective dephosphorylation of Src substrates by SHP-1. J Biol Chem. 2004;279(12): 11375–83.
15. Craggs G, Kellie S. A functional nuclear localization sequence in the C-terminal domain of SHP-1. J Biol Chem. 2001;276(26):23719–25.
16. Frank C, Keilhack H, Opitz F, Zschornig O, Böhmer FD. Binding of phosphatidic acid to the protein-tyrosine phosphatase SHP-1 as a basis for activity modulation. Biochemistry. 1999;38(37):11993–2002.
17. Sankarshanan M, Ma Z, Iype T, Lorenz U. Identification of a novel lipid raft-targeting motif in Src homology 2-containing phosphatase 1. J Immunol. 2007;179(1):483–90.
18. Poole AW, Jones ML. A SHPing tale: perspectives on the regulation of SHP-1 and SHP-2 tyrosine phosphatases by the C-terminal tail. Cell Signal. 2005;17(11):1323–32.
19. Weibrecht I, Böhmer SA, Dagnell M, Kappert K, Östman A, Böhmer FD. Oxidation sensitivity of the catalytic cysteine of the protein-tyrosine phosphatases SHP-1 and SHP-2. Free Radic Biol Med. 2007;43(1):100–10.
20. Chen CY, Willard D, Rudolph J. Redox regulation of SH2-domain-containing protein tyrosine phosphatases by two backdoor cysteines. Biochemistry. 2009;48(6):1399–409.
21. Kundu S, Fan K, Cao M, Lindner DJ, Zhao ZJ, Borden E, Yi T. Novel SHP-1 inhibitors tyrosine phosphatase inhibitor-1 and analogs with preclinical anti-tumor activities as tolerated oral agents. J Immunol. 2010;184(11):6529–36.
22. Shultz LD, Schweitzer PA, Rajan TV, Yi T, Ihle JN, Matthews RJ, Thomas ML, Beier DR. Mutations at the murine motheaten locus are within the hematopoietic cell protein-tyrosine phosphatase (Hcph) gene. Cell. 1993;73(7):1445–54.

23. Bignon JS, Siminovitch KA. Identification of PTP1C mutation as the genetic defect in motheaten and viable motheaten mice: a step toward defining the roles of protein tyrosine phosphatases in the regulation of hemopoietic cell differentiation and function. Clin Immunol Immunopathol. 1994;73(2):168–79.
24. Croker BA, Lawson BR, Rutschmann S, Berger M, Eidenschenk C, Blasius AL, Moresco EM, Sovath S, Cengia L, Shultz LD, Theofilopoulos AN, Pettersson S, Beutler BA. Inflammation and autoimmunity caused by a SHP1 mutation depend on IL-1, MyD88, and a microbial trigger. Proc Natl Acad Sci U S A. 2008;105(39):15028–33.
25. Nesterovitch AB, Szanto S, Gonda A, Bardos T, Kis-Toth K, Adarichev VA, Olasz K, Ghassemi-Najad S, Hoffman MD, Tharp MD, Mikecz K, Glant TT. Spontaneous insertion of a b2 element in the ptpn6 gene drives a systemic autoinflammatory disease in mice resembling neutrophilic dermatosis in humans. Am J Pathol. 2011;178(4):1701–14.
26. Klingmuller U, Lorenz U, Cantley LC, Neel BG, Lodish HF. Specific recruitment of SH-PTP1 to the erythropoietin receptor causes inactivation of JAK2 and termination of proliferative signals. Cell. 1995;80(5):729–38.
27. Lorenz U, Bergemann AD, Steinberg HN, Flanagan JG, Li X, Galli SJ, Neel BG. Genetic analysis reveals cell type-specific regulation of receptor tyrosine kinase c-Kit by the protein tyrosine phosphatase SHP1. J Exp Med. 1996;184(3):1111–26.
28. Paulson RF, Vesely S, Siminovitch KA, Bernstein A. Signalling by the W/Kit receptor tyrosine kinase is negatively regulated in vivo by the protein tyrosine phosphatase Shp1. Nat Genet. 1996;13(3):309–15.
29. Kozlowski M, Larose L, Lee F, Le DM, Rottapel R, Siminovitch KA. SHP-1 binds and negatively modulates the c-Kit receptor by interaction with tyrosine 569 in the c-Kit juxtamembrane domain. Mol Cell Biol. 1998;18(4):2089–99.
30. Chen HE, Chang S, Trub T, Neel BG. Regulation of colony-stimulating factor 1 receptor signaling by the SH2 domain-containing tyrosine phosphatase SHPTP1. Mol Cell Biol. 1996;16(7):3685–97.
31. Umeda S, Beamer WG, Takagi K, Naito M, Hayashi S, Yonemitsu H, Yi T, Shultz LD. Deficiency of SHP-1 protein-tyrosine phosphatase activity results in heightened osteoclast function and decreased bone density. Am J Pathol. 1999;155(1):223–33.
32. Yi T, Mui AL, Krystal G, Ihle JN. Hematopoietic cell phosphatase associates with the interleukin-3 (IL-3) receptor beta chain and down-regulates IL-3-induced tyrosine phosphorylation and mitogenesis. Mol Cell Biol. 1993;13(12):7577–86.
33. Bone H, Dechert U, Jirik F, Schrader JW, Welham MJ. SHP1 and SHP2 protein-tyrosine phosphatases associate with betac after interleukin-3-induced receptor tyrosine phosphorylation. Identification of potential binding sites and substrates. J Biol Chem. 1997;272(22):14470–6.
34. Pani G, Kozlowski M, Cambier JC, Mills GB, Siminovitch KA. Identification of the tyrosine phosphatase PTP1C as a B cell antigen receptor-associated protein involved in the regulation of B cell signaling. J Exp Med. 1995;181(6):2077–84.
35. Cyster JG, Goodnow CC. Protein tyrosine phosphatase 1C negatively regulates antigen receptor signaling in B lymphocytes and determines thresholds for negative selection. Immunity. 1995;2(1):13–24.
36. Pani G, Fischer KD, Mlinaric-Rascan I, Siminovitch KA. Signaling capacity of the T cell antigen receptor is negatively regulated by the PTP1C tyrosine phosphatase. J Exp Med. 1996;184(3):839–52.
37. Lorenz U, Ravichandran KS, Burakoff SJ, Neel BG. Lack of SHPTP1 results in src-family kinase hyperactivation and thymocyte hyperresponsiveness. Proc Natl Acad Sci U S A. 1996;93(18):9624–9.
38. Burshtyn DN, Scharenberg AM, Wagtmann N, Rajagopalan S, Berrada K, Yi T, Kinet JP, Long EO. Recruitment of tyrosine phosphatase HCP by the killer cell inhibitor receptor. Immunity. 1996;4(1):77–85.
39. Kim CH, Qu CK, Hangoc G, Cooper S, Anzai N, Feng GS, Broxmeyer HE. Abnormal chemokine-induced responses of immature and mature hematopoietic cells from motheaten mice implicate the protein tyrosine phosphatase SHP-1 in chemokine responses. J Exp Med. 1999;190(5):681–90.

40. Kruger J, Butler JR, Cherapanov V, Dong Q, Ginzberg H, Govindarajan A, Grinstein S, Siminovitch KA, Downey GP. Deficiency of Src homology 2-containing phosphatase 1 results in abnormalities in murine neutrophil function: studies in motheaten mice. J Immunol. 2000;165(10):5847–59.
41. Roach TI, Slater SE, White LS, Zhang X, Majerus PW, Brown EJ, Thomas ML. The protein tyrosine phosphatase SHP-1 regulates integrin-mediated adhesion of macrophages. Curr Biol. 1998;8(18):1035–8.
42. Zhang J, Somani AK, Siminovitch KA. Roles of the SHP-1 tyrosine phosphatase in the negative regulation of cell signalling. Semin Immunol. 2000;12(4):361–78.
43. Lowell CA. Src-family and Syk kinases in activating and inhibitory pathways in innate immune cells: signaling cross talk. Cold Spring Harb Perspect Biol. 2011;3(3).
44. Pao LI, Lam KP, Henderson JM, Kutok JL, Alimzhanov M, Nitschke L, Thomas ML, Neel BG, Rajewsky K. B cell-specific deletion of protein-tyrosine phosphatase Shp1 promotes B-1a cell development and causes systemic autoimmunity. Immunity. 2007;27(1):35–48.
45. Johnson DJ, Pao LI, Dhanji S, Murakami K, Ohashi PS, Neel BG. Shp1 regulates T cell homeostasis by limiting IL-4 signals. J Exp Med. 2013;210(7):1419–31.
46. Abram CL, Roberge GL, Pao LI, Neel BG, Lowell CA. Distinct roles for neutrophils and dendritic cells in inflammation and autoimmunity in motheaten mice. Immunity. 2013;38(3): 489–501.
47. Tomic S, Greiser U, Lammers R, Kharitonenkov A, Imyanitov E, Ullrich A, Böhmer FD. Association of SH2 domain protein tyrosine phosphatases with the epidermal growth factor receptor in human tumor cells. Phosphatidic acid activates receptor dephosphorylation by PTP1C. J Biol Chem. 1995;270(36):21277–84.
48. Keilhack H, Tenev T, Nyakatura E, Godovac-Zimmermann J, Nielsen L, Seedorf K, Böhmer FD. Phosphotyrosine 1173 mediates binding of the protein-tyrosine phosphatase SHP-1 to the epidermal growth factor receptor and attenuation of receptor signaling. J Biol Chem. 1998;273(38):24839–46.
49. Guo DQ, Wu LW, Dunbar JD, Ozes ON, Mayo LD, Kessler KM, Gustin JA, Baerwald MR, Jaffe EA, Warren RS, Donner DB. Tumor necrosis factor employs a protein-tyrosine phosphatase to inhibit activation of KDR and vascular endothelial cell growth factor-induced endothelial cell proliferation. J Biol Chem. 2000;275(15):11216–21.
50. Yu Z, Su L, Hoglinger O, Jaramillo ML, Banville D, Shen SH. SHP-1 associates with both platelet-derived growth factor receptor and the p85 subunit of phosphatidylinositol 3-kinase. J Biol Chem. 1998;273(6):3687–94.
51. Keilhack H, Muller M, Böhmer SA, Frank C, Weidner KM, Birchmeier W, Ligensa T, Berndt A, Kosmehl H, Gunther B, Muller T, Birchmeier C, Böhmer FD. Negative regulation of Ros receptor tyrosine kinase signaling. An epithelial function of the SH2 domain protein tyrosine phosphatase SHP-1. J Cell Biol. 2001;152(2):325–34.
52. Marsh HN, Dubreuil CI, Quevedo C, Lee A, Majdan M, Walsh GS, Hausdorff S, Said FA, Zoueva O, Kozlowski M, Siminovitch K, Neel BG, Miller FD, Kaplan DR. SHP-1 negatively regulates neuronal survival by functioning as a TrkA phosphatase. J Cell Biol. 2003;163(5):999–1010.
53. Dubois MJ, Bergeron S, Kim HJ, Dombrowski L, Perreault M, Fournes B, Faure R, Olivier M, Beauchemin N, Shulman GI, Siminovitch KA, Kim JK, Marette A. The SHP-1 protein tyrosine phosphatase negatively modulates glucose homeostasis. Nat Med. 2006;12(5): 549–56.
54. Xu E, Charbonneau A, Rolland Y, Bellmann K, Pao L, Siminovitch KA, Neel BG, Beauchemin N, Marette A. Hepatocyte-specific ptpn6 deletion protects from obesity-linked hepatic insulin resistance. Diabetes. 2012;61(8):1949–58.
55. Mittal Y, Pavlova Y, Garcia-Marcos M, Ghosh P. Src homology domain 2-containing protein-tyrosine phosphatase-1 (SHP-1) binds and dephosphorylates G(alpha)-interacting, vesicle-associated protein (GIV)/Girdin and attenuates the GIV-phosphatidylinositol 3-kinase (PI3K)-Akt signaling pathway. J Biol Chem. 2011;286(37):32404–15.
56. Krautwald S, Buscher D, Kummer V, Buder S, Baccarini M. Involvement of the protein tyrosine phosphatase SHP-1 in Ras-mediated activation of the mitogen-activated protein kinase pathway. Mol Cell Biol. 1996;16(11):5955–63.

57. Su L, Zhao Z, Bouchard P, Banville D, Fischer EH, Krebs EG, Shen SH. Positive effect of overexpressed protein-tyrosine phosphatase PTP1C on mitogen-activated signaling in 293 cells. J Biol Chem. 1996;271(17):10385–90.
58. Minoo P, Zadeh MM, Rottapel R, Lebrun JJ, Ali S. A novel SHP-1/Grb2-dependent mechanism of negative regulation of cytokine-receptor signaling: contribution of SHP-1 C-terminal tyrosines in cytokine signaling. Blood. 2004;103(4):1398–407.
59. Wang N, Li Z, Ding R, Frank GD, Senbonmatsu T, Landon EJ, Inagami T, Zhao ZJ. Antagonism or synergism. Role of tyrosine phosphatases SHP-1 and SHP-2 in growth factor signaling. J Biol Chem. 2006;281(31):21878–83.
60. Hsu HC, Shultz LD, Su X, Shi J, Yang PA, Relyea MJ, Zhang HG, Mountz JD. Mutation of the hematopoietic cell phosphatase (Hcph) gene is associated with resistance to gamma-irradiation-induced apoptosis in Src homology protein tyrosine phosphatase (SHP)-1-deficient "motheaten" mutant mice. J Immunol. 2001;166(2):772–80.
61. Xiao W, Hong H, Kawakami Y, Kato Y, Wu D, Yasudo H, Kimura A, Kubagawa H, Bertoli LF, Davis RS, Chau LA, Madrenas J, Hsia CC, Xenocostas A, Kipps TJ, Hennighausen L, Iwama A, Nakauchi H, Kawakami T. Tumor suppression by phospholipase C-beta3 via SHP-1-mediated dephosphorylation of Stat5. Cancer Cell. 2009;16(2):161–71.
62. Oka T, Ouchida M, Koyama M, Ogama Y, Takada S, Nakatani Y, Tanaka T, Yoshino T, Hayashi K, Ohara N, Kondo E, Takahashi K, Tsuchiyama J, Tanimoto M, Shimizu K, Akagi T. Gene silencing of the tyrosine phosphatase SHP1 gene by aberrant methylation in leukemias/lymphomas. Cancer Res. 2002;62(22):6390–4.
63. Reddy J, Shivapurkar N, Takahashi T, Parikh G, Stastny V, Echebiri C, Crumrine K, Zochbauer-Muller S, Drach J, Zheng Y, Feng Z, Kroft SH, McKenna RW, Gazdar AF. Differential methylation of genes that regulate cytokine signaling in lymphoid and hematopoietic tumors. Oncogene. 2005;24(4):732–6.
64. Chim CS, Fung TK, Cheung WC, Liang R, Kwong YL. SOCS1 and SHP1 hypermethylation in multiple myeloma: implications for epigenetic activation of the Jak/STAT pathway. Blood. 2004;103(12):4630–5.
65. Zhang Y, Zhao D, Zhao H, Wu X, Zhao W, Wang Y, Xia B, Da W. Hypermethylation of SHP-1 promoter in patient with high-risk myelodysplastic syndrome and it predicts poor prognosis. Med Oncol. 2012;29(4):2359–63.
66. Khoury JD, Rassidakis GZ, Medeiros LJ, Amin HM, Lai R. Methylation of SHP1 gene and loss of SHP1 protein expression are frequent in systemic anaplastic large cell lymphoma. Blood. 2004;104(5):1580–1.
67. Honorat JF, Ragab A, Lamant L, Delsol G, Ragab-Thomas J. SHP1 tyrosine phosphatase negatively regulates NPM-ALK tyrosine kinase signaling. Blood. 2006;107(10):4130–8.
68. Han Y, Amin HM, Franko B, Frantz C, Shi X, Lai R. Loss of SHP1 enhances JAK3/STAT3 signaling and decreases proteosome degradation of JAK3 and NPM-ALK in ALK+ anaplastic large-cell lymphoma. Blood. 2006;108(8):2796–803.
69. Bruecher-Encke B, Griffin JD, Neel BG, Lorenz U. Role of the tyrosine phosphatase SHP-1 in K562 cell differentiation. Leukemia. 2001;15(9):1424–32.
70. Liedtke M, Pandey P, Kumar S, Kharbanda S, Kufe D. Regulation of Bcr-Abl-induced SAP kinase activity and transformation by the SHPTP1 protein tyrosine phosphatase. Oncogene. 1998;17(15):1889–92.
71. Witzig TE, Hu G, Offer SM, Wellik LE, Han JJ, Stenson MJ, Dogan A, Diasio RB, Gupta M. Epigenetic mechanisms of protein tyrosine phosphatase 6 suppression in diffuse large B-cell lymphoma: implications for epigenetic therapy. Leukemia. 2014;28(1):147–54.
72. Beghini A, Ripamonti CB, Peterlongo P, Roversi G, Cairoli R, Morra E, Larizza L. RNA hyperediting and alternative splicing of hematopoietic cell phosphatase (PTPN6) gene in acute myeloid leukemia. Hum Mol Genet. 2000;9(15):2297–304.
73. Cheng J, Zhang D, Zhou C, Marasco WA. Down-regulation of SHP1 and up-regulation of negative regulators of JAK/STAT signaling in HTLV-1 transformed cell lines and freshly transformed human peripheral blood CD4+ T-cells. Leuk Res. 2004;28(1):71–82.

74. Hennige AM, Lammers R, Hoppner W, Arlt D, Strack V, Teichmann R, Machicao F, Ullrich A, Haring HU, Kellerer M. Inhibition of Ret oncogene activity by the protein tyrosine phosphatase SHP1. Endocrinology. 2001;142(10):4441–7.
75. Roccato E, Miranda C, Raho G, Pagliardini S, Pierotti MA, Greco A. Analysis of SHP-1-mediated down-regulation of the TRK-T3 oncoprotein identifies Trk-fused gene (TFG) as a novel SHP-1-interacting protein. J Biol Chem. 2005;280(5):3382–9.
76. Mok SC, Kwok TT, Berkowitz RS, Barrett AJ, Tsui FW. Overexpression of the protein tyrosine phosphatase, nonreceptor type 6 (PTPN6), in human epithelial ovarian cancer. Gynecol Oncol. 1995;57(3):299–303.
77. Yip SS, Crew AJ, Gee JM, Hui R, Blamey RW, Robertson JF, Nicholson RI, Sutherland RL, Daly RJ. Up-regulation of the protein tyrosine phosphatase SHP-1 in human breast cancer and correlation with GRB2 expression. Int J Cancer. 2000;88(3):363–8.
78. Insabato L, Amelio I, Quarto M, Zannetti A, Tolino F, de Mauro G, Cerchia L, Riccio P, Baumhoer D, Condorelli G, Terracciano L, de Franciscis V. Elevated expression of the tyrosine phosphatase SHP-1 defines a subset of high-grade breast tumors. Oncology. 2009;77(6): 378–84.
79. Montano X. Repression of SHP-1 expression by p53 leads to trkA tyrosine phosphorylation and suppression of breast cancer cell proliferation. Oncogene. 2009;28(43):3787–800.
80. Sooman L, Ekman S, Tsakonas G, Jaiswal A, Navani S, Edqvist PH, Ponten F, Bergstrom S, Johansson M, Wu X, Blomquist E, Bergqvist M, Gullbo J, Lennartsson J. PTPN6 expression is epigenetically regulated and influences survival and response to chemotherapy in high-grade gliomas. Tumour Biol. 2014;35(5):4479–88.
81. Zhang L, Wang M, Wang W, Mo J. Incidence and prognostic value of multiple gene promoter methylations in gliomas. J Neurooncol. 2014;116(2):349–56.
82. Lopez F, Esteve JP, Buscail L, Delesque N, Saint-Laurent N, Theveniau M, Nahmias C, Vaysse N, Susini C. The tyrosine phosphatase SHP-1 associates with the sst2 somatostatin receptor and is an essential component of sst2-mediated inhibitory growth signaling. J Biol Chem. 1997;272(39):24448–54.
83. Douziech N, Calvo E, Coulombe Z, Muradia G, Bastien J, Aubin RA, Lajas A, Morisset J. Inhibitory and stimulatory effects of somatostatin on two human pancreatic cancer cell lines: a primary role for tyrosine phosphatase SHP-1. Endocrinology. 1999;140(2):765–77.
84. Zatelli MC, Piccin D, Tagliati F, Bottoni A, Luchin A, degli Uberti EC. SRC homology-2-containing protein tyrosine phosphatase-1 restrains cell proliferation in human medullary thyroid carcinoma. Endocrinology. 2005;146(6):2692–8.
85. Theodoropoulou M, Zhang J, Laupheimer S, Paez-Pereda M, Erneux C, Florio T, Pagotto U, Stalla GK. Octreotide, a somatostatin analogue, mediates its antiproliferative action in pituitary tumor cells by altering phosphatidylinositol 3-kinase signaling and inducing Zac1 expression. Cancer Res. 2006;66(3):1576–82.
86. Lapalombella R, Yeh YY, Wang L, Ramanunni A, Rafiq S, Jha S, Staubli J, Lucas DM, Mani R, Herman SE, Johnson AJ, Lozanski A, Andritsos L, Jones J, Flynn JM, Lannutti B, Thompson P, Algate P, Stromatt S, Jarjoura D, Mo X, Wang D, Chen CS, Lozanski G, Heerema NA, Tridandapani S, Freitas MA, Muthusamy N, Byrd JC. Tetraspanin CD37 directly mediates transduction of survival and apoptotic signals. Cancer Cell. 2012;21(5):694–708.
87. Ahn KS, Sethi G, Sung B, Goel A, Ralhan R, Aggarwal BB. Guggulsterone, a farnesoid X receptor antagonist, inhibits constitutive and inducible STAT3 activation through induction of a protein tyrosine phosphatase SHP-1. Cancer Res. 2008;68(11):4406–15.
88. Liu SH, Wang KB, Lan KH, Lee WJ, Pan HC, Wu SM, Peng YC, Chen YC, Shen CC, Cheng HC, Liao KK, Sheu ML. Calpain/SHP-1 interaction by honokiol dampening peritoneal dissemination of gastric cancer in nu/nu mice. PLoS One. 2012;7(8), e43711.
89. Böhmer F, Friedrich K. Protein-tyrosine phosphatases as wardens of STAT signaling. JAKSTAT. 2014;3(1), e28087.
90. Esposito N, Colavita I, Quintarelli C, Sica AR, Peluso AL, Luciano L, Picardi M, Del Vecchio L, Buonomo T, Hughes TP, White D, Radich JP, Russo D, Branford S, Saglio G, Melo JV,

Martinelli R, Ruoppolo M, Kalebic T, Martinelli G, Pane F. SHP-1 expression accounts for resistance to imatinib treatment in Philadelphia chromosome-positive cells derived from patients with chronic myeloid leukemia. Blood. 2011;118(13):3634–44.
91. Tai WT, Cheng AL, Shiau CW, Huang HP, Huang JW, Chen PJ, Chen KF. Signal transducer and activator of transcription 3 is a major kinase-independent target of sorafenib in hepatocellular carcinoma. J Hepatol. 2011;55(5):1041–8.
92. Tai WT, Shiau CW, Chen PJ, Chu PY, Huang HP, Liu CY, Huang JW, Chen KF. Discovery of novel Src homology region 2 domain-containing phosphatase 1 agonists from sorafenib for the treatment of hepatocellular carcinoma. Hepatology. 2014;59(1):190–201.
93. Fan LC, Teng HW, Shiau CW, Lin H, Hung MH, Chen YL, Huang JW, Tai WT, Yu HC, Chen KF. SHP-1 is a target of regorafenib in colorectal cancer. Oncotarget. 2014;5(15): 6243–51.
94. Hanahan D, Weinberg RA. Hallmarks of cancer: the next generation. Cell. 2011;144(5): 646–74.
95. Kershaw MH, Westwood JA, Darcy PK. Gene-engineered T cells for cancer therapy. Nat Rev Cancer. 2013;13(8):525–41.
96. Monu N, Frey AB. Suppression of proximal T cell receptor signaling and lytic function in CD8+ tumor-infiltrating T cells. Cancer Res. 2007;67(23):11447–54.
97. Stromnes IM, Fowler C, Casamina CC, Georgopolos CM, McAfee MS, Schmitt TM, Tan X, Kim TD, Choi I, Blattman JN, Greenberg PD. Abrogation of SRC homology region 2 domain-containing phosphatase 1 in tumor-specific T cells improves efficacy of adoptive immunotherapy by enhancing the effector function and accumulation of short-lived effector T cells in vivo. J Immunol. 2012;189(4):1812–25.
98. Gu M, Warshawsky I, Majerus PW. Cloning and expression of a cytosolic megakaryocyte protein-tyrosine-phosphatase with sequence homology to retinaldehyde-binding protein and yeast SEC14p. Proc Natl Acad Sci U S A. 1992;89(7):2980–4.
99. Seimiya H, Sawabe T, Toho M, Tsuruo T. Phorbol ester-resistant monoblastoid leukemia cells with a functional mitogen-activated protein kinase cascade but without responsive protein tyrosine phosphatases. Oncogene. 1995;11(10):2047–54.
100. Wang X, Huynh H, Gjorloff-Wingren A, Monosov E, Stridsberg M, Fukuda M, Mustelin T. Enlargement of secretory vesicles by protein tyrosine phosphatase PTP-MEG2 in rat basophilic leukemia mast cells and Jurkat T cells. J Immunol. 2002;168(9):4612–9.
101. Kruger JM, Fukushima T, Cherepanov V, Borregaard N, Loeve C, Shek C, Sharma K, Tanswell AK, Chow CW, Downey GP. Protein-tyrosine phosphatase MEG2 is expressed by human neutrophils. Localization to the phagosome and activation by polyphosphoinositides. J Biol Chem. 2002;277(4):2620–8.
102. Huynh H, Wang X, Li W, Bottini N, Williams S, Nika K, Ishihara H, Godzik A, Mustelin T. Homotypic secretory vesicle fusion induced by the protein tyrosine phosphatase MEG2 depends on polyphosphoinositides in T cells. J Immunol. 2003;171(12):6661–71.
103. Huynh H, Bottini N, Williams S, Cherepanov V, Musumeci L, Saito K, Bruckner S, Vachon E, Wang X, Kruger J, Chow CW, Pellecchia M, Monosov E, Greer PA, Trimble W, Downey GP, Mustelin T. Control of vesicle fusion by a tyrosine phosphatase. Nat Cell Biol. 2004;6(9):831–9.
104. Gjorloff-Wingren A, Saxena M, Han S, Wang X, Alonso A, Renedo M, Oh P, Williams S, Schnitzer J, Mustelin T. Subcellular localization of intracellular protein tyrosine phosphatases in T cells. Eur J Immunol. 2000;30(8):2412–21.
105. Saito K, Williams S, Bulankina A, Honing S, Mustelin T. Association of protein-tyrosine phosphatase MEG2 via its Sec14p homology domain with vesicle-trafficking proteins. J Biol Chem. 2007;282(20):15170–8.
106. Wang Y, Vachon E, Zhang J, Cherepanov V, Kruger J, Li J, Saito K, Shannon P, Bottini N, Huynh H, Ni H, Yang H, McKerlie C, Quaggin S, Zhao ZJ, Marsden PA, Mustelin T, Siminovitch KA, Downey GP. Tyrosine phosphatase MEG2 modulates murine development and platelet and lymphocyte activation through secretory vesicle function. J Exp Med. 2005;202(11):1587–97.

107. Cho CY, Koo SH, Wang Y, Callaway S, Hedrick S, Mak PA, Orth AP, Peters EC, Saez E, Montminy M, Schultz PG, Chanda SK. Identification of the tyrosine phosphatase PTP-MEG2 as an antagonist of hepatic insulin signaling. Cell Metab. 2006;3(5):367–78.
108. Hao Q, Samten B, Ji HL, Zhao ZJ, Tang H. Tyrosine phosphatase PTP-MEG2 negatively regulates vascular endothelial growth factor receptor signaling and function in endothelial Cells. Am J Physiol Cell Physiol. 2012;303(5):C548–53.
109. Huang X, Gschweng E, Van Handel B, Cheng D, Mikkola HK, Witte ON. Regulated expression of microRNAs-126/126* inhibits erythropoiesis from human embryonic stem cells. Blood. 2011;117(7):2157–65.
110. Bu Y, Su F, Wang X, Gao H, Lei L, Chang N, Wu Q, Hu K, Zhu X, Chang Z, Meng K, Xiong JW. Protein tyrosine phosphatase PTPN9 regulates erythroid cell development through STAT3 dephosphorylation in zebrafish. J Cell Sci. 2014;127(Pt 12):2761–70.
111. Xu MJ, Sui X, Zhao R, Dai C, Krantz SB, Zhao ZJ. PTP-MEG2 is activated in polycythemia vera erythroid progenitor cells and is required for growth and expansion of erythroid cells. Blood. 2003;102(13):4354–60.
112. Yuan T, Wang Y, Zhao ZJ, Gu H. Protein-tyrosine phosphatase PTPN9 negatively regulates ErbB2 and epidermal growth factor receptor signaling in breast cancer cells. J Biol Chem. 2010;285(20):14861–70.
113. Du WW, Fang L, Li M, Yang X, Liang Y, Peng C, Qian W, O'Malley YQ, Askeland RW, Sugg SL, Qian J, Lin J, Jiang Z, Yee AJ, Sefton M, Deng Z, Shan SW, Wang CH, Yang BB. MicroRNA miR-24 enhances tumor invasion and metastasis by targeting PTPN9 and PTPRF to promote EGF signaling. J Cell Sci. 2013;126(Pt 6):1440–53.
114. Su F, Ren F, Rong Y, Wang Y, Geng Y, Feng M, Ju Y, Li Y, Zhao ZJ, Meng K, Chang Z. Protein tyrosine phosphatase Meg2 dephosphorylates signal transducer and activator of transcription 3 and suppresses tumor growth in breast cancer. Breast Cancer Res. 2012;14(2):R38.
115. Wang B, Lemay S, Tsai S, Veillette A. SH2 domain-mediated interaction of inhibitory protein tyrosine kinase Csk with protein tyrosine phosphatase-HSCF. Mol Cell Biol. 2001;21(4):1077–88.
116. Shiota M, Tanihiro T, Nakagawa Y, Aoki N, Ishida N, Miyazaki K, Ullrich A, Miyazaki H. Protein tyrosine phosphatase PTP20 induces actin cytoskeleton reorganization by dephosphorylating p190 RhoGAP in rat ovarian granulosa cells stimulated with follicle-stimulating hormone. Mol Endocrinol. 2003;17(4):534–49.
117. Spencer S, Dowbenko D, Cheng J, Li W, Brush J, Utzig S, Simanis V, Lasky LA. PSTPIP: a tyrosine phosphorylated cleavage furrow-associated protein that is a substrate for a PEST tyrosine phosphatase. J Cell Biol. 1997;138(4):845–60.
118. Kim YW, Wang H, Sures I, Lammers R, Martell KJ, Ullrich A. Characterization of the PEST family protein tyrosine phosphatase BDP1. Oncogene. 1996;13(10):2275–9.
119. Gensler M, Buschbeck M, Ullrich A. Negative regulation of HER2 signaling by the PEST-type protein-tyrosine phosphatase BDP1. J Biol Chem. 2004;279(13):12110–6.
120. Wang HM, Xu YF, Ning SL, Yang DX, Li Y, Du YJ, Yang F, Zhang Y, Liang N, Yao W, Zhang LL, Gu LC, Gao CJ, Pang Q, Chen YX, Xiao KH, Ma R, Yu X, Sun JP. The catalytic region and PEST domain of PTPN18 distinctly regulate the HER2 phosphorylation and ubiquitination barcodes. Cell Res. 2014;24(9):1067–90.
121. Yu X, Sun JP, He Y, Guo X, Liu S, Zhou B, Hudmon A, Zhang ZY. Structure, inhibitor, and regulatory mechanism of Lyp, a lymphoid-specific tyrosine phosphatase implicated in autoimmune diseases. Proc Natl Acad Sci U S A. 2007;104(50):19767–72.
122. Cohen S, Dadi H, Shaoul E, Sharfe N, Roifman CM. Cloning and characterization of a lymphoid-specific, inducible human protein tyrosine phosphatase, Lyp. Blood. 1999;93(6):2013–24.
123. Vang T, Liu WH, Delacroix L, Wu S, Vasile S, Dahl R, Yang L, Musumeci L, Francis D, Landskron J, Tasken K, Tremblay ML, Lie BA, Page R, Mustelin T, Rahmouni S, Rickert RC, Tautz L. LYP inhibits T-cell activation when dissociated from CSK. Nat Chem Biol. 2012;8(5):437–46.
124. Gjorloff-Wingren A, Saxena M, Williams S, Hammi D, Mustelin T. Characterization of TCR-induced receptor-proximal signaling events negatively regulated by the protein tyrosine phosphatase PEP. Eur J Immunol. 1999;29(12):3845–54.

125. Cloutier JF, Veillette A. Cooperative inhibition of T-cell antigen receptor signaling by a complex between a kinase and a phosphatase. J Exp Med. 1999;189(1):111–21.
126. Hasegawa K, Martin F, Huang G, Tumas D, Diehl L, Chan AC. PEST domain-enriched tyrosine phosphatase (PEP) regulation of effector/memory T cells. Science. 2004;303(5658):685–9.
127. Hermiston ML, Zikherman J, Zhu JW. CD45, CD148, and Lyp/Pep: critical phosphatases regulating Src family kinase signaling networks in immune cells. Immunol Rev. 2009;228(1):288–311.
128. Salmond RJ, Brownlie RJ, Morrison VL, Zamoyska R. The tyrosine phosphatase PTPN22 discriminates weak self peptides from strong agonist TCR signals. Nat Immunol. 2014;15(9):875–83.
129. Rhee I, Veillette A. Protein tyrosine phosphatases in lymphocyte activation and autoimmunity. Nat Immunol. 2012;13(5):439–47.
130. Rivas MA, Beaudoin M, Gardet A, Stevens C, Sharma Y, Zhang CK, Boucher G, Ripke S, Ellinghaus D, Burtt N, Fennell T, Kirby A, Latiano A, Goyette P, Green T, Halfvarson J, Haritunians T, Korn JM, Kuruvilla F, Lagace C, Neale B, Lo KS, Schumm P, Torkvist L, Dubinsky MC, Brant SR, Silverberg MS, Duerr RH, Altshuler D, Gabriel S, Lettre G, Franke A, D'Amato M, McGovern DP, Cho JH, Rioux JD, Xavier RJ, Daly MJ. Deep resequencing of GWAS loci identifies independent rare variants associated with inflammatory bowel disease. Nat Genet. 2011;43(11):1066–73.
131. Menard L, Saadoun D, Isnardi I, Ng YS, Meyers G, Massad C, Price C, Abraham C, Motaghedi R, Buckner JH, Gregersen PK, Meffre E. The PTPN22 allele encoding an R620W variant interferes with the removal of developing autoreactive B cells in humans. J Clin Invest. 2011;121(9):3635–44.
132. Habib T, Funk A, Rieck M, Brahmandam A, Dai X, Panigrahi AK, Luning Prak ET, Meyer-Bahlburg A, Sanda S, Greenbaum C, Rawlings DJ, Buckner JH. Altered B cell homeostasis is associated with type I diabetes and carriers of the PTPN22 allelic variant. J Immunol. 2012;188(1):487–96.
133. Zhang J, Zahir N, Jiang Q, Miliotis H, Heyraud S, Meng X, Dong B, Xie G, Qiu F, Hao Z, McCulloch CA, Keystone EC, Peterson AC, Siminovitch KA. The autoimmune disease-associated PTPN22 variant promotes calpain-mediated Lyp/Pep degradation associated with lymphocyte and dendritic cell hyperresponsiveness. Nat Genet. 2011;43(9):902–7.
134. Chien W, Tidow N, Williamson EA, Shih LY, Krug U, Kettenbach A, Fermin AC, Roifman CM, Koeffler HP. Characterization of a myeloid tyrosine phosphatase, Lyp, and its role in the Bcr-Abl signal transduction pathway. J Biol Chem. 2003;278(30):27413–20.
135. Negro R, Gobessi S, Longo PG, He Y, Zhang ZY, Laurenti L, Efremov DG. Overexpression of the autoimmunity-associated phosphatase PTPN22 promotes survival of antigen-stimulated CLL cells by selectively activating AKT. Blood. 2012;119(26):6278–87.
136. Yang Q, Tonks NK. Isolation of a cDNA clone encoding a human protein-tyrosine phosphatase with homology to the cytoskeletal-associated proteins band 4.1, ezrin, and talin. Proc Natl Acad Sci U S A. 1991;88(14):5949–53.
137. Itoh F, Ikuta S, Hinoda Y, Arimura Y, Ohe M, Adachi M, Ariyama T, Inazawa J, Imai K, Yachi A. Expression and chromosomal assignment of PTPH1 gene encoding a cytosolic protein tyrosine phosphatase homologous to cytoskeletal-associated proteins. Int J Cancer. 1993;55(6):947–51.
138. Sahin M, Slaugenhaupt SA, Gusella JF, Hockfield S. Expression of PTPH1, a rat protein tyrosine phosphatase, is restricted to the derivatives of a specific diencephalic segment. Proc Natl Acad Sci U S A. 1995;92(17):7859–63.
139. Gu MX, York JD, Warshawsky I, Majerus PW. Identification, cloning, and expression of a cytosolic megakaryocyte protein-tyrosine-phosphatase with sequence homology to cytoskeletal protein 4.1. Proc Natl Acad Sci U S A. 1991;88(13):5867–71.
140. Zhang SH, Kobayashi R, Graves PR, Piwnica-Worms H, Tonks NK. Serine phosphorylation-dependent association of the band 4.1-related protein-tyrosine phosphatase PTPH1 with 14-3-3beta protein. J Biol Chem. 1997;272(43):27281–7.

141. Zhang SH, Liu J, Kobayashi R, Tonks NK. Identification of the cell cycle regulator VCP (p97/CDC48) as a substrate of the band 4.1-related protein-tyrosine phosphatase PTPH1. J Biol Chem. 1999;274(25):17806–12.
142. Abaan OD, Hendriks W, Uren A, Toretsky JA, Erkizan HV. Valosin containing protein (VCP/p97) is a novel substrate for the protein tyrosine phosphatase PTPL1. Exp Cell Res. 2013;319(1):1–11.
143. Han S, Williams S, Mustelin T. Cytoskeletal protein tyrosine phosphatase PTPH1 reduces T cell antigen receptor signaling. Eur J Immunol. 2000;30(5):1318–25.
144. Sozio MS, Mathis MA, Young JA, Walchli S, Pitcher LA, Wrage PC, Bartok B, Campbell A, Watts JD, Aebersold R, Hooft van Huijsduijnen R, van Oers NS. PTPH1 is a predominant protein-tyrosine phosphatase capable of interacting with and dephosphorylating the T cell receptor zeta subunit. J Biol Chem. 2004;279(9):7760–9.
145. Zheng Y, Schlondorff J, Blobel CP. Evidence for regulation of the tumor necrosis factor alpha-convertase (TACE) by protein-tyrosine phosphatase PTPH1. J Biol Chem. 2002;277(45):42463–70.
146. Jespersen T, Gavillet B, van Bemmelen MX, Cordonier S, Thomas MA, Staub O, Abriel H. Cardiac sodium channel Na(v)1.5 interacts with and is regulated by the protein tyrosine phosphatase PTPH1. Biochem Biophys Res Commun. 2006;348(4):1455–62.
147. Bauler TJ, Hughes ED, Arimura Y, Mustelin T, Saunders TL, King PD. Normal TCR signal transduction in mice that lack catalytically active PTPN3 protein tyrosine phosphatase. J Immunol. 2007;178(6):3680–7.
148. Pilecka I, Patrignani C, Pescini R, Curchod ML, Perrin D, Xue Y, Yasenchak J, Clark A, Magnone MC, Zaratin P, Valenzuela D, Rommel C, Hooft van Huijsduijnen R. Protein-tyrosine phosphatase H1 controls growth hormone receptor signaling and systemic growth. J Biol Chem. 2007;282(48):35405–15.
149. Patrignani C, Magnone MC, Tavano P, Ardizzone M, Muzio V, Greco B, Zaratin PF. Knockout mice reveal a role for protein tyrosine phosphatase H1 in cognition. Behav Brain Funct. 2008;4:36.
150. Patrignani C, Lafont DT, Muzio V, Greco B, Hooft van Huijsduijnen R, Zaratin PF. Characterization of protein tyrosine phosphatase H1 knockout mice in animal models of local and systemic inflammation. J Inflamm (Lond). 2010;7:16.
151. Wang Z, Shen D, Parsons DW, Bardelli A, Sager J, Szabo S, Ptak J, Silliman N, Peters BA, van der Heijden MS, Parmigiani G, Yan H, Wang TL, Riggins G, Powell SM, Willson JK, Markowitz S, Kinzler KW, Vogelstein B, Velculescu VE. Mutational analysis of the tyrosine phosphatome in colorectal cancers. Science. 2004;304(5674):1164–6.
152. International Cancer Genome Consortium, Hudson TJ, Anderson W, Artez A, Barker AD, Bell C, Bernabe RR, Bhan MK, Calvo F, Eerola I, Gerhard DS, Guttmacher A, Guyer M, Hemsley FM, Jennings JL, Kerr D, Klatt P, Kolar P, Kusada J, Lane DP, Laplace F, Youyong L, Nettekoven G, Ozenberger B, Peterson J, Rao TS, Remacle J, Schafer AJ, Shibata T, Stratton MR, Vockley JG, Watanabe K, Yang H, Yuen MM, Knoppers BM, Bobrow M, Cambon-Thomsen A, Dressler LG, Dyke SO, Joly Y, Kato K, Kennedy KL, Nicolas P, Parker MJ, Rial-Sebbag E, Romeo-Casabona CM, Shaw KM, Wallace S, Wiesner GL, Zeps N, Lichter P, Biankin AV, Chabannon C, Chin L, Clement B, de Alava E, Degos F, Ferguson ML, Geary P, Hayes DN, Hudson TJ, Johns AL, Kasprzyk A, Nakagawa H, Penny R, Piris MA, Sarin R, Scarpa A, Shibata T, van de Vijver M, Futreal PA, Aburatani H, Bayes M, Botwell DD, Campbell PJ, Estivill X, Gerhard DS, Grimmond SM, Gut I, Hirst M, Lopez-Otin C, Majumder P, Marra M, McPherson JD, Nakagawa H, Ning Z, Puente XS, Ruan Y, Shibata T, Stratton MR, Stunnenberg HG, Swerdlow H, Velculescu VE, Wilson RK, Xue HH, Yang L, Spellman PT, Bader GD, Boutros PC, Campbell PJ, Flicek P, Getz G, Guigo R, Guo G, Haussler D, Heath S, Hubbard TJ, Jiang T, Jones SM, Li Q, Lopez-Bigas N, Luo R, Muthuswamy L, Ouellette BF, Pearson JV, Puente XS, Quesada V, Raphael BJ, Sander C, Shibata T, Speed TP, Stein LD, Stuart JM, Teague JW, Totoki Y, Tsunoda T, Valencia A, Wheeler DA, Wu H, Zhao S, Zhou G, Stein LD, Guigo R, Hubbard TJ, Joly Y, Jones SM, Kasprzyk A, Lathrop M, Lopez-Bigas N, Ouellette BF, Spellman PT, Teague JW, Thomas G,

Valencia A, Yoshida T, Kennedy KL, Axton M, Dyke SO, Futreal PA, Gerhard DS, Gunter C, Guyer M, Hudson TJ, McPherson JD, Miller LJ, Ozenberger B, Shaw KM, Kasprzyk A, Stein LD, Zhang J, Haider SA, Wang J, Yung CK, Cros A, Liang Y, Gnaneshan S, Guberman J, Hsu J, Bobrow M, Chalmers DR, Hasel KW, Joly Y, Kaan TS, Kennedy KL, Knoppers BM, Lowrance WW, Masui T, Nicolas P, Rial-Sebbag E, Rodriguez LL, Vergely C, Yoshida T, Grimmond SM, Biankin AV, Bowtell DD, Cloonan N, deFazio A, Eshleman JR, Etemadmoghadam D, Gardiner BB, Kench JG, Scarpa A, Sutherland RL, Tempero MA, Waddell NJ, Wilson PJ, McPherson JD, Gallinger S, Tsao MS, Shaw PA, Petersen GM, Mukhopadhyay D, Chin L, DePinho RA, Thayer S, Muthuswamy L, Shazand K, Beck T, Sam M, Timms L, Ballin V, Lu Y, Ji J, Zhang X, Chen F, Hu X, Zhou G, Yang Q, Tian G, Zhang L, Xing X, Li X, Zhu Z, Yu Y, Yu J, Yang H, Lathrop M, Tost J, Brennan P, Holcatova I, Zaridze D, Brazma A, Egevard L, Prokhortchouk E, Banks RE, Uhlen M, Cambon-Thomsen A, Viksna J, Ponten F, Skryabin K, Stratton MR, Futreal PA, Birney E, Borg A, Borresen-Dale AL, Caldas C, Foekens JA, Martin S, Reis-Filho JS, Richardson AL, Sotiriou C, Stunnenberg HG, Thoms G, van de Vijver M, van't Veer L, Calvo F, Birnbaum D, Blanche H, Boucher P, Boyault S, Chabannon C, Gut I, Masson-Jacquemier JD, Lathrop M, Pauporte I, Pivot X, Vincent-Salomon A, Tabone E, Theillet C, Thomas G, Tost J, Treilleux I, Calvo F, Bioulac-Sage P, Clement B, Decaens T, Degos F, Franco D, Gut I, Gut M, Heath S, Lathrop M, Samuel D, Thomas G, Zucman-Rossi J, Lichter P, Eils R, Brors B, Korbel JO, Korshunov A, Landgraf P, Lehrach H, Pfister S, Radlwimmer B, Reifenberger G, Taylor MD, von Kalle C, Majumder PP, Sarin R, Rao TS, Bhan MK, Scarpa A, Pederzoli P, Lawlor RA, Delledonne M, Bardelli A, Biankin AV, Grimmond SM, Gress T, Klimstra D, Zamboni G, Shibata T, Nakamura Y, Nakagawa H, Kusada J, Tsunoda T, Miyano S, Aburatani H, Kato K, Fujimoto A, Yoshida T, Campo E, Lopez-Otin C, Estivill X, Guigo R, de Sanjose S, Piris MA, Montserrat E, Gonzalez-Diaz M, Puente XS, Jares P, Valencia A, Himmelbauer H, Quesada V, Bea S, Stratton MR, Futreal PA, Campbell PJ, Vincent-Salomon A, Richardson AL, Reis-Filho JS, van de Vijver M, Thomas G, Masson-Jacquemier JD, Aparicio S, Borg A, Borresen-Dale AL, Caldas C, Foekens JA, Stunnenberg HG, van't Veer L, Easton DF, Spellman PT, Martin S, Barker AD, Chin L, Collins FS, Compton CC, Ferguson ML, Gerhard DS, Getz G, Gunter C, Guttmacher A, Guyer M, Hayes DN, Lander ES, Ozenberger B, Penny R, Peterson J, Sander C, Shaw KM, Speed TP, Spellman PT, Vockley JG, Wheeler DA, Wilson RK, Hudson TJ, Chin L, Knoppers BM, Lander ES, Lichter P, Stein LD, Stratton MR, Anderson W, Barker AD, Bell C, Bobrow M, Burke W, Collins FS, Compton CC, DePinho RA, Easton DF, Futreal PA, Gerhard DS, Green AR, Guyer M, Hamilton SR, Hubbard TJ, Kallioniemi OP, Kennedy KL, Ley TJ, Liu ET, Lu Y, Majumder P, Marra M, Ozenberger B, Peterson J, Schafer AJ, Spellman PT, Stunnenberg HG, Wainwright BJ, Wilson RK, Yang H. International network of cancer genome projects. Nature. 2010;464(7291):993–8.

153. Jung Y, Kim P, Jung Y, Keum J, Kim SN, Choi YS, Do IG, Lee J, Choi SJ, Kim S, Lee JE, Kim J, Lee S, Kim J. Discovery of ALK-PTPN3 gene fusion from human non-small cell lung carcinoma cell line using next generation RNA sequencing. Genes Chromosomes Cancer. 2012;51(6):590–7.
154. Jing M, Bohl J, Brimer N, Kinter M, Vande Pol SB. Degradation of tyrosine phosphatase PTPN3 (PTPH1) by association with oncogenic human papillomavirus E6 proteins. J Virol. 2007;81(5):2231–9.
155. Töpffer S, Müller-Schiffmann A, Matentzoglu K, Scheffner M, Steger G. Protein tyrosine phosphatase H1 is a target of the E6 oncoprotein of high-risk genital human papillomaviruses. J Gen Virol. 2007;88(Pt 11):2956–65.
156. Kranjec C, Banks L. A systematic analysis of human papillomavirus (HPV) E6 PDZ substrates identifies MAGI-1 as a major target of HPV type 16 (HPV-16) and HPV-18 whose loss accompanies disruption of tight junctions. J Virol. 2011;85(4):1757–64.
157. Warabi M, Nemoto T, Ohashi K, Kitagawa M, Hirokawa K. Expression of protein tyrosine phosphatases and its significance in esophageal cancer. Exp Mol Pathol. 2000;68(3):187–95.
158. Wu CW, Chen JH, Kao HL, Li AF, Lai CH, Chi CW, Lin WC. PTPN3 and PTPN4 tyrosine phosphatase expression in human gastric adenocarcinoma. Anticancer Res. 2006;26(2B):1643–9.

159. Gao Q, Zhao YJ, Wang XY, Guo WJ, Gao S, Wei L, Shi JY, Shi GM, Wang ZC, Zhang YN, Shi YH, Ding J, Ding ZB, Ke AW, Dai Z, Wu FZ, Wang H, Qiu ZP, Chen ZA, Zhang ZF, Qiu SJ, Zhou J, He XH, Fan J. Activating mutations in PTPN3 promote cholangiocarcinoma cell proliferation and migration and are associated with tumor recurrence in patients. Gastroenterology. 2014;146(5):1397–407.
160. Cardinale V, Alvaro D. PTPN3 mutations and HBV may exert synergistic effects in the origin of the intrahepatic cholangiocarcinoma. Gastroenterology. 2014;147(3):719–20.
161. Hsu EC, Lin YC, Hung CS, Huang CJ, Lee MY, Yang SC, Ting LP. Suppression of hepatitis B viral gene expression by protein-tyrosine phosphatase PTPN3. J Biomed Sci. 2007;14(6):731–44.
162. Zhi HY, Hou SW, Li RS, Basir Z, Xiang Q, Szabo A, Chen G. PTPH1 cooperates with vitamin D receptor to stimulate breast cancer growth through their mutual stabilization. Oncogene. 2011;30(14):1706–15.
163. Tang J, Qi X, Mercola D, Han J, Chen G. Essential role of p38gamma in K-Ras transformation independent of phosphorylation. J Biol Chem. 2005;280(25):23910–7.
164. Hou SW, Zhi HY, Pohl N, Loesch M, Qi XM, Li RS, Basir Z, Chen G. PTPH1 dephosphorylates and cooperates with p38gamma MAPK to increase ras oncogenesis through PDZ-mediated interaction. Cancer Res. 2010;70(7):2901–10.
165. Hou S, Suresh PS, Qi X, Lepp A, Mirza SP, Chen G. p38gamma Mitogen-activated protein kinase signals through phosphorylating its phosphatase ptph1 in regulating ras protein oncogenesis and stress response. J Biol Chem. 2012;287(33):27895–905.
166. Young JA, Becker AM, Medeiros JJ, Shapiro VS, Wang A, Farrar JD, Quill TA, Hooft van Huijsduijnen R, van Oers NS. The protein tyrosine phosphatase PTPN4/PTP-MEG1, an enzyme capable of dephosphorylating the TCR ITAMs and regulating NF-kappaB, is dispensable for T cell development and/or T cell effector functions. Mol Immunol. 2008;45(14):3756–66.
167. Bauler TJ, Hendriks WJ, King PD. The FERM and PDZ domain-containing protein tyrosine phosphatases, PTPN4 and PTPN3, are both dispensable for T cell receptor signal transduction. PLoS One. 2008;3(12), e4014.
168. Hironaka K, Umemori H, Tezuka T, Mishina M, Yamamoto T. The protein-tyrosine phosphatase PTPMEG interacts with glutamate receptor delta 2 and epsilon subunits. J Biol Chem. 2000;275(21):16167–73.
169. Whited JL, Robichaux MB, Yang JC, Garrity PA. Ptpmeg is required for the proper establishment and maintenance of axon projections in the central brain of Drosophila. Development. 2007;134(1):43–53.
170. Kina S, Tezuka T, Kusakawa S, Kishimoto Y, Kakizawa S, Hashimoto K, Ohsugi M, Kiyama Y, Horai R, Sudo K, Kakuta S, Iwakura Y, Iino M, Kano M, Manabe T, Yamamoto T. Involvement of protein-tyrosine phosphatase PTPMEG in motor learning and cerebellar long-term depression. Eur J Neurosci. 2007;26(8):2269–78.
171. Gu M, Meng K, Majerus PW. The effect of overexpression of the protein tyrosine phosphatase PTPMEG on cell growth and on colony formation in soft agar in COS-7 cells. Proc Natl Acad Sci U S A. 1996;93(23):12980–5.
172. Zhou J, Wan B, Shan J, Shi H, Li Y, Huo K. PTPN4 negatively regulates CrkI in human cell lines. Cell Mol Biol Lett. 2013;18(2):297–314.
173. Forbes SA, Bindal N, Bamford S, Cole C, Kok CY, Beare D, Jia M, Shepherd R, Leung K, Menzies A, Teague JW, Campbell PJ, Stratton MR, Futreal PA. COSMIC: mining complete cancer genomes in the Catalogue of Somatic Mutations in Cancer. Nucleic Acids Res. 2011;39(Database issue):D945–50.
174. Prehaud C, Wolff N, Terrien E, Lafage M, Megret F, Babault N, Cordier F, Tan GS, Maitrepierre E, Menager P, Chopy D, Hoos S, England P, Delepierre M, Schnell MJ, Buc H, Lafon M. Attenuation of rabies virulence: takeover by the cytoplasmic domain of its envelope protein. Sci Signal. 2010;3(105):ra5.
175. Trave G. PDZ-peptide complexes: as exciting as ever. Structure. 2011;19(10):1350–1.
176. Babault N, Cordier F, Lafage M, Cockburn J, Haouz A, Prehaud C, Rey FA, Delepierre M, Buc H, Lafon M, Wolff N. Peptides targeting the PDZ domain of PTPN4 are efficient inducers of glioblastoma cell death. Structure. 2011;19(10):1518–24.

177. Maisonneuve P, Caillet-Saguy C, Raynal B, Gilquin B, Chaffotte A, Perez J, Zinn-Justin S, Delepierre M, Buc H, Cordier F, Wolff N. Regulation of the catalytic activity of the human phosphatase PTPN4 by its PDZ domain. 2014. FEBS J.
178. Barr AJ, Ugochukwu E, Lee WH, King ON, Filippakopoulos P, Alfano I, Savitsky P, Burgess-Brown NA, Muller S, Knapp S. Large-scale structural analysis of the classical human protein tyrosine phosphatome. Cell. 2009;136(2):352–63.
179. Wadham C, Gamble JR, Vadas MA, Khew-Goodall Y. The protein tyrosine phosphatase Pez is a major phosphatase of adherens junctions and dephosphorylates beta-catenin. Mol Biol Cell. 2003;14(6):2520–9.
180. Cardone L, Carlucci A, Affaitati A, Livigni A, DeCristofaro T, Garbi C, Varrone S, Ullrich A, Gottesman ME, Avvedimento EV, Feliciello A. Mitochondrial AKAP121 binds and targets protein tyrosine phosphatase D1, a novel positive regulator of src signaling. Mol Cell Biol. 2004;24(11):4613–26.
181. Smith AL, Mitchell PJ, Shipley J, Gusterson BA, Rogers MV, Crompton MR. Pez: a novel human cDNA encoding protein tyrosine phosphatase- and ezrin-like domains. Biochem Biophys Res Commun. 1995;209(3):959–65.
182. Sawada M, Ogata M, Fujino Y, Hamaoka T. cDNA cloning of a novel protein tyrosine phosphatase with homology to cytoskeletal protein 4.1 and its expression in T-lineage cells. Biochem Biophys Res Commun. 1994;203(1):479–84.
183. Ogata M, Takada T, Mori Y, Uchida Y, Miki T, Okuyama A, Kosugi A, Sawada M, Oh-hora M, Hamaoka T. Regulation of phosphorylation level and distribution of PTP36, a putative protein tyrosine phosphatase, by cell-substrate adhesion. J Biol Chem. 1999;274(29):20717–24.
184. Wadham C, Gamble JR, Vadas MA, Khew-Goodall Y. Translocation of protein tyrosine phosphatase Pez/PTPD2/PTP36 to the nucleus is associated with induction of cell proliferation. J Cell Sci. 2000;113(Pt 17):3117–23.
185. Wyatt L, Wadham C, Crocker LA, Lardelli M, Khew-Goodall Y. The protein tyrosine phosphatase Pez regulates TGFbeta, epithelial-mesenchymal transition, and organ development. J Cell Biol. 2007;178(7):1223–35.
186. Yin M, Zhang L. Hippo signaling: a hub of growth control, tumor suppression and pluripotency maintenance. J Genet Genomics. 2011;38(10):471–81.
187. Poernbacher I, Baumgartner R, Marada SK, Edwards K, Stocker H. Drosophila Pez acts in Hippo signaling to restrict intestinal stem cell proliferation. Curr Biol. 2012;22(5):389–96.
188. Au AC, Hernandez PA, Lieber E, Nadroo AM, Shen YM, Kelley KA, Gelb BD, Diaz GA. Protein tyrosine phosphatase PTPN14 is a regulator of lymphatic function and choanal development in humans. Am J Hum Genet. 2010;87(3):436–44.
189. Mendola A, Schlogel MJ, Ghalamkarpour A, Irrthum A, Nguyen HL, Fastre E, Bygum A, van der Vleuten C, Fagerberg C, Baselga E, Quere I, Mulliken JB, Boon LM, Brouillard P, Vikkula M, Lymphedema Research Group. Mutations in the VEGFR3 signaling pathway explain 36% of familial lymphedema. Mol Syndromol. 2013;4(6):257–66.
190. Zhang P, Guo A, Possemato A, Wang C, Beard L, Carlin C, Markowitz SD, Polakiewicz RD, Wang Z. Identification and functional characterization of p130Cas as a substrate of protein tyrosine phosphatase nonreceptor 14. Oncogene. 2013;32(16):2087–95.
191. Niedergethmann M, Alves F, Neff JK, Heidrich B, Aramin N, Li L, Pilarsky C, Grutzmann R, Allgayer H, Post S, Gretz N. Gene expression profiling of liver metastases and tumour invasion in pancreatic cancer using an orthotopic SCID mouse model. Br J Cancer. 2007;97(10):1432–40.
192. Boiarskikh UA, Kondrakhin Iu V, Evshin IS, Sharipov RN, Komel'kov AV, Musatkina EA, Chevkina EM, Sukoian MA, Kolpakov FA, Kashkin KN, Filipenko ML. [Prediction of a non-small cell lung cancer sensitivity to cisplatin and paclitaxel based on the marker genes expression]. Mol Biol (Mosk). 2011;45(4):652–61.
193. Andersen JN, Sathyanarayanan S, Di Bacco A, Chi A, Zhang T, Chen AH, Dolinski B, Kraus M, Roberts B, Arthur W, Klinghoffer RA, Gargano D, Li L, Feldman I, Lynch B, Rush J, Hendrickson RC, Blume-Jensen P, Paweletz CP. Pathway-based identification of biomarkers for targeted therapeutics: personalized oncology with PI3K pathway inhibitors. Sci Transl Med. 2010;2(43):43ra55.

194. Liu X, Yang N, Figel SA, Wilson KE, Morrison CD, Gelman IH, Zhang J. PTPN14 interacts with and negatively regulates the oncogenic function of YAP. Oncogene. 2013;32(10):1266–73.
195. Wang W, Huang J, Wang X, Yuan J, Li X, Feng L, Park J, Chen J. PTPN14 is required for the density-dependent control of YAP1. Genes Dev. 2012;26(17):1959–71.
196. Huang JM, Nagatomo I, Suzuki E, Mizuno T, Kumagai T, Berezov A, Zhang H, Karlan B, Greene MI, Wang Q. YAP modifies cancer cell sensitivity to EGFR and survivin inhibitors and is negatively regulated by the non-receptor type protein tyrosine phosphatase 14. Oncogene. 2013;32(17):2220–9.
197. Wang W, Li X, Huang J, Feng L, Dolinta KG, Chen J. Defining the protein-protein interaction network of the human hippo pathway. Mol Cell Proteomics. 2014;13(1):119–31.
198. Wilson KE, Li YW, Yang N, Shen H, Orillion AR, Zhang J. PTPN14 forms a complex with Kibra and LATS1 proteins and negatively regulates the YAP oncogenic function. J Biol Chem. 2014;289(34):23693–700.
199. Michaloglou C, Lehmann W, Martin T, Delaunay C, Hueber A, Barys L, Niu H, Billy E, Wartmann M, Ito M, Wilson CJ, Digan ME, Bauer A, Voshol H, Christofori G, Sellers WR, Hofmann F, Schmelzle T. The tyrosine phosphatase PTPN14 is a negative regulator of YAP activity. PLoS One. 2013;8(4), e61916.
200. Möller NP, Möller KB, Lammers R, Kharitonenkov A, Sures I, Ullrich A. Src kinase associates with a member of a distinct subfamily of protein-tyrosine phosphatases containing an ezrin-like domain. Proc Natl Acad Sci U S A. 1994;91(16):7477–81.
201. Tokuchi H, Higashitsuji H, Nishiyama H, Nonoguchi K, Nagao T, Xue JH, Itoh K, Ogawa O, Fujita J. Expression of protein tyrosine phosphatase PTP-RL10 and its isoform in the mouse testis. Int J Urol. 1999;6(11):572–7.
202. Dörner C, Ciossek T, Muller S, Moller PH, Ullrich A, Lammers R. Characterization of KIF1C, a new kinesin-like protein involved in vesicle transport from the Golgi apparatus to the endoplasmic reticulum. J Biol Chem. 1998;273(32):20267–75.
203. Jui HY, Tseng RJ, Wen X, Fang HI, Huang LM, Chen KY, Kung HJ, Ann DK, Shih HM. Protein-tyrosine phosphatase D1, a potential regulator and effector for Tec family kinases. J Biol Chem. 2000;275(52):41124–32.
204. Carlucci A, Gedressi C, Lignitto L, Nezi L, Villa-Moruzzi E, Avvedimento EV, Gottesman M, Garbi C, Feliciello A. Protein-tyrosine phosphatase PTPD1 regulates focal adhesion kinase autophosphorylation and cell migration. J Biol Chem. 2008;283(16):10919–29.
205. Chen J, Lee G, Fanous AH, Zhao Z, Jia P, O'Neill A, Walsh D, Kendler KS, Chen X, International Schizophrenia Consortium. Two non-synonymous markers in PTPN21, identified by genome-wide association study data-mining and replication, are associated with schizophrenia. Schizophr Res. 2011;131(1-3):43–51.
206. Dunican DS, McWilliam P, Tighe O, Parle-McDermott A, Croke DT. Gene expression differences between the microsatellite instability (MIN) and chromosomal instability (CIN) phenotypes in colorectal cancer revealed by high-density cDNA array hybridization. Oncogene. 2002;21(20):3253–7.
207. Korff S, Woerner SM, Yuan YP, Bork P, von Knebel DM, Gebert J. Frameshift mutations in coding repeats of protein tyrosine phosphatase genes in colorectal tumors with microsatellite instability. BMC Cancer. 2008;8:329.
208. Wu ZZ, Lu HP, Chao CC. Identification and functional analysis of genes which confer resistance to cisplatin in tumor cells. Biochem Pharmacol. 2010;80(2):262–76.
209. Carlucci A, Porpora M, Garbi C, Galgani M, Santoriello M, Mascolo M, di Lorenzo D, Altieri V, Quarto M, Terracciano L, Gottesman ME, Insabato L, Feliciello A. PTPD1 supports receptor stability and mitogenic signaling in bladder cancer cells. J Biol Chem. 2010;285(50):39260–70.
210. van Eekelen M, Runtuwene V, Masselink W, den Hertog J. Pair-wise regulation of convergence and extension cell movements by four phosphatases via RhoA. PLoS One. 2012;7(4), e35913.
211. Irie S, Li Y, Kanki H, Ohyama T, Deaven LL, Somlo S, Sato TA. Identification of two Fas-associated phosphatase-1 (FAP-1) promoters in human cancer cells. DNA Seq. 2001;11(6): 519–26.

212. Yoshida S, Harada H, Nagai H, Fukino K, Teramoto A, Emi M. Head-to-head juxtaposition of Fas-associated phosphatase-1 (FAP-1) and c-Jun NH2-terminal kinase 3 (JNK3) genes: genomic structure and seven polymorphisms of the FAP-1 gene. J Hum Genet. 2002;47(11):614–9.
213. Han XJ, Xue L, Gong L, Zhu SJ, Yao L, Wang SM, Lan M, Zhang W, Li YH. Stat3 inhibits PTPN13 expression in squamous cell lung carcinoma through recruitment of HDAC5. Biomed Res Int. 2013;2013:468963.
214. Huang W, Zhu C, Wang H, Horvath E, Eklund EA. The interferon consensus sequence-binding protein (ICSBP/IRF8) represses PTPN13 gene transcription in differentiating myeloid cells. J Biol Chem. 2008;283(12):7921–35.
215. Freiss G, Chalbos D. PTPN13/PTPL1: an important regulator of tumor aggressiveness. Anticancer Agents Med Chem. 2011;11(1):78–88.
216. Abaan OD, Toretsky JA. PTPL1: a large phosphatase with a split personality. Cancer Metastasis Rev. 2008;27(2):205–14.
217. Hagemann N, Ackermann N, Christmann J, Brier S, Yu F, Erdmann KS. The serologically defined colon cancer antigen-3 interacts with the protein tyrosine phosphatase PTPN13 and is involved in the regulation of cytokinesis. Oncogene. 2013;32(39):4602–13.
218. Wansink DG, Peters W, Schaafsma I, Sutmuller RP, Oerlemans F, Adema GJ, Wieringa B, van der Zee CE, Hendriks W. Mild impairment of motor nerve repair in mice lacking PTP-BL tyrosine phosphatase activity. Physiol Genomics. 2004;19(1):50–60.
219. Nakahira M, Tanaka T, Robson BE, Mizgerd JP, Grusby MJ. Regulation of signal transducer and activator of transcription signaling by the tyrosine phosphatase PTP-BL. Immunity. 2007;26(2):163–76.
220. Lorber B, Hendriks WJ, Van der Zee CE, Berry M, Logan A. Effects of LAR and PTP-BL phosphatase deficiency on adult mouse retinal cells activated by lens injury. Eur J Neurosci. 2005;21(9):2375–83.
221. Glondu-Lassis M, Dromard M, Chavey C, Puech C, Fajas L, Hendriks W, Freiss G. Downregulation of protein tyrosine phosphatase PTP-BL represses adipogenesis. Int J Biochem Cell Biol. 2009;41(11):2173–80.
222. Sato T, Irie S, Kitada S, Reed JC. FAP-1: a protein tyrosine phosphatase that associates with Fas. Science. 1995;268(5209):411–5.
223. Ungefroren H, Kruse ML, Trauzold A, Roeschmann S, Roeder C, Arlt A, Henne-Bruns D, Kalthoff H. FAP-1 in pancreatic cancer cells: functional and mechanistic studies on its inhibitory role in CD95-mediated apoptosis. J Cell Sci. 2001;114(Pt 15):2735–46.
224. Ivanov VN, Lopez Bergami P, Maulit G, Sato TA, Sassoon D, Ronai Z. FAP-1 association with Fas (Apo-1) inhibits Fas expression on the cell surface. Mol Cell Biol. 2003;23(10): 3623–35.
225. Foehr ED, Lorente G, Vincent V, Nikolich K, Urfer R. FAS associated phosphatase (FAP-1) blocks apoptosis of astrocytomas through dephosphorylation of FAS. J Neurooncol. 2005;74(3):241–8.
226. Wieckowski E, Atarashi Y, Stanson J, Sato TA, Whiteside TL. FAP-1-mediated activation of NF-kappaB induces resistance of head and neck cancer to Fas-induced apoptosis. J Cell Biochem. 2007;100(1):16–28.
227. Ivanov VN, Ronai Z, Hei TK. Opposite roles of FAP-1 and dynamin in the regulation of Fas (CD95) translocation to the cell surface and susceptibility to Fas ligand-mediated apoptosis. J Biol Chem. 2006;281(3):1840–52.
228. Irie S, Hachiya T, Rabizadeh S, Maruyama W, Mukai J, Li Y, Reed JC, Bredesen DE, Sato TA. Functional interaction of Fas-associated phosphatase-1 (FAP-1) with p75(NTR) and their effect on NF-kappaB activation. FEBS Lett. 1999;460(2):191–8.
229. Zhang W, Tong Q, Conrad K, Wozney J, Cheung JY, Miller BA. Regulation of TRP channel TRPM2 by the tyrosine phosphatase PTPL1. Am J Physiol Cell Physiol. 2007;292(5): C1746–58.
230. Yanagisawa J, Takahashi M, Kanki H, Yano-Yanagisawa H, Tazunoki T, Sawa E, Nishitoba T, Kamishohara M, Kobayashi E, Kataoka S, Sato T. The molecular interaction of Fas and FAP-1. A tripeptide blocker of human Fas interaction with FAP-1 promotes Fas-induced apoptosis. J Biol Chem. 1997;272(13):8539–45.

231. Zhang X, Brunner T, Carter L, Dutton RW, Rogers P, Bradley L, Sato T, Reed JC, Green D, Swain SL. Unequal death in T helper cell (Th)1 and Th2 effectors: Th1, but not Th2, effectors undergo rapid Fas/FasL-mediated apoptosis. J Exp Med. 1997;185(10):1837–49.
232. Arai M, Kannagi M, Matsuoka M, Sato T, Yamamoto N, Fujii M. Expression of FAP-1 (Fas-associated phosphatase) and resistance to Fas-mediated apoptosis in T cell lines derived from human T cell leukemia virus type 1-associated myelopathy/tropical spastic paraparesis patients. AIDS Res Hum Retroviruses. 1998;14(3):261–7.
233. Zhou YW, Komada Y, Inaba H, Azuma E, Sakurai M. Down-regulation of Fas-associated phosphatase-1 (FAP-1) in interleukin-2-activated T cells. Cell Immunol. 1998;186(2):103–10.
234. Mundle SD, Mativi BY, Bagai K, Feldman G, Cheema P, Gautam U, Reza S, Cartlidge JD, Venugopal P, Shetty V, Gregory SA, Robin E, Rifkin S, Shah R, Raza A. Spontaneous down-regulation of Fas-associated phosphatase-1 may contribute to excessive apoptosis in myelodysplastic marrows. Int J Hematol. 1999;70(2):83–90.
235. Myc A, Arscott PL, Bretz JD, Thompson NW, Baker Jr JR. Characterization of FAP-1 expression and function in thyroid follicular cells. Endocrinology. 1999;140(11):5431–4.
236. Li Y, Kanki H, Hachiya T, Ohyama T, Irie S, Tang G, Mukai J, Sato T. Negative regulation of Fas-mediated apoptosis by FAP-1 in human cancer cells. Int J Cancer. 2000;87(4):473–9.
237. Elnemr A, Ohta T, Yachie A, Kayahara M, Kitagawa H, Fujimura T, Ninomiya I, Fushida S, Nishimura GI, Shimizu K, Miwa K. Human pancreatic cancer cells disable function of Fas receptors at several levels in Fas signal transduction pathway. Int J Oncol. 2001;18(2):311–6.
238. Meinhold-Heerlein I, Stenner-Liewen F, Liewen H, Kitada S, Krajewska M, Krajewski S, Zapata JM, Monks A, Scudiero DA, Bauknecht T, Reed JC. Expression and potential role of Fas-associated phosphatase-1 in ovarian cancer. Am J Pathol. 2001;158(4):1335–44.
239. Yao H, Song E, Chen J, Hamar P. Expression of FAP-1 by human colon adenocarcinoma: implication for resistance against Fas-mediated apoptosis in cancer. Br J Cancer. 2004;91(9):1718–25.
240. Xiao ZY, Wu W, Eagleton N, Chen HQ, Shao J, Teng H, Liu TH, Jiang ZM, Yao HR. Silencing Fas-associated phosphatase 1 expression enhances efficiency of chemotherapy for colon carcinoma with oxaliplatin. World J Gastroenterol. 2010;16(1):112–8.
241. Schickel R, Park SM, Murmann AE, Peter ME. miR-200c regulates induction of apoptosis through CD95 by targeting FAP-1. Mol Cell. 2010;38(6):908–15.
242. Huang W, Hu L, Bei L, Hjort E, Eklund EA. The leukemia-associated fusion protein Tel-platelet-derived growth factor receptor beta (Tel-PdgfRbeta) inhibits transcriptional repression of PTPN13 gene by interferon consensus sequence binding protein (Icsbp). J Biol Chem. 2012;287(11):8110–25.
243. Nicolini V, Cassinelli G, Cuccuru G, Bongarzone I, Petrangolini G, Tortoreto M, Mondellini P, Casalini P, Favini E, Zaffaroni N, Zunino F, Lanzi C. Interplay between Ret and Fap-1 regulates CD95-mediated apoptosis in medullary thyroid cancer cells. Biochem Pharmacol. 2011;82(7):778–88.
244. Cuppen E, Nagata S, Wieringa B, Hendriks W. No evidence for involvement of mouse protein-tyrosine phosphatase-BAS-like Fas-associated phosphatase-1 in Fas-mediated apoptosis. J Biol Chem. 1997;272(48):30215–20.
245. Houghton JA, Harwood FG, Gibson AA, Tillman DM. The fas signaling pathway is functional in colon carcinoma cells and induces apoptosis. Clin Cancer Res. 1997;3(12 Pt 1):2205–9.
246. Hayashi H, Tatebe S, Osaki M, Goto A, Sato K, Ito H. Anti-Fas antibody-induced apoptosis in human colorectal carcinoma cell lines: role of the p53 gene. Apoptosis. 1998;3(6):431–7.
247. Tillman DM, Harwood FG, Gibson AA, Houghton JA. Expression of genes that regulate Fas signalling and Fas-mediated apoptosis in colon carcinoma cells. Cell Death Differ. 1998;5(5):450–7.
248. Yamada Y, Şugahara K, Tsuruda K, Nohda K, Hata T, Maeda T, Honda M, Tawara M, Hayashibara T, Joh T, Tomonaga M, Miyazaki Y, Kamihira S. Fas-resistance in ATL cell lines not associated with HTLV-I or FAP-1 production. Cancer Lett. 1999;147(1-2):215–9.
249. Kim KM, Lee K, Hong YS, Park HY. Fas-mediated apoptosis and expression of related genes in human malignant hematopoietic cells. Exp Mol Med. 2000;32(4):246–54.
250. Castilla C, Chinchon D, Medina R, Torrubia FJ, Japon MA, Saez C. PTPL1 and PKCdelta contribute to proapoptotic signalling in prostate cancer cells. Cell Death Dis. 2013;4, e576.

251. Bompard G, Puech C, Prebois C, Vignon F, Freiss G. Protein-tyrosine phosphatase PTPL1/FAP-1 triggers apoptosis in human breast cancer cells. J Biol Chem. 2002;277(49):47861–9.
252. Dromard M, Bompard G, Glondu-Lassis M, Puech C, Chalbos D, Freiss G. The putative tumor suppressor gene PTPN13/PTPL1 induces apoptosis through insulin receptor substrate-1 dephosphorylation. Cancer Res. 2007;67(14):6806–13.
253. Miyazaki T, Atarashi Y, Yasumura S, Minatoya I, Ogawa K, Iwamoto M, Minemura M, Shimizu Y, Sato TA, Watanabe A, Sugiyama T. Fas-associated phosphatase-1 promotes Fas-mediated apoptosis in human colon cancer cells: novel function of FAP-1. J Gastroenterol Hepatol. 2006;21(1 Pt 1):84–91.
254. Winterhoff BJ, Arlt A, Duttmann A, Ungefroren H, Schafer H, Kalthoff H, Kruse ML. Characterisation of FAP-1 expression and CD95 mediated apoptosis in the A818-6 pancreatic adenocarcinoma differentiation system. Differentiation. 2012;83(3):148–57.
255. Kuchay S, Duan S, Schenkein E, Peschiaroli A, Saraf A, Florens L, Washburn MP, Pagano M. FBXL2- and PTPL1-mediated degradation of p110-free p85beta regulatory subunit controls the PI(3)K signalling cascade. Nat Cell Biol. 2013;15(5):472–80.
256. Gump JM, Staskiewicz L, Morgan MJ, Bamberg A, Riches DW, Thorburn A. Autophagy variation within a cell population determines cell fate through selective degradation of Fap-1. Nat Cell Biol. 2014;16(1):47–54.
257. Trofatter JA, MacCollin MM, Rutter JL, Murrell JR, Duyao MP, Parry DM, Eldridge R, Kley N, Menon AG, Pulaski K, et al. A novel moesin-, ezrin-, radixin-like gene is a candidate for the neurofibromatosis 2 tumor suppressor. Cell. 1993;72(5):791–800.
258. Rouleau GA, Merel P, Lutchman M, Sanson M, Zucman J, Marineau C, Hoang-Xuan K, Demczuk S, Desmaze C, Plougastel B, et al. Alteration in a new gene encoding a putative membrane-organizing protein causes neuro-fibromatosis type 2. Nature. 1993;363(6429):515–21.
259. Boedigheimer M, Laughon A. Expanded: a gene involved in the control of cell proliferation in imaginal discs. Development. 1993;118(4):1291–301.
260. Woods DF, Bryant PJ. The discs-large tumor suppressor gene of Drosophila encodes a guanylate kinase homolog localized at septate junctions. Cell. 1991;66(3):451–64.
261. Lee SH, Shin MS, Park WS, Kim SY, Kim HS, Lee JH, Han SY, Lee HK, Park JY, Oh RR, Jang JJ, Lee JY, Yoo NJ. Immunohistochemical localization of FAP-1, an inhibitor of Fas-mediated apoptosis, in normal and neoplastic human tissues. APMIS. 1999;107(12):1101–8.
262. Juric D, Sale S, Hromas RA, Yu R, Wang Y, Duran GE, Tibshirani R, Einhorn LH, Sikic BI. Gene expression profiling differentiates germ cell tumors from other cancers and defines subtype-specific signatures. Proc Natl Acad Sci U S A. 2005;102(49):17763–8.
263. Liu H, Brannon AR, Reddy AR, Alexe G, Seiler MW, Arreola A, Oza JH, Yao M, Juan D, Liou LS, Ganesan S, Levine AJ, Rathmell WK, Bhanot GV. Identifying mRNA targets of microRNA dysregulated in cancer: with application to clear cell Renal Cell Carcinoma. BMC Syst Biol. 2010;4:51.
264. Nariai Y, Mishima K, Yoshimura Y, Sekine J. FAP-1 and NF-kappaB expressions in oral squamous cell carcinoma as potential markers for chemo-radio sensitivity and prognosis. Int J Oral Maxillofac Surg. 2011;40(4):419–26.
265. Sabine VS, Sims AH, Macaskill EJ, Renshaw L, Thomas JS, Dixon JM, Bartlett JM. Gene expression profiling of response to mTOR inhibitor everolimus in pre-operatively treated post-menopausal women with oestrogen receptor-positive breast cancer. Breast Cancer Res Treat. 2010;122(2):419–28.
266. Abaan OD, Levenson A, Khan O, Furth PA, Uren A, Toretsky JA. PTPL1 is a direct transcriptional target of EWS-FLI1 and modulates Ewing's Sarcoma tumorigenesis. Oncogene. 2005;24(16):2715–22.
267. Herrmann L, Dittmar T, Erdmann KS. The protein tyrosine phosphatase PTP-BL associates with the midbody and is involved in the regulation of cytokinesis. Mol Biol Cell. 2003;14(1):230–40.
268. Huang W, Bei L, Eklund EA. Fas-associated phosphatase 1 mediates Fas resistance in myeloid progenitor cells expressing the Bcr-abl oncogene. Leuk Lymphoma. 2013;54(3):619–30.

269. Huang W, Bei L, Eklund EA. Fas-associated phosphatase 1 (Fap1) influences betacatenin activity in myeloid progenitor cells expressing the Bcr-abl oncogene. J Biol Chem. 2013;288(18):12766–76.
270. van Ham M, Hendriks W. PDZ domains-glue and guide. Mol Biol Rep. 2003;30(2):69–82.
271. Itakura M, Mori S, Park NH, Bonavida B. Both HPV and carcinogen contribute to the development of resistance to apoptosis during oral carcinogenesis. Int J Oncol. 2000;16(3): 591–7.
272. Freiss G, Puech C, Vignon F. Extinction of insulin-like growth factor-I mitogenic signaling by antiestrogen-stimulated Fas-associated protein tyrosine phosphatase-1 in human breast cancer cells. Mol Endocrinol. 1998;12(4):568–79.
273. Rey JM, Pujol P, Callier P, Cavailles V, Freiss G, Maudelonde T, Brouillet JP. Semiquantitative reverse transcription-polymerase chain reaction to evaluate the expression patterns of genes involved in the oestrogen pathway. J Mol Endocrinol. 2000;24(3):433–40.
274. Révillion F, Puech C, Rabenoelina F, Chalbos D, Peyrat JP, Freiss G. Expression of the putative tumor suppressor gene PTPN13/PTPL1 is an independent prognostic marker for overall survival in breast cancer. Int J Cancer. 2009;124(3):638–43.
275. Castilla C, Flores ML, Conde JM, Medina R, Torrubia FJ, Japon MA, Saez C. Downregulation of protein tyrosine phosphatase PTPL1 alters cell cycle and upregulates invasion-related genes in prostate cancer cells. Clin Exp Metastasis. 2012;29(4):349–58.
276. Ying J, Li H, Cui Y, Wong AH, Langford C, Tao Q. Epigenetic disruption of two proapoptotic genes MAPK10/JNK3 and PTPN13/FAP-1 in multiple lymphomas and carcinomas through hypermethylation of a common bidirectional promoter. Leukemia. 2006;20(6):1173–5.
277. Yeh SH, Wu DC, Tsai CY, Kuo TJ, Yu WC, Chang YS, Chen CL, Chang CF, Chen DS, Chen PJ. Genetic characterization of fas-associated phosphatase-1 as a putative tumor suppressor gene on chromosome 4q21.3 in hepatocellular carcinoma. Clin Cancer Res. 2006;12(4):1097–108.
278. Wei W, Jiang M, Luo L, Li Z, Wang P, Dong WQ. Colorectal cancer susceptibility variants alter risk of breast cancer in a Chinese Han population. Genet Mol Res. 2013;12(4):6268–74.
279. Niu J, Huang YJ, Wang LE, Sturgis EM, Wei Q. Genetic polymorphisms in the PTPN13 gene and risk of squamous cell carcinoma of head and neck. Carcinogenesis. 2009;30(12):2053–8.
280. Mita Y, Yasuda Y, Sakai A, Yamamoto H, Toyooka S, Gunduz M, Tanabe S, Naomoto Y, Ouchida M, Shimizu K. Missense polymorphisms of PTPRJ and PTPN13 genes affect susceptibility to a variety of human cancers. J Cancer Res Clin Oncol. 2010;136(2):249–59.
281. Zhu JH, Chen R, Yi W, Cantin GT, Fearns C, Yang Y, Yates 3rd JR, Lee JD. Protein tyrosine phosphatase PTPN13 negatively regulates Her2/ErbB2 malignant signaling. Oncogene. 2008;27(18):2525–31.
282. Hoover AC, Strand GL, Nowicki PN, Anderson ME, Vermeer PD, Klingelhutz AJ, Bossler AD, Pottala JV, Hendriks WJ, Lee JH. Impaired PTPN13 phosphatase activity in spontaneous or HPV-induced squamous cell carcinomas potentiates oncogene signaling through the MAP kinase pathway. Oncogene. 2009;28(45):3960–70.
283. Spanos WC, Hoover A, Harris GF, Wu S, Strand GL, Anderson ME, Klingelhutz AJ, Hendriks W, Bossler AD, Lee JH. The PDZ binding motif of human papillomavirus type 16 E6 induces PTPN13 loss, which allows anchorage-independent growth and synergizes with ras for invasive growth. J Virol. 2008;82(5):2493–500.
284. Maher DM, Bell MC, O'Donnell EA, Gupta BK, Jaggi M, Chauhan SC. Curcumin suppresses human papillomavirus oncoproteins, restores p53, Rb, and PTPN13 proteins and inhibits benzo[a]pyrene-induced upregulation of HPV E7. Mol Carcinog. 2011;50(1): 47–57.
285. Lucci MA, Orlandi R, Triulzi T, Tagliabue E, Balsari A, Villa-Moruzzi E. Expression profile of tyrosine phosphatases in HER2 breast cancer cells and tumors. Cell Oncol. 2010;32(5-6):361–72.
286. Palmer A, Zimmer M, Erdmann KS, Eulenburg V, Porthin A, Heumann R, Deutsch U, Klein R. EphrinB phosphorylation and reverse signaling: regulation by Src kinases and PTP-BL phosphatase. Mol Cell. 2002;9(4):725–37.

287. Vermeer PD, Bell M, Lee K, Vermeer DW, Wieking BG, Bilal E, Bhanot G, Drapkin RI, Ganesan S, Klingelhutz AJ, Hendriks WJ, Lee JH. ErbB2, EphrinB1, Src kinase and PTPN13 signaling complex regulates MAP kinase signaling in human cancers. PLoS One. 2012;7(1), e30447.
288. Vermeer PD, Colbert PL, Wieking BG, Vermeer DW, Lee JH. Targeting ERBB receptors shifts their partners and triggers persistent ERK signaling through a novel ERBB/EFNB1 complex. Cancer Res. 2013;73(18):5787–97.
289. Kawano S, Ikeda W, Kishimoto M, Ogita H, Takai Y. Silencing of ErbB3/ErbB2 signaling by immunoglobulin-like Necl-2. J Biol Chem. 2009;284(35):23793–805.
290. Yamada A, Inoue E, Deguchi-Tawarada M, Matsui C, Togawa A, Nakatani T, Ono Y, Takai Y. Necl-2/CADM1 interacts with ErbB4 and regulates its activity in GABAergic neurons. Mol Cell Neurosci. 2013;56:234–43.
291. Sugiyama H, Mizutani K, Kurita S, Okimoto N, Shimono Y, Takai Y. Interaction of Necl-4/CADM4 with ErbB3 and integrin alpha6 beta4 and inhibition of ErbB2/ErbB3 signaling and hemidesmosome disassembly. Genes Cells. 2013;18(6):519–28.
292. Zhang Y, Tu Y, Zhao J, Chen K, Wu C. Reversion-induced LIM interaction with Src reveals a novel Src inactivation cycle. J Cell Biol. 2009;184(6):785–92.
293. Scrima M, De Marco C, De Vita F, Fabiani F, Franco R, Pirozzi G, Rocco G, Malanga D, Viglietto G. The nonreceptor-type tyrosine phosphatase PTPN13 is a tumor suppressor gene in non-small cell lung cancer. Am J Pathol. 2012;180(3):1202–14.
294. Han X, Xue L, Zhou L, Gong L, Zhu S, Yao L, Wang S, Lan M, Li Y, Zhang W. The role of PTPN13 in invasion and metastasis of lung squamous cell carcinoma. Exp Mol Pathol. 2013;95(3):270–5.
295. Sun Z, Wang L, Eckloff BW, Deng B, Wang Y, Wampfler JA, Jang J, Wieben ED, Jen J, You M, Yang P. Conserved recurrent gene mutations correlate with pathway deregulation and clinical outcomes of lung adenocarcinoma in never-smokers. BMC Med Genomics. 2014;7:32.
296. Saras J, Franzen P, Aspenstrom P, Hellman U, Gonez LJ, Heldin CH. A novel GTPase-activating protein for Rho interacts with a PDZ domain of the protein-tyrosine phosphatase PTPL1. J Biol Chem. 1997;272(39):24333–8.
297. Fodero-Tavoletti MT, Hardy MP, Cornell B, Katsis F, Sadek CM, Mitchell CA, Kemp BE, Tiganis T. Protein tyrosine phosphatase hPTPN20a is targeted to sites of actin polymerization. Biochem J. 2005;389(Pt 2):343–54.
298. Ohsugi M, Kuramochi S, Matsuda S, Yamamoto T. Molecular cloning and characterization of a novel cytoplasmic protein-tyrosine phosphatase that is specifically expressed in spermatocytes. J Biol Chem. 1997;272(52):33092–9.
299. Stenzel N, Fetzer CP, Heumann R, Erdmann KS. PDZ-domain-directed basolateral targeting of the peripheral membrane protein FRMPD2 in epithelial cells. J Cell Sci. 2009;122(Pt 18):3374–84.
300. Cuppen E, Wijers M, Schepens J, Fransen J, Wieringa B, Hendriks W. A FERM domain governs apical confinement of PTP-BL in epithelial cells. J Cell Sci. 1999;112(Pt 19):3299–308.
301. Cao L, Zhang L, Ruiz-Lozano P, Yang Q, Chien KR, Graham RM, Zhou M. A novel putative protein-tyrosine phosphatase contains a BRO1-like domain and suppresses Ha-ras-mediated transformation. J Biol Chem. 1998;273(33):21077–83.
302. Toyooka S, Ouchida M, Jitsumori Y, Tsukuda K, Sakai A, Nakamura A, Shimizu N, Shimizu K. HD-PTP: A novel protein tyrosine phosphatase gene on human chromosome 3p21.3. Biochem Biophys Res Commun. 2000;278(3):671–8.
303. Doyotte A, Mironov A, McKenzie E, Woodman P. The Bro1-related protein HD-PTP/PTPN23 is required for endosomal cargo sorting and multivesicular body morphogenesis. Proc Natl Acad Sci U S A. 2008;105(17):6308–13.
304. Tanase CA. Histidine domain-protein tyrosine phosphatase interacts with Grb2 and GrpL. PLoS One. 2010;5(12), e14339.
305. Gingras MC, Zhang YL, Kharitidi D, Barr AJ, Knapp S, Tremblay ML, Pause A. HD-PTP is a catalytically inactive tyrosine phosphatase due to a conserved divergence in its phosphatase domain. PLoS One. 2009;4(4):e5105.
306. Mariotti M, Castiglioni S, Garcia-Manteiga JM, Beguinot L, Maier JA. HD-PTP inhibits endothelial migration through its interaction with Src. Int J Biochem Cell Biol. 2009;41(3):687–93.

307. Lin G, Aranda V, Muthuswamy SK, Tonks NK. Identification of PTPN23 as a novel regulator of cell invasion in mammary epithelial cells from a loss-of-function screen of the 'PTP-ome'. Genes Dev. 2011;25(13):1412–25.
308. Gingras MC, Kharitidi D, Chenard V, Uetani N, Bouchard M, Tremblay ML, Pause A. Expression analysis and essential role of the putative tyrosine phosphatase His-domain-containing protein tyrosine phosphatase (HD-PTP). Int J Dev Biol. 2009;53(7):1069–74.
309. Castiglioni S, Maier JA, Mariotti M. The tyrosine phosphatase HD-PTP: A novel player in endothelial migration. Biochem Biophys Res Commun. 2007;364(3):534–9.
310. Mariotti M, Castiglioni S, Beguinot L, Maier JA. The tyrosine phosphatase HD-PTP is regulated by FGF-2 through proteasome degradation. Front Biosci. 2006;11:2138–43.
311. Castiglioni S, Maier JA. The tyrosine phosphatase HD-PTP (PTPN23) is degraded by calpains in a calcium-dependent manner. Biochem Biophys Res Commun. 2012;421(2):380–3.
312. Popov S, Popova E, Inoue M, Gottlinger HG. Divergent Bro1 domains share the capacity to bind human immunodeficiency virus type 1 nucleocapsid and to enhance virus-like particle production. J Virol. 2009;83(14):7185–93.
313. Sette P, Mu R, Dussupt V, Jiang J, Snyder G, Smith P, Xiao TS, Bouamr F. The Phe105 loop of Alix Bro1 domain plays a key role in HIV-1 release. Structure. 2011;19(10):1485–95.
314. Miura GI, Roignant JY, Wassef M, Treisman JE. Myopic acts in the endocytic pathway to enhance signaling by the Drosophila EGF receptor. Development. 2008;135(11):1913–22.
315. Ali N, Zhang L, Taylor S, Mironov A, Urbe S, Woodman P. Recruitment of UBPY and ESCRT exchange drive HD-PTP-dependent sorting of EGFR to the MVB. Curr Biol. 2013;23(6):453–61.
316. Mariotti M, Castiglioni S, Maier JA. Inhibition of T24 human bladder carcinoma cell migration by RNA interference suppressing the expression of HD-PTP. Cancer Lett. 2009;273(1):155–63.
317. Tanaka K, Kondo K, Kitajima K, Muraoka M, Nozawa A, Hara T. Tumor-suppressive function of protein-tyrosine phosphatase non-receptor type 23 in testicular germ cell tumors is lost upon overexpression of miR142-3p microRNA. J Biol Chem. 2013;288(33):23990–9.
318. Gilbert MM, Tipping M, Veraksa A, Moberg KH. A screen for conditional growth suppressor genes identifies the Drosophila homolog of HD-PTP as a regulator of the oncoprotein Yorkie. Dev Cell. 2011;20(5):700–12.
319. Barr AJ, Knapp S. MAPK-specific tyrosine phosphatases: new targets for drug discovery? Trends Pharmacol Sci. 2006;27(10):525–30.
320. Pulido R, Zuniga A, Ullrich A. PTP-SL and STEP protein tyrosine phosphatases regulate the activation of the extracellular signal-regulated kinases ERK1 and ERK2 by association through a kinase interaction motif. EMBO J. 1998;17(24):7337–50.
321. Blanco-Aparicio C, Torres J, Pulido R. A novel regulatory mechanism of MAP kinases activation and nuclear translocation mediated by PKA and the PTP-SL tyrosine phosphatase. J Cell Biol. 1999;147(6):1129–36.
322. Muñoz JJ, Tarrega C, Blanco-Aparicio C, Pulido R. Differential interaction of the tyrosine phosphatases PTP-SL, STEP and HePTP with the mitogen-activated protein kinases ERK1/2 and p38alpha is determined by a kinase specificity sequence and influenced by reducing agents. Biochem J. 2003;372(Pt 1):193–201.
323. Bult A, Zhao F, Dirkx Jr R, Raghunathan A, Solimena M, Lombroso PJ. STEP: a family of brain-enriched PTPs. Alternative splicing produces transmembrane, cytosolic and truncated isoforms. Eur J Cell Biol. 1997;72(4):337–44.
324. Boulanger LM, Lombroso PJ, Raghunathan A, During MJ, Wahle P, Naegele JR. Cellular and molecular characterization of a brain-enriched protein tyrosine phosphatase. J Neurosci. 1995;15(2):1532–44.
325. Karim FD, Rubin GM. PTP-ER, a novel tyrosine phosphatase, functions downstream of Ras1 to downregulate MAP kinase during Drosophila eye development. Mol Cell. 1999;3(6):741–50.
326. Paul S, Lombroso PJ. Receptor and nonreceptor protein tyrosine phosphatases in the nervous system. Cell Mol Life Sci. 2003;60(11):2465–82.
327. Nguyen TH, Liu J, Lombroso PJ. Striatal enriched phosphatase 61 dephosphorylates Fyn at phosphotyrosine 420. J Biol Chem. 2002;277(27):24274–9.

328. Braithwaite SP, Adkisson M, Leung J, Nava A, Masterson B, Urfer R, Oksenberg D, Nikolich K. Regulation of NMDA receptor trafficking and function by striatal-enriched tyrosine phosphatase (STEP). Eur J Neurosci. 2006;23(11):2847–56.
329. Tashev R, Moura PJ, Venkitaramani DV, Prosperetti C, Centonze D, Paul S, Lombroso PJ. A substrate trapping mutant form of striatal-enriched protein tyrosine phosphatase prevents amphetamine-induced stereotypies and long-term potentiation in the striatum. Biol Psychiatry. 2009;65(8):637–45.
330. Kurup P, Zhang Y, Xu J, Venkitaramani DV, Haroutunian V, Greengard P, Nairn AC, Lombroso PJ. Abeta-mediated NMDA receptor endocytosis in Alzheimer's disease involves ubiquitination of the tyrosine phosphatase STEP61. J Neurosci. 2010;30(17):5948–57.
331. Paul S, Connor JA. NR2B-NMDA receptor-mediated increases in intracellular Ca^{2+} concentration regulate the tyrosine phosphatase, STEP, and ERK MAP kinase signaling. J Neurochem. 2010;114(4):1107–18.
332. Venkitaramani DV, Paul S, Zhang Y, Kurup P, Ding L, Tressler L, Allen M, Sacca R, Picciotto MR, Lombroso PJ. Knockout of striatal enriched protein tyrosine phosphatase in mice results in increased ERK1/2 phosphorylation. Synapse. 2009;63(1):69–81.
333. Venkitaramani DV, Moura PJ, Picciotto MR, Lombroso PJ. Striatal-enriched protein tyrosine phosphatase (STEP) knockout mice have enhanced hippocampal memory. Eur J Neurosci. 2011;33(12):2288–98.
334. Pelov I, Teltsh O, Greenbaum L, Rigbi A, Kanyas-Sarner K, Lerer B, Lombroso P, Kohn Y. Involvement of PTPN5, the gene encoding the striatal-enriched protein tyrosine phosphatase, in schizophrenia and cognition. Psychiatr Genet. 2012;22(4):168–76.
335. Boehm J. A 'danse macabre': tau and Fyn in STEP with amyloid beta to facilitate induction of synaptic depression and excitotoxicity. Eur J Neurosci. 2013;37(12):1925–30.
336. Goebel-Goody SM, Baum M, Paspalas CD, Fernandez SM, Carty NC, Kurup P, Lombroso PJ. Therapeutic implications for striatal-enriched protein tyrosine phosphatase (STEP) in neuropsychiatric disorders. Pharmacol Rev. 2012;64(1):65–87.
337. Fox JL, Ismail F, Azad A, Ternette N, Leverrier S, Edelmann MJ, Kessler BM, Leigh IM, Jackson S, Storey A. Tyrosine dephosphorylation is required for Bak activation in apoptosis. EMBO J. 2010;29(22):3853–68.
338. Zanke B, Suzuki H, Kishihara K, Mizzen L, Minden M, Pawson A, Mak TW. Cloning and expression of an inducible lymphoid-specific, protein tyrosine phosphatase (HePTPase). Eur J Immunol. 1992;22(1):235–9.
339. Adachi M, Sekiya M, Isobe M, Kumura Y, Ogita Z, Hinoda Y, Imai K, Yachi A. Molecular cloning and chromosomal mapping of a human protein-tyrosine phosphatase LC-PTP. Biochem Biophys Res Commun. 1992;186(3):1607–15.
340. Oh-hora M, Ogata M, Mori Y, Adachi M, Imai K, Kosugi A, Hamaoka T. Direct suppression of TCR-mediated activation of extracellular signal-regulated kinase by leukocyte protein tyrosine phosphatase, a tyrosine-specific phosphatase. J Immunol. 1999;163(3):1282–8.
341. Saxena M, Williams S, Gilman J, Mustelin T. Negative regulation of T cell antigen receptor signal transduction by hematopoietic tyrosine phosphatase (HePTP). J Biol Chem. 1998;273(25):15340–4.
342. Saxena M, Williams S, Tasken K, Mustelin T. Crosstalk between cAMP-dependent kinase and MAP kinase through a protein tyrosine phosphatase. Nat Cell Biol. 1999;1(5):305–11.
343. Nika K, Hyunh H, Williams S, Paul S, Bottini N, Tasken K, Lombroso PJ, Mustelin T. Haematopoietic protein tyrosine phosphatase (HePTP) phosphorylation by cAMP-dependent protein kinase in T-cells: dynamics and subcellular location. Biochem J. 2004;378(Pt 2):335–42.
344. Nika K, Charvet C, Williams S, Tautz L, Bruckner S, Rahmouni S, Bottini N, Schoenberger SP, Baier G, Altman A, Mustelin T. Lipid raft targeting of hematopoietic protein tyrosine phosphatase by protein kinase C theta-mediated phosphorylation. Mol Cell Biol. 2006;26(5):1806–16.
345. Wang PY, Weng J, Lee S, Anderson RG. The N terminus controls sterol binding while the C terminus regulates the scaffolding function of OSBP. J Biol Chem. 2008;283(12):8034–45.

346. Seo H, Lee IS, Park JE, Park SG, Lee Do H, Park BC, Cho S. Role of protein tyrosine phosphatase non-receptor type 7 in the regulation of TNF-alpha production in RAW 264.7 macrophages. PLoS One. 2013;8(11):e78776.
347. Gronda M, Arab S, Iafrate B, Suzuki H, Zanke BW. Hematopoietic protein tyrosine phosphatase suppresses extracellular stimulus-regulated kinase activation. Mol Cell Biol. 2001;21(20):6851–8.
348. Pettiford SM, Herbst R. The protein tyrosine phosphatase HePTP regulates nuclear translocation of ERK2 and can modulate megakaryocytic differentiation of K562 cells. Leukemia. 2003;17(2):366–78.
349. Kim HN, Lee EJ, Jung SC, Lee JY, Chung HW, Kim HL. Genetic variants that affect length/height in infancy/early childhood in Vietnamese-Korean families. J Hum Genet. 2010;55(10):681–90.
350. Fridberg M, Kjellstrom S, Anagnostaki L, Skogvall I, Mustelin T, Wiebe T, Persson JL, Dictor M, Wingren AG. Immunohistochemical analyses of phosphatases in childhood B-cell lymphoma: lower expression of PTEN and HePTP and higher number of positive cells for nuclear SHP2 in B-cell lymphoma cases compared to controls. Pediatr Hematol Oncol. 2008;25(6):528–40.
351. Zanke B, Squire J, Griesser H, Henry M, Suzuki H, Patterson B, Minden M, Mak TW. A hematopoietic protein tyrosine phosphatase (HePTP) gene that is amplified and overexpressed in myeloid malignancies maps to chromosome 1q32.1. Leukemia. 1994;8(2):236–44.
352. Eswaran J, von Kries JP, Marsden B, Longman E, Debreczeni JE, Ugochukwu E, Turnbull A, Lee WH, Knapp S, Barr AJ. Crystal structures and inhibitor identification for PTPN5, PTPRR and PTPN7: a family of human MAPK-specific protein tyrosine phosphatases. Biochem J. 2006;395(3):483–91.
353. Sergienko E, Xu J, Liu WH, Dahl R, Critton DA, Su Y, Brown BT, Chan X, Yang L, Bobkova EV, Vasile S, Yuan H, Rascon J, Colayco S, Sidique S, Cosford ND, Chung TD, Mustelin T, Page R, Lombroso PJ, Tautz L. Inhibition of hematopoietic protein tyrosine phosphatase augments and prolongs ERK1/2 and p38 activation. ACS Chem Biol. 2012;7(2):367–77.
354. Chirivi RG, Dilaver G, van de Vorstenbosch R, Wanschers B, Schepens J, Croes H, Fransen J, Hendriks W. Characterization of multiple transcripts and isoforms derived from the mouse protein tyrosine phosphatase gene Ptprr. Genes Cells. 2004;9(10):919–33.
355. van den Maagdenberg AM, Bachner D, Schepens JT, Peters W, Fransen JA, Wieringa B, Hendriks WJ. The mouse *Ptprr* gene encodes two protein tyrosine phosphatases, PTP-SL and PTPBR7, that display distinct patterns of expression during neural development. Eur J Neurosci. 1999;11(11):3832–44.
356. Hendriks WJ, Dilaver G, Noordman YE, Kremer B, Fransen JA. PTPRR protein tyrosine phosphatase isoforms and locomotion of vesicles and mice. Cerebellum. 2009;8(2):80–8.
357. Augustine KA, Silbiger SM, Bucay N, Ulias L, Boynton A, Trebasky LD, Medlock ES. Protein tyrosine phosphatase (PC12, Br7, S1) family: expression characterization in the adult human and mouse. Anat Rec. 2000;258(3):221–34.
358. Noordman YE, Jansen PA, Hendriks WJ. Tyrosine-specific MAPK phosphatases and the control of ERK signaling in PC12 cells. J Mol Signal. 2006;1:4.
359. Ogata M, Oh-hora M, Kosugi A, Hamaoka T. Inactivation of mitogen-activated protein kinases by a mammalian tyrosine-specific phosphatase, PTPBR7. Biochem Biophys Res Commun. 1999;256(1):52–6.
360. Chirivi RG, Noordman YE, Van der Zee CE, Hendriks WJ. Altered MAP kinase phosphorylation and impaired motor coordination in PTPRR deficient mice. J Neurochem. 2007;101(3):829–40.
361. Erkens M, Bakker B, van Duijn LM, Hendriks WJ, Van der Zee CE. Protein tyrosine phosphatase receptor type R deficient mice exhibit increased exploration in a new environment and impaired novel object recognition memory. Behav Brain Res. 2014;265:111–20.
362. Shi C, Zhang K, Xu Q. Gender-specific role of the protein tyrosine phosphatase receptor type R gene in major depressive disorder. J Affect Disord. 2012;136(3):591–8.

363. Nakamura F, Nakamura Y, Maki K, Sato Y, Mitani K. Cloning and characterization of the novel chimeric gene TEL/PTPRR in acute myelogenous leukemia with inv(12)(p13q13). Cancer Res. 2005;65(15):6612–21.
364. Menigatti M, Cattaneo E, Sabates-Bellver J, Ilinsky VV, Went P, Buffoli F, Marquez VE, Jiricny J, Marra G. The protein tyrosine phosphatase receptor type R gene is an early and frequent target of silencing in human colorectal tumorigenesis. Mol Cancer. 2009;8:124.
365. Laczmanska I, Karpinski P, Bebenek M, Sedziak T, Ramsey D, Szmida E, Sasiadek MM. Protein tyrosine phosphatase receptor-like genes are frequently hypermethylated in sporadic colorectal cancer. J Hum Genet. 2013;58(1):11–5.
366. Chang CC, Huang RL, Wang HC, Liao YP, Yu MH, Lai HC. High Methylation Rate of LMX1A, NKX6-1, PAX1, PTPRR, SOX1, and ZNF582 Genes in Cervical Adenocarcinoma. Int J Gynecol Cancer. 2014;24(2):201–9.
367. Su PH, Lin YW, Huang RL, Liao YP, Lee HY, Wang HC, Chao TK, Chen CK, Chan MW, Chu TY, Yu MH, Lai HC. Epigenetic silencing of PTPRR activates MAPK signaling, promotes metastasis and serves as a biomarker of invasive cervical cancer. Oncogene. 2013;32(1):15–26.
368. Lin H, Chen TC, Chang TC, Cheng YM, Chen CH, Chu TY, Hsu ST, Liu CB, Yeh LS, Wen KC, Huang CY, Yu MH. Methylated ZNF582 gene as a marker for triage of women with Pap smear reporting low-grade squamous intraepithelial lesions—A Taiwanese Gynecologic Oncology Group (TGOG) study. 2014. Gynecol Oncol.
369. Chan SW, Lim CJ, Chen L, Chong YF, Huang C, Song H, Hong W. The Hippo pathway in biological control and cancer development. J Cell Physiol. 2011;226(4):928–39.
370. Lombroso PJ, Murdoch G, Lerner M. Molecular characterization of a protein-tyrosine-phosphatase enriched in striatum. Proc Natl Acad Sci U S A. 1991;88(16):7242–6.
371. Jin YJ, Yu CL, Burakoff SJ. Human 70-kDa SHP-1L differs from 68-kDa SHP-1 in its C-terminal structure and catalytic activity. J Biol Chem. 1999;274(40):28301–7.
372. Xu Y, Banville D, Zhao HF, Zhao X, Shen SH. Transcriptional activity of the SHP-1 gene in MCF7 cells is differentially regulated by binding of NF-Y factor to two distinct CCAAT-elements. Gene. 2001;269(1-2):141–53.
373. Nystrom A, Shaik ZP, Gullberg D, Krieg T, Eckes B, Zent R, Pozzi A, Iozzo RV. Role of tyrosine phosphatase SHP-1 in the mechanism of endorepellin angiostatic activity. Blood. 2009;114(23):4897–906.
374. Yang J, Liu L, He D, Song X, Liang X, Zhao ZJ, Zhou GW. Crystal structure of human protein-tyrosine phosphatase SHP-1. J Biol Chem. 2003;278(8):6516–20.
375. Wang W, Liu L, Song X, Mo Y, Komma C, Bellamy HD, Zhao ZJ, Zhou GW. Crystal structure of human protein tyrosine phosphatase SHP-1 in the open conformation. J Cell Biochem. 2011;112(8):2062–71.
376. Critton DA, Tortajada A, Stetson G, Peti W, Page R. Structural basis of substrate recognition by hematopoietic tyrosine phosphatase. Biochemistry. 2008;47(50):13336–45.
377. Critton DA, Tautz L, Page R. Visualizing active-site dynamics in single crystals of HePTP: opening of the WPD loop involves coordinated movement of the E loop. J Mol Biol. 2011;405(3):619–29.
378. Maekawa K, Imagawa N, Nagamatsu M, Harada S. Molecular cloning of a novel protein-tyrosine phosphatase containing a membrane-binding domain and GLGF repeats. FEBS Lett. 1994;337(2):200–6.
379. Saras J, Claesson-Welsh L, Heldin CH, Gonez LJ. Cloning and characterization of PTPL1, a protein tyrosine phosphatase with similarities to cytoskeletal-associated proteins. J Biol Chem. 1994;269(39):24082–9.
380. Hendriks W, Schepens J, Brugman C, Zeeuwen P, Wieringa B. A novel receptor-type protein tyrosine phosphatase with a single catalytic domain is specifically expressed in mouse brain. Biochem J. 1995;305(Pt 2):499–504.
381. Banville D, Ahmad S, Stocco R, Shen SH. A novel protein-tyrosine phosphatase with homology to both the cytoskeletal proteins of the band 4.1 family and junction-associated guanylate kinases. J Biol Chem. 1994;269(35):22320–7.

382. van den Berk LC, van Ham MA, te Lindert MM, Walma T, Aelen J, Vuister GW, Hendriks WJ. The interaction of PTP-BL PDZ domains with RIL: an enigmatic role for the RIL LIM domain. Mol Biol Rep. 2004;31(4):203–15.
383. Walma T, Aelen J, Nabuurs SB, Oostendorp M, van den Berk L, Hendriks W, Vuister GW. A closed binding pocket and global destabilization modify the binding properties of an alternatively spliced form of the second PDZ domain of PTP-BL. Structure. 2004;12(1):11–20.
384. Walma T, Spronk CA, Tessari M, Aelen J, Schepens J, Hendriks W, Vuister GW. Structure, dynamics and binding characteristics of the second PDZ domain of PTP-BL. J Mol Biol. 2002;316(5):1101–10.
385. Cheng J, Daimaru L, Fennie C, Lasky LA. A novel protein tyrosine phosphatase expressed in lin(lo)CD34(hi)Sca(hi) hematopoietic progenitor cells. Blood. 1996;88(4):1156–67.
386. Huang K, Sommers CL, Grinberg A, Kozak CA, Love PE. Cloning and characterization of PTP-K1, a novel nonreceptor protein tyrosine phosphatase highly expressed in bone marrow. Oncogene. 1996;13(7):1567–73.
387. Gandhi TK, Chandran S, Peri S, Saravana R, Amanchy R, Prasad TS, Pandey A. A bioinformatics analysis of protein tyrosine phosphatases in humans. DNA Res. 2005;12(2):79–89.
388. Matthews RJ, Bowne DB, Flores E, Thomas ML. Characterization of hematopoietic intracellular protein tyrosine phosphatases: description of a phosphatase containing an SH2 domain and another enriched in proline-, glutamic acid-, serine-, and threonine-rich sequences. Mol Cell Biol. 1992;12(5):2396–405.
389. Cloutier JF, Veillette A. Association of inhibitory tyrosine protein kinase p50csk with protein tyrosine phosphatase PEP in T cells and other hemopoietic cells. EMBO J. 1996;15(18): 4909–18.
390. Chang HH, Tai TS, Lu B, Iannaccone C, Cernadas M, Weinblatt M, Shadick N, Miaw SC, Ho IC. PTPN22.6, a dominant negative isoform of PTPN22 and potential biomarker of rheumatoid arthritis. PLoS One. 2012;7(3), e33067.
391. Odorizzi G, Katzmann DJ, Babst M, Audhya A, Emr SD. Bro1 is an endosome-associated protein that functions in the MVB pathway in Saccharomyces cerevisiae. J Cell Sci. 2003;116(Pt 10):1893–903.
392. Sharma E, Lombroso PJ. A neuronal protein tyrosine phosphatase induced by nerve growth factor. J Biol Chem. 1995;270(1):49–53.
393. Ogata M, Sawada M, Fujino Y, Hamaoka T. cDNA cloning and characterization of a novel receptor-type protein tyrosine phosphatase expressed predominantly in the brain. J Biol Chem. 1995;270(5):2337–43.
394. Shiozuka K, Watanabe Y, Ikeda T, Hashimoto S, Kawashima H. Cloning and expression of PCPTP1 encoding protein tyrosine phosphatase. Gene. 1995;162(2):279–84.
395. Szedlacsek SE, Aricescu AR, Fulga TA, Renault L, Scheidig AJ. Crystal structure of PTP-SL/PTPBR7 catalytic domain: implications for MAP kinase regulation. J Mol Biol. 2001;311(3):557–68.
396. Floor SL, Dumont JE, Maenhaut C, Raspe E. Hallmarks of cancer: of all cancer cells, all the time? Trends Mol Med. 2012;18(9):509–15.

Role of *PTPN11* (SHP2) in Cancer

Gordon Chan and Benjamin G. Neel

Abstract Src homology-2 domain-containing phosphatase 2 (SHP2), encoded by the *PTPN11* gene, is a highly conserved, non-transmembrane protein-tyrosine phosphatase (PTP), found in all metazoans. The molecular details of SHP2 regulation by phosphotyrosyl (pTyr) peptide ligand binding are well-understood, and knowledge of these details is critical to understanding SHP2 function in health and disease. Studies using mice with gain- or loss-of-function alleles of *Ptpn11* have provided much detail about the physiological functions and signaling pathways regulated by SHP2 at the cellular and whole organism levels. Germline mutations in *PTPN11* cause Noonan syndrome, Noonan syndrome with multiple lentigines (previously, LEOPARD syndrome), as well as the cartilage tumor syndrome, metachondromatosis. Somatic *PTPN11* mutations occur in several types of hematologic malignancy, most notably juvenile myelomonocytic leukemia and, more rarely, in neuroblastoma and other solid tumors. *PTPN11* is crucial for transformation initiated by mutant receptor-tyrosine kinases (RTKs) and is an important effector of *H. pylori* virulence. However, the direct target(s) of SHP2 responsible for its physiological and pathological effects remain controversial and their identification remains a major goal for the future research.

Keywords Protein-tyrosine phosphatase (PTPs) • Receptor-tyrosine kinases (RTKs) • Cytokine receptors • Signal transduction • Cell signaling • Noonan Syndrome • Juvenile Myelomonocytic Leukemia • Metachondromatosis

---

G. Chan (✉)
Department of Pharmaceutical Sciences, University of Toronto, Toronto, ON, Canada

B.G. Neel
Laura and Isaac Perlmutter Cancer Center, New York, NY, USA

B.G. Neel, N. Tonks (eds.), *Protein Tyrosine Phosphatases in Cancer*,
DOI 10.1007/978-1-4939-3649-6_4

Introduction

Src homology-2 (SH2) domain-containing phosphatase 2 (SHP2) belongs to a small sub-family of non-receptor PTPs with two human members, SHP1 (encoded by *PTPN6*) and SHP2 (encoded by *PTPN11*). All vertebrates appear to have orthologs for both SHPs, whereas single *PTPN11*-like genes are found in *Drosophila* (*csw*) and *Caenorhabditis elegans* (*ptp-2*). An exception is provided by zebrafish, whose genome contains two genes encoding SHP2-like proteins with distinct functions and patterns of expression [1]. In mammals, SHP1 expression is largely restricted to cells of lympho-hematopoietic and epithelial origin (including neuroepithelia), whereas SHP2 is expressed more broadly [2].

SHP2 has two SH2 domains at its N-terminus (hereafter, N-SH2 and C-SH2), followed by a classical protein-tyrosine phosphatase (PTP) catalytic domain and a C-terminus containing a proline-rich stretch. Phosphorylation of SHP2 at Y^{542} and Y^{580} (and also at Y^{63} and Y^{279}) by receptor- and non-receptor protein-tyrosine kinases (PTKs) is important for some aspects of SHP2 function (see detailed discussion below). Moreover, recent evidence suggests that the C-terminal proline-rich residues can promote nuclear accumulation of SHP2 by mediating interaction with YAP/TAZ-containing complexes [3]. However, the detailed mechanism by which the C-terminus affects SHP2 function in different biological contexts remains unclear.

In vertebrates, SHP2 is expressed ubiquitously, although at variable levels in different tissues [4–7]. Splice variants within the PTP domain of SHP2 and Csw have been reported [4, 7, 8], and were shown to have different PTP activities [8], but their physiological significance has not been assessed further.

In its basal state, SHP2 activity is suppressed by intramolecular interactions between residues in the "backside loop" of the N-SH2 domain (the side opposite the phosphotyrosyl peptide binding pocket) and the catalytic surface of the PTP domain ([9]; reviewed in [2, 10]). Upon growth factor or cytokine stimulation, SHP2 is recruited, via its SH2 domains, to phosphorylated tyrosine residues on receptor-tyrosine kinases (RTKs), cytokine receptors, scaffolding adaptors, such as insulin receptor substrate (IRS), fibroblast growth factor receptor substrate (FRS), or GRB2-associated binders (GAB) proteins, and/or immune inhibitory receptors.

Engagement of the N-SH2 domain with a phosphotyrosyl (pTyr-) peptide disrupts the autoinhibitory interface, leading to exposure of the PTP domain and catalytic activation [9–11]. This elegant "molecular switch" mechanism ensures that recruitment of SHP2 to appropriate cellular locales is coupled directly to its catalytic activation. The physiological significance of this regulatory mechanism was first demonstrated by engineered mutations in the backside loop of SHP2; such mutations disrupt the SH2/PTP interface and generate mutants with enhanced catalytic and biological activity [12]. As discussed below, this mechanism is subverted in a similar way in human diseases caused by *PTPN11* mutations.

SHP2 in Receptor Signaling

Role in RAS/ERK Activation

SHP2 is required for full activation of the RAS/ERK MAP kinase (hereafter, RAS/ERK) pathway in most, if not all, RTK signaling pathways (see review [2]). For some RTKs, and in some cell types (e.g., IGF-1 signaling in fibroblasts), there is virtually no ERK activation in the absence of SHP2; in others, initial ERK activation is normal, but sustained ERK signaling is markedly diminished [13–16].

Following growth factor stimulation, SHP2 binds directly to some RTKs (e.g., PDGFR), and/or to scaffolding adapters, such as GAB1 [17, 18], IRS family proteins [19], and FRS2 [20]. Studies using fibroblasts derived from *Gab1*$^{-/-}$ [21, 22] and *Frs2*$^{-/-}$ [23] mice and experiments with chimeric and mutant forms of scaffolding adapters [24, 25] indicate that these scaffolding adapter/SHP2 complexes mediate SHP2-dependent ERK activation.

There is a general agreement that SHP2 acts upstream of RAS, as cells expressing dominant negative SHP2 [26], *Ptpn11* Ex3$^{-/-}$ fibroblasts [13, 16], and SHP2^{null} fibroblasts show defective RAS activation in response to activation of multiple RTKs [27, 28]. However, the mechanism by which SHP2 mediates RAS activation remains controversial. In one model, the C-terminus of SHP2, upon RTK-mediated tyrosyl phosphorylation, serves as a docking site for GRB2/SOS, resulting in RAS activation [29–31]. In addition to GRB2/SOS recruitment, phosphorylation at these residues is proposed to relieve basal inhibition and thereby promote SHP2 activation [32]. The latter experiments, however, employed engineered constructs of SHP2 with non-hydrolyzable pTyr mimetics, which might not accurately reproduce the effects of C-terminal tyrosyl phosphorylation *in vivo*. Furthermore, cis-interaction between the phosphorylated C-terminus and the SH2 domains is altered by GRB2 binding [33]. Thus, GRB2 recruitment might regulate catalytic and non-catalytic functions of SHP2. In any event, the "adapter" mechanism alone cannot explain how SHP2 contributes to RAS activation in all RTK pathways, as SHP2 catalytic activity is still required in these contexts. However, this model would suggest that dephosphorylation of the C-terminal tyrosines is crucial for regulating SHP2 activity. Auto-dephosphorylation is suggested to be a major mechanism of C-terminal tyrosine dephosphorylation [29, 34], although it remains unclear whether SHP2 is the only PTP that dephosphorylates these sites.

The second general model for SHP2 function in RAS/ERK activation focuses on the SHP2 catalytic domain. The structure of SHP2 and its role in regulation of RAS/ERK activation are evolutionarily conserved; hence, it seems reasonable that there is/are common substrate(s) that accounts for its role in RAS activation. However, the identity of the critical SHP2 substrate(s) remains controversial.

RASGAP binding sites on RTKs were among the earliest reported targets of SHP2. The Kazlauskas group first reported that SHP2 preferentially dephosphorylates the RASGAP binding site (Y721) on PDGFRβ, preventing precocious recruitment of RASGAP following PDGF stimulation [35]. Analogous findings, supported by genetic analysis, were reported in studies of signaling by Torso (which is structurally related to the PDGFR) in *Drosophila melanogaster* [36]. Subsequent workers reported that RASGAP sites on EGFR [37] and HER2 [38] also were targeted by SHP2. Importantly, the former study showed that mutation of the RASGAP binding site in EGFR could abrogate the effects of a dominant negative mutant of SHP2 on EGF-evoked RAS/ERK activation [37], whereas the latter report [38] showed the RASGAP binding site on HER2 is required for HER2 binding to an SHP2 substrate-trapping mutant. This finding is consistent with the previous work that showed mutation of this same site in HER2/NEU results in enhanced transforming capacity [39]. These data are, however, inconsistent with the known requirement for GAB1 and, in particular, GAB1/SHP2 binding, in signaling by EGFR family members.

Indeed, Montagner et al. reported that RASGAP can bind to multiple sites on GAB1, sites that also are targets of SHP2 [40]. Moreover, deletion of the region containing these sites was able to abrogate the requirement for SHP2 binding to GAB1 for RAS/ERK activation in their cell system. Notably, these authors did not show direct binding of GAB1 and RASGAP, leaving open the possibility of an indirect interaction. Moreover, their results conflict with a previous report showing that GAB1 regulates EGF-induced RAS-guanine nucleotide exchange (GEF), not RASGAP activity [41]. Furthermore, studies of the *Drosophila* GAB ortholog Dos indicate that its SHP2 binding sites are required for Sevenless signaling even when all other Dos tyrosines are absent [42]. Finally, as mentioned above, it is unclear how the studies of Montagner et al. can be reconciled with the results of Agazie and coworkers discussed above.

Conceivably, differences in cellular context, and, in particular, the relative level of expression of EGFR, ERBB3, and GAB1 could affect the relative requirement for SHP2 action on the EGFR versus GAB1. Consistent with this notion, Schlessinger's laboratory has reported that cells can differ in their relative requirement for GAB1 and HER3 in PI3K activation [43]. Alternatively, or in addition, SHP2 could have effects on both RAS-GEF and -GAP activity. For example, the adaptor function of SHP2 could function to enhance RAS exchange (via GRB2/SOS recruitment), while SHP2 catalytic activity prevents precocious RASGAP activation (see Fig. 1). A further possibility is suggested by a very recent report that RAS is tyrosine phosphorylated at Y^{32} by SRC; this reportedly leads to decreased RAF binding, enhanced GAP recruitment, and increased GTP hydrolysis [44]. It would be interesting to determine whether SHP2 can dephosphorylate this residue and directly regulate RAS activity.

A distinct set of candidate substrates of SHP2 are the binding proteins for CSK, a negative regulator of SRC-family kinase (SFK) activity. Cells expressing a GAB1/SHP2 fusion exhibit enhanced SRC activity, and SRC inhibitors block the ability of this fusion protein to activate ERK [45]. Furthermore, activation of SFKs in response to several growth factors is defective in *Ptpn11*-deficient cells [27, 28, 46, 47]. These effects are attributed to the failure of SHP2 to dephosphorylate binding proteins for CSK, such as CBP/PAG and PAXILLIN [27, 48], leading to enhanced negative regulation of SFK activation in $SHP2^{null}$ cells. However, it remains to be established whether this is a general model for action of SHP2 in growth factor signaling.

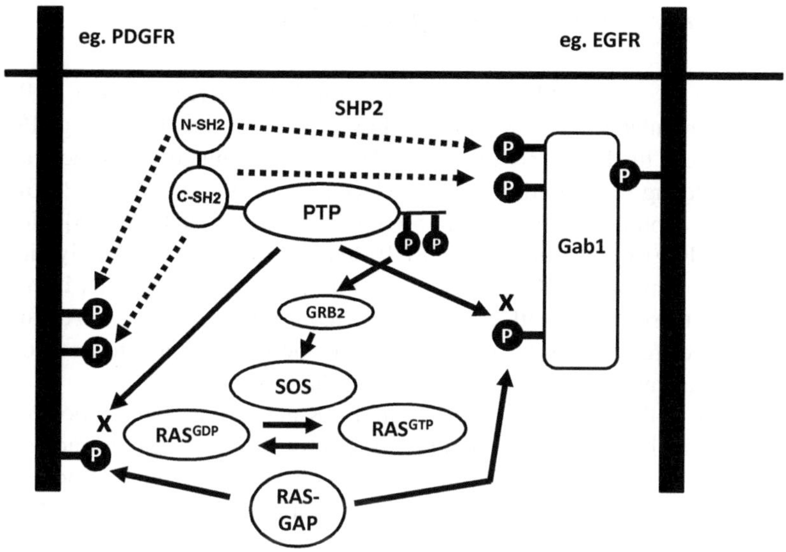

Fig. 1 Model for SHP2 action in RAS activation. SHP2 is recruited to some RTKs (e.g., PDGFR) directly and/or to one or more scaffolding adaptors (e.g., GAB proteins, IRS proteins, FRS proteins). In the example shown, the EGFR signals via GAB1. Phosphorylation of C-terminal tyrosines, catalyzed either directly by some RTKs or indirectly via Src-family kinases (SKFs) leads to recruitment of GRB2/SOS complex, which in turn facilitates RAS exchange. In parallel, SHP2 dephosphorylates RASGAP binding sites on RTKs or GAB1, thus preventing precocious RASGAP activation. The exact order of these events and/or whether they occur downstream of all RTKs remains unclear. See text for details

Other candidates for SHP2-mediated RAS/ERK activation include the SPROUTY and SPRED proteins, which are inhibitors of the RAS/ERK pathway. Transfection studies suggest that SHP2 binds to and dephosphorylates SPROUTY proteins and counteracts their inhibitory potential [49, 50]. Consistent with this idea, genetic experiments place *Spty2* downstream of, and negatively regulated by, *Ptpn11* in lens and lacrimal gland development [51]. However, the detailed effects of tyrosyl phosphorylation and the mechanism of action of the SPROUTY family are not well-understood [52]. The related SPRED proteins [53, 54] are also tyrosyl phosphorylated inhibitors implicated in negative regulation of RAS/ERK activation [53–56]. In cell expressing SPRED1, SPRED1-Y^{420}, a site shown to be important for inhibiting RAS/ERK activation, was reported to be a target of SHP2 [57]. Thus, SHP2 may promote the suppression of a negative-feedback pathway triggered by RAS/ERK activation. These observations could be consistent with the loss-of-function mutations of *SPRED1* that cause a neurofibromatosis (NF)-like syndrome in humans [58, 59]. As discussed below, *PTPN11* gain-of-function mutations cause human diseases that share many features with NF.

Roles in Other Downstream Signaling Pathways

In addition to its well-established requirement in RAS/ERK activation, SHP2 also differentially regulates the PI3K pathway downstream of many RTKs (e.g., PDGFR, IGFR, FGFR). In the EGFR pathway, SHP2 negatively regulates PI3K activation by dephosphorylating PI3K binding sites on GAB1 [14, 43]. Similarly, in PDGFR signaling, SHP2 dephosphorylates PI3K binding sites on the PDGFR [35]. However, in FGFR and IGF1R signaling, loss of SHP2 leads to impaired PI3K activation [46]. Thus, SHP2 promotes PI3K activation in some contexts, but inhibits it in others. Conceivably, these effects reflect differential requirement of active RAS for PI3K activation in different cellular contexts and/or the differential presence of PI3K binding sites in various receptors/scaffolding adaptors.

In addition to the regulation of ERK and PI3K activation, SHP2 reportedly suppresses S6K1 activation by promoting AMPK activation [60], and promotes activation of JNK [13, 61, 62], NF-κB [63], RHO [64, 65], and NFAT [66] activation in various settings. SHP2 also moderates the activation of FAK and the JAK/STAT pathway in a context-specific manner (see discussion in [2]). The effects of SHP2 on RHO regulation probably reflect its ability to dephosphorylate p190RHOGAP-B [65]. More recently, FER was suggested to be a downstream target of SHP2 in zebrafish [67]. The detailed mechanism by which SHP2 affects these pathways awaits further experimentation.

Finally, SHP2 activity is reported to be regulated by tyrosine phosphorylation at Y^{63} and Y^{279} by ABL and ARG [68]. Whereas Y^{63} resides in the N-SH2/PTP domain interface, Y^{279} is in the PTP domain. Although $SHP2^{Y63F}$ and $SHP2^{Y279F}$ showed enhanced and reduced phosphatase activity, respectively, overexpression of either mutant reportedly leads to increased basal, EGF- and PDGF-evoked ERK activation [68]. By contrast, overexpression of $SHP2^{Y279C}$, a common mutation in Noonan syndrome with multiple lentigenes (see below), inhibits EGF-induced ERK activation [69–71]. Whereas the expression of $SHP2^{Y63F}$ reduces PDGF-mediated proliferation, cells expressing $SHP2^{Y279F}$ show enhanced proliferation [68]. As $Y \rightarrow F$ substitution may not mimic the function SHP2 in its unphosphorylated state, further studies are required to elucidate the detailed mechanism by which phosphorylation at these residues regulate RAS/ERK activation and mitogen-mediated proliferation. It is also important to determine whether these roles are conserved in different cell types and in response stimulation by a wider panel of agonists.

PTPN11 in Normal Hematopoiesis

Studies of ES cell differentiation [72–75] and *Rag2*-deficient blastocyst complementation assays [76] suggested a role for SHP2 in promoting mesoderm differentiation and hematopoiesis. Interpretation of these initial studies was complicated by the use of a targeted *Ptpn11* allele ("Ex3") that encodes an N-terminally truncated SHP2 protein with increased PTP activity, which could have hypomorphic *and/or* neomorphic effects. Whereas haploinsufficiency of *Ptpn11* causes reduced self-renewal of bone marrow (BM) progenitor cells [77], homozygous loss of *Ptpn11* in hematopoietic cells leads to BM aplasia and early lethality in mice [78, 79]. *Ptpn11*-deficient mice have markedly reduced hematopoietic stem cells (HSCs) and immature progenitors of all hematopoietic lineages. Consistent with these observations, knockdown of *PTPN11* in human $CD34^+$ cord blood (CB) cells leads to diminished growth factor mediated proliferation and differentiation [80].

Ptpn11-deficient murine BM progenitors and human $CD34^+$ CB cells have defective ERK and AKT activation in response to growth factor stimulation [78–80]. Consistent with RAS as a key downstream effector of SHP2, expression of a gain-of-function *KRas* allele, $KRas^{G12D}$, can substitute for SHP2 function in BM progenitor cells [78]. *Ptpn11*-deficient HSC and progenitors undergo apoptosis with a concomitant increase in the expression of *Noxa* mRNA in a p53-independent manner [78]. However, *Noxa* deletion alone is not sufficient to restore the survival and maintenance of *Ptpn11*-deficient BM progenitors (G. C. and B. G. N., unpublished). $Ptpn11^{-/-}$BM progenitors also show reduced levels of *Kit* and *Gata2* mRNA [79]. In human CB cells, reduced *PTPN11* expression causes diminished accumulation of MCL1 and BCL_{XL} [80]. Collectively, these changes result in diminished proliferation and survival of SHP2-deficient cells.

Whereas SHP2 regulates RAS/ERK activation in the cytosol, it also can dephosphorylate nuclear proteins, such as IRF8/ICSBP [81] and RUNX1 [82], and regulate their function in hematopoietic cells. Dephosphorylation of IRF8/ICSBP inhibits *Nf1* transcription [81] and promotes RAS/ERK activation. Consistent with this notion, $Irf8^{-/-}$ mice [83] develop a myeloproliferative disorder similar to that seen in mice expressing leukemogenic SHP2 mutants (see below). However, IRF8/ICSBP deficiency also cooperates with an active *Ptpn11* allele to produce AML in mice [84]. The latter data are inconsistent with a simple enzyme/substrate interaction for SHP2 and IRF8/ISCBP and instead suggest that they have non-overlapping functions in this context.

SHP2 was also reported to dephosphorylate RUNX1 in its negative inhibitory/autoinhibitory domains, thereby promoting megakaryocyte and CD8 T cell maturation [82]. It is not clear, however, if this role is critical for the maturation of HSCs and progenitors, as *Runx1* deletion has only minimal effects on these cells [85–87]. Notably, in contrast to the effects of RUNX1 deficiency, loss of *Mapk3* (ERK1) and *Mapk1*(ERK2) in hematopoietic cells leads to rapid reduction of HSC and BM progenitors [88, 89]. Hence, the failure to activate ERK1 and ERK2 likely accounts for most, if not all, of the phenotypes in *Ptpn11*-deficient mice.

In addition to IRF8/ICSBP and RUNX1, SHP2 was reported to dephosphorylate PARAFIBROMIN/CDC73 (PF), and thereby promote enhanced β-catenin-mediated gene transcription [90]. Dephosphorylation of PF required SHP2 nuclear translocation, a process recently shown to be facilitated by SHP2 binding to dephosphorylated YAP/TAZ [3]. Despite these purported nuclear functions of SHP2, whether and how these actions contribute to roles of SHP2 in the maintenance of normal hematopoiesis and leukemogenesis (see below) awaits further investigation.

PTPN11 in NS and JMML

PTPN11 mutations occur in ~40 % of patients with Noonan Syndrome (NS), a relatively common (incidence ~1:2000 births), autosomal dominant developmental disorder characterized by facial dysmorphia, webbed neck, proportionate short stature, and cardiac abnormalities [91]. Chest and spine deformities, cognitive defects, delayed puberty, cryptorchidism, and bleeding diathesis also occur with variable penetrance [92–94], and NS is associated with increased incidence of malignancy, particularly juvenile myelomonocytic leukemia (JMML) and possibly neuroblastoma [95–98]. NS patients without *PTPN11* mutations have either gain-of-function mutation in *KRAS* (~1–2 % overall) [99, 100], *NRAS* (~1–2 %) [101, 102], *SOS1* (~10 %) [103, 104], *RAF1* (~5 %) [105, 106], *RIT1* (~9 %) [107, 108], *RASA2* (~2 %) [108], or *MAPK2K1* (rare) [108]. In addition, mutation in *SHOC*, a regulator of RAS/ERK signaling, was reported in patients with a NS-like disease [109]. These studies strongly suggest that NS is a disease caused by excessive RAS/ERK pathway activation. Although the gene(s) responsible for ~10–15 % of NS cases remain to be identified, lesions in other components of the RAS/ERK pathway will likely be uncovered by ongoing sequencing efforts.

Rare NS patients bearing specific germline *PTPN11* mutations develop JMML, a clonal pediatric myeloproliferative neoplasm (MPN) characterized by massive expansion and tissue infiltration of myeloid cells, often accompanied by macrocytic anemia with persistent fetal hemoglobinemia [110–112]. Furthermore, somatic *PTPN11* mutations are found in ~35 % of sporadic JMML patients [113, 114] and at lower incidence in other myeloid neoplasms, including acute myelogenous leukemia (AML), chronic myelomyelogenous leukemia (CMML), and myelodysplastic syndrome (MDS) [113–116], as well as in B-ALL [117, 118]. *PTPN11* mutations occur rarely in solid tumors [116, 119], although *PTPN11* is the third most commonly mutated gene in neuroblastoma, with an incidence of ~3 % [116, 120].

Most NS and leukemia mutations in *PTPN11* affect the N-SH2 or PTP domain residues in SHP2 involved in basal inhibition [2, 91, 111]. Germline *PTPN11* mutations associated with JMML are less conservative than NS mutations as a group, and generally result in higher degrees of SHP2 activation [70, 121, 122]. Somatic leukemia-associated *PTPN11* alleles also are less conservative than NS mutants and encode more highly activated versions of SHP2 [70, 121, 122].

Substantial evidence has established that *PTPN11* is a *bona fide* oncogene. Expression of leukemia-associated *PTPN11* mutants in fetal liver (FL) or bone marrow (BM) cells confers cytokine-independent myeloid colony formation, hypersensitivity to IL-3 and GM-CSF, increased colony size and survival of myeloid progenitors, and growth factor-mediated activation of ERK, AKT, and STAT5 [123–125]. Consistent with these findings, differentiation of iPS cells derived from JMML patients with a *PTPN11* mutation gives rise to myeloid cells with GM-CSF hypersensitivity and enhanced ERK and STAT5 activation compared with iPS-derived myeloid cells individuals with WT *PTPN11* alleles [126]. Interestingly, myeloid transformation evoked by the expression of the active *Ptpn11* mutant requires the C-terminal Y^{542} [124]. Hence, phosphorylation at C-terminal tyrosine residues is likely important for SHP2-mediated leukemogenesis.

Similar to NS, de-regulation of the RAS/ERK pathway is crucial for JMML pathogenesis, as patients without *PTPN11* mutations usually have either homozygous deficiency of *NF1* or somatic mutations in *CBL*, *NRAS*, or *KRAS* [110, 111], all of which regulate RAS activation [127, 128]. Also, the myeloid diseases evoked by active *PTPN11* expression are similar to those in mice with loss of *Nf1* [129–133] or expressing gain-of-function *KRas* [134–137] or *NRas* [138–140] alleles. Furthermore, gene-ablation or pharmacological inhibition of ERK1 and ERK2 blocks MPN evoked by expression of oncogenic *KRas* [88, 89, 141], *NRas* [142], or the loss of *Nf1* [143]. Collectively, these data indicate that JMML is largely a disease of excessive ERK activation in hematopoietic cells.

Transplantation of BM from BALB/c mice induced to express *Ptpn11*^{E76K} or *Ptpn11*^{D61Y} by retroviral gene transduction leads to a fatal and invasive MPN in ~60 % of the recipients, whereas ~40 % of the recipients develop T-cell leukemia/lymphoma or a mixture of the latter disease and MPN [124]. Similar retroviral transduction–transplantation experiments using transduced FL or BM cells from C57BL/6 mice failed to evoke MPN, although some mice did develop anemia with splenic infiltration along with abnormal CFU-GM [125]. The differences between the two studies suggest that mouse-strain specific genetic modifier(s) might accelerate or suppress disease development. Interestingly, *SETBP1* and *JAK3* mutations were reported recently to be secondary events in JMML patients [144]. Whether these mutations cooperate with aberrant RAS pathway activity in JMML pathogenesis, and if so, whether their mouse orthologs account for (or contribute to) strain-specific differences in myeloid phenotype, merits further investigation.

Knock-in mice bearing *Ptpn11*^{D61G} show most of the hallmark features of NS, as well as MPN that is initially well-tolerated but frequently progresses to a fatal disease [145]. Mice expressing conditional alleles for the JMML-associated mutants *Ptpn11*^{D61Y} [146] or *Ptpn11*^{E76K} [147] develop a fatal and fully penetrant MPN. HSCs and immature progenitors are reduced in the BM of these mice. Within this population, the proportion of G_0 cells is reduced, with a concomitant increase in the frequency of cells in S and G_2/M. Erythroid-specific expression of the leukemogenic *Ptpn11*^{D61Y} allele leads to ERK-dependent excess production and accumulation of erythroblasts, but not erythrocytosis, suggesting that mutant SHP2 also impairs erythroid differentiation [148].

Despite several similarities, however, mice expressing *Ptpn11*^{D61Y} and *Ptpn11*^{E76K} in their hematopoietic cells also show some distinct phenotypes. Although both lines of mice develop T-ALL in addition to MPN, a large proportion of *Ptpn11*^{E76K} mice develop B-ALL and AML [147]. The latter observation was a surprise, as B-ALL and AML have not been reported in mice expressing *KRas*^{G12D} [134–137] or *NRas*^{G12D} [138–140], which are presumably more potent activators of the RAS/ERK pathway. Moreover, mutation of genes that affect growth factor/cytokine signaling alone typically fails to cause AML; typically, cooperating mutations affecting hematapoietic cell differentiation are required [149].

The reasons for these differences in phenotype in these knock-in mice remain unclear. When tested *in vitro*, recombinant *Ptpn11*^{E76K} shows a greater increase in catalytic activity compared with *Ptpn11*^{D61Y} [113, 121], although both are highly activated. Conceivably, this difference in activity could affect certain key substrates differentially (e.g., hematopoietic transcription factors?) and/or differentially regulate modifying alleles (such as those discussed above) in disease development. In addition, chromosomal instability in *Ptpn11*^{E76K} cells (see below) might facilitate the acquisition of additional genetic aberrations that can promote B-ALL and AML development [147], although again it is unclear why *Ptpn11*^{D61Y}, which is activated to almost the same extent, would not share these effects.

When expressed under the control of its endogenous promoter, oncogenic *Ptpn11* promotes transformation through hyperactivation of ERK and AKT [146, 147]. However, *Ptpn11*^{E76K} T-ALL cells and multipotent hematopoietic progenitors also show increased aneuploidy and centrosomal amplification, suggesting the involvement of SHP2 in maintenance of chromosomal stability [147], albeit by a mechanism yet to be delineated. Expression of *Ptpn11*^{D61G} or *Ptpn11*^{E76K} reportedly leads to increased ROS production in myeloid progenitors but not in multipotent progenitor-enriched lin$^-$KIT$^+$SCA1$^+$ cells. *Ptpn11* mutant macrophages show enhanced O_2 consumption and aerobic glycolysis, either or both of which could contribute to the increase in ROS production in these cells [150].

Increased ROS levels, O_2 consumption, and mitochondrial respiration are also observed in *Ptpn11*^{E76K} primary mouse embryonic fibroblast (MEFs) [151]. However, in contrast to the enhanced proliferation of *Ptpn11*^{E76K} myeloid cells, *Ptpn11*^{E76K} MEFs show cell cycle arrest and senescence. How *Ptpn11*^{E76K} evokes enhanced mitochondrial function/ROS production, and whether these processes result from enhanced RAS/ERK activation, remain unclear. Furthermore, senescence caused by *Ptpn11*^{E76K} expression in MEFs is a sharp contrast to the enhanced proliferation of MEFs expressing an endogenous allele of *KRas*^{G12D} [152]. Further experiments are required to compare and contrast the molecular events triggered by these oncogenes.

Expression of activating mutants of *PTPN11* also causes reduced *GATA2* expression, increased JNK activation, and increased *JUN* accumulation in myeloid progenitors [62, 153]. Whether these events are caused by RAS/ERK hyperactivation remains to be established. In these myeloid cells, mutant SHP2 was reported to directly dephosphorylate STAT3, and thereby regulate its activation [154]. In erythroid progenitor cells, however, oncogenic SHP2 promotes STAT3 activation [148]. Thus, the mechanism in which STAT3 activation is regulated by SHP2 may be cell-type specific and context-dependent (see Neel *et al.* [2] for further discussion). Collectively, these observations suggest that signaling aberrations in non-RAS/ERK pathways might contribute to the cell-type specific phenotypes caused by mutant *PTPN11*.

As ERK hyperactivation is common to, and apparently critical for, myeloid diseases evoked by expression of oncogenic alleles in the RAS/ERK pathway, dampening activation of this pathway would seem to be an obvious approach for treating JMML. Indeed, MEK inhibitor (MEK-I) treatment of mice expressing oncogenic *KRas* or with homozygous loss of *Nf1* leads to amelioration of myeloid disease and improved survival [141, 143]. Nevertheless, mutant clones persist in these animals [141, 143]. These data argue that JMML stem cells are resistant to MEK inhibition, and suggest that blockade of alternative pathways may provide additive therapeutic value. Indeed, combined MEK-I and rapamycin treatment causes synergistic reduction of myeloid colony formation by BM cells expressing oncogenic *Ptpn11* [124]. Furthermore, co-expression of a catalytically inactive form of PI3K p110δ, but not p110α, normalizes MPN features in mice expressing the $Ptpn11^{D61Y}$ allele [155], and consistent with these genetic results, pharmacological blockade of p110δ cooperates with MEK-I treatment in reducing GM-CSF hypersensitivity in mouse cells and GM-CSF-mediated proliferation of BM cells from JMML patients [155]. Thus, blockade of both the ERK and PI3K pathways is likely to be a more effective therapeutic strategy in the treatment of JMML. Whether the combined MEK/PI3K blockade results in complete JMML clonal elimination *in vivo* awaits clarification.

Aberrant activation of the JAK/STAT pathway also is important in JMML. SHP2 has been reported to positively and negatively regulate JAK and/or STAT activation in pathway-specific and/or cell-context dependent manners ([2] and STAT3 discussion above). In hematopoietic cells, SHP2 positively regulates STAT5 activation. For example, immortalized yolk-sac hematopoietic cells from $Ex3^{-/-}$ mice show impaired IL-3 induced JAK2/STAT5 activation [156]. Furthermore, Ba/F3 [157] or bone marrow mast cells [124] that over-express leukemia-associated "activated" mutants of human *PTPN11* ($SHP2^{E76K}$) show enhanced IL3-evoked STAT5 phosphorylation. Also, as discussed above, myeloid cells expressing gain-of-function alleles of *PTPN11* show enhanced GM-SCF-mediated STAT5 activation [126, 146], as do BM cells from JMML patients [158–160]. Interestingly, combined pharmacologic blockade of JAK and MEK attenuate myeloid disease development in $NRas^{G12D}$ mice and prolong their survival [161]. Thus, it is possible, if not likely, that JAK inhibitors will ameliorate myeloid disease driven by SHP2 mutants.

Any effective therapy requires a complete eradication of all self-renewing cells expressing the oncogenic allele. To date, the identity of such cells in JMML remains to be established. Identification of these cells and their susceptibility to therapy invention remains an important topic for future research.

Requirement for SHP2 in Other Leukemias and Solid Tumors

Although *PTPN11* mutations are, overall, rare contributors to oncogenesis, SHP2 dysregulation plays a much broader role in cancer. Just as SHP2 is required for normal RTK signaling, it is essential for transformation initiated by RTK mutation/amplification. Loss of *Ptpn11* or treatment with a putative small molecule inhibitor of SHPs (SHP2i) reduces proliferation and survival of hematopoietic cells expressing KIT^{D814V} [162, 163] or FMS-like tyrosine kinase 3 with internal tandem duplications (FLT3-ITD) mutation [164]. KIT^{D814V} recruits p85α, GAB2, and SHP2 through KIT^{D814V}-Y^{719} [162]. In FLT3-ITD transformed cells, duplication of the region surrounding Y^{599} in FLT3-ITD allows for enhanced recruitment of SHP2, which contributes to STAT5 hyperactivation [164]. However, whether STAT5 is directly downstream of SHP2 awaits further investigation. In $FLT3\text{-}ITD^{+}$ cells, loss of *PTPN11* leads to reduced transcriptional activity of *BCL2L1*. Surprisingly, SHP2 and STAT5 were found to co-localize to the nucleus, and co-occupy the human *BCL2L1* promotor [164]. It is not clear whether this event requires the catalytic activity of SHP2; regardless, these data suggest SHP2 might have a direct role in facilitating gene expression in FLT3-ITD^{+} AML cells. Given that *PTPN11* mutations are reported in human AML ([165] and reviewed in [166]), it is likely some of these mechanisms will be involved in mediating leukemogenesis caused by oncogenic *PTPN11* alleles.

The SHP2 binding protein GAB2 is a critical determinant of the cell-lineage and latency of BCR/ABL-evoked leukemia [167]. Monobody inhibitor experiments suggest that the N-SH2 domain of SHP2, in particular, is required for binding to GAB2 in BCR–ABL-expressing cell lines [168]. *Gab2*$^{-/-}$BM cells exhibit defective activation of the PI3K/AKT and ERK pathways, and are resistant to myeloid transformation [167]. Experiments with *Gab2* mutants indicate that these results reflect loss of GAB2/SHP2 interactions. Consistent with these observations, *Ptpn11*Ex3$^{-/-}$yolk-sac (YS) cells also are refractory to BCR–ABL transformation [169]. Furthermore, knockdown of *GAB2* or *PTPN11* impairs proliferation of human CML blasts [170]. Thus, SHP2 is crucial for BCR–ABL myeloid transformation. Whether it has a similar role in BCR–ABL-transformed lymphoid cells awaits further investigation.

In addition to its role in BCR/ABL-evoked disease, GAB2 is required for Friend virus /STK-evoked erythroleukemias in mice [171]. This finding is consistent with the previous work that established a positive role for GAB2 in v-SEA (the chicken-viral ortholog of STK)-evoked transformation of mouse fibroblasts [172].

Inappropriate activation of SHP2 also is critical in solid tumor pathogenesis. *H. pylori*, the major cause of gastric ulcer and gastric carcinoma worldwide, encodes several virulence determinants, which include the *CagA* gene. SRC-family kinases phosphorylate CagA on C-terminal EPIYA motifs, resulting in recruitment and activation of SHP2 [173]. *H. pylori*-infected or CagA-transfected gastric epithelial cells exhibit morphological changes that have been termed the "hummingbird phenotype" [174, 175], and *PTPN11* knockdown impairs the development of this phenotype, highlighting a crucial role for SHP2 gastric epithelial cell transformation [173]. Transgenic mice expressing *CagA* develop gastric hyperplasia, which progresses to frank adenocarcinoma in ~10 % of mice after a long (>1 year) latent period [176]. These mice also develop MPN similar to that caused by a leukemia-associated allele of *Ptpn11* [176], suggesting that CagA, when expressed ubiquitously, can facilitate gratuitous SHP2 activation in hematopoietic cells. Consistent with the requirement for SHP2 in CagA function, neither gastric carcinoma nor MPN developed in transgenic mice expressing CagA lacking its SHP2 binding sites. Thus, these data strongly argue that CagA is effectively a "molecular mimic" of GAB family scaffolding adaptors.

The CagA-induced hummingbird phenotype is MEK-dependent [177], but whether it is RAS-dependent remains controversial [177–179]. Furthermore, it is likely that CagA/SHP2 can perturb multiple pathways that collectively promote the pathogenic effects of *H. pylori*. For example, SHP2, recruited by CagA, was reported to dephosphorylate FAK [180]. In addition, the CagA/SHP2 complex binds and inhibits the polarity kinase PAR1, leading to impaired tight junction function and defects in cell polarity [181, 182]. Interestingly, GAB1 promotes cell scattering by PAR1-mediated PAR3 phosphorylation and the dissociation of PAR3/PAR6 complex [183]. Thus, CagA and GAB1 appear to affect cellular polarity through distinct, although convergent mechanisms. Future experiments are needed to determine and compare the roles of SHP2 and its substrate(s) in these contexts.

GAB2 itself is overexpressed/amplified in 10–15 % of human breast tumors [184–190]. In an SHP2-dependent manner, GAB2 overexpression increases the proliferative potential of human MCF10A mammary epithelial cells [186, 191]. Furthermore, GAB2 cooperates with HER2/Neu overexpression to evoke an invasive-like phenotype in these cells and enhanced breast carcinogenesis in mice [186]. Cooperation with HER2/Neu requires intact SHP2 binding sites on GAB2 [186], highlighting the requirement of SHP2 in these processes. Indeed, knockdown of *PTPN11* reduces the invasiveness of HER2/3 expressing MCF10A cells [192]. Loss of *PTPN11* is accompanied by reduced expression of *ZEB1*, *LIN28b*, and an increase of *let-7a* and *let-7b* miRNAs, leading to decreased expression of *let-7* targets such as *RAS* and *MYC*. Hence, GAB2, acting through SHP2, promotes HER2-mediated proliferation, invasion, and tumorigenesis, in HER+ breast cancer.

SHP2 also plays important roles in other subtypes of breast cancer in the absence of *GAB2* amplification. For example, *PTPN11* deficiency leads to decreased expansion and cell motility of several triple-negative breast cancer (TNBC) cell lines [192, 193]. The detailed mechanism of SHP2 action in these cells awaits future investigation.

In addition to breast cancer cells, GAB2 is also amplified and overexpressed in a significant subset of serous ovarian tumors and cell lines [194, 195] (The Cancer Genome Atlas, TCGA). In these cell lines, GAB2 was reported to promote transformation, cell survival, and migration by promoting PI3K-activation [194, 195]. However, whether SHP2 is also required for GAB2-mediated oncogenesis in serous ovarian cancer remains to be established.

PTPN11 mutations are found in a small subset of lung cancer and lung cancer cell lines (TCGA, COSMIC, [116]). These mutations, though rare, are likely causal, as transgenic expression of *PTPN11*^{E76K} under the Clara cell secretory protein promoter results in adenomas and adenocarcinoma in mice [196]. Furthermore, tumor maintenance requires continuous expression of the mutant *PTPN11* allele, suggesting, at least in this model, mutant *PTPN11* expression is required for both the onset and the maintenance of solid tumors. This mouse model may be a tractable platform to test the therapeutic potential of SHP2 inhibitors in lung cancer.

Gain-of-function *PTPN11* mutations have also been reported in ~2% pilocytic astrocytoma, a childhood brain tumor [197]. All of these mutants are found in tumors that also have *FGFR1* mutations, and it remains to be established if/how they contribute to the pathogenesis of the disease. However, pilocytic astrocytomas have also been reported in NS patients [198, 199], so it seems likely that the *PTPN11* mutants make a causal contribution to this disease via an as yet undefined mechanism; conceivably they can increase the "gain" of the mutant FGFR1.

Finally, as mentioned above, *PTPN11* mutations are the third most common genetic event in neuroblastoma [116, 119, 120]. Interestingly, the RTK *ALK* also is amplified in some neuroblastomas, and *ALK* and *PTPN11* mutations are mutually exclusive (TCGA). Hence, it would not be surprising if SHP2 plays a general role in this pediatric neoplasm. Indeed, SHP2 was reported to be a direct target of NPM-ALK in anaplastic large cell lymphomas (ALCL), and was shown to be required for malignant cell proliferation [200].

The detailed roles of SHP2 in ALK signaling await further experimentation. Collectively, however, these data implicate GAB2/SHP2 is a crucial axis in transformation of both hematopoietic and epithelial cells, suggesting targeting components of this pathway may be of therapeutic value for these human diseases.

PTPN11 as a Tumor Suppressor

Hepatocyte-specific, homozygous *Ptpn11* deletion leads to increased expression of genes associated with inflammation, such *Il6* and *Tnfα*, and to enhanced IL-6-mediated STAT3 activation [201]. Aged mice (>5 months) with hepatocyte-specific SHP2 deficiency develop hepatocellular adenomas, whose frequency is enhanced when mice are exposed to the carcinogen diethylnitrosamine in a STAT3-dependent manner. However, tumors have not been observed in another strain of liver-specific *Ptpn11*$^{-/-}$ mice (Haj F. and Neel B.G., unpublished). Whether differences in genetic background and/or animal husbandry (e.g., sub-clinical infection) contribute to these discrepancies awaits future investigation.

Metachondromatosis (MC) is a rare, autosomal dominant disorder associated with the development of multiple tumors of cartilage (enchondromas) and bone (osteochondromas), and occasionally can progress to sarcoma [202]. A substantial percentage of MC patients have germline nonsense, frameshift, or splice-site mutations in *PTPN11* [203, 204], often accompanied by loss of heterozygosity in MC exostoses [204]. In mice, specific deletion of *Ptpn11* in cathepsin K (*Cstk*)-expressing cells [205] or chrondrocytes [206] produces mice with features of MC, including the presence of multiple exostoses and enchondromas. These neoplasms arise from a small population of *Ctsk*$^{+}$ cells located in the perichondrial groove of Ranvier, which have surface features of mesenchymal stem/progenitor cells and exhibit enhanced proliferation upon *Ptpn11* deletion [205]. Cartilage tumors from mutant mice show decreased ERK activation [205, 206] and increased FGF2 [206], Indian hedgehog (*Ihh*) [205, 206], and parathyroid hormone-related protein (*Pthrp*) [205] expression. Moreover, MC features in the mutant mice can be ameliorated by treatment with a smoothened inhibitor (SMOi) [205]. Thus, SHP2 activation (via ERK) apparently suppresses proliferation of these chondroid progenitors, and loss of SHP2 function relieves this suppression and results in tumorigenesis. These findings clearly show that *PTPN11* has cell context-dependent pro- and anti-oncogenic actions, albeit via the common theme of effects in RAS/ERK pathway activation.

Furthermore, they raise the possibility that MEK-I treatment could have undesirable effects on cartilage and bone, particularly in the pediatric population.

Germline heterozygous *PTPN11* mutations also are found in at least 90 % of patients with Noonan syndrome with multiple lentigenes (NS-ML). This disorder was previously termed LEOPARD syndrome (an acronym for *l*entigines, *e*lectrocardiogram abnormalities, *o*cular hypertelorism, *p*ulmonic stenosis, *a*bnormalities of genitalia, *r*etardation of growth, and *d*eafness syndrome [207, 208]), but has been renamed because the former appellation was viewed as somewhat disparaging to patients. In contrast to the mutations that cause NS, JMML, and MC, all NS-ML alleles affect residues important for phosphatase function. This results in the expression of catalytically impaired SHP2 mutants, which can inhibit growth factor-evoked ERK activation [69–71]. Although NS-ML mutants are catalytically impaired, they have increased propensity for the open confirmation and a higher affinity for pY-motifs in scaffolding proteins; as such, their net effect can be increased dephosphorylation of SHP2 substrates such as PAXILLIN [209, 210]. Thus, NS-ML mutants might have hypomorphic *and* neomorphic effects, resulting in the phenotypically distinct features in these patients. Interestingly, although NS-ML patients can exhibit bony abnormalities, they do not present with the widespread enchondromas/osteochondromas seen in MC. The distinct features caused by MC- and NS-ML mutations might reflect a combination of factors, such as tissue-specific loss of *PTPN11* in MC patients and the partial and/or neophormic effects evoked by NS-ML *PTPN11* mutants.

Summary

It is well-established that SHP2 is required for full and/or sustained RAS/ERK activation downstream of multiple growth factor and cytokine receptors. However, the resolution of the proximate targets of SHP2 and the critical sites on these substrates remains controversial. Importantly, how the spectrum of dephosphorylation at these sites differs in the normal and disease states warrant further investigation.

In humans (and also in experimental organisms), mutations of *PTPN11* cause a wide range of phenotypes/diseases, all of which are associated with perturbed RAS/ERK pathway activation. Nuances in the frequency and duration of ERK activation in specific cell types are likely crucial in the manifestation of distinct disease features. In addition to the effects of SHP2 on the RAS/ERK pathway, further investigation is needed to clarify: (a) the contribution of non-RAS/ERK pathways downstream of SHP2; (b) its non-catalytic functions; (c) its possible role in the nucleus- and (d) potential differences in feedback mechanisms in different cell types expressing *PTPN11* mutant alleles. Detecting these perturbations may only be achievable by carefully investigating signaling abnormalities caused by each type of mutation in the relevant cell type.

The development of potent and specific SHP2 inhibitors would contribute much to understanding SHP2 function, and provide potential therapeutic strategies for

PTPN11-associated diseases. Deploying such agent(s) will require consideration of the potential tumor-suppressor roles of SHP2 in some tissues, as well as its role as a negative regulator of many immune signaling pathways *in vivo* [2]. As both catalytic and non-catalytic functions of SHP2 are likely to be important, blockade of either alone might not be sufficient to neutralize the effects evoked by mutant *PTPN11* expression. However, blockade of both types of function could lead to unacceptable toxicities, as illustrated by the rapid and severe attrition of the hematopoietic system in SHP2-deficient mice. Effective therapeutic invention could well require the difficult task of normalizing the activation of signaling downstream of SHP2, as both excess and inadequate activation of these pathways contribute to human pathogenesis *in vivo*.

References

1. Bonetti M, Rodriguez-Martinez V, Paardekooper Overman J, Overvoorde J, van Eekelen M, Jopling C, Hertog J. Distinct and overlapping functions of ptpn11 genes in Zebrafish development. PLoS One. 2014;9(4), e94884. doi:10.1371/journal.pone.0094884.
2. Neel BG, Chan G, Dhanji S. SH2 domain-containing protein-tyrosine phosphatases. In: Neel BG, editor. Handbook of Cell Signaling. 2nd ed. Oxford: Academic; 2009. p. 771–810.
3. Tsutsumi R, Masoudi M, Takahashi A, Fujii Y, Hayashi T, Kikuchi I, Satou Y, Taira M, Hatakeyama M. YAP and TAZ, Hippo signaling targets, act as a rheostat for nuclear SHP2 function. Dev Cell. 2013;26(6):658–65. doi:10.1016/j.devcel.2013.08.013.
4. Freeman Jr RM, Plutzky J, Neel BG. Identification of a human src homology 2-containing protein-tyrosine-phosphatase: a putative homolog of Drosophila corkscrew. Proc Natl Acad Sci U S A. 1992;89(23):11239–43.
5. Feng GS, Hui CC, Pawson T. SH2-containing phosphotyrosine phosphatase as a target of protein-tyrosine kinases. Science. 1993;259(5101):1607–11.
6. Ahmad S, Banville D, Zhao Z, Fischer EH, Shen SH. A widely expressed human protein-tyrosine phosphatase containing src homology 2 domains. Proc Natl Acad Sci U S A. 1993;90(6):2197–201.
7. Perkins LA, Larsen I, Perrimon N. Corkscrew encodes a putative protein tyrosine phosphatase that functions to transduce the terminal signal from the receptor tyrosine kinase torso. Cell. 1992;70(2):225–36.
8. Ravetch JV, Lanier LL. Immune inhibitory receptors. Science. 2000;290(5489):84–9.
9. Hof P, Pluskey S, Dhe-Paganon S, Eck MJ, Shoelson SE. Crystal structure of the tyrosine phosphatase SHP-2. Cell. 1998;92(4):441–50.
10. Barford D, Neel BG. Revealing mechanisms for SH2 domain mediated regulation of the protein tyrosine phosphatase SHP-2. Structure. 1998;6(3):249–54.
11. Neel BG, Gu H, Pao L. The Shping news: SH2 domain-containing tyrosine phosphatases in cell signaling. Trends Biochem Sci. 2003;28(6):284–93.
12. O'Reilly AM, Pluskey S, Shoelson SE, Neel BG. Activated mutants of SHP-2 preferentially induce elongation of Xenopus animal caps. Mol Cell Biol. 2000;20(1):299–311.
13. Shi ZQ, Lu W, Feng GS. The Shp-2 tyrosine phosphatase has opposite effects in mediating the activation of extracellular signal-regulated and c-Jun NH2-terminal mitogen-activated protein kinases. J Biol Chem. 1998;273(9):4904–8.
14. Zhang SQ, Tsiaras WG, Araki T, Wen G, Minichiello L, Klein R, Neel BG. Receptor-specific regulation of phosphatidylinositol 3'-kinase activation by the protein tyrosine phosphatase Shp2. Mol Cell Biol. 2002;22(12):4062–72.

15. Saxton TM, Henkemeyer M, Gasca S, Shen R, Rossi DJ, Shalaby F, Feng GS, Pawson T. Abnormal mesoderm patterning in mouse embryos mutant for the SH2 tyrosine phosphatase Shp-2. EMBO J. 1997;16(9):2352–64.
16. Shi ZQ, Yu DH, Park M, Marshall M, Feng GS. Molecular mechanism for the Shp-2 tyrosine phosphatase function in promoting growth factor stimulation of Erk activity. Mol Cell Biol. 2000;20(5):1526–36.
17. Holgado-Madruga M, Emlet DR, Moscatello DK, Godwin AK, Wong AJ. A Grb2-associated docking protein in EGF- and insulin-receptor signalling. Nature. 1996;379(6565):560–4. doi:10.1038/379560a0.
18. Weidner KM, Di Cesare S, Sachs M, Brinkmann V, Behrens J, Birchmeier W. Interaction between Gab1 and the c-Met receptor tyrosine kinase is responsible for epithelial morphogenesis. Nature. 1996;384(6605):173–6. doi:10.1038/384173a0.
19. White MF, Yenush L. The IRS-signaling system: a network of docking proteins that mediate insulin and cytokine action. Curr Top Microbiol Immunol. 1998;228:179–208.
20. Kouhara H, Hadari YR, Spivak-Kroizman T, Schilling J, Bar-Sagi D, Lax I, Schlessinger J. A lipid-anchored Grb2-binding protein that links FGF-receptor activation to the Ras/MAPK signaling pathway. Cell. 1997;89(5):693–702.
21. Itoh M, Yoshida Y, Nishida K, Narimatsu M, Hibi M, Hirano T. Role of Gab1 in heart, placenta, and skin development and growth factor- and cytokine-induced extracellular signal-regulated kinase mitogen-activated protein kinase activation. Mol Cell Biol. 2000;20(10):3695–704.
22. Sachs M, Brohmann H, Zechner D, Muller T, Hulsken J, Walther I, Schaeper U, Birchmeier C, Birchmeier W. Essential role of Gab1 for signaling by the c-Met receptor in vivo. J Cell Biol. 2000;150(6):1375–84.
23. Hadari YR, Gotoh N, Kouhara H, Lax I, Schlessinger J. Critical role for the docking-protein FRS2 alpha in FGF receptor-mediated signal transduction pathways. Proc Natl Acad Sci U S A. 2001;98(15):8578–83. doi:10.1073/pnas.161259898.
24. Maroun CR, Naujokas MA, Holgado-Madruga M, Wong AJ, Park M. The tyrosine phosphatase SHP-2 is required for sustained activation of extracellular signal-regulated kinase and epithelial morphogenesis downstream from the met receptor tyrosine kinase. Mol Cell Biol. 2000;20(22):8513–25.
25. Schaeper U, Gehring NH, Fuchs KP, Sachs M, Kempkes B, Birchmeier W. Coupling of Gab1 to c-Met, Grb2, and Shp2 mediates biological responses. J Cell Biol. 2000;149(7):1419–32.
26. Noguchi T, Matozaki T, Horita K, Fujioka Y, Kasuga M. Role of SH-PTP2, a protein-tyrosine phosphatase with Src homology 2 domains, in insulin-stimulated Ras activation. Mol Cell Biol. 1994;14(10):6674–82.
27. Zhang SQ, Yang W, Kontaridis MI, Bivona TG, Wen G, Araki T, Luo J, Thompson JA, Schraven BL, Philips MR, Neel BG. Shp2 regulates SRC family kinase activity and Ras/Erk activation by controlling Csk recruitment. Mol Cell. 2004;13(3):341–55.
28. Yang W, Klaman LD, Chen B, Araki T, Harada H, Thomas SM, George EL, Neel BG. An Shp2/SFK/Ras/Erk signaling pathway controls trophoblast stem cell survival. Dev Cell. 2006;10(3):317–27.
29. Bennett AM, Tang TL, Sugimoto S, Walsh CT, Neel BG. Protein-tyrosine-phosphatase SHPTP2 couples platelet-derived growth factor receptor beta to Ras. Proc Natl Acad Sci U S A. 1994;91(15):7335–9.
30. Li W, Nishimura R, Kashishian A, Batzer AG, Kim WJ, Cooper JA, Schlessinger J. A new function for a phosphotyrosine phosphatase: linking GRB2-Sos to a receptor tyrosine kinase. Mol Cell Biol. 1994;14(1):509–17.
31. Welham MJ, Dechert U, Leslie KB, Jirik F, Schrader JW. Interleukin (IL)-3 and granulocyte/macrophage colony-stimulating factor, but not IL-4, induce tyrosine phosphorylation, activation, and association of SHPTP2 with Grb2 and phosphatidylinositol 3'-kinase. J Biol Chem. 1994;269(38):23764–8.
32. Lu W, Gong D, Bar-Sagi D, Cole PA. Site-specific incorporation of a phosphotyrosine mimetic reveals a role for tyrosine phosphorylation of SHP-2 in cell signaling. Mol Cell. 2001;8(4):759–69.

33. Sun J, Lu S, Ouyang M, Lin LJ, Zhuo Y, Liu B, Chien S, Neel BG, Wang Y. Antagonism between binding site affinity and conformational dynamics tunes alternative cis-interactions within Shp2. Nat Commun. 2013;4:2037. doi:10.1038/ncomms3037.
34. Vogel W, Ullrich A. Multiple in vivo phosphorylated tyrosine phosphatase SHP-2 engages binding to Grb2 via tyrosine 584. Cell Growth Differ. 1996;7(12):1589–97.
35. Klinghoffer RA, Kazlauskas A. Identification of a putative Syp substrate, the PDGF beta receptor. J Biol Chem. 1995;270(38):22208–17.
36. Cleghon V, Feldmann P, Ghiglione C, Copeland TD, Perrimon N, Hughes DA, Morrison DK. Opposing actions of CSW and RasGAP modulate the strength of Torso RTK signaling in the Drosophila terminal pathway. Mol Cell. 1998;2(6):719–27.
37. Agazie YM, Hayman MJ. Molecular mechanism for a role of SHP2 in epidermal growth factor receptor signaling. Mol Cell Biol. 2003;23(21):7875–86.
38. Zhou X, Agazie YM. Molecular mechanism for SHP2 in promoting HER2-induced signaling and transformation. J Biol Chem. 2009;284(18):12226–34. doi:10.1074/jbc.M900020200.
39. Dankort DL, Wang Z, Blackmore V, Moran MF, Muller WJ. Distinct tyrosine autophosphorylation sites negatively and positively modulate neu-mediated transformation. Mol Cell Biol. 1997;17(9):5410–25.
40. Montagner A, Yart A, Dance M, Perret B, Salles JP, Raynal P. A novel role for Gab1 and SHP2 in epidermal growth factor-induced Ras activation. J Biol Chem. 2005;280(7): 5350–60.
41. Yamasaki S, Nishida K, Yoshida Y, Itoh M, Hibi M, Hirano T. Gab1 is required for EGF receptor signaling and the transformation by activated ErbB2. Oncogene. 2003;22(10):1546–56. doi:10.1038/sj.onc.1206284.
42. Herbst R, Zhang X, Qin J, Simon MA. Recruitment of the protein tyrosine phosphatase CSW by DOS is an essential step during signaling by the sevenless receptor tyrosine kinase. EMBO J. 1999;18(24):6950–61.
43. Mattoon DR, Lamothe B, Lax I, Schlessinger J. The docking protein Gab1 is the primary mediator of EGF-stimulated activation of the PI-3K/Akt cell survival pathway. BMC Biol. 2004;2:24.
44. Bunda S, Heir P, Srikumar T, Cook JD, Burrell K, Kano Y, Lee JE, Zadeh G, Raught B, Ohh M. Src promotes GTPase activity of Ras via tyrosine 32 phosphorylation. Proc Natl Acad Sci U S A. 2014;111(36):E3785–94. doi:10.1073/pnas.1406559111.
45. Cunnick JM, Meng S, Ren Y, Desponts C, Wang HG, Djeu JY, Wu J. Regulation of the mitogen-activated protein kinase signaling pathway by SHP2. J Biol Chem. 2002;277(11):9498–504. doi:10.1074/jbc.M110547200.
46. Wu CJ, O'Rourke DM, Feng GS, Johnson GR, Wang Q, Greene MI. The tyrosine phosphatase SHP-2 is required for mediating phosphatidylinositol 3-kinase/Akt activation by growth factors. Oncogene. 2001;20(42):6018–25.
47. Bertotti A, Comoglio PM, Trusolino L. Beta4 integrin activates a Shp2-Src signaling pathway that sustains HGF-induced anchorage-independent growth. J Cell Biol. 2006;175(6): 993–1003.
48. Ren Y, Meng S, Mei L, Zhao ZJ, Jove R, Wu J. Roles of Gab1 and SHP2 in paxillin tyrosine dephosphorylation and Src activation in response to epidermal growth factor. J Biol Chem. 2004;279(9):8497–505.
49. Hanafusa H, Torii S, Yasunaga T, Matsumoto K, Nishida E. Shp2, an SH2-containing protein-tyrosine phosphatase, positively regulates receptor tyrosine kinase signaling by dephosphorylating and inactivating the inhibitor Sprouty. J Biol Chem. 2004;279(22):22992–5.
50. Jarvis LA, Toering SJ, Simon MA, Krasnow MA, Smith-Bolton RK. Sprouty proteins are in vivo targets of Corkscrew/SHP-2 tyrosine phosphatases. Development. 2006;133(6):1133–42.
51. Pan Y, Carbe C, Powers A, Feng GS, Zhang X. Sprouty2-modulated Kras signaling rescues Shp2 deficiency during lens and lacrimal gland development. Development. 2010;137(7): 1085–93.
52. Mason JM, Morrison DJ, Basson MA, Licht JD. Sprouty proteins: multifaceted negative-feedback regulators of receptor tyrosine kinase signaling. Trends Cell Biol. 2006;16(1): 45–54.

53. Wakioka T, Sasaki A, Kato R, Shouda T, Matsumoto A, Miyoshi K, Tsuneoka M, Komiya S, Baron R, Yoshimura A. Spred is a Sprouty-related suppressor of Ras signalling. Nature. 2001;412(6847):647–51. doi:10.1038/35088082.
54. Kato R, Nonami A, Taketomi T, Wakioka T, Kuroiwa A, Matsuda Y, Yoshimura A. Molecular cloning of mammalian Spred-3 which suppresses tyrosine kinase-mediated Erk activation. Biochem Biophys Res Commun. 2003;302(4):767–72.
55. Nonami A, Kato R, Taniguchi K, Yoshiga D, Taketomi T, Fukuyama S, Harada M, Sasaki A, Yoshimura A. Spred-1 negatively regulates interleukin-3-mediated ERK/mitogen-activated protein (MAP) kinase activation in hematopoietic cells. J Biol Chem. 2004;279(50):52543–51. doi:10.1074/jbc.M405189200.
56. Taniguchi K, Kohno R, Ayada T, Kato R, Ichiyama K, Morisada T, Oike Y, Yonemitsu Y, Maehara Y, Yoshimura A. Spreds are essential for embryonic lymphangiogenesis by regulating vascular endothelial growth factor receptor 3 signaling. Mol Cell Biol. 2007;27(12):4541–50. doi:10.1128/MCB.01600-06.
57. Quintanar-Audelo M, Yusoff P, Sinniah S, Chandramouli S, Guy GR. Sprouty-related Ena/vasodilator-stimulated phosphoprotein homology 1-domain-containing protein (SPRED1), a tyrosine-protein phosphatase non-receptor type 11 (SHP2) substrate in the Ras/extracellular signal-regulated kinase (ERK) pathway. J Biol Chem. 2011;286(26):23102–12. doi:10.1074/jbc.M110.212662.
58. Brems H, Chmara M, Sahbatou M, Denayer E, Taniguchi K, Kato R, Somers R, Messiaen L, De Schepper S, Fryns JP, Cools J, Marynen P, Thomas G, Yoshimura A, Legius E. Germline loss-of-function mutations in SPRED1 cause a neurofibromatosis 1-like phenotype. Nat Genet. 2007;39(9):1120–6.
59. Spurlock G, Bennett E, Chuzhanova N, Thomas N, Jim HP, Side L, Davies S, Haan E, Kerr B, Huson SM, Upadhyaya M. SPRED1 mutations (Legius syndrome): another clinically useful genotype for dissecting the neurofibromatosis type 1 phenotype. J Med Genet. 2009;46(7):431–7. doi:10.1136/jmg.2008.065474.
60. Zito CI, Qin H, Blenis J, Bennett AM. SHP-2 regulates cell growth by controlling the mTOR/S6 kinase 1 pathway. J Biol Chem. 2007;282(10):6946–53.
61. Yu M, Luo J, Yang W, Wang Y, Mizuki M, Kanakura Y, Besmer P, Neel BG, Gu H. The scaffolding adapter Gab2, via Shp-2, regulates kit-evoked mast cell proliferation by activating the Rac/JNK pathway. J Biol Chem. 2006;281(39):28615–26.
62. Yang Z, Li Y, Yin F, Chan RJ. Activating PTPN11 mutants promote hematopoietic progenitor cell-cycle progression and survival. Exp Hematol. 2008;36(10):1285–96. doi:10.1016/j.exphem.2008.04.016.
63. You M, Flick LM, Yu D, Feng GS. Modulation of the nuclear factor kappa B pathway by Shp-2 tyrosine phosphatase in mediating the induction of interleukin (IL)-6 by IL-1 or tumor necrosis factor. J Exp Med. 2001;193(1):101–10.
64. Schoenwaelder SM, Petch LA, Williamson D, Shen R, Feng GS, Burridge K. The protein tyrosine phosphatase Shp-2 regulates RhoA activity. Curr Biol. 2000;10(23):1523–6.
65. Kontaridis MI, Eminaga S, Fornaro M, Zito CI, Sordella R, Settleman J, Bennett AM. SHP-2 positively regulates myogenesis by coupling to the Rho GTPase signaling pathway. Mol Cell Biol. 2004;24(12):5340–52. doi:10.1128/MCB.24.12.5340-5352.2004.
66. Uhlen P, Burch PM, Zito CI, Estrada M, Ehrlich BE, Bennett AM. Gain-of-function/Noonan syndrome SHP-2/Ptpn11 mutants enhance calcium oscillations and impair NFAT signaling. Proc Natl Acad Sci U S A. 2006;103(7):2160–5.
67. Paardekooper Overman J, Preisinger C, Prummel K, Bonetti M, Giansanti P, Heck A, den Hertog J. Phosphoproteomics-mediated identification of Fer kinase as a target of mutant Shp2 in Noonan and LEOPARD syndrome. PLoS One. 2014;9(9), e106682. doi:10.1371/journal.pone.0106682.
68. Mitra S, Beach C, Feng GS, Plattner R. SHP-2 is a novel target of Abl kinases during cell proliferation. J Cell Sci. 2008;121(Pt 20):3335–46. doi:10.1242/jcs.035691.
69. Kontaridis MI, Swanson KD, David FS, Barford D, Neel BG. PTPN11 (Shp2) mutations in LEOPARD syndrome have dominant negative, not activating, effects. J Biol Chem. 2006;281(10):6785–92.

70. Tartaglia M, Martinelli S, Stella L, Bocchinfuso G, Flex E, Cordeddu V, Zampino G, Burgt I, Palleschi A, Petrucci TC, Sorcini M, Schoch C, Foa R, Emanuel PD, Gelb BD. Diversity and functional consequences of germline and somatic PTPN11 mutations in human disease. Am J Hum Genet. 2006;78(2):279–90.
71. Hanna N, Montagner A, Lee WH, Miteva M, Vidal M, Vidaud M, Parfait B, Raynal P. Reduced phosphatase activity of SHP-2 in LEOPARD syndrome: consequences for PI3K binding on Gab1. FEBS Lett. 2006;580(10):2477–82.
72. Zou GM, Chan RJ, Shelley WC, Yoder MC. Reduction of Shp-2 expression by small interfering RNA reduces murine embryonic stem cell-derived in vitro hematopoietic differentiation. Stem Cells. 2006;24(3):587–94.
73. Chan RJ, Johnson SA, Li Y, Yoder MC, Feng GS. A definitive role of Shp-2 tyrosine phosphatase in mediating embryonic stem cell differentiation and hematopoiesis. Blood. 2003;102(6):2074–80.
74. Qu CK, Shi ZQ, Shen R, Tsai FY, Orkin SH, Feng GS. A deletion mutation in the SH2-N domain of Shp-2 severely suppresses hematopoietic cell development. Mol Cell Biol. 1997;17(9):5499–507.
75. Qu CK, Yu WM, Azzarelli B, Cooper S, Broxmeyer HE, Feng GS. Biased suppression of hematopoiesis and multiple developmental defects in chimeric mice containing Shp-2 mutant cells. Mol Cell Biol. 1998;18(10):6075–82.
76. Qu CK, Nguyen S, Chen J, Feng GS. Requirement of Shp-2 tyrosine phosphatase in lymphoid and hematopoietic cell development. Blood. 2001;97(4):911–4.
77. Chan RJ, Li Y, Hass MN, Walter A, Voorhorst CS, Shelley WC, Yang Z, Orschell CM, Yoder MC. Shp-2 heterozygous hematopoietic stem cells have deficient repopulating ability due to diminished self-renewal. Exp Hematol. 2006;34(9):1230–9.
78. Chan G, Cheung LS, Yang W, Milyavsky M, Sanders AD, Gu S, Hong WX, Liu AX, Wang X, Barbara M, Sharma T, Gavin J, Kutok JL, Iscove NN, Shannon KM, Dick JE, Neel BG, Braun BS. Essential role for Ptpn11 in survival of hematopoietic stem and progenitor cells. Blood. 2011;117(16):4253–61. doi:10.1182/blood-2010-11-319517.
79. Zhu HH, Ji K, Alderson N, He Z, Li S, Liu W, Zhang DE, Li L, Feng GS. Kit-Shp2-Kit signaling acts to maintain a functional hematopoietic stem and progenitor cell pool. Blood. 2011;117(20):5350–61. doi:10.1182/blood-2011-01-333476.
80. Li L, Modi H, McDonald T, Rossi J, Yee JK, Bhatia R. A critical role for SHP2 in STAT5 activation and growth factor-mediated proliferation, survival, and differentiation of human CD34+ cells. Blood. 2011;118(6):1504–15. doi:10.1182/blood-2010-06-288910.
81. Huang W, Saberwal G, Horvath E, Zhu C, Lindsey S, Eklund EA. Leukemia-associated, constitutively active mutants of SHP2 protein tyrosine phosphatase inhibit NF1 transcriptional activation by the interferon consensus sequence binding protein. Mol Cell Biol. 2006;26(17):6311–32.
82. Huang H, Woo AJ, Waldon Z, Schindler Y, Moran TB, Zhu HH, Feng GS, Steen H, Cantor AB. A Src family kinase-Shp2 axis controls RUNX1 activity in megakaryocyte and T-lymphocyte differentiation. Genes Dev. 2012;26(14):1587–601. doi:10.1101/gad.192054.112.
83. Holtschke T, Lohler J, Kanno Y, Fehr T, Giese N, Rosenbauer F, Lou J, Knobeloch KP, Gabriele L, Waring JF, Bachmann MF, Zinkernagel RM, Morse 3rd HC, Ozato K, Horak I. Immunodeficiency and chronic myelogenous leukemia-like syndrome in mice with a targeted mutation of the ICSBP gene. Cell. 1996;87(2):307–17.
84. Konieczna I, Horvath E, Wang H, Lindsey S, Saberwal G, Bei L, Huang W, Platanias L, Eklund EA. Constitutive activation of SHP2 in mice cooperates with ICSBP deficiency to accelerate progression to acute myeloid leukemia. J Clin Invest. 2008;118(3):853–67.
85. Growney JD, Shigematsu H, Li Z, Lee BH, Adelsperger J, Rowan R, Curley DP, Kutok JL, Akashi K, Williams IR, Speck NA, Gilliland DG. Loss of Runx1 perturbs adult hematopoiesis and is associated with a myeloproliferative phenotype. Blood. 2005;106(2):494–504. doi:10.1182/blood-2004-08-3280.
86. Ichikawa M, Asai T, Saito T, Seo S, Yamazaki I, Yamagata T, Mitani K, Chiba S, Ogawa S, Kurokawa M, Hirai H. AML-1 is required for megakaryocytic maturation and lymphocytic

differentiation, but not for maintenance of hematopoietic stem cells in adult hematopoiesis. Nat Med. 2004;10(3):299–304. doi:10.1038/nm997.
87. Cai X, Gaudet JJ, Mangan JK, Chen MJ, De Obaldia ME, Oo Z, Ernst P, Speck NA. Runx1 loss minimally impacts long-term hematopoietic stem cells. PLoS One. 2011;6(12), e28430. doi:10.1371/journal.pone.0028430.
88. Staser K, Park SJ, Rhodes SD, Zeng Y, He YZ, Shew MA, Gehlhausen JR, Cerabona D, Menon K, Chen S, Sun Z, Yuan J, Ingram DA, Nalepa G, Yang FC, Clapp DW. Normal hematopoiesis and neurofibromin-deficient myeloproliferative disease require Erk. J Clin Invest. 2012. doi:10.1172/JCI66167.
89. Chan G, Gu S, Neel BG. Erk1 and Erk2 are required for maintenance of hematopoietic stem cells and adult hematopoiesis. Blood. 2013;121(18):3594–8. doi:10.1182/blood-2012-12-476200.
90. Takahashi A, Tsutsumi R, Kikuchi I, Obuse C, Saito Y, Seidi A, Karisch R, Fernandez M, Cho T, Ohnishi N, Rozenblatt-Rosen O, Meyerson M, Neel BG, Hatakeyama M. SHP2 tyrosine phosphatase converts parafibromin/Cdc73 from a tumor suppressor to an oncogenic driver. Mol Cell. 2011;43(1):45–56. doi:10.1016/j.molcel.2011.05.014.
91. Roberts AE, Allanson JE, Tartaglia M, Gelb BD. Noonan syndrome. Lancet. 2013;381(9863):333–42. doi:10.1016/S0140-6736(12)61023-X.
92. Noonan JA. Hypertelorism with Turner phenotype. A new syndrome with associated congenital heart disease. Am J Dis Child. 1968;116(4):373–80.
93. Noonan JA. Noonan syndrome. An update and review for the primary pediatrician. Clin Pediatr. 1994;33(9):548–55.
94. Noonan JA. Noonan syndrome revisited. J Pediatr. 1999;135(6):667–8.
95. Choong K, Freedman MH, Chitayat D, Kelly EN, Taylor G, Zipursky A. Juvenile myelomonocytic leukemia and Noonan syndrome. J Pediatr Hematol Oncol. 1999;21(6):523–7.
96. Bader-Meunier B, Tchernia G, Mielot F, Fontaine JL, Thomas C, Lyonnet S, Lavergne JM, Dommergues JP. Occurrence of myeloproliferative disorder in patients with Noonan syndrome. J Pediatr. 1997;130(6):885–9.
97. Attard-Montalto SP, Kingston JE, Eden T. Noonan's syndrome and acute lymphoblastic leukaemia. Med Pediatr Oncol. 1994;23(4):391–2.
98. Klopfenstein KJ, Sommer A, Ruymann FB. Neurofibromatosis-Noonan syndrome and acute lymphoblastic leukemia: a report of two cases. J Pediatr Hematol Oncol. 1999;21(2):158–60.
99. Schubbert S, Zenker M, Rowe SL, Boll S, Klein C, Bollag G, van der Burgt I, Musante L, Kalscheuer V, Wehner LE, Nguyen H, West B, Zhang KY, Sistermans E, Rauch A, Niemeyer CM, Shannon K, Kratz CP. Germline KRAS mutations cause Noonan syndrome. Nat Genet. 2006;38(3):331–6.
100. Carta C, Pantaleoni F, Bocchinfuso G, Stella L, Vasta I, Sarkozy A, Digilio C, Palleschi A, Pizzuti A, Grammatico P, Zampino G, Dallapiccola B, Gelb BD, Tartaglia M. Germline missense mutations affecting KRAS Isoform B are associated with a severe Noonan syndrome phenotype. Am J Hum Genet. 2006;79(1):129–35.
101. Kraoua L, Journel H, Bonnet P, Amiel J, Pouvreau N, Baumann C, Verloes A, Cave H. Constitutional NRAS mutations are rare among patients with Noonan syndrome or juvenile myelomonocytic leukemia. Am J Med Genet. 2012;158A(10):2407–11. doi:10.1002/ajmg.a.35513.
102. Cirstea IC, Kutsche K, Dvorsky R, Gremer L, Carta C, Horn D, Roberts AE, Lepri F, Merbitz-Zahradnik T, Konig R, Kratz CP, Pantaleoni F, Dentici ML, Joshi VA, Kucherlapati RS, Mazzanti L, Mundlos S, Patton MA, Silengo MC, Rossi C, Zampino G, Digilio C, Stuppia L, Seemanova E, Pennacchio LA, Gelb BD, Dallapiccola B, Wittinghofer A, Ahmadian MR, Tartaglia M, Zenker M. A restricted spectrum of NRAS mutations causes Noonan syndrome. Nat Genet. 2010;42(1):27–9. doi:10.1038/ng.497.
103. Roberts AE, Araki T, Swanson KD, Montgomery KT, Schiripo TA, Joshi VA, Li L, Yassin Y, Tamburino AM, Neel BG, Kucherlapati RS. Germline gain-of-function mutations in SOS1 cause Noonan syndrome. Nat Genet. 2007;39(1):70–4.
104. Tartaglia M, Pennacchio LA, Zhao C, Yadav KK, Fodale V, Sarkozy A, Pandit B, Oishi K, Martinelli S, Schackwitz W, Ustaszewska A, Martin J, Bristow J, Carta C, Lepri F, Neri C,

Vasta I, Gibson K, Curry CJ, Siguero JP, Digilio MC, Zampino G, Dallapiccola B, Bar-Sagi D, Gelb BD. Gain-of-function SOS1 mutations cause a distinctive form of Noonan syndrome. Nat Genet. 2007;39(1):75–9.

105. Razzaque MA, Nishizawa T, Komoike Y, Yagi H, Furutani M, Amo R, Kamisago M, Momma K, Katayama H, Nakagawa M, Fujiwara Y, Matsushima M, Mizuno K, Tokuyama M, Hirota H, Muneuchi J, Higashinakagawa T, Matsuoka R. Germline gain-of-function mutations in RAF1 cause Noonan syndrome. 2007. Nat Genet.
106. Pandit B, Sarkozy A, Pennacchio LA, Carta C, Oishi K, Martinelli S, Pogna EA, Schackwitz W, Ustaszewska A, Landstrom A, Bos JM, Ommen SR, Esposito G, Lepri F, Faul C, Mundel P, Lopez Siguero JP, Tenconi R, Selicorni A, Rossi C, Mazzanti L, Torrente I, Marino B, Digilio MC, Zampino G, Ackerman MJ, Dallapiccola B, Tartaglia M, Gelb BD. Gain-of-function RAF1 mutations cause Noonan and LEOPARD syndromes with hypertrophic cardiomyopathy. 2007. Nat Genet.
107. Aoki Y, Niihori T, Banjo T, Okamoto N, Mizuno S, Kurosawa K, Ogata T, Takada F, Yano M, Ando T, Hoshika T, Barnett C, Ohashi H, Kawame H, Hasegawa T, Okutani T, Nagashima T, Hasegawa S, Funayama R, Nagashima T, Nakayama K, Inoue S, Watanabe Y, Ogura T, Matsubara Y. Gain-of-Function Mutations in RIT1 Cause Noonan Syndrome, a RAS/MAPK Pathway Syndrome. Am J Hum Genet. 2013;93(1):173–80. doi:10.1016/j.ajhg.2013.05.021.
108. Chen PC, Yin J, Yu HW, Yuan T, Fernandez M, Yung CK, Trinh QM, Peltekova VD, Reid JG, Tworog-Dube E, Morgan MB, Muzny DM, Stein L, McPherson JD, Roberts AE, Gibbs RA, Neel BG, Kucherlapati R. Next-generation sequencing identifies rare variants associated with Noonan syndrome. Proc Natl Acad Sci U S A. 2014;111(31):11473–8. doi:10.1073/pnas.1324128111.
109. Cordeddu V, Di Schiavi E, Pennacchio LA, Ma'ayan A, Sarkozy A, Fodale V, Cecchetti S, Cardinale A, Martin J, Schackwitz W, Lipzen A, Zampino G, Mazzanti L, Digilio MC, Martinelli S, Flex E, Lepri F, Bartholdi D, Kutsche K, Ferrero GB, Anichini C, Selicorni A, Rossi C, Tenconi R, Zenker M, Merlo D, Dallapiccola B, Iyengar R, Bazzicalupo P, Gelb BD, Tartaglia M. Mutation of SHOC2 promotes aberrant protein N-myristoylation and causes Noonan-like syndrome with loose anagen hair. Nat Genet. 2009;41(9):1022–6. doi:10.1038/ng.425.
110. Emanuel PD. Juvenile myelomonocytic leukemia and chronic myelomonocytic leukemia. Leukemia. 2008;22(7):1335–42.
111. Loh ML. Recent advances in the pathogenesis and treatment of juvenile myelomonocytic leukaemia. Br J Haematol. 2011;152(6):677–87. doi:10.1111/j.1365-2141.2010.08525.x.
112. Chang TY, Dvorak CC, Loh ML. Bedside to bench in juvenile myelomonocytic leukemia: insights into leukemogenesis from a rare pediatric leukemia. Blood. 2014;124(16):2487–97. doi:10.1182/blood-2014-03-300319.
113. Tartaglia M, Niemeyer CM, Fragale A, Song X, Buechner J, Jung A, Hahlen K, Hasle H, Licht JD, Gelb BD. Somatic mutations in PTPN11 in juvenile myelomonocytic leukemia, myelodysplastic syndromes and acute myeloid leukemia. Nat Genet. 2003;34(2):148–50.
114. Loh ML, Vattikuti S, Schubbert S, Reynolds MG, Carlson E, Lieuw KH, Cheng JW, Lee CM, Stokoe D, Bonifas JM, Curtiss NP, Gotlib J, Meshinchi S, Le Beau MM, Emanuel PD, Shannon KM. Mutations in PTPN11 implicate the SHP-2 phosphatase in leukemogenesis. Blood. 2004;103(6):2325–31.
115. Loh ML, Reynolds MG, Vattikuti S, Gerbing RB, Alonzo TA, Carlson E, Cheng JW, Lee CM, Lange BJ, Meshinchi S. PTPN11 mutations in pediatric patients with acute myeloid leukemia: results from the Children's Cancer Group. Leukemia. 2004;18(11):1831–4.
116. Bentires-Alj M, Paez JG, David FS, Keilhack H, Halmos B, Naoki K, Maris JM, Richardson A, Bardelli A, Sugarbaker DJ, Richards WG, Du J, Girard L, Minna JD, Loh ML, Fisher DE, Velculescu VE, Vogelstein B, Meyerson M, Sellers WR, Neel BG. Activating mutations of the Noonan syndrome-associated SHP2/PTPN11 gene in human solid tumors and adult acute myelogenous leukemia. Cancer Res. 2004;64(24):8816–20.
117. Tartaglia M, Niemeyer CM, Shannon KM, Loh ML. SHP-2 and myeloid malignancies. Curr Opin Hematol. 2004;11(1):44–50.
118. Yamamoto T, Isomura M, Xu Y, Liang J, Yagasaki H, Kamachi Y, Kudo K, Kiyoi H, Naoe T, Kojma S. PTPN11, RAS and FLT3 mutations in childhood acute lymphoblastic leukemia. Leuk Res. 2006;30(9):1085–9.

119. Martinelli S, Carta C, Flex E, Binni F, Cordisco EL, Moretti S, Puxeddu E, Tonacchera M, Pinchera A, McDowell HP, Dominici C, Rosolen A, Di Rocco C, Riccardi R, Celli P, Picardo M, Genuardi M, Grammatico P, Sorcini M, Tartaglia M. Activating PTPN11 mutations play a minor role in pediatric and adult solid tumors. Cancer Genet Cytogenet. 2006;166(2):124–9.
120. Pugh TJ, Morozova O, Attiyeh EF, Asgharzadeh S, Wei JS, Auclair D, Carter SL, Cibulskis K, Hanna M, Kiezun A, Kim J, Lawrence MS, Lichenstein L, McKenna A, Pedamallu CS, Ramos AH, Shefler E, Sivachenko A, Sougnez C, Stewart C, Ally A, Birol I, Chiu R, Corbett RD, Hirst M, Jackman SD, Kamoh B, Khodabakshi AH, Krzywinski M, Lo A, Moore RA, Mungall KL, Qian J, Tam A, Thiessen N, Zhao Y, Cole KA, Diamond M, Diskin SJ, Mosse YP, Wood AC, Ji L, Sposto R, Badgett T, London WB, Moyer Y, Gastier-Foster JM, Smith MA, Guidry Auvil JM, Gerhard DS, Hogarty MD, Jones SJ, Lander ES, Gabriel SB, Getz G, Seeger RC, Khan J, Marra MA, Meyerson M, Maris JM. The genetic landscape of high-risk neuroblastoma. Nat Genet. 2013;45(3):279–84. doi:10.1038/ng.2529.
121. Keilhack H, David FS, McGregor M, Cantley LC, Neel BG. Diverse biochemical properties of Shp2 mutants. Implications for disease phenotypes. J Biol Chem. 2005;280(35): 30984–93.
122. Niihori T, Aoki Y, Ohashi H, Kurosawa K, Kondoh T, Ishikiriyama S, Kawame H, Kamasaki H, Yamanaka T, Takada F, Nishio K, Sakurai M, Tamai H, Nagashima T, Suzuki Y, Kure S, Fujii K, Imaizumi M, Matsubara Y. Functional analysis of PTPN11/SHP-2 mutants identified in Noonan syndrome and childhood leukemia. J Hum Genet. 2005;50(4):192–202.
123. Chan RJ, Leedy MB, Munugalavadla V, Voorhorst CS, Li Y, Yu M, Kapur R. Human somatic PTPN11 mutations induce hematopoietic-cell hypersensitivity to granulocyte-macrophage colony-stimulating factor. Blood. 2005;105(9):3737–42.
124. Mohi MG, Williams IR, Dearolf CR, Chan G, Kutok JL, Cohen S, Morgan K, Boulton C, Shigematsu H, Keilhack H, Akashi K, Gilliland DG, Neel BG. Prognostic, therapeutic, and mechanistic implications of a mouse model of leukemia evoked by Shp2 (PTPN11) mutations. Cancer Cell. 2005;7(2):179–91.
125. Schubbert S, Lieuw K, Rowe SL, Lee CM, Li X, Loh ML, Clapp DW, Shannon KM. Functional analysis of leukemia-associated PTPN11 mutations in primary hematopoietic cells. Blood. 2005;106(1):311–7.
126. Gandre-Babbe S, Paluru P, Aribeana C, Chou ST, Bresolin S, Lu L, Sullivan SK, Tasian SK, Weng J, Favre H, Choi JK, French DL, Loh ML, Weiss MJ. Patient-derived induced pluripotent stem cells recapitulate hematopoietic abnormalities of juvenile myelomonocytic leukemia. Blood. 2013;121(24):4925–9. doi:10.1182/blood-2013-01-478412.
127. Schubbert S, Shannon K, Bollag G. Hyperactive Ras in developmental disorders and cancer. Nat Rev Cancer. 2007;7(4):295–308.
128. Ward AF, Braun BS, Shannon KM. Targeting oncogenic Ras signaling in hematologic malignancies. Blood. 2012;120(17):3397–406. doi:10.1182/blood-2012-05-378596.
129. Zhang YY, Vik TA, Ryder JW, Srour EF, Jacks T, Shannon K, Clapp DW. Nf1 regulates hematopoietic progenitor cell growth and Ras signaling in response to multiple cytokines. J Exp Med. 1998;187(11):1893–902.
130. Le DT, Kong N, Zhu Y, Lauchle JO, Aiyigari A, Braun BS, Wang E, Kogan SC, Le Beau MM, Parada L, Shannon KM. Somatic inactivation of Nf1 in hematopoietic cells results in a progressive myeloproliferative disorder. Blood. 2004;103(11):4243–50.
131. Donovan S, See W, Bonifas J, Stokoe D, Shannon KM. Hyperactivation of protein kinase B and ERK have discrete effects on survival, proliferation, and cytokine expression in Nf1-deficient myeloid cells. Cancer Cell. 2002;2(6):507–14.
132. Largaespada DA, Brannan CI, Jenkins NA, Copeland NG. Nf1 deficiency causes Ras-mediated granulocyte/macrophage colony stimulating factor hypersensitivity and chronic myeloid leukaemia. Nat Genet. 1996;12(2):137–43.
133. Bollag G, Clapp DW, Shih S, Adler F, Zhang YY, Thompson P, Lange BJ, Freedman MH, McCormick F, Jacks T, Shannon K. Loss of NF1 results in activation of the Ras signaling pathway and leads to aberrant growth in haematopoietic cells. Nat Genet. 1996;12(2): 144–8.

134. Braun BS, Tuveson DA, Kong N, Le DT, Kogan SC, Rozmus J, Le Beau MM, Jacks TE, Shannon KM. Somatic activation of oncogenic *Kras* in hematopoietic cells initiates a rapidly fatal myeloproliferative disorder. Proc Natl Acad Sci U S A. 2004;101(2):597–602.
135. Chan IT, Kutok JL, Williams IR, Cohen S, Kelly L, Shigematsu H, Johnson L, Akashi K, Tuveson DA, Jacks T, Gilliland DG. Conditional expression of oncogenic K-ras from its endogenous promoter induces a myeloproliferative disease. J Clin Invest. 2004;113(4): 528–38.
136. Sabnis AJ, Cheung LS, Dail M, Kang HC, Santaguida M, Hermiston ML, Passegue E, Shannon K, Braun BS. Oncogenic Kras initiates leukemia in hematopoietic stem cells. PLoS Biol. 2009;7(3), e59. doi:10.1371/journal.pbio.1000059.
137. Zhang J, Wang J, Liu Y, Sidik H, Young KH, Lodish HF, Fleming MD. Oncogenic Kras-induced leukemogeneis: hematopoietic stem cells as the initial target and lineage-specific progenitors as the potential targets for final leukemic transformation. Blood. 2009;113(6):1304–14. doi:10.1182/blood-2008-01-134262.
138. Li Q, Haigis KM, McDaniel A, Harding-Theobald E, Kogan SC, Akagi K, Wong JC, Braun BS, Wolff L, Jacks T, Shannon K. Hematopoiesis and leukemogenesis in mice expressing oncogenic NrasG12D from the endogenous locus. Blood. 2011;117(6):2022–32. doi:10.1182/blood-2010-04-280750.
139. Wang J, Liu Y, Li Z, Du J, Ryu MJ, Taylor PR, Fleming MD, Young KH, Pitot H, Zhang J. Endogenous oncogenic Nras mutation promotes aberrant GM-CSF signaling in granulocytic/monocytic precursors in a murine model of chronic myelomonocytic leukemia. Blood. 2010;116(26):5991–6002. doi:10.1182/blood-2010-04-281527.
140. Wang J, Liu Y, Li Z, Wang Z, Tan LX, Ryu MJ, Meline B, Du J, Young KH, Ranheim E, Chang Q, Zhang J. Endogenous oncogenic Nras mutation initiates hematopoietic malignancies in a dose- and cell type-dependent manner. Blood. 2011;118(2):368–79. doi:10.1182/blood-2010-12-326058.
141. Lyubynska N, Gorman MF, Lauchle JO, Hong WX, Akutagawa JK, Shannon K, Braun BS. A MEK inhibitor abrogates myeloproliferative disease in Kras mutant mice. Sci Transl Med. 2011;3(76):76ra27. doi:10.1126/scitranslmed.3001069
142. Burgess MR, Hwang E, Firestone AJ, Huang T, Xu J, Zuber J, Bohin N, Wen T, Kogan SC, Haigis KM, Sampath D, Lowe S, Shannon K, Li Q. Preclinical efficacy of MEK inhibition in Nras mutant acute myeloid leukemia. Blood. 2014. doi:10.1182/blood-2014-05-574582.
143. Chang T, Krisman K, Theobald EH, Xu J, Akutagawa J, Lauchle JO, Kogan S, Braun BS, Shannon K. Sustained MEK inhibition abrogates myeloproliferative disease in Nf1 mutant mice. J Clin Invest. 2012. doi:10.1172/JCI63193.
144. Sakaguchi H, Okuno Y, Muramatsu H, Yoshida K, Shiraishi Y, Takahashi M, Kon A, Sanada M, Chiba K, Tanaka H, Makishima H, Wang X, Xu Y, Doisaki S, Hama A, Nakanishi K, Takahashi Y, Yoshida N, Maciejewski JP, Miyano S, Ogawa S, Kojima S. Exome sequencing identifies secondary mutations of SETBP1 and JAK3 in juvenile myelomonocytic leukemia. Nat Genet. 2013;45(8):937–41. doi:10.1038/ng.2698.
145. Araki T, Mohi MG, Ismat FA, Bronson RT, Williams IR, Kutok JL, Yang W, Pao LI, Gilliland DG, Epstein JA, Neel BG. Mouse model of Noonan syndrome reveals cell type- and gene dosage-dependent effects of Ptpn11 mutation. Nat Med. 2004;10(8):849–57.
146. Chan G, Kalaitzidis D, Usenko T, Kutok JL, Yang W, Mohi MG, Neel BG. Leukemogenic Ptpn11 causes fatal myeloproliferative disorder via cell-autonomous effects on multiple stages of hematopoiesis. Blood. 2009;113(18):4414–24.
147. Xu D, Liu X, Yu WM, Meyerson HJ, Guo C, Gerson SL, Qu CK. Non-lineage/stage-restricted effects of a gain-of-function mutation in tyrosine phosphatase Ptpn11 (Shp2) on malignant transformation of hematopoietic cells. J Exp Med. 2011;208(10):1977–88. doi:10.1084/jem.20110450.
148. Usenko T, Chan G, Torlakovic E, Klingmuller U, Neel BG. Leukemogenic ptpn11 allele causes defective erythropoiesis in mice. PLoS One. 2014;9(10), e109682. doi:10.1371/journal.pone.0109682.

149. Rampal R, Mascarenhas J. Pathogenesis and management of acute myeloid leukemia that has evolved from a myeloproliferative neoplasm. Curr Opin Hematol. 2014;21(2):65–71. doi:10.1097/MOH.0000000000000017.
150. Xu D, Zheng H, Yu WM, Qu CK. Activating mutations in protein tyrosine phosphatase Ptpn11 (Shp2) enhance reactive oxygen species production that contributes to myeloproliferative disorder. PLoS One. 2013;8(5), e63152. doi:10.1371/journal.pone.0063152.
151. Zheng H, Li S, Hsu P, Qu CK. Induction of a tumor-associated activating mutation in protein tyrosine phosphatase Ptpn11 (Shp2) enhances mitochondrial metabolism. Leading to oxidative stress and senescence. J Biol Chem. 2013. doi:10.1074/jbc.M113.462291.
152. Tuveson DA, Shaw AT, Willis NA, Silver DP, Jackson EL, Chang S, Mercer KL, Grochow R, Hock H, Crowley D, Hingorani SR, Zaks T, King C, Jacobetz MA, Wang L, Bronson RT, Orkin SH, DePinho RA, Jacks T. Endogenous oncogenic K-ras(G12D) stimulates proliferation and widespread neoplastic and developmental defects. Cancer Cell. 2004;5(4):375–87.
153. Yang Z, Kondo T, Voorhorst CS, Nabinger SC, Ndong L, Yin F, Chan EM, Yu M, Wurstlin O, Kratz CP, Niemeyer CM, Flotho C, Hashino E, Chan RJ. Increased c-Jun expression and reduced GATA2 expression promote aberrant monocytic differentiation induced by activating PTPN11 mutants. Mol Cell Biol. 2009;29(16):4376–93. doi:10.1128/MCB.01330-08.
154. Zhang W, Chan RJ, Chen H, Yang Z, He Y, Zhang X, Luo Y, Yin F, Moh A, Miller LC, Payne RM, Zhang ZY, Fu XY, Shou W. Negative regulation of Stat3 by activating PTPN11 mutants contributes to the pathogenesis of Noonan syndrome and juvenile myelomonocytic leukemia. J Biol Chem. 2009;284(33):22353–63. doi:10.1074/jbc.M109.020495.
155. Goodwin CB, Li XJ, Mali RS, Chan G, Kang M, Liu Z, Vanhaesebroeck B, Neel BG, Loh ML, Lannutti BJ, Kapur R, Chan RJ. PI3K p110delta uniquely promotes gain-of-function Shp2-induced GM-CSF hypersensitivity in a model of JMML. Blood. 2014;123(18):2838–42. doi:10.1182/blood-2013-10-535104.
156. Yu WM, Hawley TS, Hawley RG, Qu CK. Catalytic-dependent and -independent roles of SHP-2 tyrosine phosphatase in interleukin-3 signaling. Oncogene. 2003;22(38):5995–6004.
157. Yu WM, Daino H, Chen J, Bunting KD, Qu CK. Effects of a leukemia-associated gain-of-function mutation of SHP-2 phosphatase on interleukin-3 signaling. J Biol Chem. 2006;281(9):5426–34.
158. Kotecha N, Flores NJ, Irish JM, Simonds EF, Sakai DS, Archambeault S, Diaz-Flores E, Coram M, Shannon KM, Nolan GP, Loh ML. Single-cell profiling identifies aberrant STAT5 activation in myeloid malignancies with specific clinical and biologic correlates. Cancer Cell. 2008;14(4):335–43.
159. Hasegawa D, Bugarin C, Giordan M, Bresolin S, Longoni D, Micalizzi C, Ramenghi U, Bertaina A, Basso G, Locatelli F, Biondi A, Te Kronnie G, Gaipa G. Validation of flow cytometric phospho-STAT5 as a diagnostic tool for juvenile myelomonocytic leukemia. Blood Cancer J. 2013;3, e160. doi:10.1038/bcj.2013.56.
160. Gaipa G, Bugarin C, Longoni D, Cesana S, Molteni C, Faini A, Timeus F, Zecca M, Biondi A. Aberrant GM-CSF signal transduction pathway in juvenile myelomonocytic leukemia assayed by flow cytometric intracellular STAT5 phosphorylation measurement. Leukemia. 2009;23(4):791–3. doi:10.1038/leu.2008.265.
161. Kong G, Wunderlich M, Yang D, Ranheim EA, Young KH, Wang J, Chang YI, Du J, Liu Y, Tey SR, Zhang X, Juckett M, Mattison R, Damnernsawad A, Zhang J, Mulloy JC, Zhang J. Combined MEK and JAK inhibition abrogates murine myeloproliferative neoplasm. J Clin Invest. 2014;124(6):2762–73. doi:10.1172/JCI74182.
162. Mali RS, Ma P, Zeng LF, Martin H, Ramdas B, He Y, Sims E, Nabinger S, Ghosh J, Sharma N, Munugalavadla V, Chatterjee A, Li S, Sandusky G, Craig AW, Bunting KD, Feng GS, Chan RJ, Zhang ZY, Kapur R. Role of SHP2 phosphatase in KIT-induced transformation: identification of SHP2 as a druggable target in diseases involving oncogenic KIT. Blood. 2012;120(13):2669–78. doi:10.1182/blood-2011-08-375873.
163. Sharma N, Everingham S, Zeng LF, Zhang ZY, Kapur R, Craig AW. Oncogenic KIT-induced aggressive systemic mastocytosis requires SHP2/PTPN11 phosphatase for disease progression in mice. Oncotarget. 2014;5(15):6130–41.

164. Nabinger SC, Li XJ, Ramdas B, He Y, Zhang X, Zeng L, Richine B, Bowling JD, Fukuda S, Goenka S, Liu Z, Feng GS, Yu M, Sandusky GE, Boswell HS, Zhang ZY, Kapur R, Chan RJ. The protein tyrosine phosphatase, Shp2, positively contributes to FLT3-ITD-induced hematopoietic progenitor hyperproliferation and malignant disease in vivo. Leukemia. 2013;27(2):398–408. doi:10.1038/leu.2012.308.
165. Cancer Genome Atlas Research N. Genomic and epigenomic landscapes of adult de novo acute myeloid leukemia. N Engl J Med. 2013;368(22):2059–74. doi:10.1056/NEJMoa1301689.
166. Chan G, Kalaitzidis D, Neel BG. The tyrosine phosphatase Shp2 (PTPN11) in cancer. Cancer Metastasis Rev. 2008;27(2):179–92.
167. Sattler M, Mohi MG, Pride YB, Quinnan LR, Malouf NA, Podar K, Gesbert F, Iwasaki H, Li S, Van Etten RA, Gu H, Griffin JD, Neel BG. Critical role for Gab2 in transformation by BCR/ABL. Cancer Cell. 2002;1(5):479–92.
168. Sha F, Gencer EB, Georgeon S, Koide A, Yasui N, Koide S, Hantschel O. Dissection of the BCR-ABL signaling network using highly specific monobody inhibitors to the SHP2 SH2 domains. Proc Natl Acad Sci U S A. 2013;110(37):14924–9. doi:10.1073/pnas.1303640110.
169. Chen J, Yu WM, Daino H, Broxmeyer HE, Druker BJ, Qu CK. SHP-2 phosphatase is required for hematopoietic cell transformation by Bcr-Abl. Blood. 2007;109(2):778–85.
170. Scherr M, Chaturvedi A, Battmer K, Dallmann I, Schultheis B, Ganser A, Eder M. Enhanced sensitivity to inhibition of SHP2, STAT5, and Gab2 expression in chronic myeloid leukemia (CML). Blood. 2006;107(8):3279–87.
171. Teal HE, Ni S, Xu J, Finkelstein LD, Cheng AM, Paulson RF, Feng GS, Correll PH. GRB2-mediated recruitment of GAB2, but not GAB1, to SF-STK supports the expansion of Friend virus-infected erythroid progenitor cells. Oncogene. 2006;25(17):2433–43.
172. Ischenko I, Petrenko O, Gu H, Hayman MJ. Scaffolding protein Gab2 mediates fibroblast transformation by the SEA tyrosine kinase. Oncogene. 2003;22(41):6311–8.
173. Higuchi M, Tsutsumi R, Higashi H, Hatakeyama M. Conditional gene silencing utilizing the lac repressor reveals a role of SHP-2 in cagA-positive Helicobacter pylori pathogenicity. Cancer Sci. 2004;95(5):442–7.
174. Higashi H, Tsutsumi R, Muto S, Sugiyama T, Azuma T, Asaka M, Hatakeyama M. SHP-2 tyrosine phosphatase as an intracellular target of Helicobacter pylori CagA protein. Science. 2002;295(5555):683–6.
175. Tsutsumi R, Higashi H, Higuchi M, Okada M, Hatakeyama M. Attenuation of Helicobacter pylori CagA x SHP-2 signaling by interaction between CagA and C-terminal Src kinase. J Biol Chem. 2003;278(6):3664–70.
176. Ohnishi N, Yuasa H, Tanaka S, Sawa H, Miura M, Matsui A, Higashi H, Musashi M, Iwabuchi K, Suzuki M, Yamada G, Azuma T, Hatakeyama M. Transgenic expression of Helicobacter pylori CagA induces gastrointestinal and hematopoietic neoplasms in mouse. Proc Natl Acad Sci U S A. 2008;105(3):1003–8.
177. Mimuro H, Suzuki T, Tanaka J, Asahi M, Haas R, Sasakawa C. Grb2 is a key mediator of helicobacter pylori CagA protein activities. Mol Cell. 2002;10(4):745–55.
178. Suzuki M, Mimuro H, Suzuki T, Park M, Yamamoto T, Sasakawa C. Interaction of CagA with Crk plays an important role in Helicobacter pylori-induced loss of gastric epithelial cell adhesion. J Exp Med. 2005;202(9):1235–47. doi:10.1084/jem.20051027.
179. Higashi H, Nakaya A, Tsutsumi R, Yokoyama K, Fujii Y, Ishikawa S, Higuchi M, Takahashi A, Kurashima Y, Teishikata Y, Tanaka S, Azuma T, Hatakeyama M. Helicobacter pylori CagA induces Ras-independent morphogenetic response through SHP-2 recruitment and activation. J Biol Chem. 2004;279(17):17205–16.
180. Tsutsumi R, Takahashi A, Azuma T, Higashi H, Hatakeyama M. Focal adhesion kinase is a substrate and downstream effector of SHP-2 complexed with Helicobacter pylori CagA. Mol Cell Biol. 2006;26(1):261–76.
181. Saadat I, Higashi H, Obuse C, Umeda M, Murata-Kamiya N, Saito Y, Lu H, Ohnishi N, Azuma T, Suzuki A, Ohno S, Hatakeyama M. Helicobacter pylori CagA targets PAR1/MARK kinase to disrupt epithelial cell polarity. Nature. 2007;447(7142):330–3.

182. Lu H, Murata-Kamiya N, Saito Y, Hatakeyama M. Role of partitioning-defective 1/microtubule affinity-regulating kinases in the morphogenetic activity of Helicobacter pylori CagA. J Biol Chem. 2009;284(34):23024–36. doi:10.1074/jbc.M109.001008.
183. Yang Z, Xue B, Umitsu M, Ikura M, Muthuswamy SK, Neel BG. The signaling adaptor GAB1 regulates cell polarity by acting as a PAR protein scaffold. Mol Cell. 2012;47(3):469–83. doi:10.1016/j.molcel.2012.06.037.
184. Ormandy CJ, Musgrove EA, Hui R, Daly RJ, Sutherland RL. Cyclin D1, EMS1 and 11q13 amplification in breast cancer. Breast Cancer Res Treat. 2003;78(3):323–35.
185. Bekri S, Adelaide J, Merscher S, Grosgeorge J, Caroli-Bosc F, Perucca-Lostanlen D, Kelley PM, Pebusque MJ, Theillet C, Birnbaum D, Gaudray P. Detailed map of a region commonly amplified at 11q13-->q14 in human breast carcinoma. Cytogenet Cell Genet. 1997;79(1-2):125–31.
186. Bentires-Alj M, Gil SG, Chan R, Wang ZC, Wang Y, Imanaka N, Harris LN, Richardson A, Neel BG, Gu H. A role for the scaffolding adapter GAB2 in breast cancer. Nat Med. 2006;12(1):114–21.
187. Cancer Genome Atlas N. Comprehensive molecular portraits of human breast tumours. Nature. 2012;490(7418):61–70. doi:10.1038/nature11412.
188. Stephens PJ, Tarpey PS, Davies H, Van Loo P, Greenman C, Wedge DC, Nik-Zainal S, Martin S, Varela I, Bignell GR, Yates LR, Papaemmanuil E, Beare D, Butler A, Cheverton A, Gamble J, Hinton J, Jia M, Jayakumar A, Jones D, Latimer C, Lau KW, McLaren S, McBride DJ, Menzies A, Mudie L, Raine K, Rad R, Chapman MS, Teague J, Easton D, Langerod A, Oslo Breast Cancer C, Lee MT, Shen CY, Tee BT, Huimin BW, Broeks A, Vargas AC, Turashvili G, Martens J, Fatima A, Miron P, Chin SF, Thomas G, Boyault S, Mariani O, Lakhani SR, van de Vijver M, van't Veer L, Foekens J, Desmedt C, Sotiriou C, Tutt A, Caldas C, Reis-Filho JS, Aparicio SA, Salomon AV, Borresen-Dale AL, Richardson AL, Campbell PJ, Futreal PA, Stratton MR. The landscape of cancer genes and mutational processes in breast cancer. Nature. 2012;486(7403):400–4.
189. Banerji S, Cibulskis K, Rangel-Escareno C, Brown KK, Carter SL, Frederick AM, Lawrence MS, Sivachenko AY, Sougnez C, Zou L, Cortes ML, Fernandez-Lopez JC, Peng S, Ardlie KG, Auclair D, Bautista-Pina V, Duke F, Francis J, Jung J, Maffuz-Aziz A, Onofrio RC, Parkin M, Pho NH, Quintanar-Jurado V, Ramos AH, Rebollar-Vega R, Rodriguez-Cuevas S, Romero-Cordoba SL, Schumacher SE, Stransky N, Thompson KM, Uribe-Figueroa L, Baselga J, Beroukhim R, Polyak K, Sgroi DC, Richardson AL, Jimenez-Sanchez G, Lander ES, Gabriel SB, Garraway LA, Golub TR, Melendez-Zajgla J, Toker A, Getz G, Hidalgo-Miranda A, Meyerson M. Sequence analysis of mutations and translocations across breast cancer subtypes. Nature. 2012;486(7403):405–9. doi:10.1038/nature11154.
190. Shah SP, Roth A, Goya R, Oloumi A, Ha G, Zhao Y, Turashvili G, Ding J, Tse K, Haffari G, Bashashati A, Prentice LM, Khattra J, Burleigh A, Yap D, Bernard V, McPherson A, Shumansky K, Crisan A, Giuliany R, Heravi-Moussavi A, Rosner J, Lai D, Birol I, Varhol R, Tam A, Dhalla N, Zeng T, Ma K, Chan SK, Griffith M, Moradian A, Cheng SW, Morin GB, Watson P, Gelmon K, Chia S, Chin SF, Curtis C, Rueda OM, Pharoah PD, Damaraju S, Mackey J, Hoon K, Harkins T, Tadigotla V, Sigaroudinia M, Gascard P, Tlsty T, Costello JF, Meyer IM, Eaves CJ, Wasserman WW, Jones S, Huntsman D, Hirst M, Caldas C, Marra MA, Aparicio S. The clonal and mutational evolution spectrum of primary triple-negative breast cancers. Nature. 2012;486(7403):395–9. doi:10.1038/nature10933.
191. Brummer T, Schramek D, Hayes VM, Bennett HL, Caldon CE, Musgrove EA, Daly RJ. Increased proliferation and altered growth factor dependence of human mammary epithelial cells overexpressing the Gab2 docking protein. J Biol Chem. 2006;281(1):626–37.
192. Aceto N, Sausgruber N, Brinkhaus H, Gaidatzis D, Martiny-Baron G, Mazzarol G, Confalonieri S, Quarto M, Hu G, Balwierz PJ, Pachkov M, Elledge SJ, van Nimwegen E, Stadler MB, Bentires-Alj M. Tyrosine phosphatase SHP2 promotes breast cancer progression and maintains tumor-initiating cells via activation of key transcription factors and a positive feedback signaling loop. Nat Med. 2012;18(4):529–37. doi:10.1038/nm.2645.
193. Sausgruber N, Coissieux MM, Britschgi A, Wyckoff J, Aceto N, Leroy C, Stadler MB, Voshol H, Bonenfant D, Bentires-Alj M. Tyrosine phosphatase SHP2 increases cell motility

in triple-negative breast cancer through the activation of SRC-family kinases. Oncogene. 2014. doi:10.1038/onc.2014.170.

194. Wang Y, Sheng Q, Spillman MA, Behbakht K, Gu H. Gab2 regulates the migratory behaviors and E-cadherin expression via activation of the PI3K pathway in ovarian cancer cells. Oncogene. 2012;31(20):2512–20. doi:10.1038/onc.2011.435.
195. Dunn GP, Cheung HW, Agarwalla PK, Thomas S, Zektser Y, Karst AM, Boehm JS, Weir BA, Berlin AM, Zou L, Getz G, Liu JF, Hirsch M, Vazquez F, Root DE, Beroukhim R, Drapkin R, Hahn WC. In vivo multiplexed interrogation of amplified genes identifies GAB2 as an ovarian cancer oncogene. Proc Natl Acad Sci U S A. 2014;111(3):1102–7. doi:10.1073/pnas.1311909111.
196. Schneeberger VE, Luetteke N, Ren Y, Berns H, Chen L, Foroutan P, Martinez GV, Haura EB, Chen J, Coppola D, Wu J. SHP2E76K mutant promotes lung tumorigenesis in transgenic mice. Carcinogenesis. 2014;35(8):1717–25. doi:10.1093/carcin/bgu025.
197. Jones DT, Hutter B, Jager N, Korshunov A, Kool M, Warnatz HJ, Zichner T, Lambert SR, Ryzhova M, Quang DA, Fontebasso AM, Stutz AM, Hutter S, Zuckermann M, Sturm D, Gronych J, Lasitschka B, Schmidt S, Seker-Cin H, Witt H, Sultan M, Ralser M, Northcott PA, Hovestadt V, Bender S, Pfaff E, Stark S, Faury D, Schwartzentruber J, Majewski J, Weber UD, Zapatka M, Raeder B, Schlesner M, Worth CL, Bartholomae CC, von Kalle C, Imbusch CD, Radomski S, Lawerenz C, van Sluis P, Koster J, Volckmann R, Versteeg R, Lehrach H, Monoranu C, Winkler B, Unterberg A, Herold-Mende C, Milde T, Kulozik AE, Ebinger M, Schuhmann MU, Cho YJ, Pomeroy SL, von Deimling A, Witt O, Taylor MD, Wolf S, Karajannis MA, Eberhart CG, Scheurlen W, Hasselblatt M, Ligon KL, Kieran MW, Korbel JO, Yaspo ML, Brors B, Felsberg J, Reifenberger G, Collins VP, Jabado N, Eils R, Lichter P, Pfister SM, International Cancer Genome Consortium PedBrain Tumor P. Recurrent somatic alterations of FGFR1 and NTRK2 in pilocytic astrocytoma. Nat Genet. 2013;45(8):927–32. doi:10.1038/ng.2682.
198. Fryssira H, Leventopoulos G, Psoni S, Kitsiou-Tzeli S, Stavrianeas N, Kanavakis E. Tumor development in three patients with Noonan syndrome. Eur J Pediatr. 2008;167(9):1025–31. doi:10.1007/s00431-007-0636-3.
199. Schuettpelz LG, McDonald S, Whitesell K, Desruisseau DM, Grange DK, Gurnett CA, Wilson DB. Pilocytic astrocytoma in a child with Noonan syndrome. Pediatr Blood Cancer. 2009;53(6):1147–9. doi:10.1002/pbc.22193.
200. Voena C, Conte C, Ambrogio C, Boeri Erba E, Boccalatte F, Mohammed S, Jensen ON, Palestro G, Inghirami G, Chiarle R. The tyrosine phosphatase Shp2 interacts with NPM-ALK and regulates anaplastic lymphoma cell growth and migration. Cancer Res. 2007;67(9):4278–86.
201. Bard-Chapeau EA, Li S, Ding J, Zhang SS, Zhu HH, Princen F, Fang DD, Han T, Bailly-Maitre B, Poli V, Varki NM, Wang H, Feng GS. Ptpn11/Shp2 acts as a tumor suppressor in hepatocellular carcinogenesis. Cancer Cell. 2011;19(5):629–39. doi:10.1016/j.ccr.2011.03.023.
202. Fisher TJ, Williams N, Morris L, Cundy PJ. Metachondromatosis: more than just multiple osteochondromas. J Child Orthop. 2013;7(6):455–64. doi:10.1007/s11832-013-0526-3.
203. Sobreira NL, Cirulli ET, Avramopoulos D, Wohler E, Oswald GL, Stevens EL, Ge D, Shianna KV, Smith JP, Maia JM, Gumbs CE, Pevsner J, Thomas G, Valle D, Hoover-Fong JE, Goldstein DB. Whole-genome sequencing of a single proband together with linkage analysis identifies a Mendelian disease gene. PLoS Genet. 2010;6(6), e1000991. doi:10.1371/journal.pgen.1000991.
204. Bowen ME, Boyden ED, Holm IA, Campos-Xavier B, Bonafe L, Superti-Furga A, Ikegawa S, Cormier-Daire V, Bovee JV, Pansuriya TC, de Sousa SB, Savarirayan R, Andreucci E, Vikkula M, Garavelli L, Pottinger C, Ogino T, Sakai A, Regazzoni BM, Wuyts W, Sangiorgi L, Pedrini E, Zhu M, Kozakewich HP, Kasser JR, Seidman JG, Kurek KC, Warman ML. Loss-of-function mutations in PTPN11 cause metachondromatosis, but not Ollier disease or Maffucci syndrome. PLoS Genet. 2011;7(4), e1002050. doi:10.1371/journal.pgen.1002050.
205. Yang W, Wang J, Moore DC, Liang H, Dooner M, Wu Q, Terek R, Chen Q, Ehrlich MG, Quesenberry PJ, Neel BG. Ptpn11 deletion in a novel progenitor causes metachondromatosis by inducing hedgehog signalling. Nature. 2013;499(7459):491–5. doi:10.1038/nature12396.

206. Kim HK, Feng GS, Chen D, King PD, Kamiya N. Targeted disruption of Shp2 in chondrocytes leads to metachondromatosis with multiple cartilaginous protrusions. J Bone Mineral Res. 2014;29(3):761–9. doi:10.1002/jbmr.2062.
207. Digilio MC, Conti E, Sarkozy A, Mingarelli R, Dottorini T, Marino B, Pizzuti A, Dallapiccola B. Grouping of multiple-lentigines/LEOPARD and Noonan syndromes on the PTPN11 gene. Am J Hum Genet. 2002;71(2):389–94.
208. Legius E, Schrander-Stumpel C, Schollen E, Pulles-Heintzberger C, Gewillig M, Fryns JP. PTPN11 mutations in LEOPARD syndrome. J Med Genet. 2002;39(8):571–4.
209. Yu ZH, Xu J, Walls CD, Chen L, Zhang S, Zhang R, Wu L, Wang L, Liu S, Zhang ZY. Structural and mechanistic insights into LEOPARD syndrome-associated SHP2 mutations. J Biol Chem. 2013;288(15):10472–82. doi:10.1074/jbc.M113.450023.
210. Yu ZH, Zhang RY, Walls CD, Chen L, Zhang S, Wu L, Liu S, Zhang ZY. Molecular basis of gain-of-function LEOPARD syndrome-associated SHP2 mutations. Biochemistry. 2014;53(25):4136–51. doi:10.1021/bi5002695.

The Role of TCPTP in Cancer

Tony Tiganis

Abstract Human T cell protein tyrosine phosphatase (TCPTP; encoded by *PTPN2*) is a ubiquitous non-transmembrane tyrosine-specific phosphatase. TCPTP substrates include protein tyrosine kinases (PTKs) such as the epidermal growth factor receptor, Janus activated kinase (JAK) and c-Src and PTK substrates such as STAT3 and STAT5. TCPTP deficiency can enhance oncogenic tyrosine phosphorylation-dependent signalling to promote growth and survival, perturb cell cycle checkpoints and to promote tumorigenicity in vitro and in vivo. Recent studies have identified TCPTP as a *bona fide* tumour suppressor being deleted in *TLX1* overexpressing T cell acute lymphoblastic leukemias (T-ALLs) to promote cytokine signalling and to sensitise cells to transformation by oncogenic PTKs. TCPTP has also been identified as potential tumour suppressor in solid malignancies, in particular in breast cancer, promoting STAT3 signalling and tumorigenicity. This chapter will explore how perturbations in TCPTP regulation and function may contribute to the development of cancer, focusing in particular on TCPTP's roles in T-ALL and breast cancer.

Keywords Protein tyrosine phosphatase • Protein tyrosine kinase • T cell protein tyrosine phosphatase • TCPTP • PTPN2 • Cancer • Leukemia • T-ALL • Breast cancer • Tumour suppressor • STAT3 • Epidermal growth factor receptor • Src family kinase

T. Tiganis (✉)
Department of Biochemistry and Molecular Biology, Monash University, Victoria 3800, Australia

B.G. Neel, N. Tonks (eds.), *Protein Tyrosine Phosphatases in Cancer*,
DOI 10.1007/978-1-4939-3649-6_5

Abbreviations

ABC	Activated-B-cell
AMPK	AMP-activated protein kinase
c-Met	HGF receptor
DLBCL	Diffuse large B-cell lymphomas
EGF	Epidermal growth factor
EGFR	EGF receptor
ER	Oestrogen receptor
GAB2	Growth factor receptor-bound protein 2
GCB	Germinal centre B-cell
HCC	Hepatocellular carcinoma
HGF	Hepatocyte growth factor
IFN	Interferon
IL	Interleukin
IR	Insulin receptor
JAK	(Janus-activated kinase)
MAPK	Mitogen-activated protein kinase
NLS	Nuclear localisation signal
NPC	Nuclear pore complex
PDGFR	Platelet-derived growth factor receptor
PI3K	Phosphatidylinositol 3-kinase
PTK	Protein tyrosine kinase
PTP	Protein tyrosine phosphatase
PTP1B	Protein tyrosine phosphatase 1B
SFK	Src family kinase
SHP-2	Src homology phosphatase-2
SOCS3	Suppressor of cytokine signaling 3
STAT	Signal transducer and activator of transcription
T-ALL	T cell acute lymphoblastic leukemia
TCPTP	T cell PTP

Introduction

The phosphorylation of proteins on tyrosine residues is of fundamental importance to cellular signalling allowing cells to respond to varied extracellular and intracellular cues. Tyrosine phosphorylation is a reversible and dynamic process that is controlled by the opposing actions of protein tyrosine kinases (PTKs) and protein tyrosine phosphatases (PTPs). PTPs are a large and structurally diverse family of enzymes found in both eukaryotes and prokaryotes (reviewed in [1, 2]). In mammals, the PTP superfamily includes 38 classical PTPs, typified by the prototypic PTP1B, which act exclusively on tyrosine phosphorylated substrates (reviewed in [1, 2]). PTPs have been implicated in varied physiological processes and aberrations in PTP function have been linked with many human diseases including cancer (reviewed in [1, 2]).

Although generally regarded as negative regulators of PTKs, it is important to note that PTPs can function to both promote and antagonise tyrosine phosphorylation-dependent signalling (reviewed in [1, 2]) and can therefore serve as both proto-oncogenes and tumour suppressors [3, 4]. For example, activating somatic mutations in SHP-2 (Src-homology domain-containing phosphatase) that promote signalling via the Ras/mitogen-activated protein kinase (MAPK) pathway contribute to the onset/progression of several sporadic human malignancies including ~35 % of juvenile myelomonocytic leukemias [5–7]. In addition, the SHP-2-activating protein GAB2 (growth factor receptor-bound protein 2) is amplified and overexpressed in 10–15 % of human breast cancers driving SHP-2-mediated signalling to promote tumour stemness, maintenance, growth and invasion [8, 9]. Similarly, PTP1B is overexpressed in 40 % of human breast cancers coinciding with the amplification of the *ERBB2* oncogene (encoding the human epidermal growth factor receptor 2 PTK; also known as ErbB2) to promote ErbB2-mediated mammary tumorigenesis and metastases [10–15]. Thus, SHP-2 and PTP1B can serve as *bona fide* proto-oncogenes and their inhibition may be of therapeutic benefit in the treatment of human breast cancers. However, for the most part, PTPs serve as tumour suppressors, and gene deletions, inhibitory mutations or repression alleviate constraints on tyrosine phosphorylation-dependent signalling to exacerbate oncogenic PTK signalling and promote tumorigenesis. Several classical receptor-like transmembrane and cytosolic PTPs have been identified as potential tumour suppressors including *PTPRJ* [16, 17], *PTPRT* [18], *PTPRO* [19–22] and *PTPN12* [23] functioning with proto-oncogenes in the genesis and progression of human cancers. For example, in breast cancer, *PTPRO* is hypermethylated and repressed and promotes ErbB2 tyrosine phosphorylation and tumorigenicity [19–22], whereas *PTPN12* is deleted or repressed to activate proto-oncogenic PTKs such as ErbB1 to promote mammary epithelial cell transformation [23].

Understanding the role of PTPs in carcinogenesis can provide important insights into the genesis and progression of human tumours and be of prognostic and diagnostic value and may identify important new targets or approaches for therapeutic intervention [3, 4]. Human T cell protein tyrosine phosphatase (TCPTP; encoded by *PTPN2*) is a non-transmembrane tyrosine-specific PTP (reviewed in [2, 24]), so called because it was originally cloned from a T cell cDNA library [25]. Although abundant in T cells and hematopoietic cells [26, 27], TCPTP is a ubiquitous enzyme (reviewed in [2, 24]). Recent studies have identified TCPTP as a *bona fide* tumour suppressor in T cell acute lymphoblastic leukemia (T-ALL) [28] and as a potential tumour suppressor in solid malignancies, in particular in breast cancer [29]. This chapter will explore how perturbations in TCPTP regulation and function may contribute to the development of cancer, focusing in particular on TCPTP's roles in T-ALL and breast cancer.

TCPTP Localisation; Could Altered TC45 Trafficking Contribute to Cancer?

TCPTP contains a conserved PTP catalytic domain and a non-catalytic C-terminal segment that varies in length (Fig. 1) (reviewed in [2, 24]). At least two TCPTP variants are expressed in humans and rodents that arise from alternative mRNA splicing: a 48 kDa TCPTP (TC48), which is targeted to the endoplasmic reticulum (ER) and a 45 kDa TCPTP (TC45), which is targeted to the nucleus [25, 30–32] (Fig. 2). TC48 targeting to ER is mediated by hydrophobic residues (residues 396–415) within a 34-residue C-terminal tail (residues 382–415), which is replaced by a hydrophilic 6-residue sequence in TC45 (Fig. 1). Specific targeting of TC48 to the ER is also dependent on residues 346–358 [31] and might involve interactions with specific cargo receptor proteins implicated in the maintenance of the ER retention/retrieval system [33]. TC45 targeting to the nucleus is mediated by two basic clusters (residues 350–358 and 377–381; Fig. 1) that form an atypical bipartite nuclear localisation signal (NLS) that interacts with the nuclear import protein importin β1 [32]. Despite TC45 having an apparent exclusive nuclear localisation in resting cells, TC45 undergoes nucleocytoplasmic shuttling [34] and 'exits' the nucleus in response to varied stimuli (reviewed in [2, 24]) including growth factors [35], cytokines [36], hormones [37], integrin ligation [38, 39] and cellular stresses that activate the AMP-activated protein kinase (AMPK) [34]. TC45 nuclear 'exit' allows for access to substrates in the cytoplasm and at the plasma membrane for the attenuation of tyrosine phosphorylation-dependent signalling [34–38]. Interestingly, TC45 does not contain a classical nuclear export sequence and TC45 nuclear exit is not reliant on CRM1-mediated nuclear export [34]. Instead, by virtue of its small size, TC45 is capable of passively diffusing across the nuclear pore complex (NPC); tagging TC45 with GFP so that it is too large to passively diffuse across the NPC prohibits nuclear export [34]. Thus, active TC45 import is counterbalanced by passive diffusion out of the nucleus and TC45 cytoplasmic accumulation, at least in response

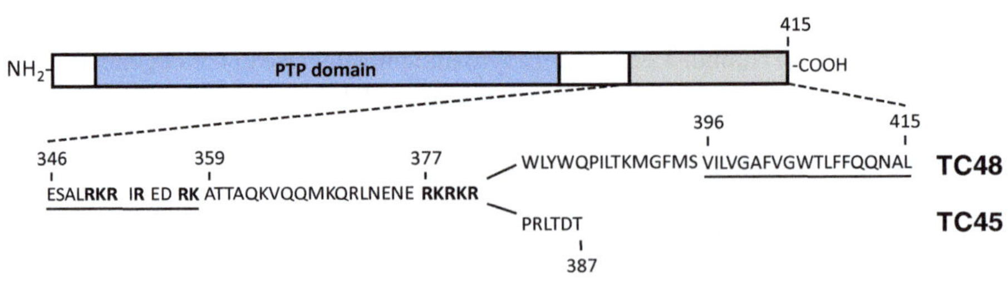

Fig. 1 Schematic representation of TCPTP. Shown are the C-terminal segments of the 45 kDa (TC45) and 48 kDa (TC48) TCPTP variants. Residues constituting the TC45 nuclear localisation sequence (NLS) are in bold and residues implicated in TC48 endoplasmic reticulum (ER) targeting are underlined

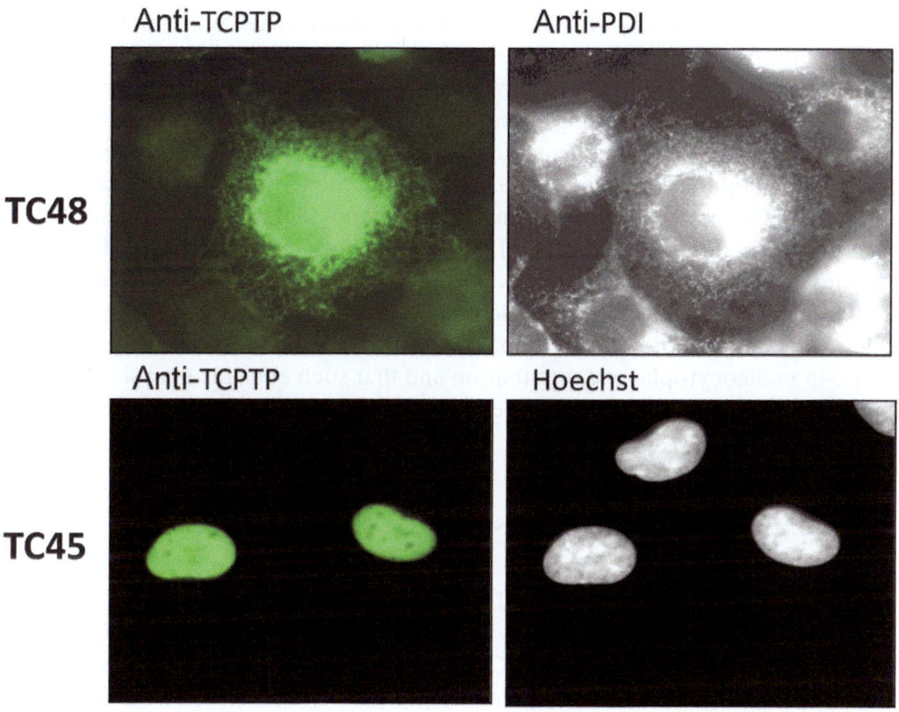

Fig. 2 TC45 and TC48 are differentially targeted to the nucleus and ER. COSs transiently overexpressing TC45 or TC48 were processed for immunofluorescence microscopy and stained with antibodies to TCPTP and either the ER marker protein disulphide isomerase (PDI), or the DNA stain Hoechst

to cellular stresses, might be mediated by the AMPK-mediated suppression of nuclear import. The regulated nucleocytoplasmic trafficking of TC45 might allow for the temporal control of cytoplasmic tyrosine phosphorylation-dependent signalling. Interestingly, AMPK signalling is thought to be diminished in certain tumours (reviewed in [40]). In particular, mTOR activity, which is inhibited by AMPK, is activated and contributes the progression of a variety of human tumours (reviewed in [40]). Also, loss of function mutations occur in the AMPK-activating kinase LKB1 and the AMPK substrate TSC2 (tuberous sclerosis complex 2) that serve as tumour suppressors (reviewed in [40]). Thus, diminished AMPK signalling in cancer may result in decreased TC45 cytoplasmic localisation to promote the phosphorylation and activation of cytoplasmic TCPTP substrates, such as c-Src [36]. Interestingly, SFK activity is increased in more than 80 % of colon tumours and this can serve as an independent negative prognostic indicator for disease progression and patient survival [41–44]. Moreover, inactivating mutations in LKB1 occur in patients with Peutz–Jeghers syndrome, which is characterised by gastrointestinal polyps and an increased risk of malignant gastrointestinal tumours [40, 45, 46]. It is possible that LKB1 mutations and diminished AMPK activation in Peutz-Jeghers syndrome may impact on TC45 nucleocytoplasmic shuttling to promote Src activation and the progression to malignant disease. In keeping with this possibility, recent studies have shown that LKB1 deficiency in melanocytes can result in the activation of the SFK Yes and the development of highly metastatic melanoma with 100 % penetrance [47]. Although evidence for alterations in TC45 nucleocytoplasmic shuttling in cancer is at present lacking, Shields et al. [29] have recently noted that TCPTP staining in human breast carcinomas was either nuclear and cytoplasmic, cytoplasmic alone, or focal nuclear. It is possible that the apparent distinct subcellular distribution of TCPTP in breast carcinomas may at least in part be attributed to changes in TC45 nucleocytoplasmic distribution and that such changes may serve to promote tyrosine phosphorylation-dependent signalling and tumorigenicity.

TCPTP Substrates: A Who's Who in Proto-oncogenes

Even a quick assessment of the TCPTP substrates identified to date (Table 1) reveals a large number of proto-oncogenes implicated in the development of both solid and hematological malignancies. TCPTP substrates implicated in cancer include: (1) receptor PTKs such as the epidermal growth factor receptor (EGFR; ErbB1)

Table 1 TCPTP Substrates

TCPTP substrates	TC45 substrates	TC48 substrates	Reference
EGFR (ErbB1)	+	+	[35]
ΔEGFR	+	?	[48]
IR	+	+	[37]
PDGFR	+	?	[49]
CSF1R	+	?	[140]
c-MET	+	+	[50]
JAK1	+	?	[52]
JAK3	+	?	[52]
c-Src	+	?	[36]
Fyn	+	?	[27]
Lck	+	?	[27]
$p52^{Shc}$	+	–	[35]
STAT1	+	+	[53]
STAT3	+	?	[55]
STAT5	+	?	[54]
STAT6	+	?	[56]
C3G	–	+	[64]
Syntaxin 17	–	+	[63]

Shown are TCPTP substrates and their recognition by the 48 kDa (TC48) and 45 kDa (TC45) TCPTP variants

[35, 38, 39, 48], the platelet-derived growth factor receptor β (PDGFR β) [49] and the hepatocyte growth factor (HGF) receptor (c-Met) [50], (2) non-receptor PTKs such as c-Src, Fyn, Lck [27, 36, 51] and JAK-1 and -3 [52] and (3) PTK substrates, such as the signal transducer and activator of transcription (STAT)-1, -3, -5 and -6 [51, 53–58]. Moreover, it is striking that a number of the identified TCPTP substrates function in concerted pathways. For example, TCPTP can dephosphorylate ErbB1 [35, 38, 39, 48], downstream cytosolic PTKs such as JAK1 [29, 52] and c-Src [29, 36] as well as their common substrate STAT3 [29, 51, 55, 57, 58] to comprehensively attenuate signalling. Similarly, TCPTP attenuates SFK and JAK PTKs as well as STAT3 in the context of the DNA replication checkpoint [51]. STAT3, which is activated in a wide range of both solid and hematological malignancies [59, 60], is in particular a prominent TCPTP substrate with *PTPN2* deletion in mice resulting in striking increases of STAT3 Y705 phosphorylation in many tissues including liver, lung, kidney, muscle, adipose tissue, heart and mammary tissue ([29]; Weide and Tiganis, unpublished observations). Moreover, TCPTP has been shown to regulate STAT3 signaling in a variety of tissues in vivo, including the liver where it regulates the STAT3-mediated repression of gluconeogenesis [57] and the hypothalamus, where TCPTP regulates leptin signaling and energy homeostasis [58]. Thus, it is conceivable that deficiencies in TCPTP or alterations in TC45 nucleocytoplasmic shuttling could have profound effects on oncogenic PTK pathway activation and tumorigenesis, in particular through the promotion of STAT3 signalling.

The TCPTP Variants Are Not Equal

In considering TCPTP's potential role in cancer, it is important to note that the two variants are not equal in their capacity to regulate PTK signalling. There is evidence that the TC48 and TC45 variants can act on distinct substrates or otherwise act on common substrates in a temporally and spatially distinct manner. Early studies using PTP 'substrate-trapping' mutants (mutants that form stable complexes with tyrosine phosphorylated substrates in a cellular context [35, 61]; for a review, see [2]) first demonstrated that TC48 and TC45 could recognise distinct substrates and could differentially affect signalling [35, 38]; whereas 'substrate-trapping' mutants of both variants bound to tyrosine phosphorylated EGFR in response to EGF, only the TC45 substrate-trapping mutant recognised to the tyrosine phosphorylated adaptor protein $p52^{Shc}$ [35] (Fig. 3) and only TC45 attenuated EGF-induced phosphatidylinositol 3-kinase (PI3K)/Akt signalling in COS1 cells [38]. The capacity of the TCPTP variants to differentially affect signalling might in part be linked to their ability to engage substrates such as the EGFR in different contexts. For example, TC48 'substrate-trapping' mutants can accumulate tyrosine phosphorylated EGFR at the ER [35], presumably reflecting a function for TC48 in preventing aberrant tyrosine phosphorylation of newly synthesised EGFR, or in inactivating the EGFR after it has undergone endocytosis as has been reported for the ER-bound PTP1B [62]. On the other hand, TC45 'substrate-trapping' mutants localise to the plasma

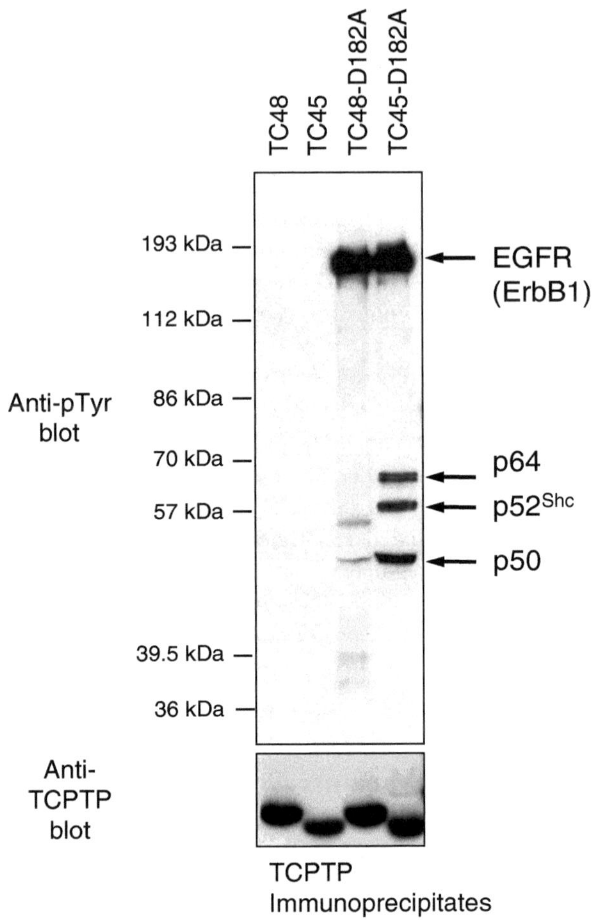

Fig. 3 Differential recognition of substrates by TC45 and TC8. COS1 cells transiently transfected with constructs for the expression of TC48 or TC45, or their corresponding D182A substrate trapping mutants (TC48D and TC45D) were serum starved and stimulated with EGF (100 ng/ml) for 15 min at 37 °C. Cells were then lysed and TCPTP immunoprecipitates (IPs) resolved by SDS-PAGE and immunoblotted with antibodies to phosphotyrosine (pTyr) or TCPTP. The major pTyr proteins coprecipitating with TC48D (EGFR) or TC45D (EGFR, p64, $p52^{Shc}$ and p50) are indicated by arrows on the right and positions of molecular size standards are shown on the left. This figure was adapted from reference [35]

membrane and prevent the endocytosis of receptor PTKs such as the EGFR or the IR ([35, 37, 38]; Tiganis and Tonks, unpublished observations), consistent with TC45 dephosphorylating receptor PTKs at the plasma membrane. In addition to exerting effects at the plasma membrane, TC45's nuclear locale affords it access to a unique subset of substrates, including STAT family members [51, 53–58]. Similarly, TC48 can act on unique substrates including both syntaxin 17 [63] and the guanine nucleotide exchange factor C3G at the Golgi [64]. Thus, the TCPTP variants have the capacity to regulate cellular signalling in different spatial and temporal contexts, with alterations in variant levels or trafficking potentially imparting distinct biological and pathological outcomes. As far as cancer is concerned, perturbations in TC45 function may be more pertinent, since recent studies demonstrate that TCPTP or TC45, but not TC48, deficiencies enhance PTK signalling and tumorigenicity in HeLa cervical adenocarcinoma cells [29].

TCPTP in Hematopoiesis and Beyond

Significant advances in our understanding of TCPTP's role in physiology have been achieved with the generation of global (*Ptpn2–/–*) [26, 65] and, more recently, tissue-specific TCPTP-knockout mice [27, 58, 66, 67]. Mice with a global deficiency in TCPTP [26] die soon after birth from severe anemia, hematopoietic defects and the development of systemic inflammatory disease, characterised by increases in circulating pro-inflammatory cytokines and lymphocytic infiltrates in non-lymphoid tissues [26, 68]. The hematopoietic defects and anemia and the overt morbidity and mortality in *Ptpn2–/–* mice are attributable to bone marrow stromal cell defects [26] associated with the abnormal production of interferon (IFN)-γ [69]. More recent studies have linked the elevated bone marrow stromal IFN-γ production in *Ptpn2–/–* mice to the increased production of interleukin (IL)-18 by hematopoietic stem cells [70]. However, further studies are necessary to elucidate the precise cellular and molecular bases for the disease and lethality in *Ptpn2–/–* mice. The recently developed *Ptpn2* floxed mice [27, 58, 67] provide a valuable resource to address TCPTP's role in biology. T cell-specific TCPTP-deficient mice (*Ptpn2$^{lox/lox}$;Lck*-Cre) have been used to establish that at least part of the inflammatory phenotype and the alterations in thymocyte development in *Ptpn2–/–* mice may be ascribed to enhanced T cell receptor signaling and T cell responses associated with elevated Lck activation [27]. TCPTP's role in T cell development and function and the implications for *PTPN2* deletion in the development of acute leukemias will be discussed later in this chapter.

In addition to affecting hematopoiesis and the immune system, TCPTP also regulates varied aspects of metabolism and body weight control. *Ptpn2+/–* heterozygous mice, which have a normal life expectancy, have been used to establish TCPTP's role in the attenuation of hepatic IR and STAT3 signalling and the regulation of hepatic glucose production [57]. More recently, mice lacking TCPTP in neuronal and glial cells (*Ptpn2$^{lox/lox}$;Nes*-Cre) have been used to establish TCPTP's role in the attenuation of leptin-induced STAT3 signalling in the hypothalamus and the central control of body weight/adiposity and glucose homeostasis [58]. Yet other studies characterising mice lacking TCPTP in osteoblasts have identified a role for TCPTP in the control of bone resorption [67]. The ongoing characterisation of conditional knockout mice will continue to define the varied roles of TCPTP in biology and disease. However, it is abundantly clear already that despite its predominant expression in immune cells, TCPTP exerts important regulatory functions outside the hematopoietic compartment. Thus, perturbations in TCPTP are likely to contribute to varied diseases including inflammatory diseases, autoimmunity, obesity, type 2 diabetes and potentially to both hematological malignancies and solid cancers by impacting on the response of both immune and non-immune cells alike.

TCPTP, A Potential Tumour Suppressor?

As discussed already, a large number of TCPTP substrates including EGFR, PDGFR, c-Met, SFKs, JAK1/3, STAT3 and STAT5 have been implicated in the genesis and progression of a wide range of human solid and hematological malignancies contributing to cellular growth, proliferation, survival, motility and invasion, angiogenesis and the development of chemotherapeutic resistance [44, 60, 71–73]. Moreover, over- and under-expression approaches have been used to establish TCPTP's capacity to regulate tumorigenic processes in vitro and in vivo. For example, TCPTP suppresses the activation of ErbB1 in response to EGF or integrin transactivation [35, 38] and dephosphorylates and suppresses the tumorigenicity associated with a constitutively active ErbB1 mutant known as ΔEGFR in vitro and in U87MG glioblastoma cell intracranial xenografts in vivo [48]. Furthermore, in response to integrin ligation, TCPTP can interact specifically with $\alpha_1\beta_1$ integrin to suppress ErbB1 transactivation and the associated tumorigenicity in HeLa cervical adenocarcinoma cells [29, 38, 39]. TCPTP has also been shown to serve as an integral regulator of the DNA replication checkpoint response [51], arguably the most important cell cycle checkpoint for preventing the genetic instability associated with cancer [74]. TCPTP-mediated inactivation of STAT3 signalling and the attenuation of cyclin D1 expression in S-phase in response to DNA replication stress serve to prevent checkpoint bypass and unscheduled cell division [51, 75]. In the absence of TCPTP, immortalised mouse embryo fibroblasts otherwise stalled in S-phase in response to replication stress progress into G2/M with incompletely replicated or damaged DNA and thereon proceed through mitosis with lagging chomosomes allowing for aneuploidy [51] and the acquisition of a transformed phenotype (Shields and Tiganis, unpublished observations). Importantly, RNA interference and the stable knockdown of TCPTP in HeLa cells can abrogate an otherwise intact DNA replication checkpoint response, whereas TC45 reconstitution into U87MG glioblastoma cells, which express little or no TC45 and have a defective checkpoint, can reinstate the checkpoint and suppress S-phase progression [51]. Thus, in addition to exacerbating tumour growth, cellular growth/proliferation and survival through the promotion of oncogenic PTK signalling, TCPTP deficiency might also contribute the cellular transformation and the genesis of certain tumours by abrogating cell cycle checkpoints.

There is growing evidence that TCPTP levels might be altered in human cancers. Lu et al. [56] reported that higher TCPTP levels in activated-B-cell (ABC)-like versus germinal centre B-cell (GCB)-like diffuse large B-cell lymphomas (DLBCL) in humans may suppress STAT6 signalling and contribute to the distinct biological characteristics of these tumour subtypes [56]. The elevated TCPTP in ABC- versus GCB-like DLBCLs was independent of tumour ontogeny consistent with it being an acquired feature [56], but it remains unclear if these findings reflect a relative increase or decrease in TCPTP expression in the respective DLBCL subtypes. Yet other studies have reported that TCPTP expression may be induced by *MYC* in lymphoma cell lines to contribute to the maintenance of *MYC*-driven B cell lymphoma xenografts [76], but it remains to be seen if this observation extends to primary human neoplasms. On the other hand, there is compelling evidence for TCPTP being decreased in certain cancers and for TCPTP serving as a tumour suppressor. Early studies demonstrated that the acquisition of resistance to the therapeutic drug Glivec (STI571, Imatinib) in chronic myelogenous leukemia cells in vitro was accompanied by the down-regulation of TCPTP and the promotion of STAT5 signalling [77]. Subsequent seminal studies by Kleppe et al. [28] reported that *PTPN2* was deleted in 6 % of all human T cell acute lymphoblastic leukemias (T-ALLs) to promote oncogenic PTK signalling, cellular proliferation and cytokine sensitivity. These and further studies identified TCPTP as a *bona fide* tumour suppressor in T-ALL demonstrating that TCPTP deficiency could cooperate with the NUP214-ABL1 translocation product, or activating *JAK1* mutations encoding for the JAK1 PTK to transform cells and promote T-ALL [28, 78]. Other studies have indicated that *PTPN2* expression may impact on the progression of hepatocellular carcinoma (HCC). In particular, gene expression profiling studies (82 patients) have indicated that *PTPN2* may be one of the top 20 of 113 genes whose presence early in HCC progression may be associated with good prognosis [79] and conversely, that decreased *PTPN2* expression may be associated with lymph node metastases [80]. Recently, TCPTP has also been identified as a putative tumour suppressor in human breast cancer, with TCPTP protein being 'lost' in a large proportion of oestrogen receptor (ER)-negative cancers, in particular in 'triple-negative' breast cancers [29].

TCPTP in T-ALL

T-ALL: T cell acute lymphoblastic leukemia (T-ALL) is an aggressive leukemia found in adults and children that constitutes roughly 10–15 % of pediatric cases and 25 % of all adult cases of acute leukemias. The prognosis of T-ALL has improved over time with intense chemotherapy achieving cure rates of 75 % in children and approximately 50 % in adults [81–83]. However, the outcome for patients with primary resistant or relapsed T-ALL remains poor [84, 85] and a greater understanding of how the underlying genetic alterations promote T-ALL is needed to develop new therapeutic strategies. T-ALL arises from the neoplastic transformation of lymphoid progenitors in the bone marrow and progresses with the stepwise accumulation of chromosomal abnormalities and gene mutations during T cell development [81–83]. Constitutive activation of NOTCH1 signalling and deletion of the *CDKN2A* locus (encompasses the *INK4A* and *ARF* tumour suppressor genes) are common in T-ALL [81, 83, 86, 87]. The translocation and aberrant expression of transcription factors, including that of homeobox genes *TLX1* and *TLX3*, are also important drivers of the transformation process [81–83]. Although *TLX1* is not normally expressed in hematopoietic cells, its aberrant expression associated with t(10;14)(q24;q11) translocation juxtaposing the *TLX1* gene downstream of T cell receptor (TCR) regulatory sequences occurs in 5–10 % of pediatric and 30 % of adults T-ALLs [81, 83]. Similarly, TLX3 is overexpressed in 20–25 % of pediatric and 5 % of adult T-ALLs harboring the t(5;14)(q35;q32) translocation [81, 83]. *TLX1* and *TLX3* overexpression occur in early T cell development and characterise the early cortical T-ALLs that are CD1a, CD4 and CD8 positive [81, 83]. *TLX1* and *TLX3* promote transformation by similar means and cooperate with other genetic lesions, including activating mutations in *JAK1* and the episomal fusion of *NUP214* to *ABL1* that generates the constitutively active NUP214-ABL1 PTK [81, 83, 88–90]. The contribution of such oncogenic PTKs to T-ALL pathogenesis may afford opportunity for therapeutic intervention with small molecule PTK inhibitors, such as the c-Abl inhibitor Imatinib, developed for BCR-ABL inhibition in CML [91, 92], or JAK PTK inhibitors that are currently in clinical development for the treatment of myeloproliferative disorders [93].

PTPN2 and T-ALL: *PTPN2* is deleted in 6% of T-ALLs with deletion being restricted to those T-ALLs overexpressing TLX1 or TLX3 [28]. Kleppe et al. [28] reported that 6 of 18 *PTPN2*-deleted T-ALLs also harbored the *NUP214-ABL1* fusion, whereas 2 of the 18 *PTPN2*-deleted T-ALLs [78] harbored mutations in *JAK1*. The JAK1 and NUP214-ABL1 PTKs can serve as *bona fide* TCPTP substrates, and short hairpin RNA (shRNA)-mediated TCPTP knockdown renders IL-3-dependent BA/F3 pro-B cells more sensitive to NUP214-ABL1- or JAK1-mediated transformation (as measured by IL-3 dependency) [28, 78]. Moreover, TCPTP knockdown enhances NUP214-ABL1-induced cellular proliferation and decreases the sensitivity of T-ALL cell lines to the c-Abl inhibitor Imatinib [28]. Similarly, TCPTP knockdown renders BA/F3 cells resistant to JAK PTK inhibition [78]. Thus, *PTPN2* deletion may exacerbate T-ALL proliferation and survival by cooperating with oncogenic PTKs to promote tyrosine phosphorylation-dependent signalling and confer resistance to small molecule PTK inhibitors that may otherwise be effective in treating T-ALL patients. Hence, it may be important to define *PTPN2* status prior to considering therapeutic approaches aimed at inhibiting oncogenic PTKs such as NUP214-ABL1 or JAK1.

At present, it remains unclear if *PTPN2* deletion represents an early genetic event that cooperates with TLX1/3 in cellular transformation. Recent studies have suggested that this may be possible, since TCPTP deficiency can alter early T cell development [27, 65]. T cells develop in the thymus from double negative (DN) thymocytes (CD4–CD8–) that undergo stepwise differentiation into double positive (DP) thymocytes ($CD4^{+}CD8^{+}$), which are thereon selected into CD4+CD8– and CD4–CD8+ single positive thymocytes based on their affinity for self-peptide and the resulting intensity and kinetics of T cell receptor (TCR) signalling [94–96]. TCPTP deficiency in thymocytes/T cells (*Ptpn2*$^{lox/lox}$;*Lck*-Cre) enhances TCR-induced SFK activation and promotes thymocyte positive selection [27]. TCPTP deficiency might also enhance cytokine signalling in particular that mediated by JAK1/3 and STAT5 to promote thymocyte survival [27, 52, 54]. Hence, *PTPN2* deletion may cooperate with TLX1 in perturbing T cell development. The need for such cooperation and additional neoplastic mutations in TLX1-mediated leukemogenesis is highlighted by the long latency to T-ALL and low penetrance following TLX1 overexpression in thymocytes and T cell progenitors [97, 98]. In *TLX1* transgenic mice, this involves the inactivation of T-ALL tumour suppressors, such as

Pten and *Ink4a-Arf*, and perturbations in the mitotic checkpoint [98]. Prior to leukemic onset, *TLX1* overexpression results in a block in DN thymocyte differentiation and the induction of apoptosis [98, 99]. Although the impact of TCPTP deficiency on DN thymocyte development has yet to be investigated, it is possible that the promotion of SFK and possibly JAK1/3-STAT5 signalling may bypass the TLX1-mediated thymocyte differentiation block to promote tumorigenicity. Moreover, *PTPN2* deletion may abrogate the DNA replication checkpoint [51] to cooperate with TLX1-induced mitotic checkpoint perturbations [98] to promote aneuploidy. Thus, PTPN2 deletion may act early in the genesis of T-ALL, as well as in the promotion of proliferation and survival of leukemic cells; however, this awaits experimental validation.

TCPTP in Breast Cancer

Breast cancer: The adult mammary epithelium is organised into ducts and lobules. Hyperplasia within the duct and lobules and the generation of abnormal cell layers is the earliest stage in breast cancer and is thought to be a precursor for the development of carcinoma in situ, most commonly ductal carcinoma in situ, a non-invasive lesion with abnormal cells. Ductal carcinoma in situ can progress to malignant invasive ductal carcinomas that ultimately account for 60–80 % of all breast tumours. The transformation of the breast epithelium to malignant and metastatic disease involves both genetic and epigenetic events and is deeply influenced by both ER and growth factor signalling, in particular that involving the EGFR/ErbB family of PTKs. Breast cancers can be subclassified according to the expression of ER, progesterone receptor (PR) and ErbB2 as well as tumour grade and transcript profiles [100]. Subtypes include (1) luminal A tumours that account for up to 60 % of breast cancers and express ER and/or PR, but not ErbB2, (2) luminal B tumours that account for 4–19 % of breast tumours, express ER or PR and are highly proliferative and/or express ErbB2, (3) highly aggressive ErbB2 positive tumours that are negative for ER and PR and account for 7–12 % of breast cancers and (4) basal-like tumours that account for 14–20 % of breast cancers and include the so-called *triple-negative* tumours that do not express ErbB2, ER or PR and are resistant to endocrine and trastuzumab-based therapies [100].

TCPTP substrates and breast cancer: Several TCPTP substrates including EGFR/ErbB1, SFKs and STAT3 have important roles in mammary gland development and have been intimately associated with breast cancer development. The EGFR family member ErbB2 is particularly important in breast cancer being overexpressed in roughly one third of all primary breast cancers and playing a causal role in tumorigenesis [101]. Expression of ErbB2 is sufficient to transform cells in vitro [102, 103] and expression of activated ErbB2 in the mouse mammary gland causes glandular hyperplasia followed by the development of adenocarcinoma [104]. Although ErbB2 has not been identified as a TCPTP substrate, ErbB2 can heterodimerise with ErbB1 to enhance EGF-induced signalling and ErbB1-mediated tumorigenicity. Moreover,

ErbB1 itself is overexpressed in triple-negative tumours and correlates with poor prognosis [105, 106]. Although ErbB1 is less transforming than ErbB2 [107], ErbB1 can cooperate with c-Src to promote the migration of breast cancer cell lines and anchorage-independent growth and aberrant human mammary epithelial cell acinar formation in three-dimensional cultures [108–112].

Elevated c-Src protein levels and/or SFK activity occur in a striking 70 % of primary breast cancers and often coincide with ErbB1 or ErbB2 overexpression [113–116]. In ductal carcinoma in situ, activated c-Src correlates with high tumour grade, high proliferation and high risk of recurrence or progression to invasive cancer [117]. SFKs cooperate with ErbB1 and ErbB2 to promote the breast cancer tumorigenicity [108–112, 118–120]. Moreover, SFKs can play an integral role in mediating ER signaling in ER+ tumours and also promote cellular proliferation, survival, invasion and chemotherapeutic resistance in ER–PR–ErbB2 basal-like tumours [116, 121, 122].

As for EGFR PTKs and SFKs, STAT3 is activated constitutively in many human breast cancers [116, 123]. Whereas STAT3 promotes cell death during normal mammary gland involution [124], in breast cancer, elevated STAT3 signalling enhances tumour cell proliferation and survival, angiogenesis, chemotherapeutic resistance and metastatic spread [116, 121, 125, 126]. During mammary gland involution, STAT3 is activated by IL-6 family cytokines, including leukemia inhibitory factor [127], which signal via gp130 [128]. Multiple factors contribute to the promotion of STAT3 signalling, including increased levels of cytokines such as IL-6, overexpression and/or activation of oncogenic PTKs such as ErbB1 and c-Src and the downregulation of negative regulators of JAK/STAT signalling, such as SOCS3 [113, 115, 116, 129–132]. In breast cancer, an increase in IL-6/gp130 signalling correlates with poor prognosis [131, 132] and promotes an invasive phenotype in mammospheres in vitro [133], whereas STAT3 knockdown attenuates xenograft growth and sensitises tumours to chemotherapeutics [121, 125].

TCPTP's capacity to attenuate ErbB1 [35, 38, 48], SFK [36] and STAT3 [51, 55, 58] signalling is consistent with the possibility that TCPTP may serve as a critical regulatory element in mammary gland development and that deficiencies in TCPTP might have a major impact on the genesis and progression of human breast cancer. Consistent with this possibility, recent studies have shown that (1) TCPTP is normally expressed in breast ductal and lobular epithelial cells [29], (2) TCPTP deficiency enhances SFK and STAT3 phosphorylation in murine mammary fat pads [29], (3) TCPTP protein is not detected in subset of human breast cancers that are ER negative [29], and (4) that TCPTP deficiency in breast cancer cell lines promotes tumorigenicity [29].

TCPTP and breast cancer: Although TCPTP's role in mammary gland physiology remains to be explored, recent studies point towards TCPTP status being important in breast cancer tumorigenicity. In particular, Shields et al. [29] have reported that TCPTP is lost in roughly half of all ER-negative breast tumours (20/45) and in 67 % (16/24) of triple-negative breast carcinomas (as assessed by immunohistochemical means; Fig. 4). Furthermore, an assessment of TCPTP status in primary breast cancer homogenates, or lysates from breast cancer cell lines by immunoblotting, not only

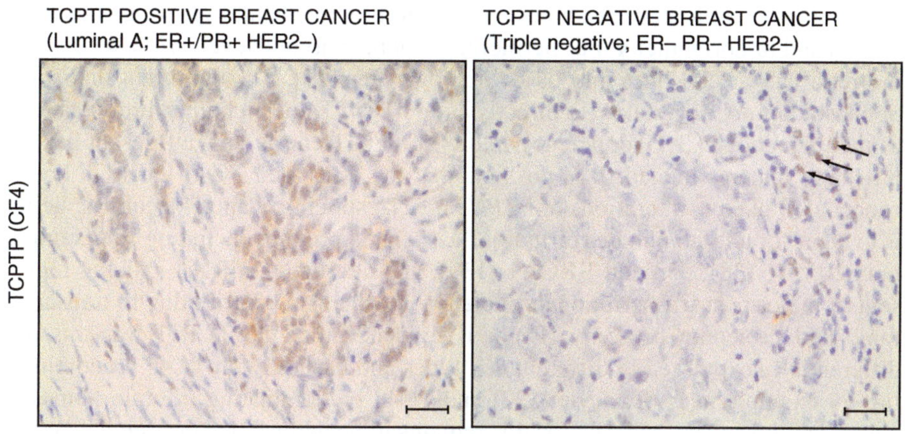

Fig. 4 TCPTP expression in human breast cancer. Formalin-fixed and paraffin-embedded breast cancers were deparaffinised and rehydrated and sections processed for immunohistochemistry and stained for TCPTP (CF4) and counterstained with hematoxylin. Representative images (400×) from TCPTP positive and negative tumours (arrows indicate TCPTP positive lymphocytes); Scale bar: 50 μm. This figure was adapted from reference [29]

reaffirms TCPTP levels being decreased, but also additionally illustrates that TC48:TC45 ratios are altered in some breast cancers so that relative TC45 levels are reduced [29]. Importantly, deficiencies in TCPTP or TC45 alone enhance PTK signalling, in particular SFK and STAT3 signalling, whereas TC45 reconstitution in TCPTP- (BT-483, MDA-MB-175 and MDA-MB-157) or TC45-deficient (MCF-7, MDA-MB-231) breast cancer cell lines dramatically suppresses cellular proliferation and survival in vitro [29]. In MDA-MB-231 cells, TC45 re-expression is also associated with a significant repression of anchorage-independent growth in vitro and xenograft growth in vivo [29]. These studies have identified TCPTP as a potential tumour suppressor in human breast cancer and have served to highlight TCPTP's capacity to act as a tumour suppressor in other solid tumours. However, it remains unclear if the decreased TCPTP levels evident in triple-negative tumours reflect *PTPN2* deletion, repression, or otherwise if they are indicative of differences in tumour cell lineage [134]. Although *PTPN2* is deleted in T-ALL [28, 78], no evidence has emerged for *PTPN2* deletion in solid tumours. An assessment of *PTPN2* expression in publicly available breast cancer data sets [135, 136] indicates that *PTPN2* mRNA levels are normally distributed [29], but these analyses do not exclude the contribution of stroma and in particular immune cells in which TCPTP is abundant to the overall *PTPN2* mRNA levels.

Further studies are needed to define the molecular basis for the decreased TCPTP levels in triple-negative breast cancers. The decreased TCPTP levels might be associated with changes in DNA methylation as reported for the receptor-like PTP *PTPRO*, which is decreased in breast cancer and is associated with elevated ErbB2 signalling and reduced breast cancer patient survival [19–22], or miR-mediated repression, as reported for *PTPN12* (encodes for PTP-PEST; [23]), which is lost in 60 % of triple-negative breast cancers to promote ErbB1 tyrosine phosphorylation and tumorigenicity. Furthermore, given the variable levels of TC45 evident in breast cancer cell lines and tissue homogenates [29], it is important that additional studies are undertaken to directly assess TC45 status in human breast cancer, as the impact of TCPTP deficiency in breast tumorigenesis may at present be significantly underestimated. It is possible that the decreased TC45 levels in breast cancer are linked to alterations in *PTPN2* mRNA splicing. Cancer-specific splicing is evident for a large number of tumour-associated proteins including for example mdm2 [137, 138], p53 and BRCA1 [139], and alternative mRNA splicing represents a major post-transcriptional regulatory mechanism contributing to transcriptome and proteome diversity. There is evidence that the TCPTP variants are differentially expressed in different cell types and tissues, with TC45 predominating for example in lymphocytes [27] and TC48 predominating in brain homogenates [58]; however, the basis for this remains to be defined. Determining whether this is linked to alterations in mRNA splicing could provide further insight into TCPTP's role in biology and delineate the basis for how TCPTP function is perturbed in breast cancer development.

Irrespective of the molecular basis for the decreased total TCPTP or TC45 in breast cancer, it will be important to determine whether TCPTP deficiency cooperates with proto-oncogenes such as ErbB1, STAT3 or SFKs for the promotion of breast cancer development and progression in vivo. This could be resolved by assessing whether mammary epithelial-specific TCPTP-deficiency exacerbates the onset and progression of tumours associated with the expression of oncogenes such as *ErbB2*. Moreover, it would be of interest to determine whether TCPTP/TC45 deficiency in human breast cancer coincides with the inactivation/repression of other PTPs such as *PTP-PEST* in triple-negative breast cancers [23], or with the overexpression/activation of PTPs such as PTP1B [10–14] or SHP2 [7–9] in ErbB2+ breast cancers, for the concerted promotion of oncogenic tyrosine phosphorylation-dependent signalling and cancer progression.

Concluding Remarks

In the last several years, strong evidence has emerged for TCPTP serving as a tumour suppressor in both hematological and solid malignancies. TCPTP's capacity to attenuate multiple key oncogenic tyrosine phosphorylation-dependent signalling pathways, in particular those involving SFKs and STAT3, raises the possibility that it may serve as a tumour suppressor in a wide variety of human tumours. Moreover, its contributions to both growth factor signalling and cellular proliferation/growth as well in cell cycle checkpoints, such as the DNA replication checkpoint [51], raise the possibility that perturbations in TCPTP function could contribute not only to the progression, but also the genesis of human cancers. This may be pertinent for example to T-ALL where TCPTP deficiency may cooperate TLX1 in bypassing DN thymocyte cell death and exacerbate oncogenic PTK signalling to promote both cellular transformation and chemotherapeutic resistance [28, 97–99]. Determining the precise contributions of TCPTP to T-ALL will require intensive and targeted study. Although to date *PTPN2* deletion has only been reported in T-ALLs [28], the decreased TCPTP versus TC45 protein levels evident in triple-negative breast cancers [29] are consistent with multiple mechanisms perturbing TCPTP function for the enhancement of oncogenic PTK signalling and the promotion of tumorigenicity. Such mechanisms might involve epigenetic modifications such as DNA methylation, as two putative CpG islands are present in the promoter region of *PTPN2* (Tiganis, unpublished observations), alterations in mRNA splicing affecting TC48:TC45 ratios and even changes in TC45 nucleocytoplasmic shuttling potentially affecting both cytoplasmic (e.g. SFKs) and nuclear (e.g. STAT3) oncogenic signalling. For the latter, it would be of interest to correlate TCPTP localisation in human tumours with the activation of AMPK, which has been implicated in the control of TC45 localisation [34]. Going forth it is important that reagents are generated that can discriminate between the TCPTP variants to discern their abundance and subcellular localisation for use in high throughput assays to systematically assess TCPTP's potential to contribute to cancer progression. Although TCPTP's precise role in the genesis and/or progression of human tumours is likely to vary for different cancer subtypes, delineating TCPTP status and the mechanisms by which it affects cancer development may ultimately be useful in stratifying cancer patients with respect to patient prognosis/survival and developing novel approaches for therapeutic intervention.

Acknowledgements T.T. is a National Health and Medical Research Council (NHMRC) of Australia Principal Research Fellow and supported by grants from the NHMRC.

References

1. Tonks NK. Protein tyrosine phosphatases: from genes, to function, to disease. Nat Rev Mol Cell Biol. 2006;7:833–46.
2. Tiganis T, Bennett AM. Protein tyrosine phosphatase function: the substrate perspective. Biochem J. 2007;402:1–15.
3. Ostman A, Hellberg C, Bohmer FD. Protein-tyrosine phosphatases and cancer. Nat Rev Cancer. 2006;6:307–20.
4. Julien SG, Dube N, Hardy S, Tremblay ML. Inside the human cancer tyrosine phosphatome. Nat Rev Cancer. 2011;11:35–49.
5. Zhang SQ, Yang W, Kontaridis MI, Bivona TG, Wen G, Araki T, Luo J, Thompson JA, Schraven BL, Philips MR, et al. Shp2 regulates SRC family kinase activity and Ras/Erk activation by controlling Csk recruitment. Mol Cell. 2004;13:341–55.
6. Mohi MG, Williams IR, Dearolf CR, Chan G, Kutok JL, Cohen S, Morgan K, Boulton C, Shigematsu H, Keilhack H, et al. Prognostic, therapeutic, and mechanistic implications of a mouse model of leukemia evoked by Shp2 (PTPN11) mutations. Cancer Cell. 2005;7:179–91.
7. Bentires-Alj M, Paez JG, David FS, Keilhack H, Halmos B, Naoki K, Maris JM, Richardson A, Bardelli A, Sugarbaker DJ, et al. Activating mutations of the noonan syndrome-associated SHP2/PTPN11 gene in human solid tumors and adult acute myelogenous leukemia. Cancer Res. 2004;64:8816–20.

8. Bentires-Alj M, Gil SG, Chan R, Wang ZC, Wang Y, Imanaka N, Harris LN, Richardson A, Neel BG, Gu H. A role for the scaffolding adapter GAB2 in breast cancer. Nat Med. 2006;12:114–21.
9. Aceto N, Sausgruber N, Brinkhaus H, Gaidatzis D, Martiny-Baron G, Mazzarol G, Confalonieri S, Quarto M, Hu G, Balwierz PJ, et al. Tyrosine phosphatase SHP2 promotes breast cancer progression and maintains tumor-initiating cells via activation of key transcription factors and a positive feedback signaling loop. Nat Med. 2012;18:529–37.
10. Bjorge JD, Pang A, Fujita DJ. Identification of protein-tyrosine phosphatase 1B as the major tyrosine phosphatase activity capable of dephosphorylating and activating c-Src in several human breast cancer cell lines. J Biol Chem. 2000;275:41439–46.
11. Wiener JR, Kerns BJ, Harvey EL, Conaway MR, Iglehart JD, Berchuck A, Bast Jr RC. Overexpression of the protein tyrosine phosphatase PTP1B in human breast cancer: association with p185c-erbB-2 protein expression. J Natl Cancer Inst. 1994;86:372–8.
12. Courjal F, Cuny M, Rodriguez C, Louason G, Speiser P, Katsaros D, Tanner MM, Zeillinger R, Theillet C. DNA amplifications at 20q13 and MDM2 define distinct subsets of evolved breast and ovarian tumours. Br J Cancer. 1996;74:1984–9.
13. Bentires-Alj M, Neel BG. Protein-tyrosine phosphatase 1B is required for HER2/Neu-induced breast cancer. Cancer Res. 2007;67:2420–4.
14. Julien SG, Dube N, Read M, Penney J, Paquet M, Han Y, Kennedy BP, Muller WJ, Tremblay ML. Protein tyrosine phosphatase 1B deficiency or inhibition delays ErbB2-induced mammary tumorigenesis and protects from lung metastasis. Nat Genet. 2007;39:338–46.
15. Tonks NK, Muthuswamy SK. A brake becomes an accelerator: PTP1B—a new therapeutic target for breast cancer. Cancer Cell. 2007;11:214–6.
16. Ruivenkamp CA, van Wezel T, Zanon C, Stassen AP, Vlcek C, Csikos T, Klous AM, Tripodis N, Perrakis A, Boerrigter L, et al. Ptprj is a candidate for the mouse colon-cancer susceptibility locus Scc1 and is frequently deleted in human cancers. Nat Genet. 2002;31:295–300.
17. Ruivenkamp C, Hermsen M, Postma C, Klous A, Baak J, Meijer G, Demant P. LOH of PTPRJ occurs early in colorectal cancer and is associated with chromosomal loss of 18q12-21. Oncogene. 2003;22:3472–4.
18. Wang Z, Shen D, Parsons DW, Bardelli A, Sager J, Szabo S, Ptak J, Silliman N, Peters BA, van der Heijden MS, et al. Mutational analysis of the tyrosine phosphatome in colorectal cancers. Science. 2004;304:1164–6.
19. Motiwala T, Kutay H, Ghoshal K, Bai S, Seimiya H, Tsuruo T, Suster S, Morrison C, Jacob ST. Protein tyrosine phosphatase receptor-type O (PTPRO) exhibits characteristics of a candidate tumor suppressor in human lung cancer. Proc Natl Acad Sci U S A. 2004;101:13844–9.
20. Motiwala T, Ghoshal K, Das A, Majumder S, Weichenhan D, Wu YZ, Holman K, James SJ, Jacob ST, Plass C. Suppression of the protein tyrosine phosphatase receptor type O gene (PTPRO) by methylation in hepatocellular carcinomas. Oncogene. 2003;22:6319–31.
21. Yu M, Lin G, Arshadi N, Kalatskaya I, Xue B, Haider S, Nguyen F, Boutros PC, Elson A, Muthuswamy LB, et al. Expression profiling during mammary epithelial cell three-dimensional morphogenesis identifies PTPRO as a novel regulator of morphogenesis and ErbB2-mediated transformation. Mol Cell Biol. 2012;32:3913–24.
22. Ramaswamy B, Majumder S, Roy S, Ghoshal K, Kutay H, Datta J, Younes M, Shapiro CL, Motiwala T, Jacob ST. Estrogen-mediated suppression of the gene encoding protein tyrosine phosphatase PTPRO in human breast cancer: mechanism and role in tamoxifen sensitivity. Mol Endocrinol. 2009;23:176–87.
23. Sun T, Aceto N, Meerbrey KL, Kessler JD, Zhou C, Migliaccio I, Nguyen DX, Pavlova NN, Botero M, Huang J, et al. Activation of multiple proto-oncogenic tyrosine kinases in breast cancer via loss of the PTPN12 phosphatase. Cell. 2011;144:703–18.
24. Tiganis T. PTP1B and TCPTP—nonredundant phosphatases in insulin signaling and glucose homeostasis. FEBS J. 2012;280(2):445–58.
25. Cool DE, Tonks NK, Charbonneau H, Walsh KA, Fischer EH, Krebs EG. cDNA isolated from a human T-cell library encodes a member of the protein-tyrosine-phosphatase family. Proc Natl Acad Sci U S A. 1989;86:5257–61.

26. You-Ten KE, Muise ES, Itie A, Michaliszyn E, Wagner J, Jothy S, Lapp WS, Tremblay ML. Impaired bone marrow microenvironment and immune function in T cell protein tyrosine phosphatase-deficient mice. J Exp Med. 1997;186:683–93.
27. Wiede F, Shields BJ, Chew SH, Kyparissoudis K, van Vliet C, Galic S, Tremblay ML, Russell SM, Godfrey DI, Tiganis T. T cell protein tyrosine phosphatase attenuates T cell signaling to maintain tolerance in mice. J Clin Invest. 2011;121:4758–74.
28. Kleppe M, Lahortiga I, El Chaar T, De Keersmaecker K, Mentens N, Graux C, Van Roosbroeck K, Ferrando AA, Langerak AW, Meijerink JP, et al. Deletion of the protein tyrosine phosphatase gene PTPN2 in T-cell acute lymphoblastic leukemia. Nat Genet. 2010;42:530–5.
29. Shields BJ, Wiede F, Gurzov EN, Wee K, Hauser C, Zhu HJ, Molloy TJ, O'Toole SA, Daly RJ, Sutherland RL, et al. TCPTP regulates SFK and STAT3 signalling and is lost in triple negative breast cancers. Mol Cell Biol. 2013;33:557–70.
30. Mosinger Jr B, Tillmann U, Westphal H, Tremblay ML. Cloning and characterization of a mouse cDNA encoding a cytoplasmic protein-tyrosine-phosphatase. Proc Natl Acad Sci U S A. 1992;89:499–503.
31. Lorenzen JA, Dadabay CY, Fischer EH. COOH-Terminal sequence motifs target the T-cell protein tyrosine phosphatase to the ER and nucleus. J Cell Biol. 1995;131:631–43.
32. Tiganis T, Flint AJ, Adam SA, Tonks NK. Association of the T-cell protein tyrosine phosphatase with nuclear import factor p97. J Biol Chem. 1997;272:21548–57.
33. Gupta V, Swarup G. Evidence for a role of transmembrane protein p25 in localization of protein tyrosine phosphatase TC48 to the ER. J Cell Sci. 2006;119:1703–14.
34. Lam MH, Michell BJ, Fodero-Tavoletti MT, Kemp BE, Tonks NK, Tiganis T. Cellular stress regulates the nucleocytoplasmic distribution of the protein tyrosine phosphatase TCPTP. J Biol Chem. 2001;276:37700–7.
35. Tiganis T, Bennett AM, Ravichandran KS, Tonks NK. Epidermal growth factor receptor and the adaptor protein $p52^{Shc}$ are specific substrates of T-cell protein tyrosine phosphatase. Mol Cell Biol. 1998;18:1622–34.
36. van Vliet C, Bukczynska PE, Puryer MA, Sadek CM, Shields BJ, Tremblay ML, Tiganis T. Selective regulation of tumor necrosis factor-induced Erk signaling by Src family kinases and the T cell protein tyrosine phosphatase. Nat Immunol. 2005;6:253–60.
37. Galic S, Klingler-Hoffmann M, Fodero-Tavoletti MT, Puryer MA, Meng TC, Tonks NK, Tiganis T. Regulation of insulin receptor signaling by the protein tyrosine phosphatase TCPTP. Mol Cell Biol. 2003;23:2096–108.
38. Tiganis T, Kemp BE, Tonks NK. The protein tyrosine phosphatase TCPTP regulates epidermal growth factor receptor-mediated and phosphatidylinositol 3-kinase-dependent signalling. J Biol Chem. 1999;274:27768–75.
39. Mattila E, Pellinen T, Nevo J, Vuoriluoto K, Arjonen A, Ivaska J. Negative regulation of EGFR signalling through integrin-alpha1beta1-mediated activation of protein tyrosine phosphatase TCPTP. Nat Cell Biol. 2005;7:78–85.
40. Luo Z, Saha AK, Xiang X, Ruderman NB. AMPK, the metabolic syndrome and cancer. Trends Pharmacol Sci. 2005;26:69–76.
41. Talamonti MS, Roh MS, Curley SA, Gallick GE. Increase in activity and level of pp60c-src in progressive stages of human colorectal cancer. J Clin Invest. 1993;91:53–60.
42. Cartwright CA, Coad CA, Egbert BM. Elevated c-Src tyrosine kinase activity in premalignant epithelia of ulcerative colitis. J Clin Invest. 1994;93:509–15.
43. Irby RB, Mao W, Coppola D, Kang J, Loubeau JM, Trudeau W, Karl R, Fujita DJ, Jove R, Yeatman TJ. Activating SRC mutation in a subset of advanced human colon cancers. Nat Genet. 1999;21:187–90.
44. Yeatman TJ. A renaissance for SRC. Nat Rev Cancer. 2004;4:470–80.
45. Boudeau J, Sapkota G, Alessi DR. LKB1, a protein kinase regulating cell proliferation and polarity. FEBS Lett. 2003;546:159–65.
46. Tiainen M, Vaahtomeri K, Ylikorkala A, Makela TP. Growth arrest by the LKB1 tumor suppressor: induction of p21(WAF1/CIP1). Hum Mol Genet. 2002;11:1497–504.

47. Liu W, Monahan KB, Pfefferle AD, Shimamura T, Sorrentino J, Chan KT, Roadcap DW, Ollila DW, Thomas NE, Castrillon DH, et al. LKB1/STK11 inactivation leads to expansion of a prometastatic tumor subpopulation in melanoma. Cancer Cell. 2012;21:751–64.
48. Klingler-Hoffmann M, Fodero-Tavoletti MT, Mishima K, Narita Y, Cavenee WK, Furnari FB, Huang HJ, Tiganis T. The protein tyrosine phosphatase TCPTP suppresses the tumorigenicity of glioblastoma cells expressing a mutant epidermal growth factor receptor. J Biol Chem. 2001;276:46313–8.
49. Persson C, Savenhed C, Bourdeau A, Tremblay ML, Markova B, Bohmer FD, Haj FG, Neel BG, Elson A, Heldin CH, et al. Site-selective regulation of platelet-derived growth factor beta receptor tyrosine phosphorylation by T-cell protein tyrosine phosphatase. Mol Cell Biol. 2004;24:2190–201.
50. Sangwan V, Paliouras GN, Abella JV, Dube N, Monast A, Tremblay ML, Park M. Regulation of the Met receptor-tyrosine kinase by the protein-tyrosine phosphatase 1B and T-cell phosphatase. J Biol Chem. 2008;283:34374–83.
51. Shields BJ, Hauser C, Bukczynska PE, Court NW, Tiganis T. DNA replication stalling attenuates tyrosine kinase signaling to suppress S phase progression. Cancer Cell. 2008;14:166–79.
52. Simoncic PD, Lee-Loy A, Barber DL, Tremblay ML, McGlade CJ. The T cell protein tyrosine phosphatase is a negative regulator of janus family kinases 1 and 3. Curr Biol. 2002;12:446–53.
53. ten Hoeve J, Ibarra-Sanchez MJ, Fu Y, Zhu W, Tremblay M, David M, Shuai K. Identification of a nuclear Stat1 protein tyrosine phosphatase. Mol Cell Biol. 2002;22:5662–8.
54. Aoki N, Matsuda T. A nuclear protein tyrosine phosphatase TC-PTP is a potential negative regulator of the PRL-mediated signaling pathway: dephosphorylation and deactivation of signal transducer and activator of transcription 5a and 5b by TC-PTP in nucleus. Mol Endocrinol. 2002;16:58–69.
55. Yamamoto T, Sekine Y, Kashima K, Kubota A, Sato N, Aoki N, Matsuda T. The nuclear isoform of protein-tyrosine phosphatase TC-PTP regulates interleukin-6-mediated signaling pathway through STAT3 dephosphorylation. Biochem Biophys Res Commun. 2002;297:811–7.
56. Lu X, Chen J, Sasmono RT, Hsi ED, Sarosiek KA, Tiganis T, Lossos IS. T-cell protein tyrosine phosphatase, distinctively expressed in activated-B-cell-like diffuse large B-cell lymphomas, is the nuclear phosphatase of STAT6. Mol Cell Biol. 2007;27:2166–79.
57. Fukushima A, Loh K, Galic S, Fam B, Shields B, Wiede F, Tremblay ML, Watt MJ, Andrikopoulos S, Tiganis T. T-cell protein tyrosine phosphatase attenuates STAT3 and insulin signaling in the liver to regulate gluconeogenesis. Diabetes. 2010;59:1906–14.
58. Loh K, Fukushima A, Zhang X, Galic S, Briggs D, Enriori PJ, Simonds S, Wiede F, Reichenbach A, Hauser C, et al. Elevated hypothalamic TCPTP in obesity contributes to cellular leptin resistance. Cell Metab. 2011;14:684–99.
59. Yu H, Pardoll D, Jove R. STATs in cancer inflammation and immunity: a leading role for STAT3. Nat Rev Cancer. 2009;9:798–809.
60. Yu H, Jove R. The STATs of cancer—new molecular targets come of age. Nat Rev Cancer. 2004;4:97–105.
61. Flint AJ, Tiganis T, Barford D, Tonks NK. Development of "substrate-trapping" mutants to identify physiological substrates of protein tyrosine phosphatases. Proc Natl Acad Sci U S A. 1997;94:1680–5.
62. Haj FG, Verveer PJ, Squire A, Neel BG, Bastiaens PI. Imaging sites of receptor dephosphorylation by PTP1B on the surface of the endoplasmic reticulum. Science. 2002;295:1708–11.
63. Muppirala M, Gupta V, Swarup G. Tyrosine phosphorylation of a SNARE protein, syntaxin 17: implications for membrane trafficking in the early secretory pathway. Biochim Biophys Acta. 2012;1823:2109–19.
64. Mitra A, Kalayarasan S, Gupta V, Radha V. TC-PTP dephosphorylates the guanine nucleotide exchange factor C3G (RapGEF1) and negatively regulates differentiation of human neuroblastoma cells. PLoS One. 2011;6, e23681.

65. Wiede F, Hui Chew S, van Vliet C, Poulton IJ, Kyparissoudis K, Sasmono T, Loh K, Tremblay ML, Godfrey DI, Sims NA, et al. Strain-dependent differences in bone development, myeloid hyperplasia, morbidity and mortality in Ptpn2-deficient mice. PLoS One. 2012;7, e36703.
66. Loh K, Merry TL, Galic S, Wu BJ, Watt MJ, Zhang S, Zhang ZY, Neel BG, Tiganis T. T cell protein tyrosine phosphatase (TCPTP) deficiency in muscle does not alter insulin signalling and glucose homeostasis in mice. Diabetologia. 2012;55:468–78.
67. Zee T, Settembre C, Levine RL, Karsenty G. T-cell protein tyrosine phosphatase regulates bone resorption and whole-body insulin sensitivity through its expression in osteoblasts. Mol Cell Biol. 2012;32:1080–8.
68. Heinonen KM, Nestel FP, Newell EW, Charette G, Seemayer TA, Tremblay ML, Lapp WS. T cell protein tyrosine phosphatase deletion results in progressive systemic inflammatory disease. Blood. 2004;103(9):3457–64.
69. Bourdeau A, Dube N, Heinonen KM, Theberge JF, Doody KM, Tremblay ML. TC-PTP-deficient bone marrow stromal cells fail to support normal B lymphopoiesis due to abnormal secretion of interferon-{gamma}. Blood. 2007;109:4220–8.
70. Bourdeau A, Trop S, Doody KM, Dumont DJ, Tremblay ML. Inhibition of T cell protein tyrosine phosphatase (TC-PTP) enhances IL-18-dependent hematopoietic stem cell expansion. Stem Cells. 2013;31(2):293–304.
71. Blume-Jensen P, Hunter T. Oncogenic kinase signalling. Nature. 2001;411:355–65.
72. Gherardi E, Birchmeier W, Birchmeier C, Vande Woude G. Targeting MET in cancer: rationale and progress. Nat Rev Cancer. 2012;12:89–103.
73. Burgess AW. EGFR family: structure physiology signalling and therapeutic targets. Growth Factors. 2008;26:263–74.
74. Bartek J, Lukas C, Lukas J. Checking on DNA damage in S phase. Nat Rev Mol Cell Biol. 2004;5:792–804.
75. Shields BJ, Tiganis T. Replication checkpoint control by a PTK/STAT3/cyclin D1 axis. Cell Cycle. 2009;8:223–30.
76. Young RM, Polsky A, Refaeli Y. TC-PTP is required for the maintenance of MYC-driven B-cell lymphomas. Blood. 2009;114:5016–23.
77. Shimizu T, Miyakawa Y, Iwata S, Kuribara A, Tiganis T, Morimoto C, Ikeda Y, Kizaki M. A novel mechanism for imatinib mesylate (STI571) resistance in CML cell line KT-1: role of TC-PTP in modulating signals downstream from the BCR-ABL fusion protein. Exp Hematol. 2004;32:1057–63.
78. Kleppe M, Soulier J, Asnafi V, Mentens N, Hornakova T, Knoops L, Constantinescu S, Sigaux F, Meijerink JP, Vandenberghe P, et al. PTPN2 negatively regulates oncogenic JAK1 in T-cell acute lymphoblastic leukemia. Blood. 2011;117:7090–8.
79. Hoshida Y, Villanueva A, Kobayashi M, Peix J, Chiang DY, Camargo A, Gupta S, Moore J, Wrobel MJ, Lerner J, et al. Gene expression in fixed tissues and outcome in hepatocellular carcinoma. N Engl J Med. 2008;359:1995–2004.
80. Lee CF, Ling ZQ, Zhao T, Fang SH, Chang WC, Lee SC, Lee KR. Genomic-wide analysis of lymphatic metastasis-associated genes in human hepatocellular carcinoma. World J Gastroenterol. 2009;15:356–65.
81. Aifantis I, Raetz E, Buonamici S. Molecular pathogenesis of T-cell leukaemia and lymphoma. Nat Rev Immunol. 2008;8:380–90.
82. Van Vlierberghe P, Ferrando A. The molecular basis of T cell acute lymphoblastic leukemia. J Clin Invest. 2012;122:3398–406.
83. Ferrando AA, Neuberg DS, Staunton J, Loh ML, Huard C, Raimondi SC, Behm FG, Pui CH, Downing JR, Gilliland DG, et al. Gene expression signatures define novel oncogenic pathways in T cell acute lymphoblastic leukemia. Cancer Cell. 2002;1:75–87.
84. Goldberg JM, Silverman LB, Levy DE, Dalton VK, Gelber RD, Lehmann L, Cohen HJ, Sallan SE, Asselin BL. Childhood T-cell acute lymphoblastic leukemia: the Dana-Farber Cancer Institute acute lymphoblastic leukemia consortium experience. J Clin Oncol. 2003;21:3616–22.

85. Oudot C, Auclerc MF, Levy V, Porcher R, Piguet C, Perel Y, Gandemer V, Debre M, Vermylen C, Pautard B, et al. Prognostic factors for leukemic induction failure in children with acute lymphoblastic leukemia and outcome after salvage therapy: the FRALLE 93 study. J Clin Oncol. 2008;26:1496–503.
86. Hebert J, Cayuela JM, Berkeley J, Sigaux F. Candidate tumor-suppressor genes MTS1 (p16INK4A) and MTS2 (p15INK4B) display frequent homozygous deletions in primary cells from T- but not from B-cell lineage acute lymphoblastic leukemias. Blood. 1994;84:4038–44.
87. Weng AP, Ferrando AA, Lee W, Morris JPT, Silverman LB, Sanchez-Irizarry C, Blacklow SC, Look AT, Aster JC. Activating mutations of NOTCH1 in human T cell acute lymphoblastic leukemia. Science. 2004;306:269–71.
88. Graux C, Cools J, Melotte C, Quentmeier H, Ferrando A, Levine R, Vermeesch JR, Stul M, Dutta B, Boeckx N, et al. Fusion of NUP214 to ABL1 on amplified episomes in T-cell acute lymphoblastic leukemia. Nat Genet. 2004;36:1084–9.
89. Asnafi V, Le Noir S, Lhermitte L, Gardin C, Legrand F, Vallantin X, Malfuson JV, Ifrah N, Dombret H, Macintyre E. JAK1 mutations are not frequent events in adult T-ALL: a GRAALL study. Br J Haematol. 2010;148:178–9.
90. Flex E, Petrangeli V, Stella L, Chiaretti S, Hornakova T, Knoops L, Ariola C, Fodale V, Clappier E, Paoloni F, et al. Somatically acquired JAK1 mutations in adult acute lymphoblastic leukemia. J Exp Med. 2008;205:751–8.
91. Quintas-Cardama A, Kantarjian H, Cortes J. Imatinib and beyond—exploring the full potential of targeted therapy for CML. Nat Rev Clin Oncol. 2009;6:535–43.
92. Quintas-Cardama A, Tong W, Manshouri T, Vega F, Lennon PA, Cools J, Gilliland DG, Lee F, Cortes J, Kantarjian H, et al. Activity of tyrosine kinase inhibitors against human NUP214-ABL1-positive T cell malignancies. Leukemia. 2008;22:1117–24.
93. Quintas-Cardama A, Kantarjian H, Cortes J, Verstovsek S. Janus kinase inhibitors for the treatment of myeloproliferative neoplasias and beyond. Nat Rev Drug Discov. 2011;10:127–40.
94. Hogquist KA, Baldwin TA, Jameson SC. Central tolerance: learning self-control in the thymus. Nat Rev Immunol. 2005;5:772–82.
95. Daniels MA, Teixeiro E, Gill J, Hausmann B, Roubaty D, Holmberg K, Werlen G, Hollander GA, Gascoigne NR, Palmer E. Thymic selection threshold defined by compartmentalization of Ras/MAPK signalling. Nature. 2006;444:724–9.
96. Palmer E, Naeher D. Affinity threshold for thymic selection through a T-cell receptor-co-receptor zipper. Nat Rev Immunol. 2009;9:207–13.
97. Rakowski LA, Lehotzky EA, Chiang MY. Transient responses to NOTCH and TLX1/HOX11 inhibition in T-cell acute lymphoblastic leukemia/lymphoma. PLoS One. 2011;6, e16761.
98. De Keersmaecker K, Real PJ, Gatta GD, Palomero T, Sulis ML, Tosello V, Van Vlierberghe P, Barnes K, Castillo M, Sole X, et al. The TLX1 oncogene drives aneuploidy in T cell transformation. Nat Med. 2010;16:1321–7.
99. Owens BM, Hawley TS, Spain LM, Kerkel KA, Hawley RG. TLX1/HOX11-mediated disruption of primary thymocyte differentiation prior to the CD4+CD8+ double-positive stage. Br J Haematol. 2006;132:216–29.
100. Reddy KB. Triple-negative breast cancers: an updated review on treatment options. Curr Oncol. 2011;18:e173–9.
101. Yarden Y. Biology of HER2 and its importance in breast cancer. Oncology. 2001;61 Suppl 2:1–13.
102. Di Marco E, Pierce JH, Knicley CL, Di Fiore PP. Transformation of NIH 3T3 cells by overexpression of the normal coding sequence of the rat neu gene. Mol Cell Biol. 1990;10:3247–52.
103. Di Fiore PP, Pierce JH, Kraus MH, Segatto O, King CR, Aaronson SA. erbB-2 is a potent oncogene when overexpressed in NIH/3T3 cells. Science. 1987;237:178–82.
104. Muller WJ, Sinn E, Pattengale PK, Wallace R, Leder P. Single-step induction of mammary adenocarcinoma in transgenic mice bearing the activated c-neu oncogene. Cell. 1988;54:105–15.

105. Corkery B, Crown J, Clynes M, O'Donovan N. Epidermal growth factor receptor as a potential therapeutic target in triple-negative breast cancer. Ann Oncol. 2009;20(5):862–7.
106. Nielsen TO, Hsu FD, Jensen K, Cheang M, Karaca G, Hu Z, Hernandez-Boussard T, Livasy C, Cowan D, Dressler L, et al. Immunohistochemical and clinical characterization of the basal-like subtype of invasive breast carcinoma. Clin Cancer Res. 2004;10:5367–74.
107. Muthuswamy SK, Li D, Lelievre S, Bissell MJ, Brugge JS. ErbB2, but not ErbB1, reinitiates proliferation and induces luminal repopulation in epithelial acini. Nat Cell Biol. 2001;3: 785–92.
108. Biscardi JS, Belsches AP, Parsons SJ. Characterization of human epidermal growth factor receptor and c-Src interactions in human breast tumor cells. Mol Carcinog. 1998;21: 261–72.
109. Biscardi JS, Maa MC, Tice DA, Cox ME, Leu TH, Parsons SJ. c-Src-mediated phosphorylation of the epidermal growth factor receptor on Tyr845 and Tyr1101 is associated with modulation of receptor function. J Biol Chem. 1999;274:8335–43.
110. Olayioye MA, Badache A, Daly JM, Hynes NE. An essential role for Src kinase in ErbB receptor signaling through the MAPK pathway. Exp Cell Res. 2001;267:81–7.
111. Olayioye MA, Beuvink I, Horsch K, Daly JM, Hynes NE. ErbB receptor-induced activation of stat transcription factors is mediated by Src tyrosine kinases. J Biol Chem. 1999;274: 17209–18.
112. Dimri M, Naramura M, Duan L, Chen J, Ortega-Cava C, Chen G, Goswami R, Fernandes N, Gao Q, Dimri GP, et al. Modeling breast cancer-associated c-Src and EGFR overexpression in human MECs: c-Src and EGFR cooperatively promote aberrant three-dimensional acinar structure and invasive behavior. Cancer Res. 2007;67:4164–72.
113. Verbeek BS, Vroom TM, Adriaansen-Slot SS, Ottenhoff-Kalff AE, Geertzema JG, Hennipman A, Rijksen G. c-Src protein expression is increased in human breast cancer. An immunohistochemical and biochemical analysis. J Pathol. 1996;180:383–8.
114. Hsieh FC, Cheng G, Lin J. Evaluation of potential Stat3-regulated genes in human breast cancer. Biochem Biophys Res Commun. 2005;335:292–9.
115. Rosen N, Bolen JB, Schwartz AM, Cohen P, DeSeau V, Israel MA. Analysis of pp60c-src protein kinase activity in human tumor cell lines and tissues. J Biol Chem. 1986;261:13754–9.
116. Diaz N, Minton S, Cox C, Bowman T, Gritsko T, Garcia R, Eweis I, Wloch M, Livingston S, Seijo E, et al. Activation of stat3 in primary tumors from high-risk breast cancer patients is associated with elevated levels of activated SRC and survivin expression. Clin Cancer Res. 2006;12:20–8.
117. Wilson GR, Cramer A, Welman A, Knox F, Swindell R, Kawakatsu H, Clarke RB, Dive C, Bundred NJ. Activated c-SRC in ductal carcinoma in situ correlates with high tumour grade, high proliferation and HER2 positivity. Br J Cancer. 2006;95:1410–4.
118. Finn RS, Dering J, Ginther C, Wilson CA, Glaspy P, Tchekmedyian N, Slamon DJ. Dasatinib, an orally active small molecule inhibitor of both the src and abl kinases, selectively inhibits growth of basal-type/"triple-negative" breast cancer cell lines growing in vitro. Breast Cancer Res Treat. 2007;105:319–26.
119. Finn RS. Targeting Src in breast cancer. Ann Oncol. 2008;19:1379–86.
120. Muthuswamy SK, Muller WJ. Direct and specific interaction of c-Src with Neu is involved in signaling by the epidermal growth factor receptor. Oncogene. 1995;11:271–9.
121. Gritsko T, Williams A, Turkson J, Kaneko S, Bowman T, Huang M, Nam S, Eweis I, Diaz N, Sullivan D, et al. Persistent activation of stat3 signaling induces survivin gene expression and confers resistance to apoptosis in human breast cancer cells. Clin Cancer Res. 2006;12:11–9.
122. Hochgrafe F, Zhang L, O'Toole SA, Browne BC, Pinese M, Porta Cubas A, Lehrbach GM, Croucher DR, Rickwood D, Boulghourjian A, et al. Tyrosine phosphorylation profiling reveals the signaling network characteristics of Basal breast cancer cells. Cancer Res. 2010;70:9391–401.
123. Alvarez JV, Febbo PG, Ramaswamy S, Loda M, Richardson A, Frank DA. Identification of a genetic signature of activated signal transducer and activator of transcription 3 in human tumors. Cancer Res. 2005;65:5054–62.

124. Chapman RS, Lourenco PC, Tonner E, Flint DJ, Selbert S, Takeda K, Akira S, Clarke AR, Watson CJ. Suppression of epithelial apoptosis and delayed mammary gland involution in mice with a conditional knockout of Stat3. Genes Dev. 1999;13:2604–16.
125. Ling X, Arlinghaus RB. Knockdown of STAT3 expression by RNA interference inhibits the induction of breast tumors in immunocompetent mice. Cancer Res. 2005;65:2532–6.
126. Barbieri I, Quaglino E, Maritano D, Pannellini T, Riera L, Cavallo F, Forni G, Musiani P, Chiarle R, Poli V. Stat3 is required for anchorage-independent growth and metastasis but not for mammary tumor development downstream of the ErbB-2 oncogene. Mol Carcinog. 2010;49:114–20.
127. Kritikou EA, Sharkey A, Abell K, Came PJ, Anderson E, Clarkson RW, Watson CJ. A dual, non-redundant, role for LIF as a regulator of development and STAT3-mediated cell death in mammary gland. Development. 2003;130:3459–68.
128. Heinrich PC, Behrmann I, Haan S, Hermanns HM, Muller-Newen G, Schaper F. Principles of interleukin (IL)-6-type cytokine signalling and its regulation. Biochem J. 2003;374:1–20.
129. Garcia R, Bowman TL, Niu G, Yu H, Minton S, Muro-Cacho CA, Cox CE, Falcone R, Fairclough R, Parsons S, et al. Constitutive activation of Stat3 by the Src and JAK tyrosine kinases participates in growth regulation of human breast carcinoma cells. Oncogene. 2001;20:2499–513.
130. Sutherland KD, Lindeman GJ, Choong DY, Wittlin S, Brentzell L, Phillips W, Campbell IG, Visvader JE. Differential hypermethylation of SOCS genes in ovarian and breast carcinomas. Oncogene. 2004;23:7726–33.
131. Selander KS, Li L, Watson L, Merrell M, Dahmen H, Heinrich PC, Muller-Newen G, Harris KW. Inhibition of gp130 signaling in breast cancer blocks constitutive activation of Stat3 and inhibits in vivo malignancy. Cancer Res. 2004;64:6924–33.
132. Bachelot T, Ray-Coquard I, Menetrier-Caux C, Rastkha M, Duc A, Blay JY. Prognostic value of serum levels of interleukin 6 and of serum and plasma levels of vascular endothelial growth factor in hormone-refractory metastatic breast cancer patients. Br J Cancer. 2003;88:1721–6.
133. Sansone P, Storci G, Tavolari S, Guarnieri T, Giovannini C, Taffurelli M, Ceccarelli C, Santini D, Paterini P, Marcu KB, et al. IL-6 triggers malignant features in mammospheres from human ductal breast carcinoma and normal mammary gland. J Clin Invest. 2007;117:3988–4002.
134. Visvader JE. Keeping abreast of the mammary epithelial hierarchy and breast tumorigenesis. Genes Dev. 2009;23:2563–77.
135. van de Vijver MJ, He YD, van't Veer LJ, Dai H, Hart AA, Voskuil DW, Schreiber GJ, Peterse JL, Roberts C, Marton MJ, et al. A gene-expression signature as a predictor of survival in breast cancer. N Engl J Med. 2002;347:1999–2009.
136. Glas AM, Floore A, Delahaye LJ, Witteveen AT, Pover RC, Bakx N, Lahti-Domenici JS, Bruinsma TJ, Warmoes MO, Bernards R, et al. Converting a breast cancer microarray signature into a high-throughput diagnostic test. BMC Genomics. 2006;7:278.
137. Fridman JS, Hernando E, Hemann MT, de Stanchina E, Cordon-Cardo C, Lowe SW. Tumor promotion by Mdm2 splice variants unable to bind p53. Cancer Res. 2003;63:5703–6.
138. Bartel F, Taubert H, Harris LC. Alternative and aberrant splicing of MDM2 mRNA in human cancer. Cancer Cell. 2002;2:9–15.
139. Okumura N, Yoshida H, Kitagishi Y, Nishimura Y, Matsuda S. Alternative splicings on p53, BRCA1 and PTEN genes involved in breast cancer. Biochem Biophys Res Commun. 2011;413:395–9.
140. Simoncic PD, Bourdeau A, Lee-Loy A, Rohrschneider LR, Tremblay ML, Stanley ER, McGlade CJ. T-cell protein tyrosine phosphatase (Tcptp) is a negative regulator of colony-stimulating factor 1 signaling and macrophage differentiation. Mol Cell Biol. 2006;26:4149–60.

PTP1B: From Metabolism to Cancer

David P. Labbé and Michel L. Tremblay

Abstract More than 25 years after its discovery, the prototypic protein-tyrosine phosphatase, PTP1B, remains at the forefront of research in the phosphatase field. In this chapter, we summarize its role in metabolic diseases that have heavily contributed to the development of targeted inhibitors. It is now clear that metabolism can be an active contributor to tumorigenesis and intriguingly some PTP1B substrates are known to play a central role in these diseases. In cancer, PTP1B acts on an array of cancer-related substrates including a variety of receptor tyrosine kinases. Because the range of PTP1B substrates generate a complex signaling network, ascertaining the net effect of PTP1B activity on tumorigenesis in a particular context proves to be a challenging task. Along this line, we discuss some of the most characterized systems in which PTP1B has been proven to act as an oncogene. Because PTP1B is ubiquitously expressed and acts on numerous substrates across all cell types, it is clear that its gene and protein regulation is critical for its specificity and in maintaining cellular homeostasis. Therefore, we have highlighted some of the important mechanisms of PTP1B regulation in relation to cancer, including its role on endocytosis, that are likely to influence a broader range of oncogenic processes. In conclusion, we present our thoughts on important questions relative to PTP1B as a target in cancer treatment.

Keywords Protein tyrosine phosphatases 1B • PTP1B • Tyrosine phosphorylation • Substrates • Tyrosine kinases • Diabetes • Obesity • Metabolism • Breast cancer • Prostate cancer • Angiogenesis • Tumor suppressor • Posttranslational modifications • Gene regulation • Endosomal trafficking

D.P. Labbé
Division of Experimental Medicine, Department of Medicine, Goodman Cancer Research Centre, McGill University, 1160 Pine Avenue, Room 601, Montréal, QC, Canada, H3A 1A3
e-mail: david.labbe@mail.mcgill.ca

M.L. Tremblay (✉)
Division of Experimental Medicine, Department of Medicine, Goodman Cancer Research Centre, McGill University, 1160 Pine Avenue, Room 601, Montréal, QC, Canada, H3A 1A3

Department of Biochemistry and Oncology, McGill University, Montréal, QC, Canada
e-mail: michel.tremblay@mcgill.ca

B.G. Neel, N. Tonks (eds.), *Protein Tyrosine Phosphatases in Cancer*,
DOI 10.1007/978-1-4939-3649-6_6

Introduction

Extracted and purified from human placenta in 1988, protein-tyrosine phosphatase 1B (PTP1B) is the founding member of the protein-tyrosine phosphatase (PTP) superfamily [1, 2]. Soon after, relatives of PTP1B (encoded by *PTPN1*) were identified and, as of today, 107 proteins are recognized as PTPs in humans [3].

As the biological counterpart of protein-tyrosine kinases (PTKs), PTPs were initially overlooked as uninteresting "off switches" for signaling pathways. Research on PTP1B contributed greatly in altering these assumptions, starting with a breakthrough study that demonstrated that its genetic ablation in mice leads to resistance to both type II diabetes and diet-induced obesity [4]. Thereafter, significant resources have been invested into the development of PTP1B inhibitors, and hundreds of scientific articles have been published on this topic.

As a PTP, and since most classical PTP1B substrates were receptor tyrosine kinases (RTKs) implicated in cancer progression, PTP1B was first hypothesized to act as a tumor suppressor in oncogenesis. However, the working hypothesis proved to be more nuanced, since in two independent in vivo studies, *Ptpn1* deletion was shown to delay the progression of an ErbB2-driven breast cancer in mice [5, 6]. Moreover, PTP1B overexpression in the mouse mammary gland led to breast tumorigenesis [5], a feature characteristic of putative oncogenes. From this point, it was clear that the predominant role of PTP1B in maintaining cellular homeostasis was not limited to the dephosphorylation of its classical RTK substrates, and that the prior housekeeping role assigned to PTP1B was inadequate.

PTP1B has been shown to provide pro-oncogenic effects by hastening growth in many different cancer types but it has also been demonstrated in other context to possess tumor suppressor characteristics. To date, over 30 PTP1B substrates have been identified [7] and additional ones that are cell and disease context specific are regularly reported. For example, coronin-1C in glioblastoma [8] and Munc18 in adipocytes [9] have been associated to important function of PTP1B in regulating, respectively, cell migration of tumor cells and glucose uptake. With this ever-growing number of putative PTP1B substrates, we chose to thoroughly discuss signaling pathways containing well-characterized PTP1B substrates involved in cell metabolism and in cancer. Finally, we include a section reviewing PTP1B transcriptional and posttranscriptional regulation, which is of utmost importance in understanding the conditions by which it is involved in neoplastic disease.

PTP1B in Metabolism

Diabetes

The tremendous increase in type II diabetes incidence in the past three decades [10] has drawn much attention to the possible heritable causes of this disease. During the 90s, genetic linkage studies identified several susceptibility loci for type II diabetes.

Interestingly, several of those identified susceptibility loci were found to map to the long arm of chromosome 20, namely the 20q13 region. Although the region identified encompasses many genes [11–13], moderate [14] to strong linkage evidence [15] of the D20S196 marker located in the 20q13.1-q13.2 [16] region hinted to the potential influence of the protein tyrosine phosphatase *PTPN1*.

Besides linkage studies, other findings also suggested a crucial role for a PTP in governing insulin signaling. In the early 70s, glucose uptake was found to be stimulated by reactive oxidative species (ROS) that had been generated by H_2O_2 induction [17]. Moreover, vanadate described as a potential sugar-lowering agent at the turn of the nineteenth century [18] was found in 1980 by Yoram Shechter and Steven J.D. Karlish to provide an insulin-mimetic action. Although the exact mechanism of action of this compound was not fully understood, they made the clever hypothesis that vanadate could possibly inhibit the activity of a cellular phosphatase, thus altering the degree of protein tyrosine phosphorylation involved in the regulation of sugar transport [19].

The identification of the prototypic PTP, named PTP1B by Nicholas K. Tonks and Edmond H. Fisher in [1], and its cloning 2 years later by Benjamin G. Neel's group [20], opened a whole new field of research (Fig. 1). Incubation of PTP1B with a phosphorylated insulin receptor (IR) preparation in vitro led to IR β subunit dephosphorylation [1]. Subsequent experiments confirmed the role of PTP1B on insulin signaling [21–23]. Higher PTP1B levels were also observed in rat models (ZDF/Drt/fa/fa and streptozotocin-diabetic rats) of diabetes [24, 25], and experiments confirmed that PTP1B interacts directly with the activated IR at specific phosphotyrosine sites (pTyr1162/pTyr1163) [23, 26]. However, it was only when genetic ablation of *Ptpn1* was found to decrease blood glucose and insulin levels in fed mice, that PTP1B was established as a critical regulator of insulin signaling (Fig. 1) [4]. Lack of PTP1B increased and prolonged IR tyrosine phosphorylation in the liver and muscles of insulin treated mice and also in the muscles of fat fed mice [4]. These results, subsequently supported by others [27], shed new light on the earlier metabolic observations on ROS and vanadate, as they were recognized as two powerful PTP inhibitors.

Obesity

Intriguingly, mice lacking PTP1B also were resistant to high-fat diet-induced obesity [4, 27] and were leaner compared with wild-type mice due to a marked reduction in fat cell mass [27]. Once again, genetic linkage studies had pointed to an obesity susceptibility locus on 20q13 (D20S196) near *PTPN1*, suggesting that this region might also be of interest in obesity (Fig. 1) [28].

Obesity is characterized by resistance to leptin, an adipocyte-derived hormone that exerts profound effects on energy expenditure, satiety, and neuroendocrine function. Introducing a *Ptpn1* deletion into leptin-deficient (*ob/ob*) mice reduced the weight and adipose tissue gain characteristic of *ob/ob* mice without modifying total food intake [29]. Hence, PTP1B genetic ablation per se was sufficient to sensitize mice to leptin and lead to a greater weight loss as compared to wild-type littermates (Fig. 1) [29, 30].

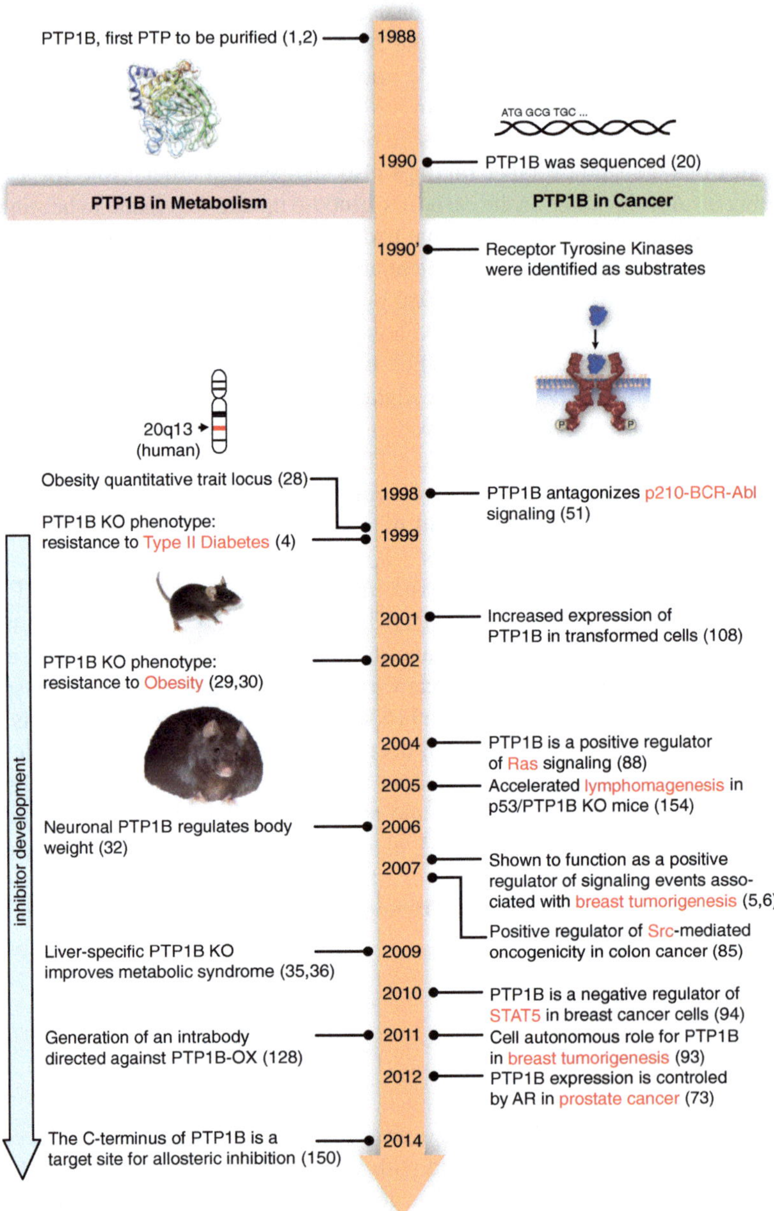

Fig. 1 Important milestones in PTP1B history

Conversely, restoration of normal leptin levels through intraperitoneal injections induced an even faster weight loss in compound *ob/ob*; *Ptpn1−/−* mice compared with *ob/ob* littermates. Thus, these results suggest that PTP1B regulates body weight both via leptin-independent and leptin-dependent pathways [29].

Activation of the leptin receptor by leptin triggers the tyrosine phosphorylation of Janus kinase 2 (JAK2) which, in turn, enables the recruitment and tyrosine phosphorylation of signal transducer and activator of transcription 3 (STAT3). Dissection of this pathway using PTP1B substrate-trapping mutants (C215S or D181A) revealed JAK2 as a putative PTP1B substrate [31]. High levels of PTP1B prevent leptin-mediated tyrosine phosphorylation of JAK2 (pTyr1007/pTyr1008) and subsequent STAT3 activation and downstream signaling [29, 30]. Consequently, PTP1B deficiency increases sensitivity to leptin. Remarkably, tissue-specific deletion of *Ptpn1* in the brain (but not in liver, skeletal muscle, or adipose tissue) largely recapitulates the phenotype observed in whole-body *Ptpn1−/−* mice, except for the serum leptin levels, which were increased instead of being reduced (Fig. 1) [32]. Further characterization led to the identification of the pro-opiomelanocortin (POMC) neurons as one neuronal subpopulation mediating the metabolic effect of PTP1B [33]. Moreover, PTP1B levels were specifically increased in the mediobasal hypothalamus of rats on a high-fat diet, and this increase was mirrored by a reduction in leptin sensitivity [34]. Additionally, liver-specific deletion of *Ptpn1* improved metabolic syndrome, which included a decrease in triglycerides and cholesterol levels [35, 36] whereas its muscle-specific deletion increased muscle glucose uptake, improved insulin sensitivity, and enhanced glucose tolerance [37]. Together with the role of PTP1B in regulating insulin sensitivity, these findings brought PTP1B to the forefront of diabetes and obesity treatments, driving the development of multiple inhibitors (Fig. 1) and leading to some initial clinical trials (discussed in Chap. 15).

At the Frontier Between Metabolism and Cancer

Nowadays, it is well documented that metabolic diseases are key factors in increasing the odds of developing several cancers. Indeed, patients with type II diabetes display an increased cancer risk and/or a worse cancer prognosis compared with healthy individuals [38]. Consistent with this notion, retrospective studies suggest that patients treated with the glucose-lowering drug metformin, which decreases insulin resistance, have decreased cancer risk [39]. Currently, dozens of prospective clinical trials are examining the protective effect of metformin in cancer. Interestingly, in contrast, patients treated with sulfonylureas (which increase circulating insulin) or exogenous insulin demonstrate an increased risk of tumor development [39], suggesting that insulin, via IR activation in tumor cells, might fuel cancer growth.

One of the most striking examples of the interrelation between metabolism and cancer is the role of the growth hormone (GH)/insulin-like growth factor (IGF) axis. GH promotes growth both directly by acting on target tissues such as liver and bones and indirectly through GH-induced production of IGF-1 and IGF-2 in the liver. Diet directly affects circulating IGF-1 protein levels. Increased energy consumption leads to higher circulating IGF-1, whereas severe caloric restriction leads to lower IGF-1 levels [40]. Epidemiological studies have linked a high circulating concentration of IGF-1 with an increased risk of prostate, breast, and colorectal cancer (Reviewed in REF [41]). At the molecular level, activation of the IGF-1 receptor (IGF-1R) by its ligand triggers both the PI3K and the Ras-MAPK pathways, which are central to cancer progression.

Additionally, GH and IGF-1 activities are controlled by the action of PTP1B [42, 43]. Upon binding of the GH to its receptor (GHR), homodimerization of the GHR leads to the recruitment and autophosphorylation of JAK2 and the subsequent recruitment and phosphorylation of STAT proteins. As it was previously observed with leptin signaling, PTP1B directly dephosphorylates JAK2, preventing further activation of the pathway. Consequently, PTP1B-deficient fibroblasts display an increase in STAT3 and STAT5 tyrosine phosphorylation. In vivo, PTP1B genetic ablation increased GH signaling in fasting mice but not in fed mice, suggesting a critical role for PTP1B under metabolic stress [42]. The IR and IGF-1R are seemingly derived from a common ancestor gene, and their signaling pathways are activated following the recruitment of the IR substrate (IRS) proteins. As observed with the IR, PTP1B expression decreases IGF-1R β-chain tyrosine kinase activity, an ability that was lost when the PTP1B D181A mutant was expressed. Mouse embryonic fibroblast (MEF) PTP1B knockouts had a greater activation of IGF-1R and phospho-Akt in response to IGF-1 stimulation compared with PTP1B wild-type MEFs. This resulted in greater cell motility, increased plating efficiency, and enhanced IGF-1-mediated survival [43]. In this line, PTP1B was found to be underexpressed in a panel of ovarian carcinoma-derived cell lines. Ectopic PTP1B expression in those cells leads to decreased cell migration/invasion, proliferation, and anchorage-independent survival, in part due to the attenuation of IGF-1R activity [44].

Therefore, as observed with the insulin and leptin signaling pathways, loss of PTP1B increases the output of the GH/IGF-1 axis. This potentiation might have little to no effect in normal circumstances. However, in the context of high IGF-1 circulating levels, as observed under high caloric consumption, it could hypothetically lead to an increase in cancer risk. Based on its negative regulation of the IGF-1 and IR signaling pathways described above, one may indeed be enticed to conclude that PTP1B acts as a tumor suppressor, an anti-oncogenic protection that would be expected to be suppressed by inhibiting PTP1B. Yet surprisingly, PTP1B knockout mice do not show a higher incidence of cancer [4], suggesting that any potential action in oncogenesis is more complex than the simple downregulation of the IGF-1R/IR signaling pathway.

PTP1B in Cancer

First identified as a kinase [45, 46], v-Src turned out in 1980 to be the first PTK to be identified [47] and, like many PTKs that were subsequently characterized [48], was revealed to be a putative oncogene. PTPs were discovered almost a decade later. As the biological counterparts of PTKs, PTPs were naturally thought of as tumor suppressors. Indeed, overexpressing PTP1B in transformed cells clearly supported such a hypothesis [49–51]. However, as new PTP1B substrates were identified, the once sharp border delimiting a tumor suppressor from an oncogene started to blur.

PTP1B Opposes Cell Transformation by Oncogenes

In order to define the role of the prototypic PTP, many early studies attempted to demonstrate that PTP1B could counteract some of the oncogenic PTKs identified at that time. Along this line, NIH 3T3 MEFs transformed with the v-Src oncogene and overexpressing PTP1, a rat homolog of PTP1B, had a 75 % reduction in their ability to form colonies in soft agar, as compared with PTP1 non-overexpressing cells [49]. Similarly, NIH 3T3 MEFs transformed with the *neu* oncogene had reduced tumorigenicity and impaired xenograft growth when overexpressing PTP1B [50]. Likewise, a tumor suppressing effect of PTP1B in p210-BCR/ABL (the p210 form of the chimeric BCR/ABL protein) transformed Rat-1 fibroblasts was also observed. Overexpression of PTP1B or PTP1B D181A impeded colony formation in soft agar, greatly diminished cell growth in reduced serum conditions, and lessened tumor formation in nude mice. Whereas PTP1B overexpression was insufficient to inhibit v-Abl-induced transformation, its transient overexpression mimicked treatment with CGP57148, a p210-BCR/ABL kinase inhibitor, and induced erythroid differentiation in p210-BCR/ABL-expressing K562 cells [51]. In fact, use of the PTP1B D181A mutant suggested that p210-BCR/ABL (but not v-Abl) is a putative PTP1B substrate [51] (Fig. 1). Overexpression of this phosphatase reduces p210-BCR/ABL tyrosine phosphorylation and prevents Grb2 binding, which is required for the full transformation mediated by this protein. However, the negative impact of PTP1B expression on the transforming capabilities of BCR/ABL expressing cells can also be attributed to a PTP-independent mechanism since overexpression of the catalytically dead PTP1B D181A mutant also demonstrated tumor suppressing capabilities [51]. Interestingly, PTP1B mRNA and protein levels were upregulated following transformation of a human myeloid cell line by p210-BCR/ABL, suggesting a possible cellular adaptation mechanism meant to prevent transformation [52]. Therefore, in the context of many overexpression studies, PTP1B acts as a tumor suppressor.

PTP1B: A Tumor Suppressor Regulating Cancer-Related RTKs?

Further evidence that supports a tumor-suppressing role of PTP1B is the unusually high number of RTKs that it negatively regulates. As described previously, PTP1B can either directly (IR, IGF-1R) or indirectly (leptin receptor, GHR) target metabolism-related RTKs and affects their signaling output. Similarly, PTP1B also regulates an array of tumor-related RTKs.

Epidermal Growth Factor Receptor

The epidermal growth factor receptor (EGFR) is the first of four members of the EGFRs family, and its amplification and/or mutation is common in some cancers such as breast, non-small cell lung cancer, or glioblastoma [53].

Using a catalytically inactive rat brain PTP1B mutant (C215S), which also acts as a trapping mutant, the EGFR was the first RTK found to associate with PTP1B in a ligand-dependent manner [54]. Development of an even more effective PTP1B substrate-trapping mutant (D181A) by the Tonks laboratory in 1997 validated EGFR as a *bona fide* substrate [55]. This was later confirmed in an unbiased quantitative mass spectrometry screen [56]. Accordingly, an increased and sustained EGFR activation and ERK phosphorylation was observed in PTP1B–/– MEFs following EGF stimulation [57].

Met Receptor

Activated by the hepatocyte growth factor (HGF), the RTK Met controls several cellular programs, such as proliferation, survival, and migration that are critical in embryogenesis and organ regeneration. Met was initially identified as an oncogene coding a constitutively active fusion protein named translocated promoter region (TPR)-Met in a chemically transformed human cell line [58]. Various mechanisms of aberrant Met activation, thought to support cell invasion and metastasis, have been observed in multiple cancer types [59]. Following HGF administration, Met was found to be hyperphosphorylated in livers from PTP1B–/– mice, compared with wild-type mice. PTP1B knockdown in cell lines treated with HGF recapitulated this finding, and substrate-trapping experiments identified the major autophosphorylation sites of Met, Tyr1234, and Tyr1235 as putative PTP1B targets. As a result, PTP1B depletion increased the invasive potential of cells in response to HGF [60]. Therefore, PTP1B negatively regulates the level of tyrosine phosphorylation and activity of two crucial RTKs present on tumor cells, clearly acting as a "tumor suppressor" when EGFR and Met signaling are isolated from the input of other pathways.

VEGFR2 and PDGFR

As PTP1B is ubiquitously expressed, it can potentially modulate the activity of other RTKs in various cell lineages. An interesting example is the regulation by PTP1B of two crucial RTKs implicated in angiogenesis: the vascular endothelial growth factor receptor 2 (VEGFR2) and the platelet-derived growth factor receptor (PDGFR). Rapid tumor outgrowth leads to poor access to oxygen in regions distant to blood vessels. This stabilizes hypoxia-inducible factors (HIFs), which translocate to the nucleus and induce various HIF-responsive genes such as VEGF-A, a ligand for VEGFR2. Secretion of VEGF-A by tumor cells triggers the recruitment of endothelial cells, which in turn secrete PDGF to recruit pericytes and stabilize the newly formed blood vessels. Surprisingly, chromatin immunoprecipitation identified *PTPN1* as a direct target of the hypoxia-inducible factor 1 alpha subunit (HIF-1α). Stabilization of HIF-1α, through loss of the von Hippel-Lindau (*VHL*) tumor suppressor gene or the expression of a constitutively stable HIFα protein, represses PTP1B expression [61].

Notably, PTP1B associates with VEGFR2. Although this association is independent of VEGFR2 tyrosine phosphorylation, high PTP1B levels (but not PTP1B C215S) inhibit VEGF-induced phosphorylation of VEGFR2 (Tyr1175) in endothelial cells [62], while *PTPN1* knockdown increases it. The knockdown results in a modulation of ERK 1/2 output as well as an increased endothelial cell proliferation. As well, overexpression of PTP1B wild-type but not the C215S mutant stabilizes VE-cadherin-mediated cell–cell adhesion following VEGF treatment. In this context, PTP1B expression would dampen VEGFR2 signaling and decrease vessel permeability, possibly preventing tumor cell intravasation towards circulation and metastasis formation [63]. However, the negative regulation of PTP1B by HIF-1α at the genomic level might be part of a built-in system meant to potentiate signaling through VEGFR2 by lowering PTP1B levels, at least when endothelial cells are in hypoxic conditions, thus favoring their survival. In this line, Lanahan et al. recently demonstrated that endothelial-specific deletions of *PTPN1* in mice result in heightened VEGF-dependent signaling, enhanced angiogenesis and increased arteriogenesis [64].

Endothelial cells that are part of a newly developed blood vessel require the recruitment of pericytes in order to be fully stabilized. In accordance with this, PTP1B was also found to interact with PDGFR [65], and treatment of PTP1B−/− MEFs with PDGF induces increased and sustained PDGFR tyrosine phosphorylation [57]. Quantitative mass spectrometry also confirmed an increased tyrosine phosphorylation of several phosphopeptides in PTP1B−/− MEFs compared to their wild-type counterparts, suggesting that PDGFR might be a putative PTP1B substrate [56]. Therefore, we can hypothesize that an increased PTP1B expression in pericytes would reduce PDGFR signaling output, reduce pericyte recruitment to newly formed blood vessels and result in immature and leaky vessels as opposed to the observed effect in endothelial cells. Indeed, whole-body *Ptpn1* knockout mice were protected from lung metastasis that are characteristic of the ErbB2-induced breast cancer model [5] suggesting that in this context, the effect of PTP1B on PDGFR signaling in pericytes might take precedence over its role on VEGFR2 activity in endothelial cell. In this report, the use of a total knockout mouse likely involves not only a cell autonomous role for PTP1B in the tumor cells, but also systemic effects affecting metabolic regulation and immune response. Taken together, the modulation of RTKs conferred by PTP1B provides crucial evidences of the different cellular context outcomes that may occur in several signaling pathways. Thus its makes the net influence of PTP1B on tumorigenesis difficult to predict without proper cell type-specific experimentation.

Regulating RTK Heterocomplexes

Predicting the net effect of PTP1B modulation might become even more challenging when we consider that many RTKs interact with each other and can form cell-specific heterocomplexes. As an example, VEGFR2 can form a VEGFR2/Met heterocomplex on tumor cells and a VEGFR2/PDGFRβ heterocomplex on pericytes. In a recent study, the Bergers' laboratory demonstrated that VEGF negatively regulates glioblastoma tumor cell invasion through the recruitment of PTP1B to a VEGFR2/Met heterocomplex [66]. This results in the suppression of HGF-dependent Met phosphorylation and, consequently, tumor cell migration. Of paramount importance, VEGF blockade restored Met activity highlighting that under those conditions, a successful anti-VEGF therapy must also target the Met receptor [66]. These experiments demonstrate a critical point: although PTP1B dephosphorylates, and thus decreases the signaling of an array of RTKs, its resulting effect depends on the specific RTKs activation status at a particular time-point. Therefore, the impact of PTP1B in a context where only the Met ligand, HGF, is available to a glioblastoma tumor cell will be different compared to situations where only VEGF or a combination of HGF/VEGF is present. However, signaling output from cancer-related RTKs converges to some critical nodes and several PTP1B substrates are part of these nodes. When taking into account the regulation of RTK's downstream signaling by PTP1B, its oncogenic potential becomes unambiguous.

Mechanisms of PTP1B Oncogenicity

Modifications that increase a protein level and/or activity specifically in a cancer tissue as compared to a matched healthy control typically suggest a tumor-promoting role. For example, with the exception of esophageal cancer, in which PTP1B mRNA levels are decreased in the affected mucosa [67], PTP1B levels are usually increased in a broad number of cancer types. Amplification of chromosome 20q13 has been observed in various malignancies, including pancreatic [68], gastric [69], breast [70], and prostate [71, 72] tumors. In a recent study, we found that *PTPN1* itself was highly amplified in metastatic prostate cancer tumors [73] and that its amplification, together with other genes located on the 20q13 common amplified region, was associated with a subset of high-risk primary tumors [73, 74]. Furthermore, an increase in PTP1B expression has been documented in prostate [73] as well as in breast [75], ovarian [76], and gastric cancer [77], with the exception of hepatocellular carcinoma in which PTP1B expression is downregulated [78]. Therefore, PTP1B modulation in human tissues suggests that this phosphatase may promote tumorigenesis.

PTP1B, a Positive Regulator of Src Activity

Under basal conditions, between 90 % and 95 % of Src proteins are phosphorylated on their Tyr529 residue located on the C-terminal tail. Phosphorylation of Tyr529 is performed by the constitutively active C-terminal Src kinase (Csk) or the Csk homologous kinase (encoded by *MATK*) [79]. The phosphorylation of Tyr529 favors an intramolecular association between the tyrosine residue and the SH2 domain, keeping the enzyme inactive under most circumstances [80]. Therefore, dephosphorylation of Src C-terminal tail, which can be triggered by a PTP, is sufficient for Src activation. It is this mechanism of regulation that is lost in v-src, since the viral version lacks the C-terminal phosphorylation site [81].

Insight that PTP1B potentially regulates Src activity came in 1998 when Arregui et al. observed a decrease in the overall Src tyrosine phosphorylation when a wild-type PTP1B but not the C215S mutant was overexpressed. Because Src activity was concurrently increased, they hypothesized that PTP1B might dephosphorylate Tyr529 and act as a positive regulator of Src [82]. Subsequently, the laboratory of Donald J. Fujita found that PTP1B protein levels were often increased in breast cancer cell lines with higher Src activity. They demonstrated that increasing levels of PTP1B (but not the C215S mutant) decreased the overall Src phospho-tyrosine levels, but more specifically Tyr529, thereby regulating Src activity [83]. Increased Src activity due to a decrease in Tyr529 phosphorylation was also observed by other groups to be associated with PTP1B overexpression [84]. In colon cancer cells, PTP1B-mediated activation of c-Src caused an acceleration of colony formation in soft agar and increased tumor growth in immunodeficient mice [85] (Fig. 1). Analogous results were obtained in breast cancer cells, where increased PTP1B activity led to the formation of invadopodia, possibly through Src activation [86]. More recently, bimolecular fluorescence complementation (BiFC) analysis suggested that PTP1B and Src interact at the plasma membrane, and this interaction is abrogated when Src Tyr529 is mutated to a phenylalanine residue [87]. Although Src has yet to be confirmed as a PTP1B substrate, the body of evidence suggests that PTP1B positively modulates Src activity and thus can act as a tumor promoter.

Regulation of the Ras-MAPK Pathway Through p62Dok

Although PTP1B depletion leads to an increase in the tyrosine phosphorylation and activation of many RTKs, their downstream signaling is not always increased to the same extent. Indeed, MAPK activation measured in terms of phospho-ERK levels was minimally augmented in PTP1B–/– MEFs treated with PDGF, while PDGFR tyrosine phosphorylation was unambiguously increased [57]. Strikingly, phospho-ERK levels were even decreased in MEFs negative for PTP1B following IGF-1 treatment even though IGF-1R was clearly activated and tyrosine phosphorylated [43]. These unexpected results advocate for the existence of a PTP1B substrate(s) that would oppose RTK signaling, dampening its effect. In order to investigate this hypothesis, we studied the role of PTP1B in PDGF signaling using MEFs immortalized by the SV40 large T antigen (TAg) [88]. Importantly, Src activity is dispensable for PDGF signaling in TAg immortalized MEFs [89, 90], which excludes PTP1B-related activity on Src in this context. In this study following PDGF treatment, PTP1B deficiency results in increased phospho-Akt levels, but decreased phospho-ERK levels, in PTP1B-knockout cells, as compared with corresponding PTP1B+/+ MEFs. Comparatively, Ras activity was diminished in PTP1B-deficient TAg immortalized MEFs. The explanation of this result is twofold. First, PTP1B deficiency induces an increase in p120RasGAP protein levels through an unknown mechanism, which promotes Ras intrinsic GTPase activity, thus attenuating its signaling. Second, the adaptor protein p62 protein downstream of tyrosine kinases (p62Dok) was identified as a putative PTP1B substrate that was confirmed later, together with p120RasGAP, by mass spectrometry analysis [56]. When phosphorylated on tyrosine, p62Dok attenuates Ras activity and consequently MAPK signaling [88] (Fig. 1). Understanding these findings is of uttermost importance. Because PTP1B can counteract its own action by directly inhibiting RTK activation while simultaneously potentiating Ras signaling, this suggests that, depending on the driving genetic alterations that lead to cancer development, PTP1B may have opposing roles. Along this line of thought, cancers harboring an activating Ras mutation are numerous and include pancreatic, colon, and lung cancer [91]. However, these cancers are resistant to modulation of p62Dok, p120RasGAP, or even Src. Therefore, inhibiting PTP1B in the context of those cancers might not be helpful and could even be deleterious since this would potentiate RTK activation and PI3K signaling, further supporting cancer growth.

PTP1B in Breast Cancer: An Oncogenic PTP

Considering both the tumor suppressing and promoting activities of PTP1B, it is challenging to predict which side the balance would tip in a specific context. Ideally, defining the role of PTP1B should be done in complex systems like murine cancer models that are driven by relevant genetic alterations previously identified in humans. Such experiments would help clarify the net effect of PTP1B on cancer development and facilitate translation from bench to bedside. Amplification and overexpression of ErbB2, a member of the EGFR RTK family, is associated with a specific breast tumor subtype occurring in about 25 % of all breast cancer [92], and results in spontaneous mammary gland tumorigenesis in transgenic mouse models. Strikingly, we and others have shown that systemic genetic ablation of PTP1B in ErbB2-transgenic mice results in an increased tumor latency (Fig. 1) [5, 6] and resistance to lung metastasis [5]. Part of this phenotype might be explained by the modulation of the Ras-MAPK-induced ErbB2 axis, which is impaired through increased p62Dok phosphorylation but not by the modulation of Src activity, which is unaltered in this model. On the other hand, specific PTP1B overexpression in the mammary gland driven by the MMTV promoter leads to spontaneous tumor formation [5].

More recently, the Bentires-Alj laboratory has suggested that the role of PTP1B in breast cancer might be cell-autonomous. Indeed, specific deletion of PTP1B in the mammary epithelium of ErbB2-transgenic mice recapitulates the delay in tumor onset observed in the whole-body PTP1B knockout mouse. However, if PTP1B is knocked down once the breast tumor is at a palpable stage, no effect on tumor growth, proliferation, or apoptosis is observed. Experiments using either MCF-10A-NeuNT or MDA-MB-231 breast cancer cells injected orthotopically in immunodeficient mice recapitulate these findings, since PTP1B knockdown ineffectively affects tumor growth once at a palpable stage. Therefore, PTP1B would not be essential for breast tumor maintenance but might be critical to tumor initiation [93] (Fig. 1). Also, STAT5 phosphorylation is frequently lost during breast tumorigenesis and interestingly, PTP1B was shown as a negative regulator of its activation in breast cancer cells [94] (Fig. 1). Taking into account that even over a year old, PTP1B-null mice do not develop tumors [4], those findings clearly demonstrate that

PTP1B acts as an oncogenic PTP in ErbB2-driven breast cancer. Furthermore, it suggests that patients treated chronically with a PTP1B inhibitor in a diabetes and/or obesity context might also benefit from a reduced risk of developing an ErbB2-positive breast cancer. However, the recent analysis of 1402 breast cancer specimens revealed that without any subtypes stratification, PTP1B protein expression is an independent predictor of improved overall survival, revealing that some breast cancer patients may be harmed by the use of PTP1B inhibitors [95].

Additionally, modulation of PTP1B was also shown to modify tumorigenicity in other cancer models. For the sake of conciseness, we have summarized the findings of those studies in Table 1. We have suggested that PTP1B can either act as a tumor suppressor or a tumor promoter. However, attempting to decipher the role of PTP1B in a particular cancer can be difficult since most studies have either not properly deconvoluted the affected signaling pathways, validated their findings in a suitable preclinical murine model or confirmed their relevance using human tissues. Future studies should address such issues and identify the specific circumstances in which PTP1B can act as an oncogene, and consequently, which cancers might benefit from PTP1B inhibitors as part of a multicomponent cancer therapy regimen.

PTP1B Specificity

As described earlier, we estimate the current number of different PTP1B substrates reported at over 30 proteins [7, 56, 96]. However, most have been briefly characterized and validated, and few have been linked to cancer-associated function of PTP1B. The ubiquitous expression of PTP1B and the variety of its substrates raises a legitimate question: "Is PTP1B a simple housekeeping enzyme, maintaining cell homeostasis by keeping a steady state level of tyrosine phosphorylation of its wide variety of substrates, or does it show specificity in its action"? In fact, PTP1B mRNA and protein levels, as well as its activity, are tightly regulated at many levels, from transcriptional to posttranscriptional and as well as posttranslational mechanisms. Moreover, cellular localization and regulation by associated proteins that depends on both cell type and specific signaling context creates all together a regulatory process that fine-tunes the dynamic function of PTP1B.

Gene Regulation

A good example of tissue-specific regulation of PTP1B is its regulation by the androgen receptor (AR) in the prostate. The AR plays a crucial role in the development/progression of prostate cancer, and its inhibition, through androgen deprivation therapy alone or in combination with AR blockade, is the standard treatment for advanced disease. Patients respond to this therapy until they develop castration-resistant metastatic prostate cancer (CRMPC) [97], a disease that still depends on

Table 1 PTP1B involvement in various cancer types

Cancer	Cell line	Overexpression/ downregulation	Onco/TS	Proliferation/ apoptosis	Soft agar/3D	Invasion	Migration	Mouse model	Human samples	References
Breast		Transgenic/KO mouse	Onco					✓		[5]
		KO mouse	Onco					✓		[6]
	MCF10A/ErbB2	Overexpression/ siRNA	Onco		✓					[109]
	MCF-7	siRNA	Onco	✓						[151]
	MCF-7	siRNA	Onco				✓			[152]
	MCF10A/ErbB2, MDA-MB-231	siRNA, KO mouse	Onco					✓		[93]
Prostate	NE (LNCaP)	Overexpression	Onco	✓					✓	[153]
	LNCaP, C4-2, DU145	siRNA, shRNA/ xenografts	Onco			✓	✓	✓	✓	[73]
Ovarian	CAOV1, CAOV4, OVK-18	Overexpression, inhibitor	TS	✓	✓	✓	✓			[44]
CML	K562	Overexpression/ xenografts	TS	✓				✓		[51]
B-cell lymphoma		KO mouse	TS					✓		[154]
Gastric	MKN45	Overexpression/ xenografts	Onco	✓	✓			✓	✓	[155]
Colon	SW48, DLD-1	Overexpression, siRNA, siRNA/ xenografts	Onco		✓			✓		[85]
Squamous cell carcinoma	KYSE150	Overexpression, siRNA, shRNA/tail vein	Onco			✓	✓	✓	✓	[156]

AR signaling for its growth and survival but that has evaded androgen deprivation therapy through various mechanisms [98]. We observed that PTP1B mRNA and protein levels were induced following androgen treatment in various prostate cancer cell lines. High-resolution location analyses on tiled array covering chromosome 20q revealed AR recruitment to two response elements within the first intron of *PTPN1* following androgen treatment. Interestingly, AR recruitment was mirrored by androgen-dependent RNA polymerase II (RNAPol II) recruitment to the *PTPN1* transcriptional start site (TSS). However, looking closely at the *PTPN1* TSS also revealed high levels of the phosphorylated form of transcriptionally active RNAPol II under basal conditions. Therefore, AR regulation only "fine-tuned" PTP1B expression in prostate cancer cells, but this increase was sufficient to promote an AR-induced cell migration dependant on the augmented PTP1B expression (Fig. 1) [73]. The steady state level of PTP1B expression is regulated by other transcription factors, such as the specificity protein 1 (SP1) or 3 (SP3) [99], that are ubiquitously expressed in mammalian cells [100].

PTP1B is also regulated by a variety of other transcription factors that are implicated in cancer progression (summarized in Fig. 2). The transcription factor Y-box binding protein 1 (YB-1), which targets the ErbB2 receptor and is itself an important oncoprotein in breast cancer [101, 102], leads the acquisition of resistance to trastuzumab therapy [103]. Interestingly, YB-1 was identified as a direct and positive modulator of PTP1B expression [104]. PTP1B expression is also regulated in the context of two different cancer hallmarks, namely sustained inflammation and tumor hypoxia. Indeed, treatment with the pro-inflammatory cytokine tumor necrosis factor α (TNFα) was found to induce PTP1B expression through the recruitment of the nuclear factor kappa-light-chain-enhancer of activated B cells (NF-κB) subunit p65 to the *PTPN1* promoter [105]. On the other hand, *PTPN1* transcription was found to be repressed by HIF-1α, which is stabilized under hypoxic conditions [61]. As previously discussed, this negative regulation under hypoxic conditions might be particularly important as it would potentiate VEGFR2 signaling in endothelial cells, promote angiogenesis and consequently tumor progression. In this particular context, PTP1B repression in tumor cells might contribute to the HIF-1α mediated cell survival/proliferation program through the potentiation of RTK signaling, even under the restricted access to nutrients that characterize poorly vascularized regions. Whether this additional regulation by HIF-1α really affects the role of PTP1B in cancer development remains to be experimentally tested. Furthermore, PTP1B is negatively regulated by early growth response protein 1 (Egr-1) transcription factor [99], which is generally considered a tumor suppressor [106]. PTP1B transcription is also repressed at the chromatin level by Sirtuin 1 (SIRT1) through the deacetylation of histone H3 in the promoter region of *PTPN1*. Strikingly, SIRT1 expression improved insulin sensitivity, a role mediated only by the repression of PTP1B transcription [107].

There are other cancer-related circumstances in which PTP1B protein levels are affected through yet uncharacterized mechanisms. For example, we reported that PTP1B expression rose significantly in primary MEFs transformed by the SV40 Tag (Fig. 1) [108]. A similar increase was also observed when p210-BCR/ABL was expressed in various cell lines [52]. We have also observed an increase in PTP1B

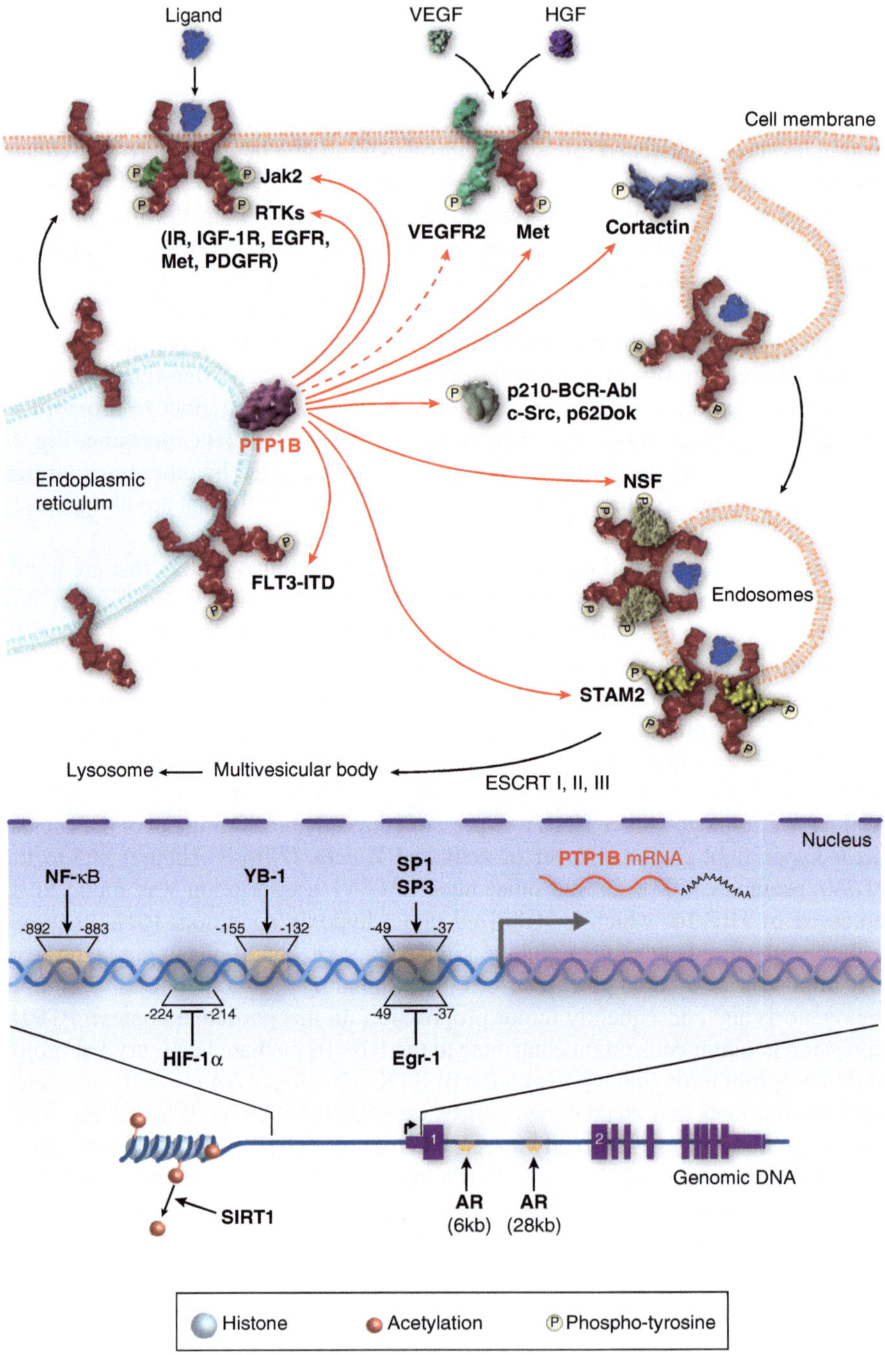

Fig. 2 Schematic representation of PTP1B transcriptional regulators and selected cancer-related substrates. *AR* androgen receptor, c-Src cellular-src, *EGFR* epidermal growth factor receptor, *Egr-1* early growth response protein 1, *FLT3-ITD* fms-related tyrosine kinase 3 internal tandem duplication, *HGF* hepatocyte growth factor, *HIF-1α* hypoxia-inducible factor 1 alpha subunit, *IGF-1R* insulin-like growth factor 1 receptor, *IR* insulin receptor, *Met* hepatocyte growth factor receptor,

protein levels in the mouse mammary gland expressing the constitutively active ErbB2 mutant [5], a finding that was also validated in cell lines by Arias-Romero et al. [109]. Treatment of a subset of B-cell lymphoma cell lines with interleukin-4 (IL-4) also led to an increase in PTP1B mRNA and protein levels [110]. Immune cell regulation of PTP1B was also recently reported by Hutchins et al. in the context of a negative feedback regulation of IL-10 signaling which triggers STAT-3 dependent increased expression of PTP1B, resulting in reduced IL-10 dependent JAK-STAT signaling [111]. It is now clear that, although PTP1B is ubiquitously expressed, a network of transcription factors and stimuli tightly regulates its levels. This certainly demonstrates that, under particular circumstances and in specific cell types, PTP1B levels need to be dynamically modulated. In addition to this complex gene regulation, PTP1B activity is also governed by a variety of posttranslational modifications.

Posttranslational Modifications

Various posttranslational modifications were found to modulate PTP1B activity either positively or negatively. An example is the modulation of PTP1B activity through its tyrosine phosphorylation upon EGF treatment. As EGFR is dephosphorylated by PTP1B following its activation, PTP1B itself gets phosphorylated on Tyr66. This tyrosine is part of the conserved YINAS motif 3 of PTPs and likely phosphorylated in many other protein-tyrosine phosphatases. Interestingly, Tyr66 phosphorylation leads to a threefold increase in PTP1B activity, which is abrogated using a Y66F mutant [65]. As it stands, the physiological dephosphorylation of Tyr66 has not been examined in detail. In contrast, insulin-induced IR activation leads to a decrease in PTP1B activity and ability to dephosphorylate the IR following its phosphorylation on Ser50 by Akt [112]. Therefore, regulation of PTP1B activity through phosphorylation has opposing roles, since it is meant to swiftly dampen EGFR signaling, while being destined to ensure a more sustained IR signaling.

Other posttranslational mechanisms, such as proteolytic cleavage by calpain [113, 114] and GP63 [115] or sumoylation [116], were found to increase or reduce PTP1B activity, respectively. Regulation of PTP1B activity through oxidation is likely to be particularly relevant in a cancer context. In tumors, high levels of ROS are present, in part because infiltrating immune cells, such as macrophages, produce

Fig. 2 (continued) *NF-κB* nuclear factor kappa-light-chain-enhancer of activated B cells, *NSF* N-ethylmaleimide-sensitive factor, *PDGFR* platelet-derived growth factor receptor, *p210-BCR-Abl* p210 form of the chimeric BRC-Abl protein, *p62Dok* p62 protein downstream of tyrosine kinases, *RTKs* receptor tyrosine kinases, *SIRT1* sirtuin 1, *SP1* specificity protein 1, *SP3* specificity protein 3, *STAM2* signal transducing adapter molecule 2, *VEGF* vascular endothelial growth factor, *VEGFR2* VEGF receptor 2, *YB-1* Y-box binding protein 1

large amounts [117], but also because tumor cells produce ROS themselves [118]. Transient and localized ROS expression within tumors regulates a number of cancer hallmarks, ranging from cell cycle progression to cell proliferation, survival, or apoptosis [118]. It is now clear that ROS, like hydrogen peroxide H_2O_2, can act as second messengers and regulate many proteins including PTKs [119], transcription factors [120], and even PTPs [121–123].

Classical PTPs share the same mechanism of action and require an embedded conserved $HC(X)_5R$ catalytic motif. Intrinsically, the cysteine residue is negatively charged, allowing nucleophilic attack on the substrate phosphate. This particularity makes this cysteine especially sensitive to oxidative stress. Depending on their degree of oxidation, PTPs can be either reversibly or irreversibly inactivated [121]. However, in a recent study by the group of Benjamin G. Neel, global assessment of oxidized PTPs in various cancer cell lines demonstrated some level of heterogeneity, while signatures were more similar in six HER2-positive breast cancer cell lines. Interestingly, transformation of NIH 3T3 cells with a constitutively active Src (Y527F) led to increased global tyrosine phosphorylation levels and ROS production with a concomitant increase in PTP oxidation, all of which were reverted to near baseline levels when cells were treated with an Src kinase inhibitor [124]. These results demonstrate that PTPs will inevitably get oxidized and inhibited in a cancer context, but that this inhibition is unequal among classical PTP members and might be variable throughout tumorigenesis.

In a recent study, Nox4-generated ROS has been shown to increase glioblastoma cells tumorigenicity through PTP1B inactivation. In this case, ROS-mediated PTP1B inactivation results in the maintenance of maximal tyrosine phosphorylation of coronin-1C, a positive regulator of cell proliferation and cell motility [8]. In this line, cells lacking a functional thioredoxin system, a major cellular reducing mechanism, displayed increased oxidation of PTP1B but unchanged oxidation of SHP2, resulting in increased PDGFR activation following PDGF treatment [125]. The impact of ROS-mediated PTP1B inhibition seems to be highly context-dependent. In primary cells, introduction of the oncogenic H-RAS^{V12} triggers ROS-induced cell senescence. Surprisingly, Yang et al. demonstrated that this ROS-mediated program was a consequence of PTP1B inactivation through oxidation that led to the hyperphosphorylation of argonaute 2 (AGO2), a putative PTP1B substrate and a master regulator of cell senescence [126]. These results suggest that RAS-mediated transformation may require increased PTP1B levels in order to bypass ROS-induced senescence. As more context-specific data are generated, a better understanding of the relative significance of oxidation-mediated PTP inactivation in cancer will be achieved.

Interestingly, PTP1B degradation was hastened when it was in its oxidized and inactive form. Indeed, Gulati et al. reported that oxidized PTP1B is overtly sensitive to calpain-mediated degradation [127]. Therefore, in a particular context, maintaining PTP1B in its oxidized form might not only inhibit its action, but may also precipitate its calpain-mediated degradation. Likewise, the laboratory of Nicholas K. Tonks developed an intracellular antibody (intrabody) meant to recognize and stabilize PTP1B in its oxidized form, therefore mimicking the action of a pharmacological inhibitor (Fig. 1) [128]. Although the delivery of an intrabody in a diabetes, obesity, or cancer context may not be achievable, this proof of concept represents a start and further underlines the crucial need to develop novel approaches to inhibit PTPs.

ER-Localized PTP1B and Its Role on Endocytosis and the Oncogenic Process

Although PTP1B was shown to associate with and dephosphorylate a plethora of phosphorylated proteins in vitro, in vivo, PTP1B localization in the cell dictates the substrates that can be readily dephosphorylated. Unlike most of other non-receptor PTPs, PTP1B is targeted to the ER through a hydrophobic C-terminal tail, leaving its catalytic domain exposed to the cytoplasm [49, 129]. Therefore, this raises a potential paradox because many well-characterized PTP1B substrates are RTKs that localize to the plasma membrane. Microscopy studies demonstrate that a subset of PTP1B can access the plasma membrane at specialized regions of cell–cell contact [130], thanks to punctate contact regions between the ER and the plasma membrane [131]. Nevertheless, being able to reach the plasma membrane is not essential for PTP1B to dephosphorylate RTKs [132]. Indeed, during biosynthesis at the ER, RTKs can get tyrosine-phosphorylated in a ligand-independent manner and initiate downstream signaling. In this case, ER-bound PTP1B is at the front line and can prevent ligand-independent RTK-mediated signaling, as was described with the RTK fms-related tyrosine kinase 3—internal tandem duplication (FLT3-ITD) [133].

PTP1B in Endocytosis

Most current evidence suggests that PTP1B dephosphorylates RTKs while they are trafficking in the different endosomal compartments. Ligand-activated RTKs are ultimately internalized and sorted for either recycling or degradation. Since their catalytic domain still faces the cytoplasm during trafficking, they can continue to initiate signaling and, conversely, are still accessible to PTPs. Fluorescence resonance energy transfer (FRET) experiments establish that the maximal interaction between ligand-activated PDGFR or EGFR and PTP1B occurs 30 minutes after ligand stimulation. Together with the fact that RTK internalization was required for interaction with PTP1B, this suggests that the interaction between ER-bound PTP1B and RTKs is dependent on endocytosis [132], a finding that was also corroborated with the IR [134–136]. More recently, Lanahan et al. demonstrated that the impaired VEGF signaling and reduction in arterial morphogenesis observed in synectin-null mice were due to a delayed trafficking of VEGFR2-containing endosomes, leading to the dephosphorylation of VEGFR2 on Tyr1175 (PLCγ/MAPK and PI3K activation site) by a PTP1B-mediated mechanism. Pharmacological inhibition of PTP1B in synectin-null mice rescues the phenotype and normalizes arterial vasculature development [62].

PTP1B in Endosomal Trafficking

However, the role of PTP1B extends beyond RTK dephosphorylation during endosomal trafficking, as recent studies identified PTP1B itself as a regulator of RTK endocytosis. The actin-regulatory protein cortactin, which is linked to cell-surface internalization and early endosomal trafficking of RTKs, is dephosphorylated by PTP1B on Tyr446. Although the exact role of this tyrosine residue for endocytosis remains elusive, studies from our laboratory demonstrated that phosphorylation on Tyr446 is required for resistance to hyperosmotic stress-induced apoptosis [137], a phenomenon that highly depends on the reorganization of the cytoskeleton. Cortactin is a substrate of the Src kinase that contributes actively to the dynamic of, which F-actin is essential for cell migration. In this line, a recent study by Radhakrishnan et al. reported that PTP1B dephosphorylation of cortactin on Tyr421 was able to reduce colon cancer cell motility [138]. Interestingly, cortactin has also been found to be frequently amplified in breast and head and neck carcinomas [139, 140].

PTP1B knockdown also leads to a delayed ligand-induced degradation of Met and EGFR by preventing their entrance to early endosomes, and subsequent trafficking through degradative pathways. In this situation, the N-ethylmaleimide-sensitive factor (NSF), a protein essential in the vesicle fusion machinery and a PTP1B substrate, was found to be hyperphosphorylated and its downregulation phenocopied PTP1B knockdown [141]. Once internalized, RTKs can either be recycled back to the cell surface or they can enter multivesicular bodies (MVBs) and be degraded once the MVBs fuse with lysosomes. Importantly, RTK cytoplasmic tails are confined to the inner MVBs, so they can no longer activate signaling or interact with PTPs [142]. Recent studies have shown that PTP1B activity promotes the internalization of the EGFR to MVB internal vesicles [143]. Work from our laboratory also uncovered that signal transducing adapter molecule 2 (STAM2), an endocytic regulator, is a putative PTP1B substrate and may shed some light on the possible mechanism responsible for this phenotype. STAM2 helps in the recognition and sorting of ubiquitinated RTKs on the surface of endosomes. Preventing STAM2 tyrosine phosphorylation on PTP1B-regulated residues results in a prolonged association between STAM2 and EGFR-positive endosomes following EGF stimulation. Therefore, these results suggest that PTP1B deficiency would lead to STAM2 hyperphosphorylation and its dissociation from ubiquitinated EGFR on endosomes, which would prevent the entrance of EGFR in MVBs and favors extended signaling [144].

Together, these studies helped clarify a puzzling observation. As described thoroughly in this chapter, PTP1B deficiency leads to RTK hyperphosphorylation, many of which are putative PTP1B substrates, but also others that were not confirmed. Following this line of thought, PTP1B deficiency leads to colony stimulating factor 1 receptor (CSF1R) phosphorylation (Tyr807) after stimulation with its ligand in bone marrow-derived macrophages, although PTP1B has not been shown to interact directly with this RTK [145]. It is tempting to speculate that, in such a case, where PTP1B seems to act in a nonspecific manner, its predominant role would be in the regulation of RTK endocytosis. In fact, most of the regulation of RTK signaling by PTP1B might be related to endosomal trafficking and endocytosis since plasma-membrane proximal PTP1B is maintained in a low-activity state, sufficient enough to maintain the basal activity of RTKs but insufficient to counteract ligand-activated RTKs [146]. Dephosphorylation of RTKs would occur as they are internalized and transported to perinuclear regions where PTP1B activity is high enough to allow signal termination [132], but only after they properly transduce their signaling.

Conclusions and Perspectives

Although PTP1B associates with and dephosphorylates a plethora of substrates (summarized in Fig. 2), its net effect on tumorigenesis is often attributed to tissue-specific substrates under particular contexts. In accordance, PTP1B genetic ablation succeeded in delaying the appearance of an ErbB2-driven mammary cancer [5, 6] but failed to modulate the development of a polyoma middle T-induced breast cancer [6]. Furthermore, specific PTP1B genetic ablation in mammary gland also successfully delayed ErbB2-driven breast cancer when done prior to tumor initiation while it had no effect on tumor growth at the stage of palpable tumors [93]. Together, these results demonstrate that PTP1B inhibitors may only be effective in specific breast cancer subtypes and probably prior to tumor initiation. Therefore, it is crucial to use mouse models that are representative of relevant cancer subtypes to identify the roles of PTP1B during tumorigenesis, and validate whether or not PTP1B inhibitors will be valuable as a part of a cancer treatment program.

Cancer treatment is on the brink of a new era of personalized medicine. Screening biopsies for driving mutations will be of utmost importance before assigning a particular drug cocktail to a patient. As described previously, on one hand, PTP1B inhibition decreases Ras signaling, but on the other hand it may drive an increase in the activity of some RTKs and PI3K downstream signaling [88]. We hypothesize that, under certain circumstances, patients harboring a tumor with a wild-type Ras may benefit from hindered MAPK activation, but patients with a constitutively activated Ras might gain no benefit from the inhibition of PTP1B, and may even experience a worse outcome since a range of RTKs might consequently become hyperactivated. As Ras is mutated in a variety of human tumors, such a hypothesis certainly deserves validation. A similar dichotomy in action could thus also be imagined in other signaling pathways where PTP1B substrates were identified to modify cancer outcome. We should also add that many other aspects of PTP1B function remain to be examined. Notably, the PTP1B modulation placed upon JAK-STAT signaling in the immune system may also play a crucial role in safeguarding from tumor initiation and expansion [147]. All of those are exciting aspects of animal homeostasis that are likely maintained by PTP1B are yet still poorly understood and additional molecular network analysis such as the one published by Miraldi et al. [148] is certainly warranted.

Finally, it is clear that the development of PTP1B small molecule inhibitors has been through a difficult journey. Yet, several companies and academic laboratories are still actively pursuing the development of original small molecules that could be used to inhibit PTP1B. One of them Kaneq Pharma (Montréal, Canada) has recently announced a new development program on PTP1B inhibitors to treat type II diabetes. Alternatively, understanding and hijacking posttranslational modifications, such as oxidation, to modulate PTP1B activity may prove to be novel and elegant approaches to target PTP1B. This was recently demonstrated by the generation of an intrabody targeting the oxidized form of PTP1B thus preventing its reduction and mimicking traditional inhibitors by maintaining a pool of inactive PTP1B [149]. Another strategy that has proven to be effective was the development of an allosteric inhibitor targeting of the intrinsically disordered C-terminus of PTP1B (Fig. 1) [150]. We trust that other original ideas to inhibit this important enzyme will continue to surface in the near future.

Although PTP1B was initially assigned an oncogenic role in cancer development, compelling evidence has demonstrated a tumor-suppressing role for PTP1B in cancer ontogeny. In such a situation, one would need to screen for small molecules that would activate PTP1B therefore optimizing its tumor-suppressing activity. Few examples of such drug activator types exist in the oncology pharmacopeia and this is without taking in account the many potential side effects that such PTP1B activators may cause in metabolism, immunity, and other biological systems. However, since predicting the net effect of PTP1B modulation has proven to be a difficult task, novel in vivo models in which PTP1B activity would be increased on a systemic level are required before further exploring this avenue.

In closing, much remains to be understood on the physiological and cellular roles of PTP1B. Ongoing research will continue to better clarify various PTP1B modes of action and in particular in cancer. We trust that in spite of the multiple PTP1B substrates and functions, these will nevertheless lead to innovative tumor-specific cancer treatments in the context of personalized medicine applications.

Acknowledgements We thank Noriko Uetani for technical assistance with figure design and drawing and Teri Hatzihristidis for a critical review of the manuscript. D.P.L. is a recipient of Canadian Institute of Health Research (CIHR) Frederick Banting and Charles Best doctoral research award and the CIHR/Fonds de la Recherche en Santé du Québec (FRSQ) training grant in cancer research FRN53888 of the McGill Integrated Cancer Research Training Program. M.L.T. is a Jeanne and Jean-Louis Lévesque chair in Cancer Research.

References

1. Tonks NK, Diltz CD, Fischer EH. Characterization of the major protein-tyrosine-phosphatases of human placenta. J Biol Chem. 1988;263(14):6731–7.
2. Tonks NK, Diltz CD, Fischer EH. Purification of the major protein-tyrosine-phosphatases of human placenta. J Biol Chem. 1988;263(14):6722–30.
3. Alonso A, Sasin J, Bottini N, Friedberg I, Friedberg I, Osterman A, Godzik A, Hunter T, Dixon J, Mustelin T. Protein tyrosine phosphatases in the human genome. Cell. 2004; 117(6):699–711.
4. Elchebly M, Payette P, Michaliszyn E, Cromlish W, Collins S, Loy AL, Normandin D, Cheng A, Himms-Hagen J, Chan CC, Ramachandran C, Gresser MJ, Tremblay ML, Kennedy BP.

Increased insulin sensitivity and obesity resistance in mice lacking the protein tyrosine phosphatase-1B gene. Science. 1999;283(5407):1544–8.

5. Julien SG, Dube N, Read M, Penney J, Paquet M, Han Y, Kennedy BP, Muller WJ, Tremblay ML. Protein tyrosine phosphatase 1B deficiency or inhibition delays ErbB2-induced mammary tumorigenesis and protects from lung metastasis. Nat Genet. 2007;39(3):338–46. doi:10.1038/ng1963. ng1963 [pii].
6. Bentires-Alj M, Neel BG. Protein-tyrosine phosphatase 1B is required for HER2/Neu-induced breast cancer. Cancer Res. 2007;67(6):2420–4. doi:10.1158/0008-5472.CAN-06-4610. 0008-5472.CAN-06-4610 [pii].
7. Ferrari E, Tinti M, Costa S, Corallino S, Nardozza AP, Chatraryamontri A, Ceol A, Cesareni G, Castagnoli L. Identification of new substrates of the protein-tyrosine phosphatase PTP1B by Bayesian integration of proteome evidence. J Biol Chem. 2011;286(6):4173–85. doi:10.1074/jbc.M110.157420.
8. Mondol AS, Tonks NK, Kamata T. Nox4 redox regulation of PTP1B contributes to the proliferation and migration of glioblastoma cells by modulating tyrosine phosphorylation of coronin-1C. Free Radic Biol Med. 2013;67C:285–91. doi:10.1016/j.freeradbiomed.2013.11.005.
9. Bakke J, Bettaieb A, Nagata N, Matsuo K, Haj FG. Regulation of the SNARE-interacting protein Munc18c tyrosine phosphorylation in adipocytes by protein-tyrosine phosphatase 1B. Cell Commun Signal. 2013;11:57. doi:10.1186/1478-811X-11-57.
10. Danaei G, Finucane MM, Lu Y, Singh GM, Cowan MJ, Paciorek CJ, Lin JK, Farzadfar F, Khang YH, Stevens GA, Rao M, Ali MK, Riley LM, Robinson CA, Ezzati M. National, regional, and global trends in fasting plasma glucose and diabetes prevalence since 1980: systematic analysis of health examination surveys and epidemiological studies with 370 country-years and 2.7 million participants. Lancet. 2011;378(9785):31–40. doi:10.1016/S0140-6736(11)60679-X.
11. Ji L, Malecki M, Warram JH, Yang Y, Rich SS, Krolewski AS. New susceptibility locus for NIDDM is localized to human chromosome 20q. Diabetes. 1997;46(5):876–81.
12. Bowden DW, Sale M, Howard TD, Qadri A, Spray BJ, Rothschild CB, Akots G, Rich SS, Freedman BI. Linkage of genetic markers on human chromosomes 20 and 12 to NIDDM in Caucasian Sib pairs with a history of diabetic nephropathy. Diabetes. 1997;46(5):882–6. doi:10.2337/diab.46.5.882.
13. Ghosh S, Watanabe RM, Hauser ER, Valle T, Magnuson VL, Erdos MR, Langefeld CD, Balow Jr J, Ally DS, Kohtamaki K, Chines P, Birznieks G, Kaleta HS, Musick A, Te C, Tannenbaum J, Eldridge W, Shapiro S, Martin C, Witt A, So A, Chang J, Shurtleff B, Porter R, Kudelko K, Unni A, Segal L, Sharaf R, Blaschak-Harvan J, Eriksson J, Tenkula T, Vidgren G, Ehnholm C, Tuomilehto-Wolf E, Hagopian W, Buchanan TA, Tuomilehto J, Bergman RN, Collins FS, Boehnke M. Type 2 diabetes: evidence for linkage on chromosome 20 in 716 Finnish affected sib pairs. Proc Natl Acad Sci U S A. 1999;96(5):2198–203.
14. Zouali H, Hani EH, Philippi A, Vionnet N, Beckmann JS, Demenais F, Froguel P. A susceptibility locus for early-onset non-insulin dependent (type 2) diabetes mellitus maps to chromosome 20q, proximal to the phosphoenolpyruvate carboxykinase gene. Hum Mol Genet. 1997;6(9):1401–8.
15. Klupa T, Malecki MT, Pezzolesi M, Ji L, Curtis S, Langefeld CD, Rich SS, Warram JH, Krolewski AS. Further evidence for a susceptibility locus for type 2 diabetes on chromosome 20q13.1-q13.2. Diabetes. 2000;49(12):2212–6.
16. Brown-Shimer S, Johnson KA, Lawrence JB, Johnson C, Bruskin A, Green NR, Hill DE. Molecular cloning and chromosome mapping of the human gene encoding protein phosphotyrosyl phosphatase 1B. Proc Natl Acad Sci U S A. 1990;87(13):5148–52.
17. Czech MP, Lawrence Jr JC, Lynn WS. Evidence for electron transfer reactions involved in the Cu2+ -dependent thiol activation of fat cell glucose utilization. J Biol Chem. 1974;249(4):1001–6.
18. Thompson KH, Orvig C. Vanadium in diabetes: 100 years from Phase 0 to Phase I. J Inorg Biochem. 2006;100(12):1925–35. doi:10.1016/j.jinorgbio.2006.08.016.
19. Shechter Y, Karlish SJ. Insulin-like stimulation of glucose oxidation in rat adipocytes by vanadyl (IV) ions. Nature. 1980;284(5756):556–8.

20. Chernoff J, Schievella AR, Jost CA, Erikson RL, Neel BG. Cloning of a cDNA for a major human protein-tyrosine-phosphatase. Proc Natl Acad Sci U S A. 1990;87(7):2735–9.
21. Cicirelli MF, Tonks NK, Diltz CD, Weiel JE, Fischer EH, Krebs EG. Microinjection of a protein-tyrosine-phosphatase inhibits insulin action in Xenopus oocytes. Proc Natl Acad Sci U S A. 1990;87(14):5514–8.
22. Kenner KA, Anyanwu E, Olefsky JM, Kusari J. Protein-tyrosine phosphatase 1B is a negative regulator of insulin- and insulin-like growth factor-I-stimulated signaling. J Biol Chem. 1996;271(33):19810–6.
23. Seely BL, Staubs PA, Reichart DR, Berhanu P, Milarski KL, Saltiel AR, Kusari J, Olefsky JM. Protein tyrosine phosphatase 1B interacts with the activated insulin receptor. Diabetes. 1996;45(10):1379–85.
24. Ahmad F, Goldstein BJ. Increased abundance of specific skeletal muscle protein-tyrosine phosphatases in a genetic model of insulin-resistant obesity and diabetes mellitus. Metabolism. 1995;44(9):1175–84.
25. Ahmad F, Goldstein BJ. Alterations in specific protein-tyrosine phosphatases accompany insulin resistance of streptozotocin diabetes. Am J Physiol. 1995;268(5 Pt 1):E932–40.
26. Bandyopadhyay D, Kusari A, Kenner KA, Liu F, Chernoff J, Gustafson TA, Kusari J. Protein-tyrosine phosphatase 1B complexes with the insulin receptor in vivo and is tyrosine-phosphorylated in the presence of insulin. J Biol Chem. 1997;272(3):1639–45. doi:10.1074/jbc.272.3.1639.
27. Klaman LD, Boss O, Peroni OD, Kim JK, Martino JL, Zabolotny JM, Moghal N, Lubkin M, Kim YB, Sharpe AH, Stricker-Krongrad A, Shulman GI, Neel BG, Kahn BB. Increased energy expenditure, decreased adiposity, and tissue-specific insulin sensitivity in protein-tyrosine phosphatase 1B-deficient mice. Mol Cell Biol. 2000;20(15):5479–89.
28. Lee JH, Reed DR, Li WD, Xu W, Joo EJ, Kilker RL, Nanthakumar E, North M, Sakul H, Bell C, Price RA. Genome scan for human obesity and linkage to markers in 20q13. Am J Hum Genet. 1999;64(1):196–209. doi:10.1086/302195.
29. Cheng A, Uetani N, Simoncic PD, Chaubey VP, Lee-Loy A, McGlade CJ, Kennedy BP, Tremblay ML. Attenuation of leptin action and regulation of obesity by protein tyrosine phosphatase 1B. Dev Cell. 2002;2(4):497–503.
30. Zabolotny JM, Bence-Hanulec KK, Stricker-Krongrad A, Haj F, Wang Y, Minokoshi Y, Kim YB, Elmquist JK, Tartaglia LA, Kahn BB, Neel BG. PTP1B regulates leptin signal transduction in vivo. Dev Cell. 2002;2(4):489–95.
31. Myers MP, Andersen JN, Cheng A, Tremblay ML, Horvath CM, Parisien JP, Salmeen A, Barford D, Tonks NK. TYK2 and JAK2 are substrates of protein-tyrosine phosphatase 1B. J Biol Chem. 2001;276(51):47771–4. doi:10.1074/jbc.C100583200.
32. Bence KK, Delibegovic M, Xue B, Gorgun CZ, Hotamisligil GS, Neel BG, Kahn BB. Neuronal PTP1B regulates body weight, adiposity and leptin action. Nat Med. 2006;12(8):917–24. doi:10.1038/nm1435.
33. Banno R, Zimmer D, De Jonghe BC, Atienza M, Rak K, Yang W, Bence KK. PTP1B and SHP2 in POMC neurons reciprocally regulate energy balance in mice. J Clin Invest. 2010;120(3):720–34. doi:10.1172/JCI39620.
34. White CL, Whittington A, Barnes MJ, Wang Z, Bray GA, Morrison CD. HF diets increase hypothalamic PTP1B and induce leptin resistance through both leptin-dependent and -independent mechanisms. Am J Physiol Endocrinol Metab. 2009;296(2):E291–9. doi:10.1152/ajpendo.90513.2008.
35. Delibegovic M, Zimmer D, Kauffman C, Rak K, Hong EG, Cho YR, Kim JK, Kahn BB, Neel BG, Bence KK. Liver-specific deletion of protein-tyrosine phosphatase 1B (PTP1B) improves metabolic syndrome and attenuates diet-induced endoplasmic reticulum stress. Diabetes. 2009;58(3):590–9. doi:10.2337/db08-0913.
36. Owen C, Lees EK, Grant L, Zimmer DJ, Mody N, Bence KK, Delibegovic M. Inducible liver-specific knockdown of protein tyrosine phosphatase 1B improves glucose and lipid homeostasis in adult mice. Diabetologia. 2013;56(10):2286–96. doi:10.1007/s00125-013-2992-z.

37. Delibegovic M, Bence KK, Mody N, Hong EG, Ko HJ, Kim JK, Kahn BB, Neel BG. Improved glucose homeostasis in mice with muscle-specific deletion of protein-tyrosine phosphatase 1B. Mol Cell Biol. 2007;27(21):7727–34. doi:10.1128/MCB.00959-07.
38. Giovannucci E, Harlan DM, Archer MC, Bergenstal RM, Gapstur SM, Habel LA, Pollak M, Regensteiner JG, Yee D. Diabetes and cancer: a consensus report. CA Cancer J Clin. 2010;60(4):207–21. doi:10.3322/caac.20078.
39. Bowker SL, Yasui Y, Veugelers P, Johnson JA. Glucose-lowering agents and cancer mortality rates in type 2 diabetes: assessing effects of time-varying exposure. Diabetologia. 2010;53(8):1631–7. doi:10.1007/s00125-010-1750-8.
40. Giovannucci E, Pollak M, Liu Y, Platz EA, Majeed N, Rimm EB, Willett WC. Nutritional predictors of insulin-like growth factor I and their relationships to cancer in men. Cancer Epidemiol Biomarkers Prevention. 2003;12(2):84–9.
41. Clayton PE, Banerjee I, Murray PG, Renehan AG. Growth hormone, the insulin-like growth factor axis, insulin and cancer risk. Nat Rev Endocrinol. 2011;7(1):11–24. doi:10.1038/nrendo.2010.171.
42. Gu F, Dube N, Kim JW, Cheng A, Ibarra-Sanchez Mde J, Tremblay ML, Boisclair YR. Protein tyrosine phosphatase 1B attenuates growth hormone-mediated JAK2-STAT signaling. Mol Cell Biol. 2003;23(11):3753–62.
43. Buckley DA, Cheng A, Kiely PA, Tremblay ML, O'Connor R. Regulation of insulin-like growth factor type I (IGF-I) receptor kinase activity by protein tyrosine phosphatase 1B (PTP-1B) and enhanced IGF-I-mediated suppression of apoptosis and motility in PTP-1B-deficient fibroblasts. Mol Cell Biol. 2002;22(7):1998–2010.
44. Fan G, Lin G, Lucito R, Tonks NK. Protein-tyrosine phosphatase 1B antagonized signaling by insulin-like growth factor-1 receptor and kinase BRK/PTK6 in ovarian cancer cells. J Biol Chem. 2013;288(34):24923–34. doi:10.1074/jbc.M113.482737.
45. Collett MS, Erikson RL. Protein kinase activity associated with the avian sarcoma virus src gene product. Proc Natl Acad Sci U S A. 1978;75(4):2021–4.
46. Levinson AD, Oppermann H, Levintow L, Varmus HE, Bishop JM. Evidence that the transforming gene of avian sarcoma virus encodes a protein kinase associated with a phosphoprotein. Cell. 1978;15(2):561–72.
47. Hunter T, Sefton BM. Transforming gene product of Rous sarcoma virus phosphorylates tyrosine. Proc Natl Acad Sci U S A. 1980;77(3):1311–5.
48. Lemmon MA, Schlessinger J. Cell signaling by receptor tyrosine kinases. Cell. 2010;141(7):1117–34. doi:10.1016/j.cell.2010.06.011.
49. Woodford-Thomas TA, Rhodes JD, Dixon JE. Expression of a protein tyrosine phosphatase in normal and v-src-transformed mouse 3T3 fibroblasts. J Cell Biol. 1992;117(2):401–14.
50. Brown-Shimer S, Johnson KA, Hill DE, Bruskin AM. Effect of protein tyrosine phosphatase 1B expression on transformation by the human neu oncogene. Cancer Res. 1992;52(2):478–82.
51. LaMontagne Jr KR, Hannon G, Tonks NK. Protein tyrosine phosphatase PTP1B suppresses p210 bcr-abl-induced transformation of rat-1 fibroblasts and promotes differentiation of K562 cells. Proc Natl Acad Sci U S A. 1998;95(24):14094–9.
52. LaMontagne Jr KR, Flint AJ, Franza Jr BR, Pandergast AM, Tonks NK. Protein tyrosine phosphatase 1B antagonizes signalling by oncoprotein tyrosine kinase p210 bcr-abl in vivo. Mol Cell Biol. 1998;18(5):2965–75.
53. Sergina NV, Moasser MM. The HER family and cancer: emerging molecular mechanisms and therapeutic targets. Trends Mol Med. 2007;13(12):527–34. doi:10.1016/j.molmed.2007.10.002.
54. Milarski KL, Zhu G, Pearl CG, McNamara DJ, Dobrusin EM, MacLean D, Thieme-Sefler A, Zhang ZY, Sawyer T, Decker SJ, et al. Sequence specificity in recognition of the epidermal growth factor receptor by protein tyrosine phosphatase 1B. J Biol Chem. 1993;268(31):23634–9.
55. Flint AJ, Tiganis T, Barford D, Tonks NK. Development of "substrate-trapping" mutants to identify physiological substrates of protein tyrosine phosphatases. Proc Natl Acad Sci U S A. 1997;94(5):1680–5.
56. Mertins P, Eberl HC, Renkawitz J, Olsen JV, Tremblay ML, Mann M, Ullrich A, Daub H. Investigation of protein-tyrosine phosphatase 1B function by quantitative proteomics. Mol Cell Proteomics. 2008;7(9):1763–77. doi:10.1074/mcp.M800196-MCP200.

57. Haj FG, Markova B, Klaman LD, Bohmer FD, Neel BG. Regulation of receptor tyrosine kinase signaling by protein tyrosine phosphatase-1B. J Biol Chem. 2003;278(2):739–44. doi:10.1074/jbc.M210194200.
58. Cooper CS, Park M, Blair DG, Tainsky MA, Huebner K, Croce CM, Vande Woude GF. Molecular cloning of a new transforming gene from a chemically transformed human cell line. Nature. 1984;311(5981):29–33.
59. Gherardi E, Birchmeier W, Birchmeier C, Vande Woude G. Targeting MET in cancer: rationale and progress. Nat Rev Cancer. 2012;12(2):89–103. doi:10.1038/nrc3205.
60. Sangwan V, Paliouras GN, Abella JV, Dube N, Monast A, Tremblay ML, Park M. Regulation of the Met receptor-tyrosine kinase by the protein-tyrosine phosphatase 1B and T-cell phosphatase. J Biol Chem. 2008;283(49):34374–83. doi:10.1074/jbc.M805916200.
61. Suwaki N, Vanhecke E, Atkins KM, Graf M, Swabey K, Huang P, Schraml P, Moch H, Cassidy AM, Brewer D, Al-Lazikani B, Workman P, De-Bono J, Kaye SB, Larkin J, Gore ME, Sawyers CL, Nelson P, Beer TM, Geng H, Gao L, Qian DZ, Alumkal JJ, Thomas G, Thomas GV. A HIF-regulated VHL-PTP1B-Src signaling axis identifies a therapeutic target in renal cell carcinoma. Sci Transl Med. 2011;3(85):85ra47. doi:10.1126/scitranslmed.3002004
62. Lanahan AA, Hermans K, Claes F, Kerley-Hamilton JS, Zhuang ZW, Giordano FJ, Carmeliet P, Simons M. VEGF receptor 2 endocytic trafficking regulates arterial morphogenesis. Dev Cell. 2010;18(5):713–24. doi:10.1016/j.devcel.2010.02.016.
63. Nakamura Y, Patrushev N, Inomata H, Mehta D, Urao N, Kim HW, Razvi M, Kini V, Mahadev K, Goldstein BJ, McKinney R, Fukai T, Ushio-Fukai M. Role of protein tyrosine phosphatase 1B in vascular endothelial growth factor signaling and cell-cell adhesions in endothelial cells. Circ Res. 2008;102(10):1182–91. doi:10.1161/CIRCRESAHA.107.167080.
64. Lanahan AA, Lech D, Dubrac A, Zhang J, Zhuang ZW, Eichmann A, Simons M. PTP1b is a physiologic regulator of vascular endothelial growth factor signaling in endothelial cells. Circulation. 2014;130(11):902–9. doi:10.1161/CIRCULATIONAHA.114.009683.
65. Liu F, Chernoff J. Protein tyrosine phosphatase 1B interacts with and is tyrosine phosphorylated by the epidermal growth factor receptor. Biochem J. 1997;327(Pt 1):139–45.
66. Lu KV, Chang JP, Parachoniak CA, Pandika MM, Aghi MK, Meyronet D, Isachenko N, Fouse SD, Phillips JJ, Cheresh DA, Park M, Bergers G. VEGF inhibits tumor cell invasion and mesenchymal transition through a MET/VEGFR2 complex. Cancer Cell. 2012;22(1):21–35. doi:10.1016/j.ccr.2012.05.037.
67. Warabi M, Nemoto T, Ohashi K, Kitagawa M, Hirokawa K. Expression of protein tyrosine phosphatases and its significance in esophageal cancer. Exp Mol Pathol. 2000;68(3):187–95. 10.1006/exmp.2000.2303. S0014-4800(00)92303-7 [pii].
68. Mahlamaki EH, Barlund M, Tanner M, Gorunova L, Hoglund M, Karhu R, Kallioniemi A. Frequent amplification of 8q24, 11q, 17q, and 20q-specific genes in pancreatic cancer. Genes Chromosomes Cancer. 2002;35(4):353–8. doi:10.1002/gcc.10122.
69. Yang SH, Seo MY, Jeong HJ, Jeung HC, Shin J, Kim SC, Noh SH, Chung HC, Rha SY. Gene copy number change events at chromosome 20 and their association with recurrence in gastric cancer patients. Clin Cancer Res. 2005;11(2 pt 1):612–20.
70. Tanner MM, Tirkkonen M, Kallioniemi A, Isola J, Kuukasjarvi T, Collins C, Kowbel D, Guan XY, Trent J, Gray JW, Meltzer P, Kallioniemi OP. Independent amplification and frequent co-amplification of three nonsyntenic regions on the long arm of chromosome 20 in human breast cancer. Cancer Res. 1996;56(15):3441–5.
71. Bar-Shira A, Pinthus JH, Rozovsky U, Goldstein M, Sellers WR, Yaron Y, Eshhar Z, Orr-Urtreger A. Multiple genes in human 20q13 chromosomal region are involved in an advanced prostate cancer xenograft. Cancer Res. 2002;62(23):6803–7.
72. Tabach Y, Kogan-Sakin I, Buganim Y, Solomon H, Goldfinger N, Hovland R, Ke XS, Oyan AM, Kalland KH, Rotter V, Domany E. Amplification of the 20q chromosomal arm occurs early in tumorigenic transformation and may initiate cancer. PLoS One. 2011;6(1), e14632. doi:10.1371/journal.pone.0014632.

73. Lessard L, Labbé DP, Deblois G, Bégin LR, Hardy S, Mes-Masson AM, Saad F, Trotman LC, Giguère V, Tremblay ML. PTP1B is an androgen receptor-regulated phosphatase that promotes the progression of prostate cancer. Cancer Res. 2012;72(6):1529–37. doi:10.1158/0008-5472.CAN-11-2602.
74. Labbé DP, Nowak DG, Deblois G, Lessard L, Giguère V, Trotman LC, Tremblay ML. Prostate cancer genetic-susceptibility locus on chromosome 20q13 is amplified and coupled to androgen receptor-regulation in metastatic tumors. Mol Cancer Res. 2014;12(2):184–9. doi:10.1158/1541-7786.MCR-13-0477.
75. Wiener JR, Kerns BJ, Harvey EL, Conaway MR, Iglehart JD, Berchuck A, Bast Jr RC. Overexpression of the protein tyrosine phosphatase PTP1B in human breast cancer: association with p185c-erbB-2 protein expression. J Natl Cancer Inst. 1994;86(5):372–8.
76. Wiener JR, Hurteau JA, Kerns BJ, Whitaker RS, Conaway MR, Berchuck A, Bast Jr RC. Overexpression of the tyrosine phosphatase PTP1B is associated with human ovarian carcinomas. Am J Obstet Gynecol. 1994;170(4):1177–83. doi:S0002937894005740 [pii].
77. Wang J, Liu B, Chen X, Su L, Wu P, Wu J, Zhu Z. PTP1B expression contributes to gastric cancer progression. Med Oncol. 2011. doi:10.1007/s12032-011-9911-2.
78. Zheng LY, Zhou DX, Lu J, Zhang WJ, Zou DJ. Down-regulated expression of the protein-tyrosine phosphatase 1B (PTP1B) is associated with aggressive clinicopathologic features and poor prognosis in hepatocellular carcinoma. Biochem Biophys Res Commun. 2012;420(3):680–4. doi:10.1016/j.bbrc.2012.03.066.
79. Brown MT, Cooper JA. Regulation, substrates and functions of src. Biochim Biophys Acta. 1996;1287(2-3):121–49.
80. Roskoski Jr R. Src kinase regulation by phosphorylation and dephosphorylation. Biochem Biophys Res Commun. 2005;331(1):1–14. doi:10.1016/j.bbrc.2005.03.012.
81. Martin GS. The hunting of the Src. Nat Rev Mol Cell Biol. 2001;2(6):467–75. doi:10.1038/35073094.
82. Arregui CO, Balsamo J, Lilien J. Impaired integrin-mediated adhesion and signaling in fibroblasts expressing a dominant-negative mutant PTP1B. J Cell Biol. 1998;143(3):861–73.
83. Bjorge JD, Pang A, Fujita DJ. Identification of protein-tyrosine phosphatase 1B as the major tyrosine phosphatase activity capable of dephosphorylating and activating c-Src in several human breast cancer cell lines. J Biol Chem. 2000;275(52):41439–46. doi:10.1074/jbc.M004852200.
84. Dadke S, Chernoff J. Interaction of protein tyrosine phosphatase (PTP) 1B with its substrates is influenced by two distinct binding domains. Biochem J. 2002;364(Pt 2):377–83. doi:10.1042/BJ20011372.
85. Zhu S, Bjorge JD, Fujita DJ. PTP1B contributes to the oncogenic properties of colon cancer cells through Src activation. Cancer Res. 2007;67(21):10129–37.
86. Cortesio CL, Chan KT, Perrin BJ, Burton NO, Zhang S, Zhang ZY, Huttenlocher A. Calpain 2 and PTP1B function in a novel pathway with Src to regulate invadopodia dynamics and breast cancer cell invasion. J Cell Biol. 2008;180(5):957–71. doi:10.1083/jcb.200708048.
87. Monteleone MC, Gonzalez Wusener AE, Burdisso JE, Conde C, Caceres A, Arregui CO. ER-bound protein tyrosine phosphatase PTP1B interacts with Src at the plasma membrane/substrate interface. PLoS One. 2012;7(6), e38948. doi:10.1371/journal.pone.0038948.
88. Dube N, Cheng A, Tremblay ML. The role of protein tyrosine phosphatase 1B in Ras signaling. Proc Natl Acad Sci U S A. 2004;101(7):1834–9. doi:10.1073/pnas.0304242101.
89. Klinghoffer RA, Sachsenmaier C, Cooper JA, Soriano P. Src family kinases are required for integrin but not PDGFR signal transduction. EMBO J. 1999;18(9):2459–71. doi:10.1093/emboj/18.9.2459.
90. Broome MA, Courtneidge SA. No requirement for src family kinases for PDGF signaling in fibroblasts expressing SV40 large T antigen. Oncogene. 2000;19(24):2867–9. doi:10.1038/sj.onc.1203608.
91. Schubbert S, Shannon K, Bollag G. Hyperactive Ras in developmental disorders and cancer. Nat Rev Cancer. 2007;7(4):295–308. doi:10.1038/nrc2109.

92. Hynes NE, MacDonald G. ErbB receptors and signaling pathways in cancer. Curr Opin Cell Biol. 2009;21(2):177–84. doi:S0955-0674(09)00014-3 [pii] 10.1016/j.ceb.2008.12.010.
93. Balavenkatraman KK, Aceto N, Britschgi A, Mueller U, Bence KK, Neel B, Bentires-Alj M. Epithelial protein-tyrosine phosphatase 1B (PTP1B) contributes to the induction of mammary tumors by HER2/Neu but is not essential for tumor maintenance. Mol Cancer Res. 2011. doi:10.1158/1541-7786.MCR-11-0198.
94. Johnson KJ, Peck AR, Liu C, Tran TH, Utama FE, Sjolund AB, Schaber JD, Witkiewicz AK, Rui H. PTP1B suppresses prolactin activation of Stat5 in breast cancer cells. Am J Pathol. 2010;177(6):2971–83. doi:10.2353/ajpath.2010.090399.
95. Soysal S, Obermann EC, Gao F, Oertli D, Gillanders WE, Viehl CT, Muenst S. PTP1B expression is an independent positive prognostic factor in human breast cancer. Breast Cancer Res Treat. 2013;137(2):637–44. doi:10.1007/s10549-012-2373-1.
96. Li X, Wilmanns M, Thornton J, Kohn M. Elucidating human phosphatase-substrate networks. Sci Signal. 2013;6(275):10. doi:10.1126/scisignal.2003203.
97. Damber JE, Aus G. Prostate cancer. Lancet. 2008;371(9625):1710–21. doi:10.1016/S0140-6736(08)60729-1.
98. Shen MM, Abate-Shen C. Molecular genetics of prostate cancer: new prospects for old challenges. Genes Dev. 2010;24(18):1967–2000. doi:10.1101/gad.1965810.
99. Fukada T, Tonks NK. The reciprocal role of Egr-1 and Sp family proteins in regulation of the PTP1B promoter in response to the p210 Bcr-Abl oncoprotein-tyrosine kinase. J Biol Chem. 2001;276(27):25512–9. doi:10.1074/jbc.M101354200.
100. Li L, He S, Sun JM, Davie JR. Gene regulation by Sp1 and Sp3. Biochem Cell Biol. 2004;82(4):460–71. doi:10.1139/o04-045.
101. Lee C, Dhillon J, Wang MY, Gao Y, Hu K, Park E, Astanehe A, Hung MC, Eirew P, Eaves CJ, Dunn SE. Targeting YB-1 in HER-2 overexpressing breast cancer cells induces apoptosis via the mTOR/STAT3 pathway and suppresses tumor growth in mice. Cancer Res. 2008;68(21):8661–6. doi:10.1158/0008-5472.CAN-08-1082.
102. Habibi G, Leung S, Law JH, Gelmon K, Masoudi H, Turbin D, Pollak M, Nielsen TO, Huntsman D, Dunn SE. Redefining prognostic factors for breast cancer: YB-1 is a stronger predictor of relapse and disease-specific survival than estrogen receptor or HER-2 across all tumor subtypes. Breast Cancer Res. 2008;10(5):R86. doi:10.1186/bcr2156.
103. Dhillon J, Astanehe A, Lee C, Fotovati A, Hu K, Dunn SE. The expression of activated Y-box binding protein-1 serine 102 mediates trastuzumab resistance in breast cancer cells by increasing CD44+ cells. Oncogene. 2010;29(47):6294–300. doi:10.1038/onc.2010.365.
104. Fukada T, Tonks NK. Identification of YB-1 as a regulator of PTP1B expression: implications for regulation of insulin and cytokine signaling. EMBO J. 2003;22(3):479–93. doi:10.1093/emboj/cdg067.
105. Zabolotny JM, Kim YB, Welsh LA, Kershaw EE, Neel BG, Kahn BB. Protein-tyrosine phosphatase 1B expression is induced by inflammation in vivo. J Biol Chem. 2008;283(21):14230–41. doi:10.1074/jbc.M800061200.
106. Baron V, Adamson ED, Calogero A, Ragona G, Mercola D. The transcription factor Egr1 is a direct regulator of multiple tumor suppressors including TGFbeta1, PTEN, p53, and fibronectin. Cancer Gene Ther. 2006;13(2):115–24. doi:10.1038/sj.cgt.7700896.
107. Sun C, Zhang F, Ge X, Yan T, Chen X, Shi X, Zhai Q. SIRT1 improves insulin sensitivity under insulin-resistant conditions by repressing PTP1B. Cell Metab. 2007;6(4):307–19. doi:10.1016/j.cmet.2007.08.014.
108. Cheng A, Bal GS, Kennedy BP, Tremblay ML. Attenuation of adhesion-dependent signaling and cell spreading in transformed fibroblasts lacking protein tyrosine phosphatase-1B. J Biol Chem. 2001;276(28):25848–55. doi:10.1074/jbc.M009734200.
109. Arias-Romero LE, Saha S, Villamar-Cruz O, Yip SC, Ethier SP, Zhang ZY, Chernoff J. Activation of Src by protein tyrosine phosphatase 1B Is required for ErbB2 transformation of human breast epithelial cells. Cancer Res. 2009;69(11):4582–8. doi:0008-5472.CAN-08-4001 [pii] 10.1158/0008-5472.CAN-08-4001.

110. Lu X, Malumbres R, Shields B, Jiang X, Sarosiek KA, Natkunam Y, Tiganis T, Lossos IS. PTP1B is a negative regulator of interleukin 4-induced STAT6 signaling. Blood. 2008;112(10):4098–108. doi:10.1182/blood-2008-03-148726.
111. Hutchins AP, Diez D, Takahashi Y, Ahmad S, Jauch R, Tremblay ML, Miranda-Saavedra D. Distinct transcriptional regulatory modules underlie STAT3's cell type-independent and cell type-specific functions. Nucleic Acids Res. 2013;41(4):2155–70. doi:10.1093/nar/gks1300.
112. Ravichandran LV, Chen H, Li Y, Quon MJ. Phosphorylation of PTP1B at Ser(50) by Akt impairs its ability to dephosphorylate the insulin receptor. Mol Endocrinol. 2001;15(10):1768–80.
113. Frangioni JV, Oda A, Smith M, Salzman EW, Neel BG. Calpain-catalyzed cleavage and subcellular relocation of protein phosphotyrosine phosphatase 1B (PTP-1B) in human platelets. EMBO J. 1993;12(12):4843–56.
114. Kuchay SM, Kim N, Grunz EA, Fay WP, Chishti AH. Double knockouts reveal that protein tyrosine phosphatase 1B is a physiological target of calpain-1 in platelets. Mol Cell Biol. 2007;27(17):6038–52. doi:10.1128/MCB.00522-07.
115. Gomez MA, Contreras I, Halle M, Tremblay ML, McMaster RW, Olivier M. Leishmania GP63 alters host signaling through cleavage-activated protein tyrosine phosphatases. Sci Signal. 2009;2(90):58.
116. Dadke S, Cotteret S, Yip SC, Jaffer ZM, Haj F, Ivanov A, Rauscher 3rd F, Shuai K, Ng T, Neel BG, Chernoff J. Regulation of protein tyrosine phosphatase 1B by sumoylation. Nat Cell Biol. 2007;9(1):80–5. doi:10.1038/ncb1522.
117. Hussain SP, Hofseth LJ, Harris CC. Radical causes of cancer. Nat Rev Cancer. 2003;3(4):276–85. doi:10.1038/nrc1046.
118. Storz P, Liou GY. Reactive oxygen species in cancer. Free Radic Res. 2010;44(5):479–96. doi:10.3109/10715761003667554.
119. Chiarugi P, Giannoni E, Taddei ML. Src redox regulation: again in the front line. Free Radic Bio Med. 2010;49(4):516–27. doi:10.1016/j.freeradbiomed.2010.04.025.
120. Li N, Karin M. Is NF-kappaB the sensor of oxidative stress? FASEB J. 1999; 13(10):1137–43.
121. Salmeen A, Andersen JN, Myers MP, Meng TC, Hinks JA, Tonks NK, Barford D. Redox regulation of protein tyrosine phosphatase 1B involves a sulphenyl-amide intermediate. Nature. 2003;423(6941):769–73. doi:10.1038/nature01680.
122. Salmeen A, Barford D. Functions and mechanisms of redox regulation of cysteine-based phosphatases. Antioxid Redox Sign. 2005;7(5-6):560–77. doi:10.1089/ars.2005.7.560.
123. Yang J, Groen A, Lemeer S, Jans A, Slijper M, Roe SM, den Hertog J, Barford D. Reversible oxidation of the membrane distal domain of receptor PTPalpha is mediated by a cyclic sulfenamide. Biochemistry. 2007;46(3):709–19. doi:10.1021/bi061546m.
124. Karisch R, Fernandez M, Taylor P, Virtanen C, St-Germain JR, Jin LL, Harris IS, Mori J, Mak TW, Senis YA, Ostman A, Moran MF, Neel BG. Global proteomic assessment of the classical protein-tyrosine phosphatome and "Redoxome". Cell. 2011;146(5):826–40. doi:10.1016/j.cell.2011.07.020.
125. Dagnell M, Frijhoff J, Pader I, Augsten M, Boivin B, Xu J, Mandal PK, Tonks NK, Hellberg C, Conrad M, Arner ES, Ostman A. Selective activation of oxidized PTP1B by the thioredoxin system modulates PDGF-beta receptor tyrosine kinase signaling. Proc Natl Acad Sci U S A. 2013;110(33):13398–403. doi:10.1073/pnas.1302891110.
126. Yang M, Haase AD, Huang FK, Coulis G, Rivera KD, Dickinson BC, Chang CJ, Pappin DJ, Neubert TA, Hannon GJ, Boivin B, Tonks NK. Dephosphorylation of tyrosine 393 in argonaute 2 by protein tyrosine phosphatase 1B regulates gene silencing in oncogenic RAS-induced senescence. Mol Cell. 2014;55(5):782–90. doi:10.1016/j.molcel.2014.07.018.
127. Gulati P, Markova B, Gottlicher M, Bohmer FD, Herrlich PA. UVA inactivates protein tyrosine phosphatases by calpain-mediated degradation. EMBO Rep. 2004;5(8):812–7. doi:10.1038/sj.embor.7400190.
128. Haque A, Andersen JN, Salmeen A, Barford D, Tonks NK. Conformation-sensing antibodies stabilize the oxidized form of PTP1B and inhibit its phosphatase activity. Cell. 2011;147(1):185–98. doi:10.1016/j.cell.2011.08.036.

129. Frangioni JV, Beahm PH, Shifrin V, Jost CA, Neel BG. The nontransmembrane tyrosine phosphatase PTP-1B localizes to the endoplasmic reticulum via its 35 amino acid C-terminal sequence. Cell. 1992;68(3):545–60.
130. Haj FG, Sabet O, Kinkhabwala A, Wimmer-Kleikamp S, Roukos V, Han HM, Grabenbauer M, Bierbaum M, Antony C, Neel BG, Bastiaens PI. Regulation of signaling at regions of cell-cell contact by endoplasmic reticulum-bound protein-tyrosine phosphatase 1B. PLoS One. 2012;7(5), e36633. doi:10.1371/journal.pone.0036633.
131. Anderie I, Schulz I, Schmid A. Direct interaction between ER membrane-bound PTP1B and its plasma membrane-anchored targets. Cell Signal. 2007;19(3):582–92. doi:10.1016/j.cellsig.2006.08.007.
132. Haj FG, Verveer PJ, Squire A, Neel BG, Bastiaens PI. Imaging sites of receptor dephosphorylation by PTP1B on the surface of the endoplasmic reticulum. Science. 2002;295(5560):1708–11. doi:10.1126/science.1067566.
133. Choudhary C, Olsen JV, Brandts C, Cox J, Reddy PN, Bohmer FD, Gerke V, Schmidt-Arras DE, Berdel WE, Muller-Tidow C, Mann M, Serve H. Mislocalized activation of oncogenic RTKs switches downstream signaling outcomes. Mol Cell. 2009;36(2):326–39. doi:10.1016/j.molcel.2009.09.019.
134. Boute N, Boubekeur S, Lacasa D, Issad T. Dynamics of the interaction between the insulin receptor and protein tyrosine-phosphatase 1B in living cells. EMBO Rep. 2003;4(3):313–9. doi:10.1038/sj.embor.embor767.
135. Romsicki Y, Reece M, Gauthier JY, Asante-Appiah E, Kennedy BP. Protein tyrosine phosphatase-1B dephosphorylation of the insulin receptor occurs in a perinuclear endosome compartment in human embryonic kidney 293 cells. J Biol Chem. 2004;279(13):12868–75. doi:10.1074/jbc.M309600200.
136. Cromlish WA, Tang M, Kyskan R, Tran L, Kennedy BP. PTP1B-dependent insulin receptor phosphorylation/residency in the endocytic recycling compartment of CHO-IR cells. Biochem Pharmacol. 2006;72(10):1279–92. doi:10.1016/j.bcp.2006.07.038.
137. Stuible M, Dube N, Tremblay ML. PTP1B regulates cortactin tyrosine phosphorylation by targeting Tyr446. J Biol Chem. 2008;283(23):15740–6.
138. Radhakrishnan VM, Kojs P, Young G, Ramalingam R, Jagadish B, Mash EA, Martinez JD, Ghishan FK, Kiela PR. PTyr(421) cortactin is overexpressed in colon cancer and is dephosphorylated by curcumin: involvement of Non-receptor type 1 protein tyrosine phosphatase (PTPN1). PLoS One. 2014;9(1), e85796. doi:10.1371/journal.pone.0085796.
139. Ormandy CJ, Musgrove EA, Hui R, Daly RJ, Sutherland RL. Cyclin D1, EMS1 and 11q13 amplification in breast cancer. Breast Cancer Res Treat. 2003;78(3):323–35.
140. Schuuring E. The involvement of the chromosome 11q13 region in human malignancies: cyclin D1 and EMS1 are two new candidate oncogenes—a review. Gene. 1995;159(1):83–96.
141. Sangwan V, Abella J, Lai A, Bertos N, Stuible M, Tremblay ML, Park M. Protein-tyrosine phosphatase 1B modulates early endosome fusion and trafficking of Met and epidermal growth factor receptors. J Biol Chem. 2011;286(52):45000–13. doi:10.1074/jbc.M111.270934.
142. Sorkin A, von Zastrow M. Endocytosis and signalling: intertwining molecular networks. Nat Rev Mol Cell Biol. 2009;10(9):609–22. doi:10.1038/nrm2748.
143. Eden ER, White IJ, Tsapara A, Futter CE. Membrane contacts between endosomes and ER provide sites for PTP1B-epidermal growth factor receptor interaction. Nat Cell Biol. 2010;12(3):267–72. doi:10.1038/ncb2026.
144. Stuible M, Abella JV, Feldhammer M, Nossov M, Sangwan V, Blagoev B, Park M, Tremblay ML. PTP1B targets the endosomal sorting machinery: dephosphorylation of regulatory sites on the endosomal sorting complex required for transport component STAM2. J Biol Chem. 2010;285(31):23899–907. doi:10.1074/jbc.M110.115295.
145. Heinonen KM, Dube N, Bourdeau A, Lapp WS, Tremblay ML. Protein tyrosine phosphatase 1B negatively regulates macrophage development through CSF-1 signaling. Proc Natl Acad Sci U S A. 2006;103(8):2776–81. doi:10.1073/pnas.0508563103.

146. Yudushkin IA, Schleifenbaum A, Kinkhabwala A, Neel BG, Schultz C, Bastiaens PI. Live-cell imaging of enzyme-substrate interaction reveals spatial regulation of PTP1B. Science. 2007;315(5808):115–9. doi:10.1126/science.1134966.
147. Pike KA, Hutchins AP, Vinette V, Theberge JF, Sabbagh L, Tremblay ML, Miranda-Saavedra D. Protein tyrosine phosphatase 1B is a regulator of the interleukin-10-induced transcriptional program in macrophages. Sci Signal. 2014;7(324):43.
148. Miraldi ER, Sharfi H, Friedline RH, Johnson H, Zhang T, Lau KS, Ko HJ, Curran TG, Haigis KM, Yaffe MB, Bonneau R, Lauffenburger DA, Kahn BB, Kim JK, Neel BG, Saghatelian A, White FM. Molecular network analysis of phosphotyrosine and lipid metabolism in hepatic PTP1b deletion mice. Integr Biol. 2013;5(7):940–63. doi:10.1039/c3ib40013a.
149. Haque A, Tonks NK. The use of phage display to generate conformation-sensor recombinant antibodies. Nat Protoc. 2012;7(12):2127–43. doi:10.1038/nprot.2012.132.
150. Krishnan N, Koveal D, Miller DH, Xue B, Akshinthala SD, Kragelj J, Jensen MR, Gauss CM, Page R, Blackledge M, Muthuswamy SK, Peti W, Tonks NK. Targeting the disordered C terminus of PTP1B with an allosteric inhibitor. Nat Chem Biol. 2014;10(7):558–66. doi:10.1038/nchembio.1528.
151. Blanquart C, Karouri SE, Issad T. Implication of protein tyrosine phosphatase 1B in MCF-7 cell proliferation and resistance to 4-OH tamoxifen. Biochem Biophys Res Commun. 2009;387(4):748–53.
152. Blanquart C, Karouri SE, Issad T. Protein tyrosine phosphatase-1B and T-cell protein tyrosine phosphatase regulate IGF-2-induced MCF-7 cell migration. Biochem Biophys Res Commun. 2010;392(1):83–8. doi:10.1016/j.bbrc.2009.12.176.
153. Wu C, Zhang L, Bourne PA, Reeder JE, di Sant'Agnese PA, Yao JL, Na Y, Huang J. Protein tyrosine phosphatase PTP1B is involved in neuroendocrine differentiation of prostate cancer. Prostate. 2006;66(11):1125–35. doi:10.1002/pros.20412.
154. Dube N, Bourdeau A, Heinonen KM, Cheng A, Loy AL, Tremblay ML. Genetic ablation of protein tyrosine phosphatase 1B accelerates lymphomagenesis of p53-null mice through the regulation of B-cell development. Cancer Res. 2005;65(21):10088–95. doi:10.1158/0008-5472.CAN-05-1353.
155. Wang J, Liu B, Chen X, Su L, Wu P, Wu J, Zhu Z. PTP1B expression contributes to gastric cancer progression. Med Oncol. 2012;29(2):948–56. doi:10.1007/s12032-011-9911-2.
156. Wang XM, Shang L, Zhang Y, Hao JJ, Shi F, Luo W, Zhang TT, Wang BS, Yang Y, Liu ZH, Zhan QM, Wang MR. PTP1B contributes to calreticulin-induced metastatic phenotypes in esophageal squamous cell carcinoma. Mol Cancer Res. 2013;11(9):986–94. doi:10.1158/1541-7786.MCR-12-0704.

Dual-Specificity Map Kinase (MAPK) Phosphatases (MKPs) and Their Involvement in Cancer

Andrew M. Kidger and Stephen M. Keyse

Abstract Mitogen-activated protein kinase (MAPK) cascades regulate diverse cellular functions including proliferation, differentiation, migration and apoptosis. The abnormal regulation or activity of MAPK signalling pathways has been implicated as a key event in the initiation and progression of many human cancers. Furthermore, MAPK signalling also plays a key role in the cellular response to many cancer treatments, including radiation and chemotherapeutic drugs. Dual-specificity mitogen-activated protein kinase phosphatases (DUSPs or MKPs) dephosphorylate the key threonine and tyrosine residues within the activation loop of the MAPKs and act as negative regulators to modulate the spatiotemporal dynamics of MAPK activity and thus the biological outcome of signalling. There is accumulating evidence that alterations in the expression and activities of MKPs are found in a wide range of cancers. However, this data is complex and often contradictory with evidence for both oncogenic and tumour suppressive roles, depending on disease and cellular context. This review summarises our current understanding of the role(s) that these enzymes might play in the oncogenic process and suggests avenues for further studies. Such work will lead to novel insights into the abnormal regulation of MAPK activity in cancer and reveal novel therapeutic approaches.

Keywords MAPK • Ras-ERK • DUSP • MAPK phosphatase • MKP • Cancer • Signal transduction

---

A.M. Kidger • S.M. Keyse (✉)
Division of Cancer Research, University of Dundee, Jacqui Wood Cancer Centre, James Arrott Drive, Ninewells Hospital & Medical School, Dundee DD1 9SY, UK
e-mail: s.m.keyse@dundee.ac.uk

B.G. Neel, N. Tonks (eds.), *Protein Tyrosine Phosphatases in Cancer*,
DOI 10.1007/978-1-4939-3649-6_7

MAP Kinase Signalling

Mitogen-activated protein kinase (MAPK) cascades are a group of highly conserved signal transduction pathways which regulate diverse cellular functions including cell proliferation, differentiation, migration, inflammation and stress responses such as growth arrest or apoptotic cell death [1–3]. A wide range of extracellular and intracellular stimuli including growth factors, cytokines, cellular stress and DNA damage induce the sequential activation of a three-tiered kinase cascade which constitutes the core MAPK signalling module [4]. Active MAPK kinase kinase (MAPKKK) phosphorylates and activates MAPK kinase (MAPKK or MEK), a dual-specificity (Thr/Tyr) protein kinase, thus facilitating the phosphorylation of both the threonine and tyrosine residues of a conserved T-X-Y motif, which is required to activate the MAPK itself [5]. There are four major MAPK signalling pathways in mammalian cells. These are the classical Ras-activated extracellular signal activated kinases ERK1 and ERK2, the p38 family of four MAPK isoforms (p38α, β, δ and γ), the three c-Jun amino-terminal kinases (JNKs 1,2 and 3) and ERK5. In addition, there are less well-characterised or "atypical" MAPKs such as ERK3 and ERK4 [6]. The best characterised MAPKs in terms of their involvement in the aetiology of human cancers are the ERK1/2, JNK and p38 signalling pathways (Fig. 1).

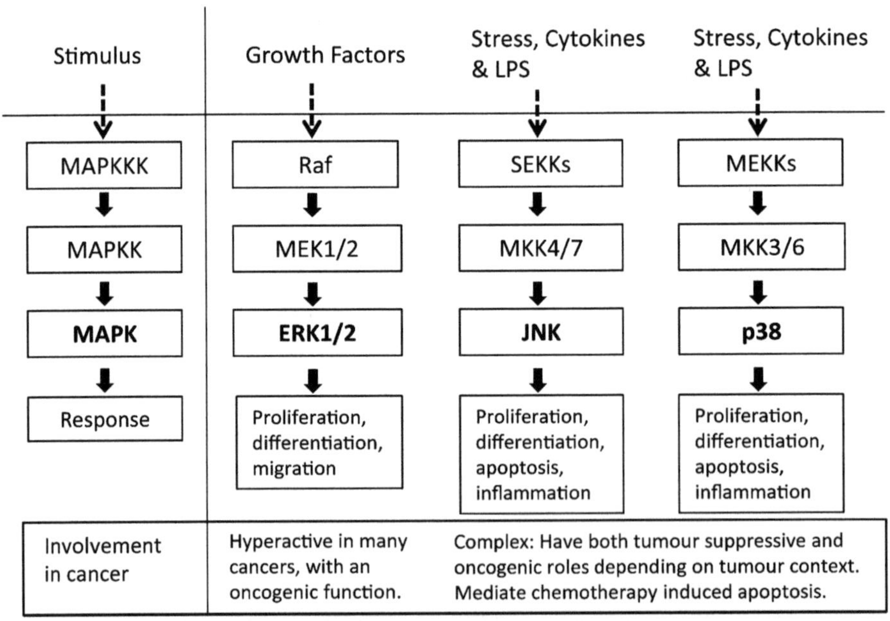

Fig. 1 MAPK signalling pathways and cancer. Schematic showing the three major MAPK modules in mammalian cells, some of the biological endpoints associated with pathway activity and a description of their putative or characterised roles in cancer. For details see text

The Classical Ras-ERK Pathway in Cancer

The Ras-ERK pathway is the most widely studied and well-characterised MAPK pathway in terms of its involvement in the oncogenic transformation of cells and the subsequent stages of tumour development. ERK signalling is deregulated in approximately 30 % of human cancers, either due to increased levels of growth factors and/or their receptors, or activating mutations in upstream components of the Ras-MAPK pathway itself (Fig. 2.). The latter include receptor tyrosine kinases and the Ras family of cellular proto-oncogenes. More recently, the finding that the

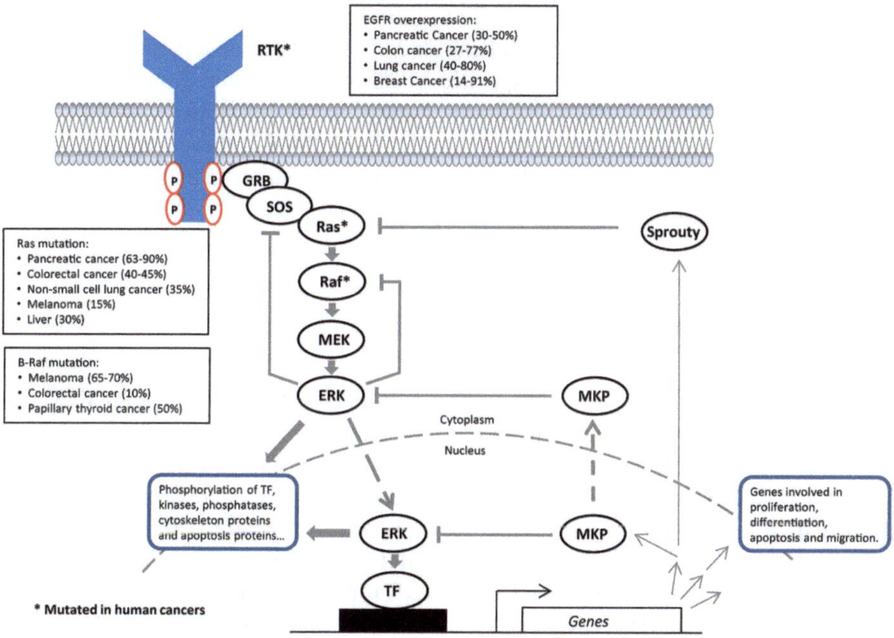

Fig. 2 Oncogenic activation and feedback control mechanisms acting on the Ras–ERK MAPK cascade. Schematic showing the architecture of the Ras-extracellular signal regulated kinase (ERK) pathway from the receptor tyrosine kinase (RTK) at the plasma membrane, through the adaptor protein GRB (growth receptor bound protein), the guanine nucleotide exchange factor SOS (son of sevenless) to the small GTPase Ras. Once GTP bound and activated, Ras activates the MAPK module consisting of the Raf MAPK kinase kinase (MAPKKK), MEK (MAPKK) and ERK itself. Once activated in the cytoplasm, ERK can phosphorylate and inhibit both SOS and Raf, thus exerting negative feedback control of the pathway. Activated ERK also translocates to the cell nucleus where it phosphorylates and activates transcription factors (TF) and induces the expression of ERK target genes. These encode many proteins involved in the regulation of cell proliferation and/or differentiation, but also include negative feedback regulators of the Ras-ERK pathway, including the Sprouty proteins and MAPK phosphatases (MKPs). Oncogenic activation of the Ras-ERK pathway can occur by overexpression or mutation of a number of components. These are indicated with asterisks, along with the cancer type and approximate frequency at which these changes have been observed

MAPKKK B-Raf is mutated in 40–60 % of malignant melanomas and is also associated with papillary thyroid cancer, colorectal cancer, non-small-cell lung cancer and hairy cell leukaemia has served to underscore the importance of the ERK pathway in human malignancy. The most common mutation in B-Raf is a valine to glutamic acid substitution at codon 600 (V600E), which leads to increased kinase activity and thus activates MKK/MEK and ERK [7]. However, while many of the less common B-Raf mutations also result in increased kinase activity, a subset actually reduce the activity of B-Raf towards MKK/MEK. However, these mutants adopt a conformation that causes activation of endogenous C-Raf and this again leads to activation of MKK/MEK and hence ERK [7].

Much excitement has been generated by the ability of vemurafenib/PLX4032, a specific inhibitor of B-Raf, to suppress ERK activation and inhibit tumour cell growth in preclinical models and to elicit clinical responses in B-Raf^{V600E}-positive melanoma patients [8, 9]. Significant benefits have also been observed in B-Raf^{V600E} melanoma patients treated with a specific MEK inhibitor when compared to conventional chemotherapy [10]. Despite the emergence of resistance and significant side effects in patients treated with vemurafenib/PLX4032, there remains significant hope that B-Raf inhibitors, either alone or in combination with other drugs, will prove useful in melanoma and other cancers [11, 12].

JNK Signalling and Cancer

The role of stress-activated MAPK isoforms in cancer is complex, with JNK and p38 demonstrating both oncogenic and tumour suppressive functions in a variety of experimental systems and animal models [13, 14]. Initial reports suggested that JNK activity was oncogenic, at least in the context of Ras and Met-induced cell transformation in vitro [15, 16]. However, experiments in mice lacking different JNK isoforms paint a more complex picture of the relationship between JNK signalling and cancer. For instance, mice lacking JNK1, but not JNK2, are less susceptible to diethylnitrosamine (DEN)-induced liver tumours than wild-type animals, implying that JNK1 is essential for tumour development in this model [17, 18]. Conversely, mice lacking JNK1 were more susceptible than wild-type animals to DMBA-TPA-induced skin papilloma formation whereas mice lacking JNK2 were significantly protected against tumour development [19, 20]. Thus, while JNK1 is oncogenic in the liver, it is a tumour suppressor in skin. In the mouse liver and also in human liver cancer cell lines, loss or downregulation of JNK1 correlates with reduced expression of c-myc and increased expression of the cyclin-dependent kinase inhibitor p21waf1. The latter effect is possibly a reflection of the ability of JNK1 to antagonise p53-dependent pro-apoptotic signalling and thus promote tumour growth and proliferation [17]. In skin, the differential effects of JNK1 and JNK2 deficiency were associated with changes in ERK activation, PI3-kinase/AKT signalling and sequence-specific DNA binding by the AP-1 family of transcription factors [19, 20]. However, it is unclear whether these changes or the modulation of apoptosis by JNK are specifically related to the isoform and organ-specific effects of JNK deletion on carcinogenesis.

Another endpoint of JNK signalling with possible links to cancer is its role in the induction of paracrine factors, including Wnt, BMPs and Il-6. Such signals activate the JAK/STAT3 and Wnt signalling pathways in surrounding cells inducing the expression of genes including cyclin D1 (CCND1), octamer-binding transcription factor 4 (OCT4), SRY-box 2 (Sox2) and c-Myc which promote proliferation and self-renewal of stem cells or cancer stem cells (CSCs) thus driving tumourigenesis. In this way, cells subject to stress signalling and apoptosis can help to stimulate compensatory growth of the tumour as a whole [21].

In humans, large-scale sequencing of the coding regions of protein kinase genes and global exome sequencing in a diverse range of cancers have revealed mutations clustering in the JNK pathway in lung, colorectal, pancreatic and other tumours. Genes affected include those encoding MAP4K3/germinal centre kinase (GLK), MKK4, JNK1 and JNK2 indicating that mutations in this pathway are very likely to be involved in cancer development [22, 23].

The p38 MAPKs and Cancer

As in the case of the JNKs, the involvement of the p38 family of MAPKs in cancer may be complex. Evidence of a tumour suppressive role for p38α derives from a variety of studies. Firstly, mouse embryo fibroblasts from p38α$^{-/-}$ mice are more susceptible to transformation by activated H-Ras than wild-type cells and form more tumours when injected into nude mice [24]. More directly, conditional ablation of p38α in the mouse lung sensitises these animals to Kras^{G12V}-induced lung tumours [25], while hepatocyte-specific ablation of p38α sensitises mice to diethylnitrosamine (DEN)/phenobarbital-induced liver cancer [26]. As is the case for JNK signalling, different p38 isoforms may also have quite different, and opposing, roles in tumour development. Mice lacking the p38δ MAPK are extremely resistant to both DMBA/TPA-induced skin tumours and K-Ras-induced lung cancers, indicating that this MAPK is oncogenic in these tissues and could oppose the tumour suppressive function of p38α in the lung [27].

Possible mechanisms for the tumour suppressive function of p38α include its ability to negatively regulate cell proliferation and block cell cycle progression at both the G1/S and G2/M transitions. Mechanisms for the former include the negative regulation of the JNK pathway, the induction of premature senescence, particularly in the context of activated oncogenes, and downregulation of epidermal growth factor receptor (EGFR) expression. The effects of p38α on the cell cycle may involve the up-regulation of cyclin dependent kinase (CDK) inhibitors, downregulation of cyclins and modulation of p53 activity [28]. However, not all studies are consistent with a purely suppressive effect of p38α in cancer. p38 signalling may also promote tumour development in a number of ways. Firstly, it may increase tumour survival, either by the induction of tumour dormancy or increased levels of autophagy [29, 30]. Furthermore, p38 signalling can lead to increased inflammation, primarily via its ability to regulate pro-inflammatory mediators such as cyclooxygenase-2 (COX2) and cytokines including interleukin-1(IL1), IL6 and TNFα. The latter effects are largely mediated by the activation of MAPK-activated kinase-2 (MK2), a downstream effector of p38α, which acts to both stabilise cytokine mRNAs and increase levels of translation [31–33]. Finally, p38 signalling may also positively influence tumour cell invasion and angiogenesis. In addition to its pro-inflammatory role, MK2 can also play a positive role in cell migration by phosphorylation of the small heat shock protein Hsp27. This modification has been implicated in the remodelling of the actin cytoskeleton associated with cell movement [34]. Activation of p38 also results in increased expression of matrix metalloproteinases (MMPs), which mediate degradation and remodelling of the extracellular matrix (ECM) during cancer cell invasion and increased levels of vascular endothelial growth factor (VEGF), a potent inducer of tumour angiogenesis. All of these events may combine to increase the invasive and metastatic potential of tumour cells [35].

Unlike the JNK pathway, mutations in components of the p38 pathway have not been detected in human cancers. As mentioned previously, MKK4 is inactivated by mutation with low frequency (5 %) in a variety of human tumours. However, as this MAPKK can activate both JNK and p38 it is unclear whether significant changes in p38 activity are associated with its loss. Finally, WIP1/PPM1D is an oncogenic, p53-inducible type 2C Ser/Thr protein phosphatase, which is amplified in approximately 11 % of primary breast cancers [36]. The ability of WIP1/PPM1D to dephosphorylate and inactivate p38 MAPK and thus oppose the tumour suppressive functions of the pathway has been linked with its oncogenic function [37]. However, WIP1/PPM1D also dephosphorylates many substrates of the DNA damage sensor kinases ATM and ATR, as well as suppressing the activities of a range of tumour suppressors including p53, p16(INK4a) and ARF. Thus it is unclear how important the ability of WIP1/PPM1D to regulate p38 signalling is in mediating its oncogenic functions [38].

The sheer diversity of these JNK and p38-mediated responses is probably due to the wide range of stresses and stimuli, which are able to activate these pathways. One important aspect of this is that many stimuli will engage with both pathways simultaneously and it is the duration and intensity of signalling through the individual modules coupled with various mechanisms of crosstalk between the JNK and p38 pathways which will result in the integration of information and the resulting biological outcome. Consequently, such responses will be both cell context and cell-type dependent thus accounting for the variable outcome of p38 or JNK signalling in the context of cell growth, differentiation, apoptosis or senescence and the contribution of these process to tumourigenesis [14].

Protein Phosphatases and the Regulation of MAPK Signalling

The duration, magnitude and subcellular localisation of MAPK activation are all critical factors in determining the biological outcome of signalling [39–41]. These parameters can be modified in a number of ways. MAPK Scaffold proteins can regulate the spatial activation of MAPKs and thus modulate the signalling outcome by controlling which cellular substrates the MAPK is able to interact with. Furthermore, by tethering individual MAPK module components (typically MAPKKK, MAPKK and MAPK) they can facilitate more efficient sequential activation, thereby increasing the efficiency of MAPK signalling [42]. The overall duration and magnitude of MAPK is controlled by a balance between the opposing actions of upstream pathway activators and a number of negative regulatory mechanisms, which can act at multiple levels within the pathway [43] (Figure 2 shows the negative feedback regulation of the ERK pathways). However, it is now clear that a major point of control over pathway activity is the dephosphorylation and inactivation of the MAPKs themselves. Due to the unique mode of MAPK activation, this task can be performed by all three 3 classes of protein phosphatases: type 1/2 Ser/Thr phosphatases, tyrosine-specific protein phosphatases or dual-specificity (Thr/Tyr) phosphatases (DUSPs) [44]. The latter group constitutes the largest family of protein phosphatases dedicated entirely to the regulation of MAPK signalling.

Dual-Specificity MAPK Phosphatases

Mitogen-activated protein kinase phosphatases (MKPs) form a distinct subgroup of the DUSPs, a heterogeneous family of cysteine-dependent protein phosphatases which are able to dephosphorylate both phosphotyrosine and phosphothreonine/phosphoserine residues [45]. There are ten catalytically active MKPs in mammalian cells, which act to specifically regulate the phosphorylation, activity and localisation of MAPKs [46]. MKPs share a common structure comprising an N-terminal regulatory domain containing a kinase interaction motif (KIM) to facilitate binding to specific MAPKs and a highly conserved C-terminal catalytic domain [43] (Fig. 3). Furthermore, MKPs contain nuclear localisation or export signals (NLS/NES) that determine their subcellular localisation [46].

The ten MKPs have been classified into three subfamilies based on amino acid sequence homology, subcellular localisation and substrate specificity [44, 47, 48]. The first group comprises DUSP1/MKP-1, DUSP2, DUSP4/MKP-2 and DUSP5. These are mitogen- or stress-inducible nuclear MKPs. In the case of DUSP1/MKP-1, DUSP2, and DUSP4/MKP-2, these enzymes can interact with and inactivate more than one MAPK isoform with DUSP1/MKP-1 able to regulate ERK, p38 and JNK signalling. In contrast, DUSP5 displays a highly specific activity towards ERK1/2 showing little or no activity towards either p38 or JNK MAPK isoforms. The second group includes the cytoplasmic ERK-selective MKPs DUSP6/MKP-3, DUSP7/MKP-X and DUSP9/MKP-4. The third group consists of DUSP8, DUSP10/MKP-5 and DUSP16/MKP-7, which are JNK/p38-specific MKPs and are localised in both the cytoplasm and the nucleus.

Most MKPs are expressed at relatively low levels in resting (unstimulated) cells and are transcriptionally induced as early response genes in response to a wide variety of mitogenic and/or stress stimuli, many of which also result in MAPK pathway activation. The former observation led to the idea that these enzymes would act as

Gene/MKP	Other Names	Subcellular Localisation	Substrate Specificity	Inducible by MAPKs:	Domain Structure
DUSP1	MKP-1, CL100, erp, 3CH134, hVH1	Nuclear	JNK = p38 > ERK	ERK, p38	
DUSP2	PAC-1	Nuclear	ERK = p38 > JNK	ERK, JNK	
DUSP4	MKP-2, Typ1, Sty8, hVH2	Nuclear	ERK = JNK > p38	ERK	
DUSP5	hVH-3, B23	Nuclear	ERK	ERK	
DUSP6	MKP-3, Pyst1, rVH6	Cytoplasmic	ERK	ERK	
DUSP7	MKP-X, Pyst2, B59	Cytoplasmic	ERK	N/D	
DUSP9	MKP-4, Pyst3	Cytoplasmic	ERK > p38	N/D	
DUSP8	M3/6, hVH5, HB5	Cytoplasmic/ nuclear	JNK = p38	N/D	
DUSP10	MKP-5	Cytoplasmic/ nuclear	JNK = p38	N/D	
DUSP16	MKP-7	Cytoplasmic/ nuclear	JNK = p38	N/D	

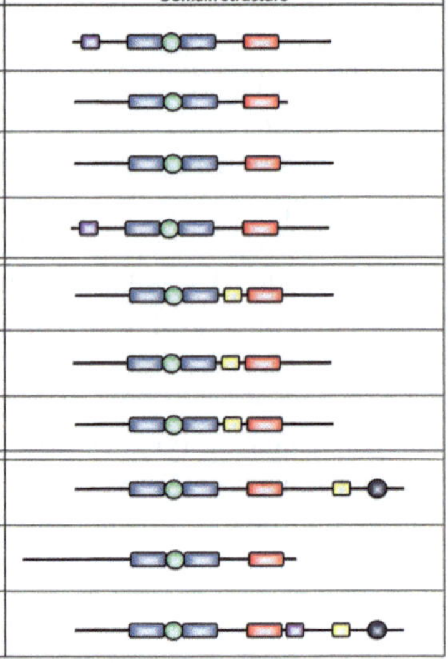

Fig. 3 Classification, properties and domain structure of the dual-specificity MAPK phosphatases (MKPs). Classification and basic properties of the MKPs including trivial names, subcellular localisation, substrate specificity, the involvement of MAPK pathways in transcriptional regulation and the domain structure of the proteins. See main text for details

part of negative feedback loops to control MAPK activity. However, it is now clear that many other signalling pathways can also increase the expression levels of the MKPs. For example, DUSP1/MKP-1 is transcriptionally regulated by p53 during responses to oxidative stress [49], DUSP6/MKP-3 has been shown to be induced by wnt-β-catenin signalling during zebrafish development and in mouse hepatoma cells [50, 51], and transforming growth factor-β (TGFβ) can induce DUSP4/ MKP-2 in a SMAD3-dependent manner in lymphocytes [52]. Thus the MKPs can also act as mediators of crosstalk between a number of distinct signalling pathways and MAPK modules. Following a transient stimulus, the downregulation of DUSP expression is mediated by multiple mechanisms, both at the transcriptional and post-transcriptional level. The latter include changes in rates of both mRNA synthesis and degradation [53].

Post-translational modification also plays a key role in the control of MKP expression and activity. Phosphorylation of C-terminal residues within DUSP1/ MKP-1 and DUSP4/MKP-2 by ERK1/2 has been shown to stabilise these proteins in response to growth factor stimulation, thus reinforcing the negative feedback control of MAPK signalling [54]. In contrast, the phosphorylation of DUSP6/ MKP-3 by ERK1/2 promotes ubiquitin-mediated degradation of the protein [55] and this results in the initial disappearance of DUSP6/MKP-3 in cells exposed to platelet-derived growth factor (PDGF) [56]. Phosphorylation of the KIM motif within DUSP9/MKP-4 by protein kinase-A (PKA) prevents the phosphatase from physically interacting with and inactivating both ERK1/2 and p38 MAPKs thus representing another mechanism of crosstalk between distinct signalling pathways [57]. Finally, while the substrate specificity of MKPs is largely determined by their ability to specifically recognise and bind to individual MAPKs, certain of these enzymes also undergo a conformational (allosteric) change on MAPK binding, which results in the reordering of the active site residues into the optimum conformation for catalysis [58–60]. This mechanism of catalytic activation is thought to reinforce substrate selectivity and ensure that these enzymes do not display activity towards non-MAPK substrates [43].

Spatio-temporal Regulation of MAPK Signalling by MKPs

The differing subcellular localisation and substrate specificity of MKPs allows them to regulate the magnitude and duration of activity of specific MAPKs and also to spatially restrict MAPK activity. Furthermore, MKPs are able to bind their MAPK substrates regardless of their phosphorylation status, giving them the ability to anchor inactive MAPKs in distinct subcellular compartments following dephosphorylation [43, 61, 62]. Together these properties indicate that MKPs are more than simply negative regulators of MAPKs and are able to regulate MAPK signalling in ways, which are both complex and incompletely understood [43].

The regulation of MAPK signalling by MKPs can mediate multiple functions including autoregulation, a signalling "memory" and crosstalk between signalling pathways [43]. Transcription of the ERK-specific phosphatases DUSP5 and DUSP6/MKP-3 is induced in an ERK-dependent manner following mitogenic stimulation [63, 64]. This forms a delayed autoregulatory negative feedback loop; in this case MKP expression might be expected to prevent sustained, but not transient, ERK responses. However, if the MKP protein is still present when cells are exposed to a subsequent signal, it will modulate this response and therefore the cell will have retained a temporal memory of the initial signal [43].

The ability of MKPs to anchor dephosphorylated MAPKs in a particular subcellular location could also modify signalling by controlling access to the upstream components of the pathway. For instance, high levels of DUSP6/MKP-3 expression may lead to the cytoplasmic accumulation of dephosphorylated (inactive) ERK allowing rapid reactivation of this pool in the presence of activated MEK. In contrast, the nuclear sequestration of inactive ERK by DUSP5 might restrict the access of the inactive ERK to activated MEK in the cytoplasm thus preventing rapid reactivation [43].

Given the above, it is not surprising that the levels and activities of the MKPs have been implicated in modulating both the oncogenic and anti-oncogenic properties of the MAP signalling pathways. The evidence for this involvement is discussed in detail in the following sections of this chapter.

MKPs and Cancer

The Inducible Nuclear MKPs and Cancer

DUSP1/MKP-1

Of the four inducible nuclear MKPs, the first to be identified and the most studied is DUSP1/MKP-1, which is able to interact with and inactivate ERK, JNK and p38 MAPKs [59]. Soon after the discovery that DUSP1/MKP-1 was a MAPK phosphatase, reports appeared in the literature describing the abnormal regulation of DUSP1/MKP-1 in human tumours (Table 1). These include cases of overexpression in early stage prostate, stomach, colon and bladder cancers, with lower expression levels observed in late stage disease and in metastases [65–67]. In breast cancer, the situation is more confused with reports of increased expression maintained throughout the course of the disease [68] but also reduced expression of DUSP1/MKP-1 correlating with a poor outcome [69]. The latter result led to the suggestion that methylation of the DUSP1/MKP-1 gene might be a useful biomarker of malignancy in this disease [70]. However, in ovarian cancer, DUSP1/MKP-1 is expressed at moderate to strong levels in almost 60 % of tumours, where it correlated with poor prognosis [71]. In non-small-cell lung cancer (NSCLC), levels of DUSP1/MKP-1 also were reported to be higher when compared to normal tissue or small-cell lung cancers and DUSP1/MKP-1 appeared to promote angiogenesis, tumour cell invasion and metastasis, indicating that it is a positive regulator of malignancy in this disease [72, 73]. Finally, DUSP1/MKP-1 expression was reduced in a small but significant number of human hepatocellular carcinomas where it correlated with a poor prognosis [74]. The loss of expression was attributed either to increased ubiquitin-mediated degradation of DUSP1/MKP-1 protein or promoter methylation associated with loss of heterozygosity at the DUSP1/MKP-1 locus, again suggesting a tumour suppressive function for this phosphatase in liver cancer [75]. Overall, the expression levels of DUSP1/MKP-1 in human tumours present a very complex picture with examples of increased or reduced expression depending on the tissue of origin, the tumour type and grade of disease.

Given its ability to inactivate both JNK and p38 MAPKs, early work showed that by reducing the activity of these pathways the overexpression of DUSP1/MKP-1 could suppress apoptosis in response to a variety of stress conditions including exposure to UV radiation and cis-platin [76, 77]. This led to the idea that the overexpression of DUSP1/MKP-1 might correlate with and contribute to chemoresistance in human tumours [78–83]. This concept is reinforced by the observation that cells derived from DUSP1/MKP-1$^{-/-}$ mice are acutely sensitive to UV, cisplatin and hydrogen peroxide [84–86]. At least in mouse embryo fibroblasts, the induction of DUSP1/MKP-1 mRNA and protein is mediated by p38α signalling, while the resulting protection against apoptosis is mediated by inactivation of JNK. Thus the response to oxidative stress and DNA damaging agents may involve DUSP1/MKP-1-mediated crosstalk between the p38 and JNK pathways [86].

Table 1 Summary of reported expression changes for individual MKPs in cancer

MKP	Cancer	Change in expression	References
DUSP1/MKP-1	Human epithelial tumours: Colon, prostate, bladder	Overexpression found in the early phases of tumours. Expression subsequently decreases in tumours of higher histological grade or which have begun to metastasise	[65, 66, 78]
	Ovarian	Moderate to strong DUSP1/MKP-1 expression was observed in 60 % of invasive carcinomas, which correlated with shorter progression free survival	[71]
	Breast	Overexpressed in breast cancer cell lines and tumour samples undergoing malignant transformation, compared to non-malignant controls	[68]
		DUSP1/MKP-1 protein levels were decreased in 63.7 % of breast cancer tissues, and decreased DUSP1 levels correlate with poor prognosis in breast cancer	[69]
		The DUSP1/MKP-1 gene is a potential methylation biomarker for malignancy of breast cancer	[70]
	Non-small-cell lung cancer (NSCLC)	DUSP1/MKP-1 promotes angiogenesis, invasion and metastasis	[72]
		Elevated expression of DUSP1 in NSCLC compared to normal lung cells or SCLC. Increased DUSP1/MKP-1 expression correlated with increased survival	[73]
		The therapeutic efficacy of Notch pathway inhibition in NSCLC may be mediated by the up-regulation of DUSP1	[89]
		Ectopic expression of DUSP1/MKP-1 reduces the growth rate of NSCLC	[90]
		Gastric	DUSP1/MKP-1 is overexpressed in gastric adenocarcinomas
	[67]	Liver	DUSP1/MKP-1 expression is decreased in 15 % of hepatocellular carcinoma (HCC), correlating with poor prognosis in HCC patients

(continued)

Table 1 (continued)

MKP	Cancer	Change in expression	References
	[74]		DUSP1 inactivation in HCC is due to either ERK-dependent ubiquitination or promoter hypermethylation and correlates with poor prognosis
	[75]	Chemoresistance	DUSP1/MKP-1 modulates JNK activation and apoptosis in response to cisplatin. Demonstrated in NSCLC, osteosarcoma, breast and ovarian cancer
	[77, 79–81, 85]		Cells derived from DUSP1/MKP-1$^{-/-}$ mice are acutely sensitive to UV, cisplatin and hydrogen peroxide
	[84–86]		Overexpression of DUSP1/MKP-1 confers multidrug resistance in human glioblastoma cells
[83]	DUSP2	Colon	DUSP2 expression is inversely correlated with tumour status, and stage of progression. DUSP2 is suppressed by Hypoxia inducible factor–1 (HIF-1)
[104]		Breast	Decreased expression
[104]		Cervical	Decreased expression
[104]		Ovary	Decreased expression
[104]			Expression in serous carcinoma correlates with poor overall survival
[103]		Kidney	Decreased expression
[104]		Prostate	Decreased expression

[104]		Acute Leukaemia	Downregulated significantly correlates with increased pERK activity
[105]	DUSP4/MKP-2	Ovarian	Overexpression in ovarian serous borderline tumours, but it is downregulated following tumour progression to serous carcinomas
[93]		Breast	Overexpression in breast cancer cell lines and tumour samples undergoing malignant transformation, compared to non-malignant controls
[68]			DUSP4/MKP-2 overexpression sensitises breast cancer cells to chemotherapy-induced apoptosis, whereas DUSP4 depletion decreases the response
[101]		Colorectal (CRC)	DUSP4/MKP-2 up-regulated in CRC, and causes increased proliferation
[97]			DUSP4/MKP-2 up-regulated in CRC
[95]			In CRC cell lines, oncogenic KRas/BRaf activation promotes the overexpression of DUSP4/MKP-2 and DUSP5 in a MEK-dependent manner
[96]		Pancreatic	Overexpression in many pancreatic cancer cell lines and tumours. Pancreatic tumour cells with mutant Kras suppress ERK activity by MEK-dependent induction of DUSP4/MKP-2
[94]		Lung adenocarcinoma	Decreased expression in EGFR-mutant lung adenocarcinoma

(continued)

Table 1 (continued)

MKP	Cancer	Change in expression	References
[99, 100]		Astrocytic gliomas	DUSP4/MKP-2 promoter hypermethylation and corresponding reducing DUSP4 expression identified in glioma tissue and cell lines. Overexpression of DUSP4/MKP-2 inhibits glioma growth
[98]	DUSP5	Colorectal (CRC)	Transcriptomic analysis of CRC cell lines revealed DUSP5 to be one of the 3 most up-regulated genes
[107]			In CRC cell lines, oncogenic KRas/BRaf activation promotes the overexpression of DUSP4 and DUSP5 in a MEK-dependent manner
[96]		Gastric	DUSP5 is downregulated in gastric cancer cell lines and tissue samples by promoter CpG island hypermethylation. DUSP5 hypermethylation correlated with shorter survival and poor prognosis
[108]	DUSP6/MKP-3	Pancreatic	Overexpression in neoplasia and early carcinoma, but downregulation in invasive carcinoma
[113]			The loss of DUSP6/MKP-3 enables progression from pancreatic intraepithelial neoplasia to invasive ductal carcinoma
[115]			DUSP6/MKP-3 transcriptionally suppressed by hypermethylation in pancreatic cancer cell lines
[114]		Lung adenocarcinoma	Decreased expression

[119]		Colorectal Cancer (CRC)	Overexpression in colorectal cancer cell lines expressing mutant Kras
[118]			Oesophageal squamous cell carcinoma and nasopharyngeal carcinoma
DUSP6 downregulated. DUSP6/MKP-3 loss promotes cell invasion and epithelial-mesenchymal transition (EMT)-associated phenotypes		[121]	Melanoma
Low expression observed in the majority of melanomas, but DUSP6/MKP-3 is overexpressed in a distinct subset of melanoma		[124]	Breast
DUSP6 overexpression is a mechanism of tamoxifen resistance in breast cancer		[128]	Ovarian
Downregulation in primary human ovarian cancer cells. DUSP6/MKP-3 loss increases resistance of ovarian cancer cells to cisplatin, and forced expression of DUSP6 reduces ERK activity resensitising cells to cisplatin-induced apoptosis		[129]	Thyroid

(continued)

Table 1 (continued)

MKP	Cancer	Change in expression	References
Overexpression in both cell lines and patient samples from papillary and poorly differentiated thyroid carcinoma. DUSP6/MKP-3 silencing reduced the proliferation rate of papillary thyroid carcinoma cell lines, indicating that DUSP6 may have an oncogenic role in thyroid carcinogenesis		[126]	
DUSP6/MKP-3 overexpression correlates with poor prognosis of thyroid carcinomas. DUSP6/MKP-3 silencing decreased invasiveness and viability	[125]	DUSP7/MKP-X	Acute Leukaemias
Overexpression of DUSP7/MKP-X in acute leukaemia	[130–132]	DUSP9/MKP-4	Renal cell carcinoma
Decreased expression correlates with poor prognosis	[136, 137]		
mRNA profiling revealed DUSP9/MKP-4 downregulation	[135]	DUSP8	No associations

		DUSP10/MKP-5	Prostate
Vitamin D activates DUSP10/MKP-5 in prostate cancer cells and has an anti-inflammatory effect, slowing prostate cancer progression	[139, 140]	DUSP16/MKP-7	Burkitt's lymphoma
DUSP16/MKP-7 is subject to methylation-dependent transcriptional silencing in Burkitt's Lymphoma cell lines and tissues, but unmethylated in other non-Hodgkin's lymphomas (NHLs) and epithelial malignancies	[143, 146]		ETV6-RUNX1 B-cell precursor acute lymphoblastic leukaemia (BCP-ALL)
DUSP16/MKP-7 is one of several genes lost following a centromeric deletion in ETV6-RUNX1 BCP-ALL. This deletion is associated with increased survival in BCP-ALL	[145]		

The many correlations between DUSP1/MKP-1 expression and tumour drug resistance have led to the suggestion that DUSP1/MKP-1 might be a viable therapeutic target in cancer therapy [87, 88]. However, recent work indicates that the therapeutic efficacy of Notch pathway inhibition using pharmacological inhibitors of γ-secretase activity in non-small-cell lung cancer (NSCLC) may at least in part be mediated by the up-regulation of DUSP1/MKP-1 and a resulting decrease in activity of the ERK signalling pathway [89]. While this might seem counter-intuitive in light of the previous links between DUSP1/MKP-1 overexpression, inactivation of JNK and cancer drug resistance, it is possible that this effect reflects an undue reliance on the ERK pathway for tumour survival in cancers such as NSCLC in which tumour growth is driven by activated Kras. This is supported by the observation that expression of DUSP1/MKP-1 reduced the growth rate of NSCLC cells in vitro and impaired NSCLC tumour formation and metastasis in a mouse xenograft model [90]. Overall, these results serve to emphasise that the effects of inhibiting or elevating DUSP1/MKP-1 activity may be both complex and tumour type specific. Certainly far more knowledge of the relevant cellular targets affected by manipulation of DUSP1/MKP-1 levels will be required before a rationale for targeting DUSP1/MKP-1 in cancer can be further developed. This process is further complicated by the profound effects of DUSP1/MKP-1 deletion on the regulation of immune function and the possible effects these might have in cancer initiation or development [91, 92].

DUSP4/MKP-2

DUSP4/MKP-2 is most closely related to DUSP1/MKP-1 and abnormalities in DUSP4/MKP-2 expression have also been reported in a number of human tumours (Table 1). In ovarian cancer, DUSP4/MKP-2 expression is associated with serous borderline tumours which have a benign phenotype when compared with serous carcinomas [93]. The overexpression of DUSP4/MKP-2 has also been associated with mutations in either Kras or Braf in pancreatic, and colorectal cancers, where both its transcriptional induction and biological effect seem to be dependent on regulation of the ERK MAPK pathway rather than on modulation of JNK or p38 signalling [94–96]. In colorectal cancer, DUSP4/MKP-2 expression has been associated with both microsatellite instability and increased cell proliferation [97] indicating that DUSP4/MKP-2 may promote tumour growth. In contrast, a tumour suppressive role for DUSP4/MKP-2 has been proposed based on epigenetic silencing of DUSP4/MKP-2 in a significant number of secondary glioblastomas and the ability of DUSP4/MKP-2 to suppress glioblastoma cell growth in vitro [98]. Interestingly, a high incidence of DUSP4/MKP-2 gene loss is associated with lung cancers carrying mutations in the epidermal growth factor receptor (EGFR) [99]. However, others have reported high levels of DUSP4/MKP-2 protein in EGFR mutant lung cancer cell lines [100].

Finally, DUSP4/MKP-2 is expressed in breast tumours [68] and a recent study has identified low levels of DUSP4/MKP-2 expression as correlating with residual disease in patients receiving neoadjuvant chemotherapy, with these drug resistant tumours exhibiting the characteristics of basal-like breast cancer (BLBC). The lower levels of

DUSP4/MKP-2 expression would appear to result from promoter methylation and also correlate with gene expression patterns indicative of Ras-ERK pathway activation [101]. The authors present data indicating that DUSP4/MKP-2 levels show a positive correlation with docetaxel-induced apoptosis and that inhibition of MEK could synergise with docetaxel to achieve increased cell kill in BLBC with low levels of DUSP4/MKP-2. Clearly the idea of combining conventional chemotherapy with inhibition of the Ras-MAPK pathway may be worthy of further study.

In conclusion, as is the case for DUSP1/MKP-1, the overall picture with respect to DUSP4/MKP-2 and cancer is complex, but there are grounds to believe that this phosphatase may be playing a significant role in modulating tumour growth and could influence anticancer drug responses. The recent generation of mice in which the DUSP4/MKP-2 gene has been deleted by homologous recombination should facilitate mechanistic studies into the role of this phosphatase in a range of murine cancer models [102].

Other Nuclear MKPs and Cancer

Of the remaining two members of the inducible nuclear MKPs DUSP2 (also known as Pac-1) and DUSP5, relatively little is known in terms of possible roles in cancer. DUSP2 expression was observed in serous carcinoma of the ovary where its presence correlated with a poor outcome in terms of patient survival, and loss of DUSP2 expression, has been associated with increased levels of ERK activation in acute leukaemias [103–105] (Table 1). DUSP5 is unique amongst the nuclear MKPs in that it is an ERK-specific phosphatase and can also act as a nuclear anchoring protein for the inactive MAPK [61]. Given this biochemical profile, it might be expected to regulate nuclear phospho-ERK levels and thus exert some activity in cancers in which the Ras-Erk pathway is activated. Indeed, elevated levels of DUSP5 mRNA are seen in Kras mutant cancer cell lines and have been postulated to exert negative feedback control of nuclear ERK activity [106, 107]. A recent study also reports promoter methylation and downregulation of DUSP5 in gastric cancers, but not in normal gastric mucosa. Furthermore, in a set of surgically resected gastric cancer cases, patients with DUSP5 promoter hypermethylation exhibited significantly shortened survival, compared with those lacking this modification and restoring DUSP5 expression in DUSP5-silenced gastric cancer cell lines decreased their growth and colony-forming ability [108]. These results were interpreted as suggesting that the silencing of DUSP5 by promoter hypermethylation causes increased maintenance of phosphorylated ERK1/2, driving cell proliferation and thus contributing to gastric carcinogenesis.

Cytoplasmic MKPs and Cancer

There are three cytoplasmic dual-specificity MKPs, which are highly homologous at the amino acid sequence level. The prototypic member of this family is DUSP6/MKP-3, which is notable for its absolute substrate selectivity for ERK1/2 over all other MAPK isoforms. DUSP7/MKP-X is also highly selective for ERK1/2 while the third member of the subfamily, DUSP9/MKP-4, seems to be able to act on both ERK1/2 and p38 MAPK, but has little or no activity towards JNK.

DUSP6/MKP-3 and Cancer

The expression of DUSP6/MKP-3 mRNA is highly regulated during early development with high levels of expression found at sites of fibroblast growth factor (FGF) signalling in mouse embryos [109]. Furthermore, DUSP6/MKP-3 expression was demonstrated to be both FGF and ERK dependent in the developing chicken limb and neural plate, clearly showing that DUSP6/MKP-3 acts as part of a negative feedback loop to regulate ERK activity downstream of the FGF receptor [110, 111]. Finally, the molecular details of FGF-dependent DUSP6/MKP-3 transcriptional regulation were elucidated with the discovery of a highly conserved binding site for the E-twenty six (Ets) family of transcription factors in the DUSP6/MKP-3 proximal promoter and the demonstration that this is essential for ERK-dependent up-regulation of the gene in response to FGF in vivo [63, 112].

A link between DUSP6/MKP-3 and cancer was first indicated by the observation that DUSP6/MKP-3 is overexpressed in low- and high-grade pancreatic intraepithelial neoplasia (PanIN), but its expression is subsequently suppressed or lost as these lesions developed into pancreatic ductal adenocarcinomas (PDAC). In poorly differentiated, invasive PDAC, DUSP6/MKP-3 expression is lost in approximately 80 % of tumours [113]. The loss of DUSP6/MKP-3 expression has been associated with epigenetic silencing by aberrant hypermethylation of the DUSP6/MKP-3 promoter in pancreatic cancer cell lines [114]. Furthermore, ectopic expression of DUSP6/MKP-3 in these cells results in a decrease in phosphorylated ERK coupled with an inhibition of proliferation and the induction of apoptosis [113]. The loss of DUSP6/MKP-3 expression is only seen in invasive carcinoma, whereas other tumour suppressors implicated in pancreatic cancer development, such as p53 or SMAD4, are lost earlier during PanIN development [115]. These observations have led to the proposal that DUSP6/MKP-3 expression constrains the progression of PanINs to invasive carcinomas in a manner which is independent of other major tumour suppressors, making DUSP6/MKP-3 suppression a key player in the malignancy of PDAC [116, 117].

Over 90 % of pancreatic cancers are initiated by mutation of the Kras cellular proto-oncogene and overexpression of DUSP6/MKP-3 has also been reported in colorectal cancer cell lines expressing mutant Kras where it appears to be responsible for suppression of ERK activity [118]. More recently, loss of DUSP6/MKP-3 expression has also been reported in non-small-cell lung cancer (NSCLC), a tumour which also has a significant frequency (15–25 %) of Kras mutation [119]. As was the case in pancreatic cancers, early stage or low-grade neoplastic lesions expressed DUSP6/MKP-3 at high levels compared with the normal bronchial epithelia and these levels decreased as tumours progressed to late stage disease. A recent study in NSCLC cell lines noted an inverse correlation between levels of DUSP6/MKP-3 expression and ERK activation and observed that ERK-dependent up-regulation of DUSP6/MKP-3 was mediated at the promoter level by the same conserved Ets factor binding site responsible for FGF responses during early development [120]. Again, knockdown of DUSP6/MKP-3 using siRNA in H441 lung cancer cells increased both ERK activation and cellular proliferation, whereas overexpression in H1975 lung cancer cells, which express low levels of DUSP6/MKP-3, reduced both ERK activation and cellular proliferation and promoted apoptosis. A tumour suppressor role for DUSP6/MKP-3 is also suggested by the observation that the protein is downregulated in oesophageal squamous cell carcinomas and nasopharyngeal carcinomas where its loss seems to promote both cell invasion and the epithelial to mesenchymal transition (EMT), both of which are hallmarks of malignant progression and tumour spread [121]. Taken together, these studies support the notion that DUSP6/MKP-3 could be a tumour suppressor whose downregulation is involved in the progression of certain cancers.

In addition to a link between tumours with Kras mutations and DUSP6/MKP-3, ERK-dependent overexpression of this phosphatase has also been seen in melanoma cell lines with activating mutations in Braf (a MAPK kinase kinase or MKKK) [122, 123]. This could reflect an attempt to restrain the activation of ERK2 in response to constitutive Braf activity in these tumours. However, recent work has raised the possibility that DUSP6/MKP-3 may play differing roles in melanoma and that this may reflect tumour heterogeneity. This is based on the observation that while DUSP6/MKP-3 expression appears to promote the growth and invasive properties of tumourigenic mouse melanocytes, which lack mutant Braf, it inhibits the growth of human melanoma cells with Braf mutations. The former cells seem to have key characteristics in common with a subset of thick primary human melanomas with high DUSP6/MKP-3 expression and a poor prognosis [124]. Interestingly, DUSP6/MKP-3 expression has also been observed in papillary and poorly differentiated thyroid cancers [125]. These tumours often carry mutations in the receptor tyrosine kinase RET/PTC (rearranged in transformation/papillary thyroid cancer), Ras or Braf indicating a role for the ERK pathway in pathogenesis. However, DUSP6/MKP-3 expression seems to correlate with the invasive properties of these tumours and siRNA-mediated knockdown of DUSP6/MKP-3 led to a reduction in tumour cell growth and invasion suggesting that DUSP6/MKP-3 may promote thyroid carcinogenesis [126]. Finally, while the anti-diabetic drug metformin exhibits antitumour activity by inhibiting the mTORC1 complex, melanoma cells are refractory to this drug because Braf-ERK-RSK1-dependent TORC1 activation overcomes the inhibitory effects of AMPK activation. In these cells, AMPK activation also leads to the degradation of DUSP6/MKP-3 protein, resulting in increased ERK activity and higher expression of VEGF-A, thus promoting angiogenesis and tumour growth. Surprisingly, VEGF inhibitors synergise with metformin to suppress melanoma growth, suggesting a novel combination therapy that may be effective in this disease [127].

In addition to links with cancers in which the Ras-ERK pathway is dysregulated, elevated expression of DUSP6/MKP-3 has also been observed in breast cancers where it has been linked to tamoxifen resistance [128]. In contrast, DUSP6/MKP-3 expression levels are low in ovarian cancer cells and correlate with increased ERK activity and cisplatin resistance, both of which could be reversed by ectopic expression of DUSP6/MKP-3 [129].

Other Cytoplasmic MKPs and Cancer

Compared with DUSP6/MKP-3, relatively little is known about the physiological roles of either DUSP7/MKP-X or DUSP9/MKP-4 or their involvement in human cancer. There are reports that DUSP7/MKP-X is overexpressed in acute leukaemias. Initially this was revealed by a microarray analysis [130], but was subsequently confirmed by RT-PCR and Northern blot analysis [131]. Subsequent work showed that DUSP7/MKP-X was highly expressed in leucocytes from acute myeloid leukaemia (AML) patients and high levels of mRNA were also detected in bone marrow and peripheral leucocytes from patients with acute lymphoblastic leukaemia (ALL) [132]. To date, it is unclear if these observations have any direct link with either pathogenesis or prognosis in these cancers.

DUSP9/MKP-4 is the third member of this subfamily and compared with DUSP6/MKP-3 and DUSP7/MKP-X it seems to exhibit a more relaxed substrate selectivity having significant activity towards both ERK1/2 and p38 MAPKs. Gene targeting experiments have revealed that this phosphatase plays an essential role in placental development and function [133] and recent work has implicated DUSP9/MKP-4 as regulator of ERK activation in pluripotent mouse embryonic stem cells exposed to bone morphogenetic protein-4 (BMP4), where it seems critical for cell fate determination [134].

In clear cell renal cell carcinomas (ccRCC), DUSP9/MKP-4 was identified one of a group of genes associated with normal differentiated kidney cells whose expression was disrupted in tumours [135]. In a larger study, the expression of DUSP9/MKP-4 mRNA and protein was determined in paired samples of ccRCC and normal tissue and found to be significantly reduced in tumours. Furthermore, in patients with lower DUSP9/MKP-4 expression, survival was significantly shorter, suggesting that DUSP9/MKP-4 might represent a useful prognostic marker in this disease [136]. Interestingly, DUSP9/MKP-4 levels are also found to be lower in hepatocellular carcinomas (HCC) when compared to normal liver tissue and siRNA-mediated knockdown of DUSP9/MKP-4 in HCC cells both enhanced cell survival and inhibited apoptosis. Furthermore, patients with high DUSP9/MKP-4 expression exhibited longer overall survival and significantly lower recurrence rates than those with low MKP-4 expression, and multivariate analysis indicated DUSP9/MKP-4 as an independent prognostic factor for HCC after curative resection [137]. Although preliminary, these studies would seem to indicate that DUSP9/MKP-4 may play a tumour suppressive role in kidney and liver cancer and warrant further work to confirm and extend these observations.

p38/JNK Selective MKPs and Cancer

This subfamily consists of three enzymes DUSP8, DUSP10/MKP-5 and DUSP16/MKP-7. Compared with the inducible nuclear or cytoplasmic ERK-specific groups of MKPs, these proteins are relatively understudied. In the case of DUSP8, there is very little known about the physiological function of the endogenous phosphatase and there is no clear evidence of abnormal regulation or differential expression in any human cancers. Studies of knockout mice revealed that DUSP10/MKP-5 deficient cells produced greatly enhanced levels of pro-inflammatory cytokines during innate immune responses and showed changes in T-cell activation and proliferation, indicating that DUSP10/MKP-5 regulates both innate and adaptive immune systems in vivo [138]. There are some hints from in vitro experiments that up-regulation of DUSP10/MKP-5 might be related to the ability of 1,25-Dihydroxyvitamin D(3) to inhibit growth and promote apoptosis in prostate cancer cell lines [139]. Interestingly, the effects of 1,25-Dihydroxyvitamin D(3) in vivo might be due to the ability of DUSP10/MKP-5 to inactivate p38 MAPK and thus downregulate production of the pro-inflammatory cytokine IL-6. This would implicate a DUSP10/MKP-5 mediated reduction in inflammation as a potential mechanism for prostate cancer prevention by 1,25-Dihydroxyvitamin D(3) [140]. A similar mechanism involving up-regulation of DUSP10/MKP-5 has also been proposed to operate in the chemopreventive and anti-inflammatory actions of curcumin, resveratrol and other phytochemicals in prostate cancer [141].

The third member of this subfamily DUSP16/MKP-7 is one of several candidate genes lying within a tumour suppressor locus on chromosome 12p12 [142]. Overexpression of DUSP16/MKP-7 in BCR-ABL-transformed Rat-1 fibroblasts reduces their transforming capacity in vitro and in vivo via downregulation of BCR-ABL-induced JNK activation [143] and DUSP16/MKP-7 was one of three genes downregulated in prostate cancers with deletions lying within chromosome 12p12-13 [144]. Aberrations involving chromosome 12 are also seen in childhood B-cell precursor acute lymphoblastic leukaemia (BCP-ALL) caused by chromosome translocations encoding the ETV6-RUNX1 fusion protein [t(12;21)(p13;q22)]. A centromeric deletion on chromosome 12 associated with this translocation contained DUSP16/MKP-7 and four other loci. At present it is uncertain if this additional chromosome abnormality modulates the expression of the ETV6-RUNX1 fusion or leads to loss of a putative tumour suppressor [145]. Finally, DUSP16/MKP-7 is subject to methylation-dependent transcriptional silencing in both Burkitt's Lymphoma (BL) cell lines and primary tumours. Activation of JNK was increased in BL cell lines with DUSP16/MKP-7 methylation and these cell lines showed increased sensitivity to anthracyclines, which are known to activate JNK. These results suggest that DUSP16/MKP-7 is an epigenetically regulated determinant of JNK activation in BL and may thus influence the outcome of chemotherapy [146].

Conclusions and Perspectives

The MAPK signalling pathways play diverse and often contradictory roles in human cancer and therefore it should be no surprise that this complexity is mirrored in the biology of the phosphatases that act to regulate their activity. At present, we have a wealth of data indicating that MKPs are abnormally regulated in a wide variety of human tumours, but there is little or no evidence that these changes reflect cancer associated mutations or alterations in the primary sequence or structure of these genes. Instead it is more likely to be either a consequence of the dysregulation of the signalling pathways that regulate MKP expression or epigenetic modifications of regulatory regions, which control gene transcription. The challenge of future work is to determine if changes in MKP levels and activity are simply a consequence of the cancer process or if these changes make a significant contribution to the progression or maintenance of the malignant phenotype. Recent work using mouse knockout models has revealed key roles for MKPs in early development and particularly in the regulation of immune and metabolic homeostasis. These models should be instrumental in revealing the roles of the various MKPs in the initiation and progression of cancers in response to activated oncogenes and chemical carcinogens. They should also be of use in determining the role of MKPs in mediating sensitivity or resistance to cancer chemotherapeutic agents, particularly those that specifically target elements of the MAPK signalling pathways themselves. Taken together, an increased understanding of the ways in which MKPs act to regulate the biological consequences of MAPK signalling should give us important insights into new ways that these pathways can be manipulated for therapeutic gain in the treatment of cancer.

Acknowledgements Work in the Keyse laboratory is supported by a Cancer Research-UK programme grant (CR-UK Stress Response Laboratory C8227/A12053). Andrew Kidger is supported by a Cancer Research UK funded Ph.D studentship through the auspices of the Dundee Cancer Centre.

References

1. Chang L, Karin M. Mammalian MAP kinase signalling cascades. Nature. 2001;410:37–40.
2. Wada T, Penninger JM. Mitogen-activated protein kinases in apoptosis regulation. Oncogene. 2004;23:2838–49.
3. Kyriakis JM, Avruch J. Mammalian MAPK signal transduction pathways activated by stress and inflammation: a 10-year update. Physiol Rev. 2012;92:689–737.
4. Raman M, Chen W, Cobb MH. Differential regulation and properties of MAPKs. Oncogene. 2007;26:3100–12.
5. Marshall CJ. MAP kinase kinase kinase, MAP kinase kinase and MAP kinase. Curr Opin Genet Dev. 1994;4:82–9.
6. Coulombe P, Meloche S. Atypical mitogen-activated protein kinases: structure, regulation and functions. Biochim Biophys Acta. 2007;1773:1376–87.
7. Wan PT, Garnett MJ, Roe SM, et al. Mechanism of activation of the RAF-ERK signaling pathway by oncogenic mutations of B-RAF. Cell. 2004;116:855–67.
8. Joseph EW, Pratilas CA, Poulikakos PI, et al. The RAF inhibitor PLX4032 inhibits ERK signaling and tumor cell proliferation in a V600E BRAF-selective manner. Proc Natl Acad Sci. 2010;107:14903–8.
9. Bollag G, Hirth P, Tsai J, et al. Clinical efficacy of a RAF inhibitor needs broad target blockade in BRAF-mutant melanoma. Nature. 2010;467:596–9.
10. Flaherty KT, Robert C, Hersey P, et al. Improved survival with MEK inhibition in BRAF-mutated melanoma. N Engl J Med. 2012;367:107–14.
11. Das Thakur M, Salangsang F, Landman AS, Sellers WR, Pryer NK, Levesque MP, Dummer R, McMahon M, Stuart DD. Modelling vemurafenib resistance in melanoma reveals a strategy to forestall drug resistance. Nature. 2013. doi:10.1038/nature11814.

12. Lito P, Pratilas CA, Joseph EW, et al. Relief of profound feedback inhibition of mitogenic signaling by RAF inhibitors attenuates their activity in BRAFV600E melanomas. Cancer Cell. 2012;22:668–82.
13. Weston CR, Davis RJ. The JNK signal transduction pathway. Curr Opin Cell Biol. 2007;19:142–9.
14. Wagner EF, Nebreda ÁR. Signal integration by JNK and p38 MAPK pathways in cancer development. Nat Rev Cancer. 2009;9:537–49.
15. Clark GJ, Westwick JK, Der CJ. p120 GAP modulates Ras activation of Jun kinases and transformation. J Biol Chem. 1997;272:1677–81.
16. Rodrigues GA, Park M, Schlessinger J. Activation of the JNK pathway is essential for transformation by the Met oncogene. EMBO J. 1997;16:2634–45.
17. Hui L, Zatloukal K, Scheuch H, Stepniak E, Wagner EF. Proliferation of human HCC cells and chemically induced mouse liver cancers requires JNK1-dependent p21 downregulation. J Clin Invest. 2008;118:3943–53.
18. Sakurai T, Maeda S, Chang L, Karin M. Loss of hepatic NF-κB activity enhances chemical hepatocarcinogenesis through sustained c-Jun N-terminal kinase 1 activation. Proc Natl Acad Sci. 2006;103:10544–51.
19. She Q-B, Chen N, Bode AM, Flavell RA, Dong Z. Deficiency of c-Jun-NH2-terminal kinase-1 in mice enhances skin tumor development by 12-O-tetradecanoylphorbol-13-acetate. Cancer Res. 2002;62:1343–8.
20. Chen N, Nomura M, She Q-B, Ma W-Y, Bode AM, Wang L, Flavell RA, Dong Z. Suppression of skin tumorigenesis in c-Jun NH2-terminal kinase-2-deficient mice. Cancer Res. 2001;61:3908–12.
21. Chen F. JNK-induced apoptosis, compensatory growth, and cancer stem cells. Cancer Res. 2012;72:379–86.
22. Greenman C, Stephens P, Smith R, et al. Patterns of somatic mutation in human cancer genomes. Nature. 2007;446:153–8.
23. Jones S, Zhang X, Parsons DW, et al. Core signaling pathways in human pancreatic cancers revealed by global genomic analyses. Science. 2008;321:1801–6.
24. Dolado I, Swat A, Ajenjo N, De Vita G, Cuadrado A, Nebreda AR. p38α MAP kinase as a sensor of reactive oxygen species in tumorigenesis. Cancer Cell. 2007;11:191–205.
25. Ventura JJ, Tenbaum S, Perdiguero E, Huth M, Guerra C, Barbacid M, Pasparakis M, Nebreda AR. p38α MAP kinase is essential in lung stem and progenitor cell proliferation and differentiation. Nat Genet. 2007;39:750–8.
26. Hui L, Bakiri L, Mairhorfer A, Schweifer N, Haslinger C, Kenner L, Komnenovic V, Scheuch H, Beug H, Wagner EF. p38α suppresses normal and cancer cell proliferation by antagonizing the JNK–c-Jun pathway. Nat Genet. 2007;39:741–9.
27. Schindler EM, Hindes A, Gribben EL, et al. p38δ mitogen-activated protein kinase is essential for skin tumor development in mice. Cancer Res. 2009;69:4648–55.
28. Thornton TM. Non-classical P38 Map kinase functions: cell cycle checkpoints and survival. Int J Biol Sci. 2009;44.
29. Aguirre-Ghiso JA. Models, mechanisms and clinical evidence for cancer dormancy. Nat Rev Cancer. 2007;7:834–46.
30. Comes F, Matrone A, Lastella P, et al. A novel cell type-specific role of p38α in the control of autophagy and cell death in colorectal cancer cells. Cell Death Differ. 2007;14:693–702.
31. Kumar S, Boehm J, Lee JC. p38 MAP kinases: key signalling molecules as therapeutic targets for inflammatory diseases. Nat Rev Drug Discov. 2003;2:717–26.
32. Schieven G. The p38alpha kinase plays a central role in inflammation. Curr Top Med Chem. 2009;9:1038–48.
33. Kotlyarov A, Gaestel M. Is MK2 (mitogen-activated protein kinase-activated protein kinase 2) the key for understanding post-transcriptional regulation of gene expression? Biochem Soc Trans. 2002;30:959–63.
34. Rousseau S, Dolado I, Beardmore V, et al. CXCL12 and C5a trigger cell migration via a PAK1/2-p38α MAPK-MAPKAP-K2-HSP27 pathway. Cell Signal. 2006;18:1897–905.

35. Dolado I, Nebreda AR. Regulation of tumorigenesis by p38α MAP kinase. In: Posas F, Nebreda AR, editors. Stress-Act. Protein kinases. Berlin Heidelberg: Springer; 2008. p. 99–128.
36. Bulavin DV, Demidov ON, Saito S, et al. Amplification of PPM1D in human tumors abrogates p53 tumor-suppressor activity. Nat Genet. 2002;31:210–5.
37. Bulavin DV, Phillips C, Nannenga B, Timofeev O, Donehower LA, Anderson CW, Appella E, Fornace AJ. Inactivation of the Wip1 phosphatase inhibits mammary tumorigenesis through p38 MAPK–mediated activation of the p16Ink4a-p19Arf pathway. Nat Genet. 2004;36:343–50.
38. Lu X, Nguyen T-A, Moon S-H, Darlington Y, Sommer M, Donehower LA. The type 2C phosphatase Wip1: an oncogenic regulator of tumor suppressor and DNA damage response pathways. Cancer Metastasis Rev. 2008;27:123–35.
39. Marshall C. Specificity of receptor tyrosine kinase signaling: Transient versus sustained extracellular signal-regulated kinase activation. Cell. 1995;80:179–85.
40. Ebisuya M, Kondoh K, Nishida E. The duration, magnitude and compartmentalization of ERK MAP kinase activity: mechanisms for providing signaling specificity. J Cell Sci. 2005;118:2997–3002.
41. Kholodenko BN, Hancock JF, Kolch W. Signalling ballet in space and time. Nat Rev Mol Cell Biol. 2010;11:414–26.
42. Kolch W. Coordinating ERK/MAPK signalling through scaffolds and inhibitors. Nat Rev Mol Cell Biol. 2005;6:827–37.
43. Caunt CJ, Keyse SM. Dual-specificity MAP kinase phosphatases (MKPs). FEBS J. 2013;280:489–504.
44. Keyse SM. Protein phosphatases and the regulation of mitogen-activated protein kinase signalling. Curr Opin Cell Biol. 2000;12:186–92.
45. Patterson KI, Brummer T, O'Brien PM, Daly RJ. Dual-specificity phosphatases: critical regulators with diverse cellular targets. Biochem J. 2009;418:475–89.
46. Owens DM, Keyse SM. Differential regulation of MAP kinase signalling by dual-specificity protein phosphatases. Oncogene. 2007;26:3203–13.
47. Theodosiou A, Ashworth A. MAP kinase phosphatases. Genome Biol. 2002;3:1–10.
48. Camps M, Nichols A, Arkinstall S. Dual specificity phosphatases: a gene family for control of MAP kinase function. FASEB J. 2000;14:6–16.
49. Liu Y-X, Wang J, Guo J, Wu J, Lieberman HB, Yin Y. DUSP1 is controlled by p53 during the cellular response to oxidative stress. Mol Cancer Res. 2008;6:624–33.
50. Tsang M, Maegawa S, Kiang A, Habas R, Weinberg E, Dawid IB. A role for MKP3 in axial patterning of the zebrafish embryo. Development. 2004;131:2769–79.
51. Zeller E, Mock K, Horn M, Colnot S, Schwarz M, Braeuning A. Dual-specificity phosphatases are targets of the Wnt/β-catenin pathway and candidate mediators of β-catenin/Ras signaling interactions. Biol Chem. 2012;393:1183–91.
52. Ramesh S, Qi X-J, Wildey GM, Robinson J, Molkentin J, Letterio J, Howe PH. TGFβ-mediated BIM expression and apoptosis are regulated through SMAD3-dependent expression of the MAPK phosphatase MKP2. EMBO Rep. 2008;9:990–7.
53. Kuwano Y, Kim HH, Abdelmohsen K, Pullmann R, Martindale JL, Yang X, Gorospe M. MKP-1 mRNA stabilization and translational control by RNA-binding proteins HuR and NF90. Mol Cell Biol. 2008;28:4562–75.
54. Brondello J-M, Pouysségur J, McKenzie FR. Reduced MAP kinase phosphatase-1 degradation after p42/p44MAPK-dependent phosphorylation. Science. 1999;286:2514–7.
55. Marchetti S, Gimond C, Chambard J-C, Touboul T, Roux D, Pouysségur J, Pagès G. Extracellular signal-regulated kinases phosphorylate mitogen-activated protein kinase phosphatase 3/DUSP6 at serines 159 and 197, two sites critical for its proteasomal degradation. Mol Cell Biol. 2005;25:854–64.
56. Jurek A, Amagasaki K, Gembarska A, Heldin C-H, Lennartsson J. Negative and positive regulation of MAPK phosphatase 3 controls platelet-derived growth factor-induced Erk activation. J Biol Chem. 2009;284:4626–34.

57. Dickinson RJ, Delavaine L, Cejudo-Marín R, Stewart G, Staples CJ, Didmon MP, Trinidad AG, Alonso A, Pulido R, Keyse SM. Phosphorylation of the kinase interaction motif in mitogen-activated protein (MAP) kinase phosphatase-4 mediates cross-talk between protein kinase a and MAP kinase signaling pathways. J Biol Chem. 2011;286:38018–26.
58. Camps M, Nichols A, Gillieron C, Antonsson B, Muda M, Chabert C, Boschert U, Arkinstall S. Catalytic activation of the phosphatase MKP-3 by ERK2 mitogen-activated protein kinase. Science. 1998;280:1262–5.
59. Slack DN, Seternes O-M, Gabrielsen M, Keyse SM. Distinct binding determinants for ERK2/p38α and JNK MAP kinases mediate catalytic activation and substrate selectivity of MAP kinase phosphatase-1. J Biol Chem. 2001;276:16491–500.
60. Stewart AE, Dowd S, Keyse SM, McDonald NQ. Crystal structure of the MAPK phosphatase Pyst1 catalytic domain and implications for regulated activation. Nat Struct Mol Biol. 1999;6:174–81.
61. Mandl M, Slack DN, Keyse SM. Specific inactivation and nuclear anchoring of extracellular signal-regulated kinase 2 by the inducible dual-specificity protein phosphatase DUSP5. Mol Cell Biol. 2005;25:1830–45.
62. Karlsson M, Mathers J, Dickinson RJ, Mandl M, Keyse SM. Both nuclear-cytoplasmic shuttling of the dual specificity phosphatase MKP-3 and its ability to anchor MAP kinase in the cytoplasm are mediated by a conserved nuclear export signal. J Biol Chem. 2004;279:41882–91.
63. Ekerot M, Stavridis MP, Delavaine L, Mitchell MP, Staples C, Owens DM, Keenan ID, Dickinson RJ, Storey KG, Keyse SM. Negative-feedback regulation of FGF signalling by DUSP6/MKP-3 is driven by ERK1/2 and mediated by Ets factor binding to a conserved site within the DUSP6/MKP-3 gene promoter. Biochem J. 2008;412:287.
64. Kucharska A, Rushworth LK, Staples C, Morrice NA, Keyse SM. Regulation of the inducible nuclear dual-specificity phosphatase DUSP5 by ERK MAPK. Cell Signal. 2009;21:1794–805.
65. Loda M, Capodieci P, Mishra R, Yao H, Corless C, Grigioni W, Wang Y, Magi-Galluzzi C, Stork PJ. Expression of mitogen-activated protein kinase phosphatase-1 in the early phases of human epithelial carcinogenesis. Am J Pathol. 1996;149:1553–64.
66. Rauhala HE, Porkka KP, Tolonen TT, Martikainen PM, Tammela TLJ, Visakorpi T. Dual-specificity phosphatase 1 and serum/glucocorticoid-regulated kinase are downregulated in prostate cancer. Int J Cancer. 2005;117:738–45.
67. Bang YJ, Kwon JH, Kang SH, Kim JW, Yang YC. Increased MAPK activity and MKP-1 overexpression in human gastric adenocarcinoma. Biochem Biophys Res Commun. 1998;250:43–7.
68. Wang H, Cheng Z, Malbon CC. Overexpression of mitogen-activated protein kinase phosphatases MKP1, MKP2 in human breast cancer. Cancer Lett. 2003;191:229–37.
69. Hou M-F, Chang C-W, Chen F-M, Wang S-N, Yang S-F, Chen P-H, Su J-H, Yeh Y-T. Decreased total MKP-1 protein levels predict poor prognosis in breast cancer. World J Surg. 2012;36:1922–32.
70. Chen F-M, Chang H-W, Yang S-F, Huang Y-F, Nien P-Y, Yeh Y-T, Hou M-F. The mitogen-activated protein kinase phosphatase-1 (MKP-1) gene is a potential methylation biomarker for malignancy of breast cancer. Exp Mol Med. 2012;44:356–62.
71. Denkert C, Schmitt WD, Berger S, Reles A, Pest S, Siegert A, Lichtenegger W, Dietel M, Hauptmann S. Expression of mitogen-activated protein kinase phosphatase-1 (MKP-1) in primary human ovarian carcinoma. Int J Cancer. 2002;102:507–13.
72. Moncho-Amor V, de Cáceres II, Bandres E, et al. DUSP1/MKP1 promotes angiogenesis, invasion and metastasis in non-small-cell lung cancer. Oncogene. 2011;30:668–78.
73. Vicent S, Garayoa M, López-Picazo JM, Lozano MD, Toledo G, Thunnissen FBJM, Manzano RG, Montuenga LM. Mitogen-activated protein kinase phosphatase-1 is overexpressed in non-small cell lung cancer and is an independent predictor of outcome in patients. Clin Cancer Res. 2004;10:3639–49.
74. Tsujita E, Taketomi A, Gion T, Kuroda Y, Endo K, Watanabe A, Nakashima H, Aishima S, Kohnoe S, Maehara Y. Suppressed MKP-1 is an independent predictor of outcome in patients with hepatocellular carcinoma. Oncology. 2005;69:342–7.

75. Calvisi DF, Pinna F, Meloni F, et al. Dual-specificity phosphatase 1 ubiquitination in extracellular signal-regulated kinase—mediated control of growth in human hepatocellular carcinoma. Cancer Res. 2008;68:4192–200.
76. Franklin CC, Srikanth S, Kraft AS. Conditional expression of mitogen-activated protein kinase phosphatase-1, MKP-1, is cytoprotective against UV-induced apoptosis. Proc Natl Acad Sci. 1998;95:3014–9.
77. Sanchez-Perez I, Martínez-Gomariz M, Williams D, Keyse SM, Perona R. CL100/MKP-1 modulates JNK activation and apoptosis in response to cisplatin. Oncogene. 2000;19(45): 5142–52.
78. Magi-Galluzzi C, Mishra R, Fiorentino M, Montironi R, Yao H, Capodieci P, Wishnow K, Kaplan I, Stork PJ, Loda M. Mitogen-activated protein kinase phosphatase 1 is overexpressed in prostate cancers and is inversely related to apoptosis. Lab Investig J Tech Methods Pathol. 1997;76:37–51.
79. Chattopadhyay S, Machado-Pinilla R, Manguan-García C, et al. MKP1/CL100 controls tumor growth and sensitivity to cisplatin in non-small-cell lung cancer. Oncogene. 2006;25: 3335–45.
80. Small GW, Shi YY, Higgins LS, Orlowski RZ. Mitogen-activated protein kinase phosphatase-1 is a mediator of breast cancer chemoresistance. Cancer Res. 2007;67:4459–66.
81. Wang Z, Zhou J-Y, Kanakapalli D, Buck S, Wu GS, Ravindranath Y. High level of mitogen-activated protein kinase phosphatase-1 expression is associated with cisplatin resistance in osteosarcoma. Pediatr Blood Cancer. 2008;51:754–9.
82. Montagut C, Iglesias M, Arumi M, et al. Mitogen-activated protein kinase phosphatase-1 (MKP-1) impairs the response to anti-epidermal growth factor receptor (EGFR) antibody cetuximab in metastatic colorectal cancer patients. Br J Cancer. 2010;102:1137–44.
83. Yu H, Park J, Lee J, Choi K, Choi C. Constitutive expression of MAP kinase phosphatase-1 confers multi-drug resistance in human glioblastoma cells. Cancer Res Treat. 2012;44:195.
84. Zhou J-Y, Liu Y, Wu GS. The role of mitogen-activated protein kinase phosphatase-1 in oxidative damage—induced cell death. Cancer Res. 2006;66:4888–94.
85. Wang Z, Xu J, Zhou J-Y, Liu Y, Wu GS. Mitogen-activated protein kinase phosphatase-1 is required for cisplatin resistance. Cancer Res. 2006;66:8870–7.
86. Staples CJ, Owens DM, Maier JV, Cato ACB, Keyse SM. Cross-talk between the p38α and JNK MAPK pathways mediated by MAP kinase phosphatase-1 determines cellular sensitivity to UV radiation. J Biol Chem. 2010;285:25928–40.
87. Nunes-Xavier C, Romá-Mateo C, Ríos P, Tárrega C, Cejudo-Marín R, Tabernero L, Pulido R. Dual-specificity MAP kinase phosphatases as targets of cancer treatment. Anticancer Agents Med Chem. 2011;11:109–32.
88. Vogt A, McDonald PR, Tamewitz A, Sikorski RP, Wipf P, Skoko JJ, Lazo JS. A cell-active inhibitor of mitogen-activated protein kinase phosphatases restores paclitaxel-induced apoptosis in dexamethasone-protected cancer cells. Mol Cancer Ther. 2008;7:330–40.
89. Maraver A, Fernandez-Marcos PJ, Herranz D, et al. Therapeutic effect of γ-secretase inhibition in KrasG12V-driven non-small cell lung carcinoma by derepression of DUSP1 and inhibition of ERK. Cancer Cell. 2012;22:222–34.
90. Tai C-J, Wu AT, Chiou J-F, et al. The investigation of mitogen-activated protein kinase phosphatase-1 as a potential pharmacological target in non-small cell lung carcinomas, assisted by non-invasive molecular imaging. BMC Cancer. 2010;10:95.
91. Wang X, Liu Y. Regulation of innate immune response by MAP kinase phosphatase-1. Cell Signal. 2007;19:1372–82.
92. Lawan A, Shi H, Gatzke F, Bennett AM. Diversity and specificity of the mitogen-activated protein kinase phosphatase-1 functions. Cell Mol Life Sci. 2012;70(2):223–37.
93. Sieben NLG, Oosting J, Flanagan AM, Prat J, Roemen GMJM, Kolkman-Uljee SM, van Eijk R, Cornelisse CJ, Fleuren GJ, van Engeland M. Differential gene expression in ovarian tumors reveals dusp 4 and serpina 5 as key regulators for benign behavior of serous borderline tumors. J Clin Oncol. 2005;23:7257–64.

94. Yip-Schneider MT, Lin A, Marshall MS. Pancreatic tumor cells with mutant K-ras suppress ERK activity by MEK-dependent induction of MAP kinase phosphatase-2. Biochem Biophys Res Commun. 2001;280:992–7.
95. Gaedcke J, Grade M, Jung K, et al. Mutated KRAS results in overexpression of DUSP4, a MAP-kinase phosphatase, and SMYD3, a histone methyltransferase, in rectal carcinomas. Genes Chromosomes Cancer. 2010;49:1024–34.
96. Cagnol S, Rivard N. Oncogenic KRAS and BRAF activation of the MEK/ERK signaling pathway promotes expression of dual-specificity phosphatase 4 (DUSP4/MKP2) resulting in nuclear ERK1/2 inhibition. Oncogene. 2012. doi:10.1038/onc.2012.88.
97. Gröschl B, Bettstetter M, Giedl C, Woenckhaus M, Edmonston T, Hofstädter F, Dietmaier W. Expression of the MAP kinase phosphatase DUSP4 is associated with microsatellite instability in colorectal cancer (CRC) and causes increased cell proliferation. 2012. Int J Cancer n/a–n/a.
98. Waha A, Felsberg J, Hartmann W, et al. Epigenetic downregulation of mitogen-activated protein kinase phosphatase MKP-2 relieves its growth suppressive activity in glioma cells. Cancer Res. 2010;70:1689–99.
99. Chitale D, Gong Y, Taylor BS, et al. An integrated genomic analysis of lung cancer reveals loss of DUSP4 in EGFR-mutant tumors. Oncogene. 2009;28:2773–83.
100. Britson JS, Barton F, Balko JM, Black EP. Deregulation of DUSP activity in EGFR-mutant lung cancer cell lines contributes to sustained ERK1/2 signaling. Biochem Biophys Res Commun. 2009;390:849–54.
101. Balko JM, Cook RS, Vaught DB, et al. Profiling of residual breast cancers after neoadjuvant chemotherapy identifies DUSP4 deficiency as a mechanism of drug resistance. Nat Med. 2012;18:1052–9.
102. Lawan A, Torrance E, Al-Harthi S, Shweash M, Alnasser S, Neamatallah T, Schroeder J, Plevin R. MKP-2: out of the DUSP-bin and back into the limelight. Biochem Soc Trans. 2012;40:235–9.
103. Givant-Horwitz V, Davidson B, Goderstad JM, Nesland JM, Tropé CG, Reich R. The PAC-1 dual specificity phosphatase predicts poor outcome in serous ovarian carcinoma. Gynecol Oncol. 2004;93:517–23.
104. Lin S-C, Chien C-W, Lee J-C, Yeh Y-C, Hsu K-F, Lai Y-Y, Lin S-C, Tsai S-J. Suppression of dual-specificity phosphatase—2 by hypoxia increases chemoresistance and malignancy in human cancer cells. J Clin Invest. 2011;121:1905–16.
105. Kim S-C, Hahn J-S, Min Y-H, Yoo N-C, Ko Y-W, Lee W-J. Constitutive activation of extracellular signal-regulated kinase in human acute leukemias: combined role of activation of MEK, hyperexpression of extracellular signal-regulated kinase, and downregulation of a phosphatase, PAC1. Blood. 1999;93:3893–9.
106. Kreeger PK, Mandhana R, Alford SK, Haigis KM, Lauffenburger DA. RAS mutations affect tumor necrosis factor–induced apoptosis in colon carcinoma cells via ERK-modulatory negative and positive feedback circuits along with non-ERK pathway effects. Cancer Res. 2009;69:8191–9.
107. Yun J, Rago C, Cheong I, et al. Glucose deprivation contributes to the development of KRAS pathway mutations in tumor cells. Science. 2009;325:1555–9.
108. Shin S-H, Park S-Y, Kang GH. Down-regulation of dual-specificity phosphatase 5 in gastric cancer by promoter CpG island hypermethylation and its potential role in carcinogenesis. Am J Pathol. 2013. doi:10.1016/j.ajpath.2013.01.004.
109. Dickinson RJ, Eblaghie MC, Keyse SM, Morriss-Kay GM. Expression of the ERK-specific MAP kinase phosphatase PYST1/MKP3 in mouse embryos during morphogenesis and early organogenesis. Mech Dev. 2002;113:193–6.
110. Eblaghie MC, Lunn JS, Dickinson RJ, Münsterberg AE, Sanz-Ezquerro J-J, Farrell ER, Mathers J, Keyse SM, Storey K, Tickle C. Negative feedback regulation of FGF signaling levels by Pyst1/MKP3 in chick embryos. Curr Biol. 2003;13:1009–18.
111. Smith TG, Karlsson M, Lunn JS, Eblaghie MC, Keenan ID, Farrell ER, Tickle C, Storey KG, Keyse SM. Negative feedback predominates over cross-regulation to control ERK MAPK activity in response to FGF signalling in embryos. FEBS Lett. 2006;580:4242–5.

112. Znosko WA, Yu S, Thomas K, Molina GA, Li C, Tsang W, Dawid IB, Moon AM, Tsang M. Overlapping functions of Pea3 ETS transcription factors in FGF signaling during zebrafish development. Dev Biol. 2010;342:11–25.
113. Furukawa T, Sunamura M, Motoi F, Matsuno S, Horii A. Potential tumor suppressive pathway involving DUSP6/MKP-3 in pancreatic cancer. Am J Pathol. 2003;162:1807–15.
114. Xu S, Furukawa T, Kanai N, Sunamura M, Horii A. Abrogation of DUSP6 by hypermethylation in human pancreatic cancer. J Hum Genet. 2005;50:159–67.
115. Furukawa T, Fujisaki R, Yoshida Y, Kanai N, Sunamura M, Abe T, Takeda K, Matsuno S, Horii A. Distinct progression pathways involving the dysfunction of DUSP6/MKP-3 in pancreatic intraepithelial neoplasia and intraductal papillary-mucinous neoplasms of the pancreas. Mod Pathol. 2005;18:1034–42.
116. Furukawa T, Sunamura M, Horii A. Molecular mechanisms of pancreatic carcinogenesis. Cancer Sci. 2006;97:1–7.
117. Furukawa T. Molecular pathology of pancreatic cancer: implications for molecular targeting therapy. Clin Gastroenterol Hepatol. 2009;7:S35–9.
118. Haigis KM, Kendall KR, Wang Y, et al. Differential effects of oncogenic K-Ras and N-Ras on proliferation, differentiation and tumor progression in the colon. Nat Genet. 2008;40:600–8.
119. Okudela K, Yazawa T, Woo T, et al. Down-regulation of DUSP6 expression in lung cancer: its mechanism and potential role in carcinogenesis. Am J Pathol. 2009;175:867–81.
120. Zhang Z, Kobayashi S, Borczuk AC, Leidner RS, LaFramboise T, Levine AD, Halmos B. Dual specificity phosphatase 6 (DUSP6) is an ETS-regulated negative feedback mediator of oncogenic ERK signaling in lung cancer cells. Carcinogenesis. 2010;31:577–86.
121. Wong VCL, Chen H, Ko JMY, et al. Tumor suppressor dual-specificity phosphatase 6 (DUSP6) impairs cell invasion and epithelial-mesenchymal transition (EMT)-associated phenotype. Int J Cancer. 2012;130:83–95.
122. Bloethner S, Chen B, Hemminki K, Müller-Berghaus J, Ugurel S, Schadendorf D, Kumar R. Effect of common B-RAF and N-RAS mutations on global gene expression in melanoma cell lines. Carcinogenesis. 2005;26:1224–32.
123. Packer LM, East P, Reis-Filho JS, Marais R. Identification of direct transcriptional targets of V600EBRAF/MEK signalling in melanoma. Pigment Cell Melanoma Res. 2009;22: 785–98.
124. Li W, Song L, Ritchie A-M, Melton DW. Increased levels of DUSP6 phosphatase stimulate tumourigenesis in a molecularly distinct melanoma subtype. Pigment Cell Melanoma Res. 2012;25:188–99.
125. Lee JU, Huang S, Lee MH, et al. Dual specificity phosphatase 6 as a predictor of invasiveness in papillary thyroid cancer. Eur J Endocrinol. 2012;167:93–101.
126. Degl'Innocenti D, Romeo P, Tarantino E, et al. DUSP6/MKP3 is overexpressed in papillary and poorly differentiated thyroid carcinoma and contributes to neoplastic properties of thyroid cancer cells. Endocr Relat Cancer. 2013;20:23–37.
127. Martin MJ, Hayward R, Viros A, Marais R. Metformin accelerates the growth of BRAF V600E-driven melanoma by upregulating VEGF-A. Cancer Discov. 2012;2:344–55.
128. Cui Y, Parra I, Zhang M, Hilsenbeck SG, Tsimelzon A, Furukawa T, Horii A, Zhang Z-Y, Nicholson RI, Fuqua SAW. Elevated expression of mitogen-activated protein kinase phosphatase 3 in breast tumors: a mechanism of tamoxifen resistance. Cancer Res. 2006;66: 5950–9.
129. Chan DW, Liu VWS, Tsao GSW, Yao K-M, Furukawa T, Chan KKL, Ngan HYS. Loss of MKP3 mediated by oxidative stress enhances tumorigenicity and chemoresistance of ovarian cancer cells. Carcinogenesis. 2008;29:1742–50.
130. Levy-Nissenbaum O, Sagi-Assif O, Raanani P, Avigdor A, Ben-Bassat I, Witz IP. cDNA microarray analysis reveals an overexpression of the dual-specificity MAPK phosphatase PYST2 in acute leukemia. Methods Enzymol. 2003;366:103–13.
131. Levy-Nissenbaum O, Sagi-Assif O, Raanani P, Avigdor A, Ben-Bassat I, Witz IP. Overexpression of the dual-specificity MAPK phosphatase PYST2 in acute leukaemia. Cancer Lett. 2003;199:185–92.

132. Levy-Nissenbaum O, Sagi-Assif O, Kapon D, Hantisteanu S, Burg T, Raanani P, Avigdor A, Ben-Bassat I, Witz IP. Dual-specificity phosphatase Pyst2-L is constitutively highly expressed in myeloid leukemia and other malignant cells. Oncogene. 2003;22:7649–60.
133. Christie GR, Williams DJ, MacIsaac F, Dickinson RJ, Rosewell I, Keyse SM. The dual-specificity protein phosphatase DUSP9/MKP-4 is essential for placental function but is Not required for normal embryonic development. Mol Cell Biol. 2005;25:8323–33.
134. Li Z, Fei T, Zhang J, et al. BMP4 signaling acts via dual-specificity phosphatase 9 to control ERK activity in mouse embryonic stem cells. Cell Stem Cell. 2012;10:171–82.
135. Zhou L, Chen J, Li Z, et al. Integrated profiling of microRNAs and mRNAs: microRNAs located on Xq27.3 associate with clear cell renal cell carcinoma. PLoS One. 2010;5(12):e15224.
136. Wu S, Wang Y, Sun L, et al. Decreased expression of dual-specificity phosphatase 9 is associated with poor prognosis in clear cell renal cell carcinoma. BMC Cancer. 2011;11:413.
137. Liu J, Ni W, Xiao M, Jiang F, Ni R. Decreased expression and prognostic role of mitogen-activated protein kinase phosphatase 4 in hepatocellular carcinoma. J Gastrointest Surg. 2013;17:756–65.
138. Zhang Y, Blattman JN, Kennedy NJ, Duong J, Nguyen T, Wang Y, Davis RJ, Greenberg PD, Flavell RA, Dong C. Regulation of innate and adaptive immune responses by MAP kinase phosphatase 5. Nature. 2004;430:793–7.
139. Peehl DM, Shinghal R, Nonn L, Seto E, Krishnan AV, Brooks JD, Feldman D. Molecular activity of 1,25-dihydroxyvitamin D3 in primary cultures of human prostatic epithelial cells revealed by cDNA microarray analysis. J Steroid Biochem Mol Biol. 2004;92:131–41.
140. Nonn L, Peng L, Feldman D, Peehl DM. Inhibition of p38 by vitamin D reduces interleukin-6 production in normal prostate cells via mitogen-activated protein kinase phosphatase 5: implications for prostate cancer prevention by vitamin D. Cancer Res. 2006;66:4516–24.
141. Krishnan AV, Feldman D. Molecular pathways mediating the anti-inflammatory effects of calcitriol: implications for prostate cancer chemoprevention and treatment. Endocr Relat Cancer. 2010;17:R19–38.
142. Montpetit A, Boily G, Sinnett D. A detailed transcriptional map of the chromosome 12p12 tumour suppressor locus. Eur J Hum Genet. 2002;10:62–71.
143. Hoornaert I, Marynen P, Goris J, Sciot R, Baens M. MAPK phosphatase DUSP16/MKP-7, a candidate tumor suppressor for chromosome region 12p12–13, reduces BCR-ABL-induced transformation. Oncogene. 2003;22:7728–36.
144. Kibel AS, Huagen J, Guo C, Isaacs WB, Yan Y, Pienta KJ, Goodfellow PJ. Expression mapping at 12p12-13 in advanced prostate carcinoma. Int J Cancer. 2004;109:668–72.
145. Al-Shehhi H, Konn ZJ, Schwab CJ, Erhorn A, Barber KE, Wright SL, Gabriel AS, Harrison CJ, Moorman AV. Abnormalities of the der(12)t(12;21) in ETV6-RUNX1 acute lymphoblastic leukemia. Genes Chromosomes Cancer. 2013;52:202–13.
146. Lee S, Syed N, Taylor J, et al. DUSP16 is an epigenetically regulated determinant of JNK signalling in Burkitt's lymphoma. Br J Cancer. 2010;103:265–74.

PTEN

Dawid G. Nowak and Lloyd C. Trotman

Abstract The PTEN tumor suppressor gene is among the most frequently targeted genes in cancer. It inhibits cellular growth, proliferation, and survival functions and is essential for epithelial cell polarity and migration. Mechanistically, PTEN achieves these various functions by antagonizing PI-3 kinase-mediated production of the PIP_3 lipid second messenger which in turn leads to differential recruitment of key signaling molecules to the plasma membrane, and to inactivation of the protein kinase AKT.

Modeling of Pten deficiency in mice has firmly established that losing only one copy of the gene is sufficient for the process of tumor initiation, while losing both copies triggers cellular senescence. These two features control tumorigenesis in several tissues, which highlights the critical role of PTEN-regulatory mechanisms, especially at the transcriptional and posttranslational levels.

Understanding and interfering with these processes will therefore greatly impact our success in cancer treatment.

Keyword PTEN • Cancer • Cancer genetics • Mouse models • Tumor suppressor • PI-3 kinase • PIP_3 • Proliferation • Apoptosis • Migration

Introduction

The PTEN phosphatase has an active site motif characteristic of the protein tyrosine phosphatase superfamily, yet its major substrate is the membrane lipid phosphatidylinositol (3,4,5) trisphosphate (PIP_3) [1]. PTEN is thus the major antagonist of PI-3 kinase signaling, which explains its supreme role in tumor suppression, development, and growth control. One and a half decades of research on PTEN has provided us with fundamental insights into many aspects of biology: phospholipid metabolism and the downstream signaling networks, regulation of basic cellular processes such as migration and senescence and, finally, it has allowed us to redefine genetic principles of tumor suppression. As the first mechanisms of PTEN regulation and the role of PTEN in disease recurrence after targeted therapy have

D.G. Nowak (✉) • L.C. Trotman
Cold Spring Harbor Laboratory, Cancer Center,
One Bungtown Road, Cold Spring Harbor, NY 11724, USA

B.G. Neel, N. Tonks (eds.), *Protein Tyrosine Phosphatases in Cancer*,
DOI 10.1007/978-1-4939-3649-6_8

become clear, a conceptual framework for successful cancer therapy is emerging. Since such efforts will require a comprehensive understanding of its biology, this chapter will integrate various facets of PTEN research.

PTEN Discovery and Function

Analysis of the genomic 10q23 region, which frequently displays loss of heterozygosity in human tumors, simultaneously led two research teams to the cloning of the tumor suppressor gene named *PTEN* (Phosphatase and TENsin homolog on chromosome TEN) [2] or *MMAC1* (Mutated in Multiple Advanced Cancers 1) [3] (see also timeline in Fig. 1). This discovery yielded the first direct evidence of the long

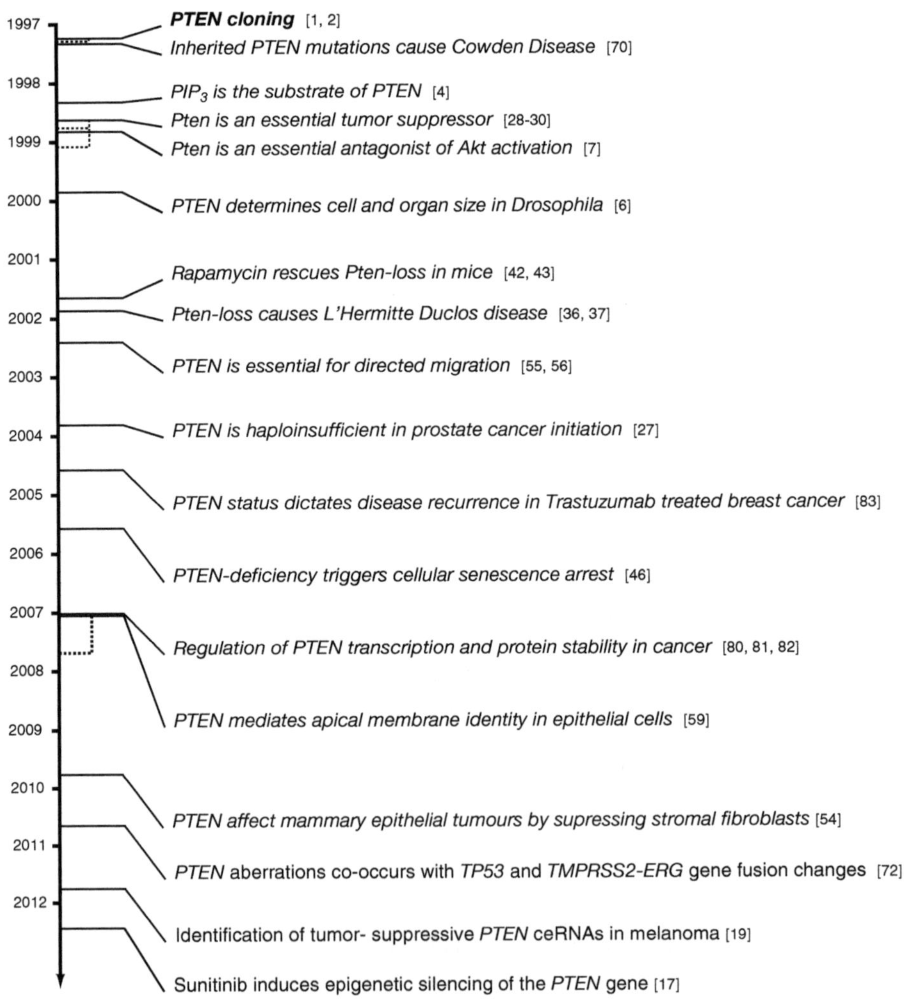

Fig. 1 A timeline of landmark discoveries in PTEN biology. References are given in brackets

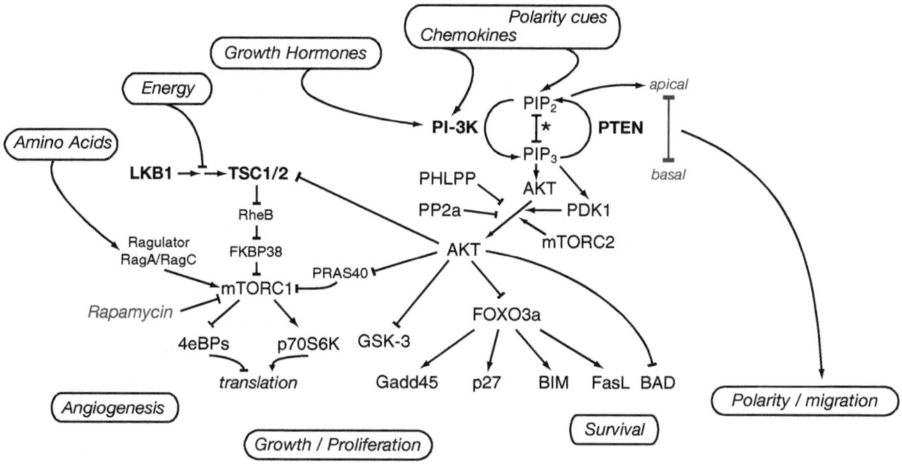

Fig. 2 The PTEN/PI3-kinase signaling pathway connects extracellular inputs (*boxed, top*) with basic cellular functions (*boxed, bottom*). Only a selection of molecular mediators is shown (see also text). The most frequently targeted cancer genes are indicated in *bold print*, *arrows* denote stimulation, *blunted arrows* inhibition

suspected role of phosphatases in tumor suppression through their ability to reverse oncogenic kinase function. While initial experiments soon established PTEN target preferences on synthetic peptides, [4] the highly acidic nature of the optimal substrate (polyGlu$_4$Tyr$_1$) and the unusual dual specificity for both phosphorylated Tyr and Ser/Thr residues prompted research on the ability of PTEN to target other substrates, especially the highly charged phosphatidylinositol second messengers. This effort led to the identification of the major PTEN substrate, PIP$_3$, an essential second messenger for recruitment and activation of PH domain-containing kinases such as AKT (also known as PKB) [1]. PTEN specifically dephosphorylates the three-position on the inositol ring, converting PIP$_3$ to PI (4,5)P$_2$, and thus effectively reverses PI-3 kinase function (see Fig. 2). The discovery of this target revealed that a phosphatase is at the top control level of a signaling pathway for intercellular communication, a highly conserved trait among metazoans [5]. *PTEN* mutation in Drosophila, for example, provided the first link between PTEN and the control of cell and organ size [6], which can be rescued by mutations in the drosophila *dAKT gene*. The first genetic proof of this signaling hierarchy between PI3-kinase, PTEN, and AKT came with the availability of *Pten*-deficient Mouse Embryonic Fibroblasts and activation-specific phospho-AKT antibodies [7]. To date, phospho-Akt-specific western blotting and immunohistochemistry remain the most common readouts of PTEN activity, as quantification of PIP$_3$ levels requires HPLC or radiolabelling and thin layer chromatography.

PTEN and Cancer

Together with the discovery of the *PTEN* gene in a loss of heterozygosity (LOH) prone region of chromosome 10, mutational analysis immediately revealed frequent PTEN point mutations in commonly used glioblastoma, prostate and breast cancer cell lines and primary tumor samples [2, 3], confirming the notion that PTEN is the critical target gene of the 10q23 region. Homozygous or heterozygous deletion is found in up to 80 % of glioblastoma samples and in endometrial cancers its mutation frequency reaches 65 % [8]. Other epithelial tumors that commonly present with *PTEN* deletions are prostate cancer (up to 50 % in metastasis [9, 10]), melanoma (57 % heterozygous loss), and bladder cancer (39 % heterozygous loss). Note, however, that these numbers are subject to change as large-scale sequencing based efforts deliver ever better resolution of the cancer genomes, which are curated online [8]. Immunohistochemical (IHC) analysis of PTEN levels and activity in tumor tissue has so far provided the best comprehensive analysis needed to complement the genetic and expression data. For example, while *PTEN* LOH frequency is relatively low in breast cancer and mutation averages below 5 % [8], loss of PTEN protein is found by IHC analysis in close to 50 % of cases [11]. Similar results have been found in lung [12], colon [13], and prostate cancer [10].

This discrepancy between low protein level and intact gene copy number could for example be explained by changes in methylation of the *PTEN* promoter that leads to *PTEN* silencing in cancer. In patients with metastatic melanoma, over 60 % of serum-derived DNA was found to have significant PTEN promoter methylation, and this change was correlated with very low *PTEN* expression [14]. *PTEN* promoter methylation has been shown as a mechanism by which cancers evade the effects of anticancer drugs. In cells derived from metastatic prostate cancer patients, *PTEN* silencing was increased after treatment with the antiandrogenic drug bicalutamide. This silencing could be reversed by treatment with methylating agents [15]. Imatinib-resistant leukemia cells isolated from individuals with Chronic Myeloid Leukemia (CML), Chronic Eosinophilic Leukemia (CEL), and Philadelphia-positive Acute Lymphoblastic Leukemia (ALL) also show hypermethylation of the *PTEN* promoter in association with the downregulation of the gene transcript [16]. Similarly, gastrointestinal stromal tumor cell lines that are resistant to sunitinib treatment show increased PTEN promoter methylation, and interestingly, forced expression of PTEN in these cells led to resensitization to sunitinib treatment [17].

Beyond methylation of promoter, multiple microRNAs have been shown to target the 3′ UTR of PTEN, allowing for DICER to degrade the mRNA transcript. An example is the miR-106b~25 cluster, which is found at abnormally high levels in prostate cancer cell lines and prostate tumor samples [18]. Infection of DU145 prostate cancer cells with this miRNA cluster resulted in reduced PTEN protein levels and increased proliferation [18].

Interestingly *PTENP1*, a PTEN pseudogene that shares a similar 3′ UTR with PTEN, has been found to regulate PTEN in a microRNA dependent manner and shows copy number loss in patient samples of colon cancer [19] and melanoma

[20]. Similar miRNAs target *PTENP1* and PTEN. Introduction of *PTENP1* or the *PTENP1* 3′ UTR was shown to increase PTEN levels, decrease Akt phosphorylation, and reduce proliferation of HCT116 colon cancer cells. These effects were shown to be microRNA dependent, as expression of *PTENP1* or its 3′ UTR did not exert these effects in DICER$^{-/-}$ HCT116 cells [19].

Once the miRNA dependent effects of *PTENP1* were discovered, other transcripts with similar miRNA binding sites in their 3′ UTR were found to have a similar effect (termed competing endogenous RNAs, ceRNAs). Several ceRNAs have been found to have tumor suppressive properties through sequestering *PTEN* targeting microRNAs, thus allowing for enhanced PTEN expression. Expression levels of the VAPA, CNOT6L, SERINC1, and ZEB2 ceRNAs have been found to correlate with PTEN expression in a wide variety of human cancers, and to stimulate PTEN expression in a DICER and microRNA dependent manner. The ceRNA regulation is reciprocal: as these PTEN ceRNAs can sequester microRNAs and allow for enhanced PTEN expression, so too can PTEN sequester microRNAs and enhance expression of these transcripts [21, 22].

PTEN Haploinsufficiency in Cancer

Early reports on PTEN loss in epithelial cancers of breast and prostate presented researchers with a paradox: while LOH analysis of the *PTEN* locus revealed frequent deletion of one copy, mutation of the remaining allele was rarely found. Cowden Disease patients also frequently showed retention of their remaining normal *PTEN* copy in tumor tissues [23]. These observations raised the question of whether PTEN heterozygosity might be insufficient for prevention of certain tumor types, a principle termed haploinsufficiency, which does not conform to the classic "two-hit hypothesis." Originally defined for the Retinoblastoma tumor suppressor (and essential for the realization of the role of tumor opposing genes in cancer), the two-hit hypothesis states that tumor suppressor nullizygosity is required for a cancer to form [24]. The question if *Pten* is haploinsufficient in cancer initiation was solved as mouse models allowed precise functional genetic analysis of disease progression. Initial results clearly demonstrated *Pten* haploinsufficiency for the lethal autoimmunity phenotype in heterozygous lymphocytes. Haploinsufficiency of *Pten* in prostate cancer was also seen if *Pten* heterozygosity was either combined with expression of an oncogenic transgene (resulting in accelerated tumor formation) or with loss of $p27^{\mathrm{kip1}}$ [25, 26]. The first evidence that *Pten* is haploinsufficient on its own and that its hemizygous loss suffices to initiate cancer formation came with the establishment of a hypomorphic series of *Pten* expressing mice, which revealed that Pten protein levels and the corresponding Akt activity determine tumor onset and rate of cancer progression in the prostate without full loss of function [27]. Collectively, these findings reveal that the frequent cases of tumors with heterozygous *PTEN* loss indeed represent PTEN pathway-driven disease (search "PTEN: HETLOSS" at http://www.cbioportal.org/public-portal/ [8]).

Modeling the PTEN Signaling Pathway in Cancer

Our understanding of the role of PTEN is ever expanding and has greatly profited from its high conservation in model organisms such as Dictyostelium, Caenorhabditis, and Drosophila. The major impact on dissecting the complex role of PTEN in cancer was achieved thanks to mouse modeling. Murine knockout models for *Pten* unambiguously demonstrate its potential to trigger cancer. While *Pten* null animals display embryonic lethality, *Pten* heterozygous animals are readily obtained and develop lesions in a wide variety of tissues including breast, prostate, colon, endometrium, and the thyroid as well as adrenal glands [28–30]. In addition, complete conditional inactivation of *Pten* using the *Cre/Lox* technology has confirmed roles for *Pten* in regulating the proliferation of skin [31], heart [32], and brain tissues [33] as well as cell and organ sizes in the latter two. Summarily, the in vivo analysis of *Pten* activity reveals four core biological functions: the ability of Pten to curb proliferation, as seen in epithelial tumors, the essential function of Pten in mediating apoptosis (as seen in B- and T-lymphocytes) [34], the control of cell size, polarity and migration to maintain tissue architecture and brain function (reviewed in [35]), which is most strikingly seen in the brain. The example of L'Hermitte Duclos disease is particularly instructive for demonstrating the impact that mouse models can have on our understanding of human disease: mice revealed that abnormal size and migration-defects of *Pten*-deficient neurons in the developing cerebellum and not hyper-proliferation underlie human L'Hermitte Duclos disease [36, 37].

Taken together, loss of *PTEN* activity affects major physiological responses that shape cancer initiation and progression including (and not limited to) proliferation, apoptosis, senescence, migration, and angiogenesis (Fig. 2).

PTEN and Proliferation

PTEN negatively regulates cell growth and proliferation, which is chiefly mediated by AKT. AKT inhibits key targets through their phosphorylation; GSK3-phosphorylation thus links PTEN to metabolism and cell cycle control (via Glycogen synthase and MYC, respectively), FOXO3a-phosphorylation leads to decreased transcription of cell cycle inhibitors such as $p27^{Kip1}$ [38] and GADD45 [39], and activation of the mTOR complex 1 (mTORC1) indirectly via phosphorylation of the Tuberous Sclerosis Complex tumor suppressor TSC2 or more directly via phosphorylation of the mTORC1 inhibitor PRAS40 leads to increased protein synthesis [40]. As illustrated in Fig. 2, mTOR kinase therefore couples protein translation with the sensing of growth factor signals (PTEN-AKT-dependent sensor) and amino acid availability through its activation on lysosomes [41]. It thus appears as if the requirement for multicellular organisms to conditionally integrate nutrient availability with growth hormone-controlled proliferation is achieved by introduction of the PIP_3 signaling pathway of which PTEN is the major antagonist. This realization and the discovery of the naturally occurring mTORC1 inhibitor Rapamycin resulted in tremendous interest in the therapeutic effects of that compound. Indeed, successful preclinical trials with Rapamycin in *Pten* mutant mice fostered high hopes for the potential of this drug in clinical cancer therapy [42, 43].

However, results from ongoing clinical trials seem to indicate high variability of results with specific effects in only few cancer types, which is in contrast to trials using Rapamycin in Tuberous Sclerosis Complex patients, that are more promising [44]. The potential benefits of blocking PI3K/mTOR signaling are clear; however, there are great concerns about the tumor's ability of developing resistance to single agents. Thus, complementary therapies to counter the evolutionary adaptation of cancers are currently under intense investigation [45].

PTEN and Senescence

Interestingly, complete loss of *Pten* leads to a p53-dependent cellular senescence response that blocks tumorigenesis as long as the p53-pathway is intact. This demonstrates existence of an intrinsic response in the cell that protects it from tumorigenesis by the "two-hit" paradigm [46]. Combined inactivation of *Pten* and *Trp53* (a hallmark of metastatic disease) causes an invasive prostate cancer as early as 2 weeks after puberty and is consistently lethal by 7 months of age [46]. These data support the hypothesis that *Pten*-loss Induced Cellular Senescence (PICS) acts as a natural barrier to tumor development and that progression to metastasis requires evading senescence and cooperation of these genes. Therefore, discovering other genes responsible for bypassing PICS will be crucial to design future therapeutics.

Notably, the senescence response, which has been found to counteract tumor formation in various tissues [47], can also be bypassed in cancers of mice where one or both copies of $p27^{kip1}$ or the Promyelocytic Leukemia (*Pml*) gene are lost in combination with *Pten*. These genetic combinations cause prostate and lethal colon carcinomas, neither of which are caused by *Pml*- or *p27*-loss alone [25, 48]. Since PML and p27 degradation at the protein level have been associated with cancer progression in tissues that show frequent *PTEN* LOH [49, 50], it is possible that the senescence and haploinsufficiency paradigms direct the formation of tumors in tissues that are most likely to combine *PTEN* heterozygosity with the loss of such secondary tumor suppressors (reviewed in [51]).

PTEN and Apoptosis

PTEN signaling converges on its ability to induce apoptosis, a characteristic that is most strikingly visible in *Pten* heterozygous mice, which mostly succumb to an autoimmune disorder due to failure of lymphocyte apoptosis. Collectively, PTEN seems to mediate pro-apoptotic functions via inactivation of AKT, which can directly inactivate the pro-apoptotic BAD protein as well as inhibit FOXO3a-dependent transcription of apoptosis genes such as Fas-ligand and Bim [52]. It will therefore be of particular interest to see whether in *PTEN*-deficient epithelial tumorigenesis, signaling through the AKT-FOXO3a axis is more important than that through the TSC-mTOR axis (see Fig. 2). Mouse modeling so far highlights the importance of signaling through the former axis in *Pten* heterozygous prostate and colon cancer upon the introduction of mutations that affect FOXO3a and p27 [25, 48].

PTEN and the Tumor Microenvironment

Emerging pieces of evidence support the idea that stroma can contribute to malignancy of epithelial cells (reviewed in [53]). Conditional loss of *Pten* in stromal fibroblasts of mouse mammary glands accelerated the initiation, progression, and malignant transformation of mammary epithelial tumors [54]. Inactivation of *Pten* in the stroma was related to enriched vascular networks, innate immune system infiltration, and activation of the Ets2 transcription factor. Increased binding of Ets2 to Mmp9 raises the bioavailability of VEGF-A through ECM remodeling, and through the chemotactic and pro-inflammatory molecule Ccl3, thus explaining the effect of stromal *Pten*-loss on epithelia [54]. It will be interesting to further investigate whether there is cross-talk between stromal and epithelial cells via auto- and paracrine mechanisms in this scenario (Fig. 3).

PTEN and Migration

To enable directional migration, small G-proteins such as Ras and Rho regulate localized production and hydrolysis of PIP_3 through their effects on PI3-kinase and PTEN, respectively, in a coordinated process, which was discovered by studying chemoattractant responses in the slime mold *Dictyostelium discoideum* [55, 56]. This results in PIP_3-dependent accumulation of actin remodeling factors to the chemoattractant-facing "front" side of the cell and PTEN dependent establishment of the cell's back-side identity in migration [57, 58]. Similarly, to establish and maintain cell polarity in epithelial glands during organogenesis, it has been shown that PTEN and PIP_2 define the apical plasma membrane via recruitment of Cdc42 and aPKC; however, in contrast to migration, the upstream mediators of this event are still unknown [59]. In addition to PIP_3-mediated migration, PTEN also dephosphorylates focal adhesion kinase (FAK), which leads to the inhibition of cell migration, decreasing integrin-mediated cell spreading, and reduction in the formation of focal contacts [60]. In glioblastoma cells, restoration of PTEN results in blocking the phenotypical features of tumor progression including cell invasion and migration. These effects were correlated with decreased tyrosine phosphorylation levels of FAK and p130Cas [61].

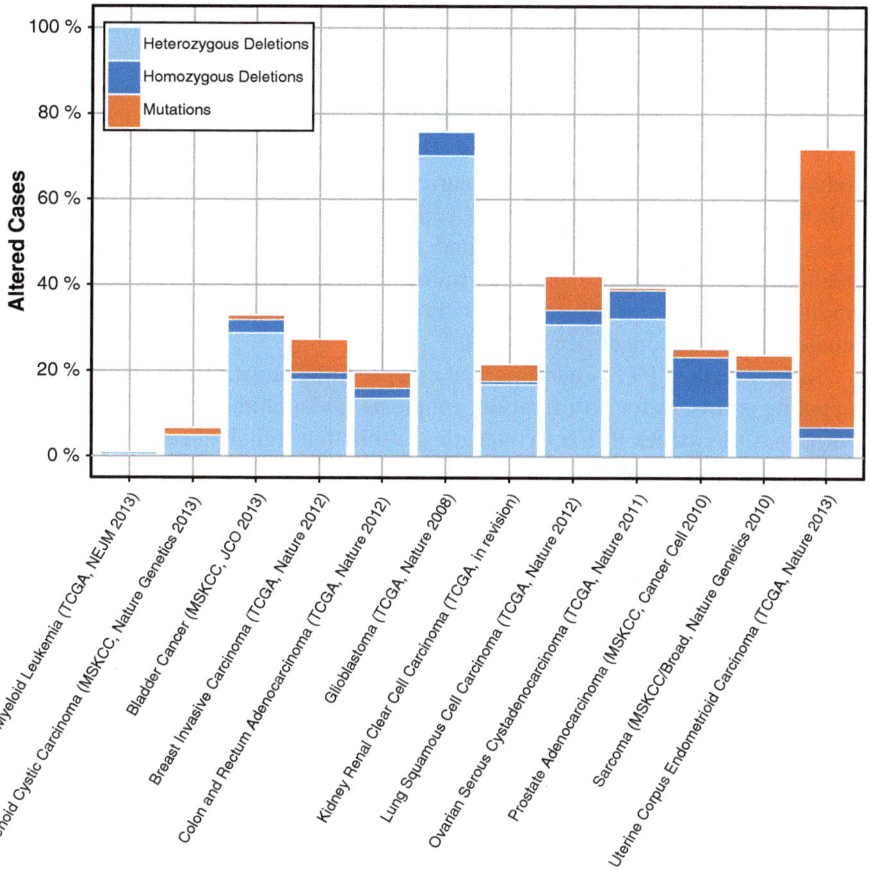

Fig. 3 Genomic changes of PTEN gene in different types of tumors

PTEN and Angiogenesis

PTEN downregulation induces angiogenesis via activating PI3K/AKT signaling, which mediates the effects of proteins such as Vascular Endothelial Growth Factor-A (VEGF-A) [62] and Angiopoietin-1 (Ang-1) [63]. Loss of PTEN function enhances VEGF-A-mediated Akt phosphorylation and increases angiogenesis in in vitro assays, as opposed to while PTEN expression inhibits these effects [64]. Additionally, results from GEMs carrying heterozygous deletion of *Pten* specifically in endothelial cells are hypersensitized to various pro-angiogenic molecules including different combinations of VEGF-A, Ang-1, and bFGF (basic Fibroblast Growth Factor), which leads to significantly increased vascularization [65]. Moreover, reintroduction of the PTEN gene into human prostate PC-3 and LNCaP cells leads to decreased VEGF-A secretion and also causes inhibition of endothelial cell growth and migration [66]. Neo-angiogenesis can be promoted through both mTOR activity [67] and inactivation of Forkhead transcription factors [68], which links PTEN to yet another characteristic of cancer.

Overall, the current picture of cellular PTEN function attributes a central role to its regulation of AKT kinase. However, the frequency of *AKT* alterations in cancer is far below that of *PTEN* alterations. Therefore, notwithstanding additional AKT-independent PTEN functions, our current knowledge strongly suggests that the PIP_3/PIP_2-dependent maintenance of cellular architecture, which is achieved independently of AKT by PI3-kinase and PTEN, is one essential factor in PTEN-mediated suppression of cancerous transformation. This notion is consistent with findings in an in vitro breast cancer model [69] and has also been confirmed by studies of human cancer genetics [5].

Taken together, PTEN has emerged as a principle negative regulator of the cell's signaling pathways for proliferation, apoptosis, and architecture. Its ability to antagonize mTOR enables PTEN to overrule sensors that signal nutrient availability and recruitment of vasculature.

PTEN in Hereditary Diseases

Another major line of evidence for the important tumor suppressive function of PTEN was published almost simultaneously with its discovery, namely the realization that *PTEN* germline mutations are linked to cancer susceptibility syndromes [70]. Cowden Disease (CD, MIM 158350) is an autosomal dominant cancer susceptibility syndrome that confers a strong predisposition for breast, thyroid, and skin tumors featuring large numbers of initially benign hamartomas in these tissues. Another inherited disorder, L'Hermitte Duclos disease, which is characterized by macrocephaly, seizures, and ataxia that cosegregates with CD, was also found to harbor *PTEN* mutations. Indeed, the linkage with *PTEN* mutation has proven so strong that several phenotypically related syndromes are today collectively referred to as PTEN Hamartoma Tumor Syndromes (PTHS), with CD, Proteus-like- and Bannayan–Riley–Ruvalcaba syndromes showing the highest PTEN mutation frequency (80 %, 60 %, and 60 %, respectively) [71].

Even though a firm connection between *PTEN* and cancer could soon be established, there appeared to be seemingly contradictory data regarding the ability of *PTEN*-loss to initiate tumorigenesis as opposed to a proposed role only in advanced cancers. In certain tumor types (e.g., gliomas), loss mainly occurs at late stages, while germline mutations clearly are able to initiate neoplasia in other tissues. Faithful model systems for cancer were found in mouse and they recapitulated the tissue-dependent nature of *Pten* alteration. Now, the emerging cancer genomes confirm the striking tissue-specific differences in PTEN alteration type: uterine cancers show 67 % mutation and 4 % homozygous deletion [8], while prostate cancers show the reverse scenario with 1 % mutation and 42 % deletion [72].

A number of reports in mouse have pointed to a role of PTEN in autism spectrum disorders [73–75]. Links between patients with inherited PTEN and TSC mutation and autism have been reported [76, 77], and some 10 % of autistic patients were found to harbor PTEN mutation [78]. This has spawned the first trials on TSC mutant autistic children and young adults with rapamycin [79]—the results are eagerly awaited.

PTEN and Outcome of Cancer Therapy

The above insights into tumor suppression strongly suggest that understanding PTEN regulation harbors great therapeutic potential since many tumors are diagnosed while they still retain one copy of the gene. Thus, the identification of the first Ubiquitin E3-ligase of PTEN, NEDD4-1, represented a significant step towards elucidating how tumor cells degrade PTEN and how this posttranslational modification could be inhibited therapeutically [80]. Intriguingly, mono-ubiquitination of PTEN is essential for its nuclear import and stability, as revealed in a patient family where mutation of one ubiquitination site leads to nuclear PTEN exclusion and familial Cowden Disease. PTEN degradation and nuclear transport are closely linked processes and interventions favoring the latter should be of great therapeutic potential as nuclear PTEN localization has been generally associated with better disease outcomes (see [81] and references therein). Inhibition of *PTEN* at the transcriptional level has also been associated with T-cell leukemia, as a NOTCH1 target was found to repress *PTEN* in a manner sensitive to the clinical NOTCH1 inhibitor gamma-secretase (GSI) [82].

Indeed, GSI resistance in *NOTCH1*-driven leukemia was also found to be associated with *PTEN* mutation, demonstrating the importance of the PTEN pathway in such malignancies and highlighting the need for means to rescue PTEN function in order to prevent resistance to targeted therapy.

Additional examples for a similar role of PTEN in targeted therapy have been reported, collectively linking PTEN/PI-3 K alterations with resistance to the HER2 inhibitors Herceptin/Trastuzumab in breast cancer [83] and resistance to the EGFR inhibitors Erlotinib and Gefitinib in glioblastoma patients [84].

In conclusion, a picture is emerging in which the neoplastic process requires fine-tuned downregulation of PTEN that dictates tumor formation, tumor progression to malignancy, and tumor sensitivity to targeted therapy. Our continued efforts to understand PTEN regulation are therefore of critical impact to multilayered control of PI 3-Kinase pathway therapy in future cancer treatments.

References

1. Maehama T, Dixon JE. The tumor suppressor, PTEN/MMAC1, dephosphorylates the lipid second messenger, phosphatidylinositol 3,4,5-trisphosphate. J Biol Chem. 1998;273(22):13375–8.
2. Li J, et al. PTEN, a putative protein tyrosine phosphatase gene mutated in human brain, breast, and prostate cancer. Science. 1997;275(5308):1943–7.

3. Steck PA, et al. Identification of a candidate tumour suppressor gene, MMAC1, at chromosome 10q23.3 that is mutated in multiple advanced cancers. Nat Genet. 1997;15(4):356–62.
4. Myers MP, et al. P-TEN, the tumor suppressor from human chromosome 10q23, is a dual-specificity phosphatase. Proc Natl Acad Sci U S A. 1997;94(17):9052–7.
5. Engelman JA, Luo J, Cantley LC. The evolution of phosphatidylinositol 3-kinases as regulators of growth and metabolism. Nat Rev Genet. 2006;7(8):606–19.
6. Huang H, et al. PTEN affects cell size, cell proliferation and apoptosis during Drosophila eye development. Development. 1999;126(23):5365–72.
7. Stambolic V, et al. Negative regulation of PKB/Akt-dependent cell survival by the tumor suppressor PTEN. Cell. 1998;95(1):29–39.
8. Cerami E, et al. The cBio cancer genomics portal: an open platform for exploring multidimensional cancer genomics data. Cancer Discov. 2012;2(5):401–4.
9. Liu W, et al. Copy number analysis indicates monoclonal origin of lethal metastatic prostate cancer. Nat Med. 2009;15(5):559–65.
10. Chen M, et al. Identification of PHLPP1 as a tumor suppressor reveals the role of feedback activation in PTEN-mutant prostate cancer progression. Cancer Cell. 2011;20(2):173–86.
11. Saal LH, et al. Poor prognosis in carcinoma is associated with a gene expression signature of aberrant PTEN tumor suppressor pathway activity. Proc Natl Acad Sci U S A. 2007;104(18): 7564–9.
12. Yanagawa N, et al. Loss of phosphatase and tensin homolog protein expression is an independent poor prognostic marker in lung adenocarcinoma. J Thorac Oncol. 2012;7(10):1513–21.
13. Kim SS, et al. Expression of NEDD4-1, a PTEN regulator, in gastric and colorectal carcinomas. APMIS. 2008;116(9):779–84.
14. Mirmohammadsadegh A, et al. Epigenetic silencing of the PTEN gene in melanoma. Cancer Res. 2006;66(13):6546–52.
15. Gravina GL, et al. Epigenetic modulation of PTEN expression during antiandrogenic therapies in human prostate cancer. Int J Oncol. 2009;35(5):1133–9.
16. Nishioka C, et al. Long-term exposure of leukemia cells to multi-targeted tyrosine kinase inhibitor induces activations of AKT, ERK and STAT5 signaling via epigenetic silencing of the PTEN gene. Leukemia. 2010;24(9):1631–40.
17. Yang J, et al. Long-term exposure of gastrointestinal stromal tumor cells to sunitinib induces epigenetic silencing of the PTEN gene. Int J Cancer. 2012;130(4):959–66.
18. Poliseno L, et al. Identification of the miR-106b~25 microRNA cluster as a proto-oncogenic PTEN-targeting intron that cooperates with its host gene MCM7 in transformation. Sci Signal. 2010;3(117):ra29.
19. Poliseno L, et al. A coding-independent function of gene and pseudogene mRNAs regulates tumour biology. Nature. 2010;465(7301):1033–8.
20. Poliseno L, et al. Deletion of PTENP1 pseudogene in human melanoma. J Invest Dermatol. 2011;131(12):2497–500.
21. Tay Y, et al. Coding-independent regulation of the tumor suppressor PTEN by competing endogenous mRNAs. Cell. 2011;147(2):344–57.
22. Karreth FA, et al. In vivo identification of tumor- suppressive PTEN ceRNAs in an oncogenic BRAF-induced mouse model of melanoma. Cell. 2011;147(2):382–95.
23. Sansal I, Sellers WR. The biology and clinical relevance of the PTEN tumor suppressor pathway. J Clin Oncol. 2004;22(14):2954–63.
24. Knudson Jr AG. Mutation and cancer: statistical study of retinoblastoma. Proc Natl Acad Sci U S A. 1971;68(4):820–3.
25. Di Cristofano A, et al. Pten and p27KIP1 cooperate in prostate cancer tumor suppression in the mouse. Nat Genet. 2001;27(2):222–4.
26. Kwabi-Addo B, et al. Haploinsufficiency of the Pten tumor suppressor gene promotes prostate cancer progression. Proc Natl Acad Sci U S A. 2001;98(20):11563–8.
27. Trotman LC, et al. Pten dose dictates cancer progression in the prostate. PLoS Biol. 2003;1(3):E59.
28. Di Cristofano A, et al. Pten is essential for embryonic development and tumour suppression. Nat Genet. 1998;19(4):348–55.

29. Podsypanina K, et al. Mutation of Pten/Mmac1 in mice causes neoplasia in multiple organ systems. Proc Natl Acad Sci U S A. 1999;96(4):1563–8.
30. Suzuki A, et al. High cancer susceptibility and embryonic lethality associated with mutation of the PTEN tumor suppressor gene in mice. Curr Biol. 1998;8(21):1169–78.
31. Suzuki A, et al. Keratinocyte-specific Pten deficiency results in epidermal hyperplasia, accelerated hair follicle morphogenesis and tumor formation. Cancer Res. 2003;63(3):674–81.
32. Oudit GY, et al. Loss of PTEN attenuates the development of pathological hypertrophy and heart failure in response to biomechanical stress. Cardiovasc Res. 2008;78(3):505–14.
33. Fraser MM, et al. Pten loss causes hypertrophy and increased proliferation of astrocytes in vivo. Cancer Res. 2004;64(21):7773–9.
34. Di Cristofano A, et al. Impaired Fas response and autoimmunity in Pten+/- mice. Science. 1999;285(5436):2122–5.
35. Naguib A, Trotman LC. PTEN plasticity: how the taming of a lethal gene can go too far. Trends Cell Biol. 2013;23:374–9.
36. Kwon CH, et al. Pten regulates neuronal soma size: a mouse model of Lhermitte-Duclos disease. Nat Genet. 2001;29(4):404–11.
37. Backman SA, et al. Deletion of Pten in mouse brain causes seizures, ataxia and defects in soma size resembling Lhermitte-Duclos disease. Nat Genet. 2001;29(4):396–403.
38. Medema RH, et al. AFX-like Forkhead transcription factors mediate cell-cycle regulation by Ras and PKB through p27kip1. Nature. 2000;404(6779):782–7.
39. Tran H, et al. DNA repair pathway stimulated by the forkhead transcription factor FOXO3a through the Gadd45 protein. Science. 2002;296(5567):530–4.
40. Guertin DA, Sabatini DM. Defining the role of mTOR in cancer. Cancer Cell. 2007;12(1):9–22.
41. Guertin DA, et al. mTOR complex 2 is required for the development of prostate cancer induced by Pten loss in mice. Cancer Cell. 2009;15(2):148–59.
42. Podsypanina K, et al. An inhibitor of mTOR reduces neoplasia and normalizes p70/S6 kinase activity in Pten+/- mice. Proc Natl Acad Sci U S A. 2001;98(18):10320–5.
43. Neshat MS, et al. Enhanced sensitivity of PTEN-deficient tumors to inhibition of FRAP/mTOR. Proc Natl Acad Sci U S A. 2001;98(18):10314–9.
44. Franz DN, et al. Rapamycin causes regression of astrocytomas in tuberous sclerosis complex. Ann Neurol. 2006;59(3):490–8.
45. Wong KK, Engelman JA, Cantley LC. Targeting the PI3K signaling pathway in cancer. Curr Opin Genet Dev. 2010;20(1):87–90.
46. Chen Z, et al. Crucial role of p53-dependent cellular senescence in suppression of Pten-deficient tumorigenesis. Nature. 2005;436(7051):725–30.
47. Narita M, Lowe SW. Senescence comes of age. Nat Med. 2005;11(9):920–2.
48. Trotman LC, et al. Identification of a tumour suppressor network opposing nuclear Akt function. Nature. 2006;441(7092):523–7.
49. Cordon-Cardo C, et al. Distinct altered patterns of p27KIP1 gene expression in benign prostatic hyperplasia and prostatic carcinoma. J Natl Cancer Inst. 1998;90(17):1284–91.
50. Gurrieri C, et al. Loss of the tumor suppressor PML in human cancers of multiple histologic origins. J Natl Cancer Inst. 2004;96(4):269–79.
51. Berger AH, Knudson AG, Pandolfi PP. A continuum model for tumour suppression. Nature. 2011;476(7359):163–9.
52. Franke TF, et al. PI3K/Akt and apoptosis: size matters. Oncogene. 2003;22(56):8983–98.
53. Friedl P, Alexander S. Cancer invasion and the microenvironment: plasticity and reciprocity. Cell. 2011;147(5):992–1009.
54. Trimboli AJ, et al. Pten in stromal fibroblasts suppresses mammary epithelial tumours. Nature. 2009;461(7267):1084–91.
55. Funamoto S, et al. Spatial and temporal regulation of 3-phosphoinositides by PI 3-kinase and PTEN mediates chemotaxis. Cell. 2002;109(5):611–23.
56. Iijima M, Devreotes P. Tumor suppressor PTEN mediates sensing of chemoattractant gradients. Cell. 2002;109(5):599–610.
57. Charest PG, Firtel RA. Big roles for small GTPases in the control of directed cell movement. Biochem J. 2007;401(2):377–90.

58. Li Z, et al. Regulation of PTEN by Rho small GTPases. Nat Cell Biol. 2005;7(4):399–404.
59. Martin-Belmonte F, et al. PTEN-mediated apical segregation of phosphoinositides controls epithelial morphogenesis through Cdc42. Cell. 2007;128(2):383–97.
60. Tamura M, et al. Inhibition of cell migration, spreading, and focal adhesions by tumor suppressor PTEN. Science. 1998;280(5369):1614–7.
61. Tamura M, et al. Tumor suppressor PTEN inhibition of cell invasion, migration, and growth: differential involvement of focal adhesion kinase and p130Cas. Cancer Res. 1999;59(2):442–9.
62. Gerber HP, et al. Vascular endothelial growth factor regulates endothelial cell survival through the phosphatidylinositol 3'-kinase/Akt signal transduction pathway. Requirement for Flk-1/KDR activation. J Biol Chem. 1998;273(46):30336–43.
63. Kim I, et al. Angiopoietin-1 regulates endothelial cell survival through the phosphatidylinositol 3'-Kinase/Akt signal transduction pathway. Circ Res. 2000;86(1):24–9.
64. Huang J, Kontos CD. PTEN modulates vascular endothelial growth factor-mediated signaling and angiogenic effects. J Biol Chem. 2002;277(13):10760–6.
65. Hamada K, et al. The PTEN/PI3K pathway governs normal vascular development and tumor angiogenesis. Genes Dev. 2005;19(17):2054–65.
66. Koul D, et al. MMAC/PTEN tumor suppressor gene regulates vascular endothelial growth factor-mediated angiogenesis in prostate cancer. Int J Oncol. 2002;21(3):469–75.
67. Majumder PK, et al. mTOR inhibition reverses Akt-dependent prostate intraepithelial neoplasia through regulation of apoptotic and HIF-1-dependent pathways. Nat Med. 2004;10(6):594–601.
68. Paik JH, et al. FoxOs are lineage-restricted redundant tumor suppressors and regulate endothelial cell homeostasis. Cell. 2007;128(2):309–23.
69. Aranda V, et al. Par6-aPKC uncouples ErbB2 induced disruption of polarized epithelial organization from proliferation control. Nat Cell Biol. 2006;8(11):1235–45.
70. Liaw D, et al. Germline mutations of the PTEN gene in Cowden disease, an inherited breast and thyroid cancer syndrome. Nat Genet. 1997;16(1):64–7.
71. Zbuk KM, Eng C. Cancer phenomics: RET and PTEN as illustrative models. Nat Rev Cancer. 2007;7(1):35–45.
72. Taylor BS, et al. Integrative genomic profiling of human prostate cancer. Cancer Cell. 2010;18(1):11–22.
73. Kwon CH, et al. Pten regulates neuronal arborization and social interaction in mice. Neuron. 2006;50(3):377–88.
74. Zhou J, Parada LF. PTEN signaling in autism spectrum disorders. Curr Opin Neurobiol. 2012;22(5):873–9.
75. Xiong Q, et al. PTEN regulation of local and long-range connections in mouse auditory cortex. J Neurosci. 2012;32(5):1643–52.
76. Smalley SL, et al. Autism and tuberous sclerosis. J Autism Dev Disord. 1992;22(3):339–55.
77. Goffin A, et al. PTEN mutation in a family with Cowden syndrome and autism. Am J Med Genet. 2001;105(6):521–4.
78. Butler MG, et al. Subset of individuals with autism spectrum disorders and extreme macrocephaly associated with germline PTEN tumour suppressor gene mutations. J Med Genet. 2005;42(4):318–21.
79. Sahin M. Targeted treatment trials for tuberous sclerosis and autism: no longer a dream. Curr Opin Neurobiol. 2012;22(5):895–901.
80. Wang X, et al. NEDD4-1 is a proto-oncogenic ubiquitin ligase for PTEN. Cell. 2007;128(1):129–39.
81. Trotman LC, et al. Ubiquitination regulates PTEN nuclear import and tumor suppression. Cell. 2007;128(1):141–56.
82. Palomero T, et al. Mutational loss of PTEN induces resistance to NOTCH1 inhibition in T-cell leukemia. Nat Med. 2007;13(10):1203–10.
83. Nagata Y, et al. PTEN activation contributes to tumor inhibition by trastuzumab, and loss of PTEN predicts trastuzumab resistance in patients. Cancer Cell. 2004;6(2):117–27.
84. Mellinghoff IK, et al. Molecular determinants of the response of glioblastomas to EGFR kinase inhibitors. N Engl J Med. 2005;353(19):2012–24.

Phosphatase-Independent Functions of the Tumor Suppressor PTEN

Antonella Papa and Pier Paolo Pandolfi

Abstract PTEN (Phosphatase and tensin homolog deleted on Chromosome 10) is a major tumor suppressor with both phosphatase-dependent and -independent functions. As it dephosphorylates the second messenger phosphatidylinositol-3,4,5-triphosphate (PtdIns3*P* or PI3P), PTEN regulates the activation of the proto-oncogenic PI3K-AKT signaling pathway and thereby inhibits cell proliferation and cell growth. In addition, through protein phosphatase activity, PTEN can regulate the phosphorylation of protein substrates in order to inhibit cell migration and invasion. Recently several studies have highlighted the importance of PTEN localization in cell nuclei, where it can function as a tumor suppressor in a phosphatase-independent manner. For instance, nuclear PTEN function has been associated with a PTEN scaffolding role favoring the activation of the DNA-repair machinery and ensuring chromosome stability. Further, it has been discovered that through phosphatase-independent functions PTEN also regulates cell cycle progression and key tumor-suppressive metabolic states in the nucleus. In this chapter we focus our attention on the scaffolding role of PTEN. We summarize the mechanisms involved in the regulation of nuclear and cytoplasmic PTEN shuttling, and discuss the potential clinical relevance of reported nuclear PTEN targets. Importantly, we also discuss the implications of nuclear PTEN functions to tumorigenesis, and outline their potential as targets in the development of new therapeutic approaches to cancer.

A. Papa
Monash Biomedicine Discovery Institute and Department of Biochemistry and Molecular Biology, Monash University, Clayton, VIC, Australia

Departments of Medicine and Pathology, Cancer Research Institute, Beth Israel Deaconess Cancer Center, Beth IsraelDeaconess Medical Center, Harvard Medical School, Boston, MA, USA
e-mail: apapa@bidmc.harvard.edu

P.P. Pandolfi (✉)
Departments of Medicine and Pathology, Cancer Research Institute, Beth Israel Deaconess Cancer Center, Beth IsraelDeaconess Medical Center, Harvard Medical School, Boston, MA, USA
e-mail: ppandolf@bidmc.harvard.edu

B.G. Neel, N. Tonks (eds.), *Protein Tyrosine Phosphatases in Cancer*,
DOI 10.1007/978-1-4939-3649-6_9

Keywords PTEN phosphatase-independent activity • PTEN nuclear localization

Subcellular Localization of the Tumor Suppressor PTEN

In 1997, two independent groups identified *PTEN* as a new candidate tumor suppressor located on chromosome 10q23, and found that it was highly mutated in several forms of human cancer [1, 2]. Initial biochemical studies revealed that PTEN shares structural similarities with members of the Dual Specificity Protein Phosphatase (DUSPs) family [3]; and in 1998, Maehama and Dixon discovered that PTEN catalyzes the dephosphorylation of the second messenger PIP3 at the 3-position on the inositol ring [4].

Ever since, PTEN has been an object of intense investigations due to its prominent role in controlling cell biology and tissue homeostasis. Indeed, through its lipid phosphatase function PTEN was found to inhibit the activation of several PIP3-dependent kinases, including the phosphatidylinositol-3-kinase (PI3K)-Akt-mTOR signaling pathway, which regulates cell proliferation, cell growth, and cell metabolism [5, 6]. In vivo studies performed in mice likewise revealed that Pten plays an essential role during embryonic development, as its complete loss leads to early embryonic lethality due to severe hyperproliferation and an inability to trigger apoptosis [7–9]. In addition, it has been consistently reported that heterozygous reductions of Pten levels promotes tumorigenesis in several organs and epithelial tissues, thus identifying Pten as a haplo-insufficient tumor suppressor [10, 11].

PTEN was initially described as distributed in the cytoplasm, where it functions toward the PIP3 through lipid phosphatase activity [12]. Surprisingly, later investigations revealed that this tumor suppressor was also easily detectable in the nucleus, although its nuclear role was at first unclear [13]. While initial reports suggested that PTEN was localized to the nucleus as a consequence of the cell cycle [14], further investigations concluded that PTEN was actively entering the nucleus in order to fully achieve its tumor-suppressive functions (by favoring DNA-damage response (DDR) and controlling cell cycle progression and metabolism rather than exerting any catalytic activity) [15–17].

The discovery of nuclear PTEN also helped illuminate the relevance of the many *PTEN* mutations located outside the protein's catalytic core, which is specifically encoded by the *exon5*. Remarkably, analysis of the status of *PTEN* in human cancer made clear that while almost 40 % of PTEN mutations were found located at the structural active site, and were therefore directly connected to its phosphatase functions, several sporadic and inherited missense and nonsense mutations were found across the entire enzyme sequence [18, 19]. This observation led to speculation that beyond its catalytic activity, PTEN might use additional protein domains encoded along its enzyme sequence in order to fully exert its tumor-suppressive function, in a phosphatase-independent manner [10, 16].

Regulation of PTEN Nuclear Entry

The discovery of the nuclear compartmentalization of PTEN suggested it might be possible to identify new nuclear PTEN interactors that could potentially play a role in the development of alternative strategies for cancer treatment. However, in turn this also begged another key question: what were the mechanisms governing the cytoplasmic and nuclear shuttling of PTEN?

Cargo Proteins Transport PTEN into the Nucleus

Nuclear proteins can be redirected into the nucleus by different mechanisms. One common and efficient system for the selective translocation of nuclear proteins is achieved through the expression of nuclear localization signal sequences (NLSs). NLSs are either single or repeated clusters of basic amino-acid residues with partially conserved sequences which are recognized by protein import receptors [20]. PTEN does not express typical NLSs, such as that expressed by the SV40 large-T antigen; however, Chung and coworkers eventually identified four redundant NLSs-like sequences in PTEN, two apiece in the N-terminal and C-terminal domains [21]. Strikingly, the authors found that while a single mutation of any of these NLSs-like sequences did not yield an effect on PTEN localization, a combined mutation in two out of four sequences prevented its nuclear translocation, and furthermore, inhibited Akt activation. Notably in 293 T and HeLa cells, it was found that the nuclear import of PTEN was dependent on its binding to the vault protein, Major Vault Protein (MVP). Thus the authors concluded that MVP is the transport protein that carries PTEN into the nucleus [21].

Different conclusions were however reported by Gil et al. By performing experiments in the U87-MG human glioblastoma cell line, the authors found that PTEN nuclear translocation was dependent on its binding to Ran-GTPase, a protein associated with importin-β and required for importin-mediated nuclear transport [22] (Fig. 1).

Moreover, these authors concluded that PTEN nuclear accumulation increased upon treatment with sorbitol and TNFα, suggesting that stress conditions and apoptotic stimuli may favor PTEN nuclear translocation and function. Notably, PTEN-driven nuclear apoptosis was shown to be independent of the protein's phosphatase functions, as the expression of either PTENC124S, a catalytically dead PTEN, or a lipid-dead (but protein phosphatase active) PTENG129E mutant was still able to signal apoptosis [22].

Finally, and in contrast to what had been previously reported, Liu et al. described that in HeLa cells rather than being actively transported through the nuclear pores, PTEN could enter the nucleus simply by diffusion [23, 24]. Because PTEN is a phosphatase of approximately 50–55 kDa in size, it may be able to translocate into the nucleus by passive diffusion through the nuclear pores, with no need of interacting with protein carriers (limiting protein size for diffusion through the pores is <60,000 Da) [24]. In addition, the authors found that selected mutations in the very N-terminal

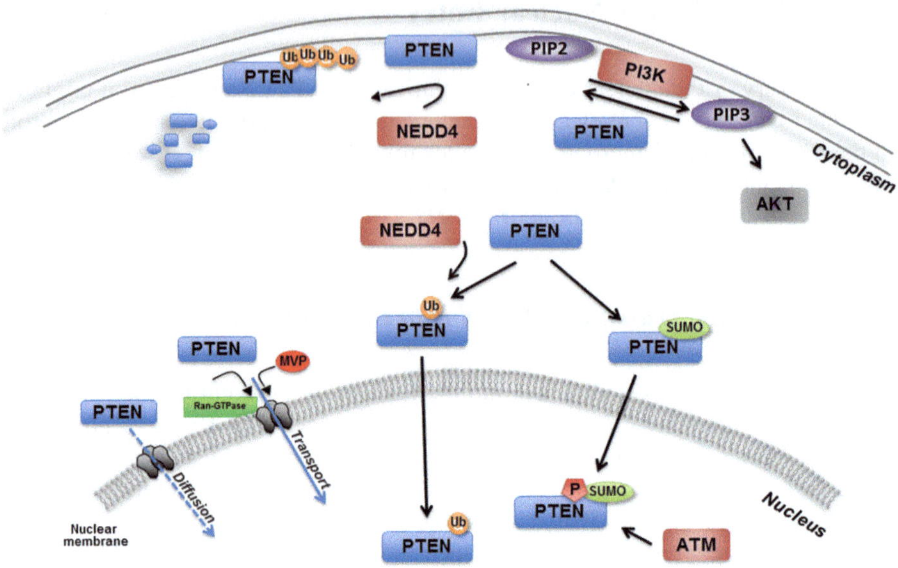

Fig. 1 PTEN at the plasma membrane dephosphorylates the second messenger PIP3 and regulates the activation of the PI3K-AKT pathway. Further polyubiquitination leads to PTEN degradation mediated by the Ubiquitin- Proteasome Pathway (UPP). SUMOylation and NEDD4-dependent monoubiquitination favor PTEN shuttling and nuclear translocation. Across the nuclear membrane, carrier proteins such as MVP or Ran-GTPase transport PTEN into the nucleus. In addition, passive diffusion may further increase PTEN nuclear localization

sequence of PTEN, mainly PTENK13 and PTENR14, precluded nuclear PTEN accumulation and induced its cytoplasmic sequestration [23] (Fig. 1).

Taken together, these findings suggest that while a complete description of the mechanisms responsible for PTEN translocation into the nucleus is still missing, it is doubtless fair to conclude that nuclear localization plays a role in its essential functional activity. Furthermore, it is legitimate to speculate that the discrepancies found between different reports may reflect the distinct and diverse genetic backgrounds of the respective cell lines used in the studies. For instance, retention of endogenous PTEN in HeLa and 293T cells may have influenced the localization of mutated PTEN, which differed from experiments performed in U87-MG which, on the contrary, exhibited lost endogenous PTEN expression.

Clearly further investigations into the mechanisms that bring PTEN into the nucleus are still required.

Posttranslational Modifications Regulate PTEN Nuclear Shuttling

While active transport and passive diffusion explain the routes by which PTEN enters the nucleus, there is an additional level of complexity to be elucidated when it comes to the signals that activate PTEN translocation and the molecular features that favor its nuclear relocation.

To this end, it is well established that posttranslational modifications can regulate PTEN functions at different levels [11]. While phosphorylation, oxidation, and acetylation have been mainly linked to the regulation of PTEN enzymatic activity [25], ubiquitination and sumoylation have been more directly connected to its subcellular localization [26–28]. Commonly, polyubiquitination regulates protein levels and stability by signaling for protein degradation to the proteasome [29]. For instance, PTEN recruitment to the plasma membrane also favors its subsequent degradation in a polyubiquitination-dependent manner [28, 30]. However, PTEN can also be monoubiquitinated, and PTEN monoubiquitination has been found to be essential to its nuclear translocation [27].

Along its entire amino-acid sequence PTEN contains several lysines (Lys, K) that are targeted by E3-ligases. To date, NEDD4-1, WWP2 and RFP are the only reported E3-ligases for PTEN [28, 31]. But while WWP2 and RFP seems to regulate PTEN polyubiquitination and membrane recruitment [31], NEDD4-1 can regulate both poly- and monoubiquitination of PTEN [27, 28].

In 2007 Trotman et al. reported that two major lysines in PTEN, PTENK13 and PTENK289, were mutated in cancer. But importantly, while Lys13 was mutated in somatic cancer [32], Lys289 was identified in a patient diagnosed with the cancer predisposition syndrome Cowden Disease (CD). CD, together with Bannayan–Riley–Ruvalcaba syndrome (BRRS), *PTEN*-related Proteus syndrome (PS), and Proteus-like syndrome, is one of the closely related autosomal dominant cancer predisposition syndromes commonly associated with inherited mutations of *PTEN* [33]. This patient developed recurrent intestinal polyps in which, surprisingly, PTEN nuclear localization was found to be decreased. Furthermore, when the patient developed overt carcinoma, the wild-type allele was found lost hence fulfilling the "two hit hypothesis" paradigm [27]. In vitro experiments confirmed that cells expressing PTENK289E, a mutated PTEN (Lys289 (K) to Glu (E), as found in human cancer), exhibited lower levels of nuclear localization due to reduced nuclear entry, while the mutant protein still retained full catalytic activity. In addition, generation of a double PTENK13E-K289E mutant displayed an even more dramatic PTEN nuclear exclusion, leading to a reduced apoptotic signal. Thus it was concluded that impairment of PTEN nuclear accumulation through regulation of its monoubiquitination reduces PTEN function and favors tumorigenesis [27]. In a parallel study, Wang et al. also identified NEDD-4 as the E3-ligase responsible for monoubiquitinating PTEN [28].

An additional report further highlighted the importance of another PTEN posttranslational modification as a means of nuclear entry. Bassi and co-authors identified a site of SUMOylation in the C-terminal region of PTEN [26]. Critically, the authors found that mutation of Lys254 to Arg (R) in PTEN (PTENK254R) abrogated PTEN sumoylation and, importantly, prevented its nuclear localization. Consistent with lipid phosphatase-independent activity expected of nuclear PTEN, the authors reported that while PTENK254R was still fully capable of inhibiting Akt through PIP3, this mutation was also accompanied by a concomitant hypersensitivity to genotoxic stress. Specifically, upon exposure to ionizing radiation (IR), cells with complete loss of PTEN and engineered to express only PTENK254R were unable to repair double-strand breaks (DSBs), unlike their PTEN wild-type counterpart, which indicated a defect in the activation of the DNA repair machinery. Furthermore, authors identified ataxia telangiectasia mutated (ATM) as the kinase responsible for the SUMO-PTEN phosphorylation, which consequently regulates PTEN import in response to IR. Thus, in agreement with previous reports, the authors concluded that nuclear PTEN functions may be responsible for the activation of the DNA damage response process, with potential implications for cancer therapy [26].

Notably, the inability of PTENK254R to accumulate in the nucleus depended on its increased nuclear export rather than on decreased nuclear entry, suggesting that the mechanisms regulating PTEN subcellular localization may involve a functional network whereby several players and pathways are regulated and integrated in a coordinated fashion; and that cellular stresses and suboptimal cellular conditions (such as hypoxia or IR) may favor or inhibit PTEN nuclear localization and functions.

Collectively, these reports support a functional role for nuclear PTEN which may be partially independent on its phosphatase activity.

PTEN Nuclear Functions

Although several studies had noted the presence of nuclear PTEN in normal breast, pancreas, and thyroid tissues (as opposed to their derived cancer cells) [34], the first compelling evidence of nuclear PTEN accumulation and its functional implications was reported by Ginn-Pease and Eng in 2003 [14]. By performing in vitro fractionation of a MCF-7 breast cancer cell line, the authors found that the ratio between cytoplasmic and nuclear PTEN changed over cell cycle progression. It was reported that when synchronized in G0-G1 phase, cells exhibited increased nuclear PTEN localization. In contrast, cells going through the S-phase, and with a higher proliferative index, displayed mainly cytoplasmic PTEN. Thus the authors concluded that nuclear PTEN may favor cell cycle arrest in the G0-G1 phase, and thus may regulate cell cycle progression [14, 35].

One of the first nuclear PTEN functions to be characterized was associated with the tumor suppressor p53. Freeman and coworkers studied the functional cross-talk between Pten and p53 in mice [35]. They found that loss of *Pten* led to an acceleration of p53-driven tumorigenesis due to both phosphatase-dependent and -independent functions of Pten. Specifically, while lowering of Pten induces Akt activation, MDM2 phosphorylation and nuclear translocation to induce p53 degradation, the authors also reported that through physical interaction, PTEN stabilizes p53 protein to favor its stability and transcriptional activity. Importantly, the PTEN-p53 binding was found to be dependent on the non-catalytic C-terminal domain of PTEN, and the use of a catalytically dead PTEN, PTENC124S, or the single PTEN C-terminal domain with no phosphatase domain, generated the same functional effects. Therefore the authors concluded that PTEN can regulate p53 activity with a combination of phosphatase-dependent and -independent functions [35]. These findings are, however, at odd with the fact that PTEN loss triggers a potent p53-dependent cellular senescence response that needs to be evaded in order for the tumor to progress to malignancy [36]. Hence the functional cross-talk of PTEN with p53 seems to be redundant, at least in the prostate.

Taken together, these early and descriptive reports propose only correlative observations regarding the potential function of PTEN in the nucleus, while the possible PTEN-p53 connection only suggests a redundant role for nuclear PTEN. Hence, the central questions regarding the key functions of PTEN in the nucleus, and the contribution of the phosphatase-independent functions to tumor suppression, still remained to be answered.

PTEN Preserves Chromosome Stability

One of the first nuclear functions attributed to PTEN refers to its ability to maintain chromosome stability [37]. Starting from the observation that lack of PTEN expression was associated with aneuploidy in human breast cancers [38], Shen and colleagues performed in vitro studies on primary mouse embryonic fibroblasts (MEFs) and embryonic stem cells (ES cells) derived from Pten *wild-type* and Pten *null* mice. In these settings, they noticed that even in steady state condition, cells *null* for Pten displayed a significant increase in chromosomal anomalies. Specifically, the authors reported consistent centromere breakages and a number of centromere-associated chromosomal instabilities, with reciprocal and non-reciprocal chromosomal translocations. Through immunostaining analysis, they also discovered that nuclear Pten localization overlapped with the staining of a major component of the centromere, CENP-C. The binding between PTEN and CENP-C was then further confirmed by co-immunoprecipitation experiments and was found to be essential for the stability of the centromere itself. Indeed, a specific nonsense mutation in PTEN, R189X, that was associated with CD and encoded a deletion of the C-terminus region, was no longer able to bind CENP-C. In addition, by mapping the binding domain between the two proteins, it was found that CENP-C binds PTEN through its C-terminal region, and that the phosphatase-dead PTENC124S mutant was still able to bind CENP-C. Thus, the authors concluded that the centromere-associated function of PTEN is accomplished through scaffolding effects and is not dependent on its catalytic activity [16].

In addition, together with centromere breakages and chromosomal translocations, the authors also noticed an increase in spontaneous chromosomal breaks and random fragmentations which were more likely to be associated with defects in the DNA repair machinery. They likewise observed in Pten *null* MEFs a higher incidence of DNA double-strand breaks (DSBs), alongside an increase in the phosphorylation of the DSB marker H2AX, γ-H2AX, as compared to Pten *wild-type* cells. To further investigate the mechanisms by which Pten preserves genomic stability, the authors utilized microarray analysis to identify the DNA-repair family member Rad51 as one of the altered genes whose expression was inhibited upon loss of PTEN. Importantly, Rad51 modulation was dependent on the phosphatase functions of PTEN as the catalytically dead mutant PTENC124S had no effect on transcript levels.

The authors concluded that through a combination of functions both phosphatase independent (PTEN binding to CENP-C) and dependent (PTEN activation of Rad51 expression), PTEN regulates chromosome stability and favors the activation of Homologous recombination-directed repair (HDR) for the repair of DSBs [16].

Nuclear PTEN Regulates Cell Cycle Progression and Metabolism

Early studies suggested that nuclear PTEN may be connected to the regulation of cell cycle progression. In agreement with this, Song et al. recently demonstrated that nuclear PTEN can directly enhance the activation of the anaphase-promoting complex/cyclosome APC/C by favoring its association with the specific target CDH1 [17]. Critically, in this study the authors characterized the induction of cellular senescence induced by PTEN loss as a fail-safe mechanism opposing tumor progression. While previous reports have pointed to the functional interaction between PTEN loss and p53 activation through the activation of Akt-mTOR signaling [36, 39], this new study demonstrated that through scaffolding effects, PTEN loss also favors cellular senescence in a p53- and Akt-independent manner. Specifically, the APC/C complex is an E3-ubiquitin ligase whose activation is restricted to the late mitosis and G1 phase, where it is responsible for the ubiquitination of cell cycle components which are tightly regulated in expression and stability [40]. In the study, the authors found a direct nuclear binding between PTEN and the APC family members (APC3/CDC27, APC4, APC5, and APC7), and reported that through physical interaction, PTEN favors the binding of APC3 to its substrate CDH1, thereby promoting formation of the complex. Critically, it was found that the PTEN and APC/C-CDH1 binding was mediated by the PTEN C-terminal domain, suggesting that the phosphatase function of PTEN was dispensable at this point. Notably, PTEN loss induced a reduction of the APC/C-CDH1 function with consequent accumulation of its known targets, such as Ets2 and p16, Cyclin A2, Aurora A, and PLK1. Taken together, the authors concluded that through the CDH1-Ets2-p16 axis, PTEN modulates the cellular senescence response in a phosphatase-independent manner. In addition, because loss of PTEN may lead to the accumulation of Aurora A and PLK1 kinases in the nucleus, and concomitant activation of Akt signaling in the cytoplasm, combined inhibition of these pathways in PTEN null cancers may lead to new strategies for more effective cancer therapy, as we discuss below.

Moreover, it has been recently reported that systemic elevation of Pten in the mouse can induce a tumor-suppressive “anti-Warburg” metabolic state through coordination of both phosphatase-dependent and -independent PTEN functions [15]. Garcia-Cao et al. generated a series of transgenic mouse lines with constitutive elevated expression of Pten. Taking advantage of these transgenic lines and primary cells derived from these mice, they discovered that Pten can regulate both glucose and glutamine metabolism by inhibiting the PI3K activation on one hand,

and promoting the APC-CDH1 functions toward the PFKFB3 (6-phosphofructo-2-kinase/fructose-2,6-biphosphatse isoform 3) and GLS (glutaminase) enzymes on the other. Therefore, Pten triggers a tumor-suppressive cell metabolic state both in the cytoplasm in a lipid-phosphatase-dependent manner toward the PI3K suppression, and in the nucleus, in a phosphatase-independent manner toward the activation of APC-CDH1 functions.

Nuclear PTEN Functions in Cancer Therapy

Loss of PTEN expression is a hallmark of cancer progression, and is considered a molecular marker for the use of PI3K inhibitors in clinics [10]. The selection of resistant cells for this treatment, however, can lead to tumor relapse and recurrence of the disease. To this end, the discovery of nuclear PTEN and its phosphatase-independent activity toward genomic stability have opened new possibilities for developing highly targeted treatments for cancer eradication, and have prompted many scientists to renew their investigations of the protein's functional roles.

A concept long pursued in the clinic is the synthetic lethal approach, in which targeted inhibition of the DNA-damage repair machinery in the cancer cells leads to the emergence of deleterious genetic mutations, synthetic sickness, and eventual cell death [41].

The prototypical example of this is treatment of cells harboring mutations in *BRCA1* or *BRCA2* genes with poly (ADP-ribose) polymerase (PARP) inhibitors, or PARPi. As the *BRCA* genes encode proteins responsible for the repair of DSBs through Homologous Recombination (HR), and as inhibition of PARP activity leads to an accumulation of DSBs, the use of PARPi, such as Olaparib, has been highly effective in breast, ovarian, and pancreatic cancer cells with BRCA1 and BRCA2 mutations. These initial in vitro findings led to the establishment of ongoing clinical trials in patients harboring mutations in their *BRCA* genes [42]. Following a similar rationale, cells that have undergone loss of PTEN may be sensitized to PARPi, as they would also be defective in HR through impaired Rad51 function and increased chromosome instability. To investigate this hypothesis, Mendes-Pereira et al. generated two cell lines in which they specifically targeted a truncated mutation to both copies of the PTEN *exon2*, thus generating cell lines expressing only the first 24 amino acids of PTEN [43]. First, the authors verified that loss of PTEN expression caused Rad51-dependent impairment in DNA repair. Second, and more importantly, the authors tested the sensitivity of these cells to a specific PARPi (KU0058948) and also to the less selective chemotherapy drug, cisplatin. Strikingly, the authors found that cells with loss of PTEN displayed sensitivity to PARPi 20 times higher, and sensitivity to the DNA cross-linker cisplatin five times higher than the respective responses of the wild-type counterpart. In addition, the authors studied the *PTEN*-deficient prostate cancer cell line PC3, and found that reintroduction of an expression vector encoding PTEN wild-type reverted the sensitivity to PARPi, rendering these cells resistant to treatment. Finally, to assess the contribution of the phosphatase-dependent and -independent functions of PTEN to this process, the authors tested sensitivity to PARPi by introducing PTEN phosphatase-dead mutant PTENC124, as well as the nuclear-excluded PTENK289E. Here they found that while PTENC124S was resistant to treatment, expression of PTENK289E, a mutant with impaired nuclear import, showed sensitivity to Olaparib. Thus, nuclear PTEN function can predict sensitivity to PARPi [43]. Collectively, these data support combinatorial treatment with PI3K and PARPi, such as Olaparib, as it may be highly effective in the treatment of cancer harboring loss of PTEN.

In addition, together with the regulation of the DNA-damage response, PTEN loss in the nucleus was also found to impair the formation of the APC-CDH1 complex [17]. As loss of APC-CDH1 functions leads to accumulation of downstream oncogenic proteins such as PLK1 and Aurora A, tumors with loss of PTEN may lead to an oncogenic addiction to these kinases and hence hypersensitivity to their pharmacological inhibition. Indeed, as demonstrated by Song et al., loss of PTEN in human and mouse lines is associated with growth inhibition and induction of apoptosis upon treatment with either BI 2536 (PLK1 inhibitor) or VX680 (Aurora A inhibitor) inhibitors. Thus, while loss of PTEN may predict sensitivity to PI3K inhibition, it can at the same time also predict sensitivity to the use of PLK1 and Aurora A inhibitors that are currently being tested in clinical trials.

Concluding Remarks

The tumor suppressor PTEN is a key regulator of several biological processes, many of which are fulfilled through inhibition of the proto-oncogenic PI3K-AKT pathway, which for several years has been a prime focus of research into various forms of cancer.

In vivo studies have recently pointed, however, to the existence of AKT-independent mechanisms contributing to Pten loss-driven tumorigenesis. For instance, Blanco-Aparicio et al. found that in the breast and prostate, constitutive Akt activation achieved through the expression of a myristoylated Akt (myrAkt1), only led to development of mammary or prostatic hyperplasia, rather than the more advanced small adenocarcinomas found in mice with actual loss of Pten [44]. In addition, Vivanco et al. reported that loss of PTEN may lead to activation of the JNK pathway through Akt-independent signaling, and thus cooperate with Akt activation for tumorigenesis [45].

In light of these findings, investigators are exploring whether the nuclear PTEN functions executed through phosphatase-independent effects may offer a straightforward explanation for the discrepancies between these phenotypes—and more importantly, whether they point to additional pathways potentially deregulated by loss of PTEN. Indeed, many investigators believe that a number of recent discoveries and experiments, including the profound defects found in the DSB repair process of Pten null cells, together with their differential sensitivity to the PARP inhibitor Olaparib, and Aurora kinase inhibitors, may coalesce into a turning point

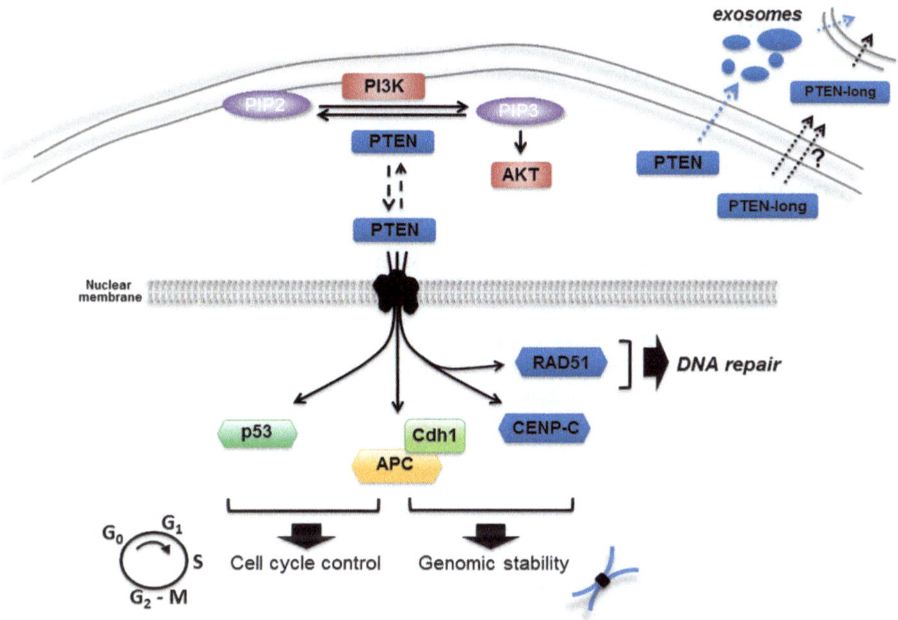

Fig. 2 Nuclear PTEN was initially reported to favor p53 functions and activity in controlling cell cycle check-points. Later studies acknowledged the essential role of PTEN in maintaining chromosome stability by exerting scaffolding effects on the centromere protein CENP-C. In addition, by inducing the expression of Rad51, PTEN favors the activation of the DNA-repair machinery to fix DSBs. In the nucleus, PTEN also stabilizes the formation of the APC/C-Cdh1 complex to regulate cell cycle progression and senescence. In addition, secreted PTEN, either through exosomes or through unidentified mechanisms, can function as a tumor suppressor in a cell non-autonomous manner. Collectively, the functions of nuclear PTEN are executed through its scaffolding properties in a phosphatase-independent manner

in our ability to efficiently treat PTEN-deficient cancers. Notably, clinical trials are currently under investigations for all these inhibitors. In this scenario, tumors displaying PTEN mutations leading to either defective nuclear accumulation or interfering with the mechanisms that regulate PTEN subcellular localization may illuminate new combinatorial approaches for novel, personalized, and highly effective cancer treatments.

In addition, it has been recently proposed that through independent mechanisms not yet fully defined, PTEN can also be secreted into the extracellular space for uptake by recipient cells where it can function as tumor suppressor in a cell non-autonomous manner [46, 47] (Fig. 2). In line with this, both phosphatase-dependent and -independent functions of PTEN should be taken into consideration as together they could alter the physiology of recipient cells. More provocatively, PTEN scaffold effects may also exert biological functions in the extracellular matrix. Thus, while over the last decade we have greatly improved our knowledge of the mechanisms of action of this essential tumor suppressor, recent and exciting discoveries have opened new avenues and unexpected directions for cancer research and therapy.

Acknowledgments The authors would like to thank SuJung Song for critical comments and Thomas Garvey for editing this manuscript. A.P. was supported in part by the American-Italian Cancer Foundation Post-Doctoral Fellowship.

References

1. Li J, Simpson L, Takahashi M, Miliaresis C, Myers MP, Tonks N, Parsons R. The PTEN/MMAC1 tumor suppressor induces cell death that is rescued by the AKT/protein kinase B oncogene. Cancer Res. 1998;58:5667–72.
2. Steck PA, Pershouse MA, Jasser SA, Yung WK, Lin H, Ligon AH, Langford LA, Baumgard ML, Hattier T, Davis T, Frye C, Hu R, Swedlund B, Teng DH, Tavtigian SV. Identification of a candidate tumour suppressor gene, MMAC1, at chromosome 10q23.3 that is mutated in multiple advanced cancers. Nat Genet. 1997;15:356–62.
3. Myers MP, Stolarov JP, Eng C, Li J, Wang SI, Wigler MH, Parsons R, Tonks NK. P-TEN, the tumor suppressor from human chromosome 10q23, is a dual-specificity phosphatase. Proc Natl Acad Sci U S A. 1997;94:9052–7.
4. Maehama T, Dixon JE. The tumor suppressor, PTEN/MMAC1, dephosphorylates the lipid second messenger, phosphatidylinositol 3,4,5-trisphosphate. J Biol Chem. 1998;273:13375–8.
5. Cantley LC, Neel BG. New insights into tumor suppression: PTEN suppresses tumor formation by restraining the phosphoinositide 3-kinase/AKT pathway. Proc Natl Acad Sci U S A. 1999;96:4240–5.
6. Myers MP, Pass I, Batty IH, van der kaay J, Stolarov JP, Hemmings BA, Wigler MH, Downes CP, Tonks NK. The lipid phosphatase activity of PTEN is critical for its tumor suppressor function. Proc Natl Acad Sci U S A. 1998;95:13513–8.
7. DI Cristofano A, Pesce B, Cordon-Cardo C, Pandolfi PP. Pten is essential for embryonic development and tumour suppression. Nat Genet. 1998;19:348–55.
8. Podsypanina K, Ellenson LH, Nemes A, Gu J, Tamura M, Yamada KM, Cordon-Cardo C, Catoretti G, Fisher PE, Parsons R. Mutation of Pten/Mmac1 in mice causes neoplasia in multiple organ systems. Proc Natl Acad Sci U S A. 1999;96:1563–8.
9. Suzuki A, de la Pompa JL, Stambolic V, Elia AJ, Sasaki T, Del Barco Barrantes I, Ho A, Wakeham A, Itie A, Khoo W, Fukumoto M, Mak TW. High cancer susceptibility and embryonic lethality associated with mutation of the PTEN tumor suppressor gene in mice. Curr Biol. 1998;8:1169–78.
10. Hollander MC, Blumenthal GM, Dennis PA. PTEN loss in the continuum of common cancers, rare syndromes and mouse models. Nat Rev Cancer. 2011;11:289–301.
11. Song MS, Salmena L, Pandolfi PP. The functions and regulation of the PTEN tumour suppressor. Nat Rev Mol Cell Biol. 2012;13:283–96.
12. Leevers SJ, Vanhaesebroeck B, Waterfield MD. Signalling through phosphoinositide 3-kinases: the lipids take centre stage. Curr Opin Cell Biol. 1999;11:219–25.
13. Planchon SM, Waite KA, Eng C. The nuclear affairs of PTEN. J Cell Sci. 2008;121:249–53.
14. Ginn-Pease ME, Eng C. Increased nuclear phosphatase and tensin homologue deleted on chromosome 10 is associated with G0-G1 in MCF-7 cells. Cancer Res. 2003;63:282–6.
15. Garcia-Cao I, Song MS, Hobbs RM, Laurent G, Giorgi C, de Boer VC, Anastasiou D, Ito K, Sasaki AT, Rameh L, Carracedo A, Vander Heiden MG, Cantley LC, Pinton P, Haigis MC, Pandolfi PP. Systemic elevation of PTEN induces a tumor-suppressive metabolic state. Cell. 2012;149:49–62.
16. Shen WH, Balajee AS, Wang J, Wu H, Eng C, Pandolfi PP, Yin Y. Essential role for nuclear PTEN in maintaining chromosomal integrity. Cell. 2007;128:157–70.
17. Song MS, Carracedo A, Salmena L, Song SJ, Egia A, Malumbres M, Pandolfi PP. Nuclear PTEN regulates the APC-CDH1 tumor-suppressive complex in a phosphatase-independent manner. Cell. 2011;144:187–99.
18. Ali IU, Schriml LM, Dean M. Mutational spectra of PTEN/MMAC1 gene: a tumor suppressor with lipid phosphatase activity. J Natl Cancer Inst. 1999;91:1922–32.

19. Marsh DJ, Coulon V, Lunetta KL, Rocca-Serra P, Dahia PL, Zheng Z, Liaw D, Caron S, Duboue B, Lin AY, Richardson AL, Bonnetblanc JM, Bressieux JM, Cabarrot-Moreau A, Chompret A, Demange L, Eeles RA, Yahanda AM, Fearon ER, Fricker JP, Gorlin RJ, Hodgson SV, Huson S, Lacombe D, Eng C, et al. Mutation spectrum and genotype-phenotype analyses in Cowden disease and Bannayan-Zonana syndrome, two hamartoma syndromes with germline PTEN mutation. Hum Mol Genet. 1998;7:507–15.
20. Hodel MR, Corbett AH, Hodel AE. Dissection of a nuclear localization signal. J Biol Chem. 2001;276:1317–25.
21. Chung JH, Ginn-Pease ME, Eng C. Phosphatase and tensin homologue deleted on chromosome 10 (PTEN) has nuclear localization signal-like sequences for nuclear import mediated by major vault protein. Cancer Res. 2005;65:4108–16.
22. GIL A, Andres-Pons A, Fernandez E, Valiente M, Torres J, Cervera J, Pulido R. Nuclear localization of PTEN by a Ran-dependent mechanism enhances apoptosis: Involvement of an N-terminal nuclear localization domain and multiple nuclear exclusion motifs. Mol Biol Cell. 2006;17:4002–13.
23. Liu F, Wagner S, Campbell RB, Nickerson JA, Schiffer CA, Ross AH. PTEN enters the nucleus by diffusion. J Cell Biochem. 2005;96:221–34.
24. Peters R. Fluorescence microphotolysis to measure nucleocytoplasmic transport and intracellular mobility. Biochim Biophys Acta. 1986;864:305–59.
25. Leslie NR, Foti M. Non-genomic loss of PTEN function in cancer: not in my genes. Trends Pharmacol Sci. 2011;32:131–40.
26. Bassi C, Ho J, Srikumar T, Dowling RJ, Gorrini C, Miller SJ, Mak TW, Neel BG, Raught B, Stambolic V. Nuclear PTEN controls DNA repair and sensitivity to genotoxic stress. Science. 2013;341:395–9.
27. Trotman LC, Wang X, Alimonti A, Chen Z, Teruya-Feldstein J, Yang H, Pavletich NP, Carver BS, Cordon-Cardo C, Erdjument-Bromage H, Tempst P, Chi SG, Kim HJ, Misteli T, Jiang X, Pandolfi PP. Ubiquitination regulates PTEN nuclear import and tumor suppression. Cell. 2007;128:141–56.
28. Wang X, Trotman LC, Koppie T, Alimonti A, Chen Z, Gao Z, Wang J, Erdjument-Bromage H, Tempst P, Cordon-Cardo C, Pandolfi PP, Jiang X. NEDD4-1 is a proto-oncogenic ubiquitin ligase for PTEN. Cell. 2007;128:129–39.
29. Ciechanover A, Finley D, Varshavsky A. The ubiquitin-mediated proteolytic pathway and mechanisms of energy-dependent intracellular protein degradation. J Cell Biochem. 1984;24:27–53.
30. Cheung LW, Hennessy BT, Li J, Yu S, Myers AP, Djordjevic B, Lu Y, Stemke-Hale K, Dyer MD, Zhang F, Ju Z, Cantley LC, Scherer SE, Liang H, Lu KH, Broaddus RR, Mills GB. High frequency of PIK3R1 and PIK3R2 mutations in endometrial cancer elucidates a novel mechanism for regulation of PTEN protein stability. Cancer Discov. 2011;1:170–85.
31. Lee JT, Shan J, Zhong J, Li M, Zhou B, Zhou A, Parsons R, Gu W. RFP-mediated ubiquitination of PTEN modulates its effect on AKT activation. Cell Res. 2013;23:552–64.
32. Duerr EM, Rollbrocker B, Hayashi Y, Peters N, Meyer-Puttlitz B, Louis DN, Schramm J, WIESTLER OD, Parsons R, Eng C, von Deimling A. PTEN mutations in gliomas and glioneuronal tumors. Oncogene. 1998;16:2259–64.
33. Eng C. PTEN: one gene, many syndromes. Hum Mutat. 2003;22:183–98.
34. Salmena L, Carracedo A, Pandolfi PP. Tenets of PTEN tumor suppression. Cell. 2008;133:403–14.
35. Freeman DJ, Li AG, Wei G, Li HH, Kertesz N, Lesche R, Whale AD, Martinez-Diaz H, Rozengurt N, Cardiff RD, Liu X, Wu H. PTEN tumor suppressor regulates p53 protein levels and activity through phosphatase-dependent and -independent mechanisms. Cancer Cell. 2003;3:117–30.
36. Chen Z, Trotman LC, Shaffer D, Lin HK, Dotan ZA, Niki M, Koutcher JA, Scher HI, Ludwig T, Gerald W, Cordon-Cardo C, Pandolfi PP. Crucial role of p53-dependent cellular senescence in suppression of Pten-deficient tumorigenesis. Nature. 2005;436:725–30.
37. Yin Y, Shen WH. PTEN: a new guardian of the genome. Oncogene. 2008;27:5443–53.
38. Puc J, Keniry M, Li HS, Pandita TK, Choudhury AD, Memeo L, Mansukhani M, Murty VV, Gaciong Z, Meek SE, Piwnica-Worms H, Hibshoosh H, Parsons R. Lack of PTEN sequesters CHK1 and initiates genetic instability. Cancer Cell. 2005;7:193–204.

39. Alimonti A, Nardella C, Chen Z, Clohessy JG, Carracedo A, Trotman LC, Cheng K, Varmeh S, Kozma SC, Thomas G, Rosivatz E, Woscholski R, Cognetti F, Scher HI, Pandolfi PP. A novel type of cellular senescence that can be enhanced in mouse models and human tumor xenografts to suppress prostate tumorigenesis. J Clin Invest. 2010;120:681–93.
40. Sullivan M, Morgan DO. Finishing mitosis, one step at a time. Nat Rev Mol Cell Biol. 2007;8:894–903.
41. Canaani D. Application of the concept synthetic lethality toward anticancer therapy: A promise fulfilled? Cancer Lett. 2014;352:59–65.
42. Ashworth A. A synthetic lethal therapeutic approach: poly(ADP) ribose polymerase inhibitors for the treatment of cancers deficient in DNA double-strand break repair. J Clin Oncol. 2008;26:3785–90.
43. Mendes-Pereira AM, Martin SA, Brough R, Mccarthy A, Taylor JR, Kim JS, Waldman T, Lord CJ, Ashworth A. Synthetic lethal targeting of PTEN mutant cells with PARP inhibitors. EMBO Mol Med. 2009;1:315–22.
44. Blanco-Aparicio C, Renner O, Leal JF, Carnero A. PTEN, more than the AKT pathway. Carcinogenesis. 2007;28:1379–86.
45. Vivanco I, Palaskas N, Tran C, Finn SP, Getz G, Kennedy NJ, Jiao J, Rose J, Xie W, Loda M, Golub T, Mellinghoff IK, Davis RJ, Wu H, Sawyers CL. Identification of the JNK signaling pathway as a functional target of the tumor suppressor PTEN. Cancer Cell. 2007;11:555–69.
46. Hopkins BD, Fine B, Steinbach N, Dendy M, Rapp Z, Shaw J, Pappas K, Yu JS, Hodakoski C, Mense S, Klein J, Pegno S, Sulis ML, Goldstein H, Amendolara B, Lei L, Maurer M, Bruce J, Canoll P, Hibshoosh H, Parsons R. A secreted PTEN phosphatase that enters cells to alter signaling and survival. Science. 2013;341:399–402.
47. Putz U, Howitt J, Doan A, Goh CP, Low LH, Silke J, Tan SS. The tumor suppressor PTEN is exported in exosomes and has phosphatase activity in recipient cells. Sci Signal. 2012;5:ra70.

The PRL PTPs: Regulating Gene Expression to Reprogram the Cancer Cell

Cameron A. Rogers and Catherine J. Pallen

Abstract Since the discovery of the PRLs in the 1990s, a large body of work has established that PRL overexpression endows cells with malignant properties and in human cancers is associated with disease progression and poor outcome. How the PRLs exert these effects at the molecular level remains unknown despite an intensive search for PRL substrates, including amongst components of key cancer-associated receptor-proximal signaling pathways. We discuss evidence to support a developing theme that the main action of the PRLs is to alter programs of gene expression. By so doing, the PRLs drive changes in molecular networks and cellular architecture to transform cell properties and advance cancer. We propose that defining and functionally investigating the PRL "interactome" could inform the mechanistic basis of PRL actions by revealing effectors and targeting partners to facilitate substrate identification. We also highlight recent findings on the actions of the PRLs gained from the first few studies in whole animal models, on regulation of the PRLs, and on progress towards therapeutic targeting of PRL-overexpressing tumors.

C.A. Rogers
Department of Medicine, University of British Columbia, Vancouver, BC, Canada

Child and Family Research Institute, 3102-950 West 28th Ave., Vancouver, BC, Canada, V5Z 4H4

C.J. Pallen (✉)
Department of Pediatrics, University of British Columbia, Vancouver, BC, Canada

Child and Family Research Institute, 3102-950 West 28th Ave., Vancouver, BC, Canada, V5Z 4H4
e-mail: cpallen@mail.ubc.ca

B.G. Neel, N. Tonks (eds.), *Protein Tyrosine Phosphatases in Cancer*,
DOI 10.1007/978-1-4939-3649-6_10

The PRLs: Cancer-Associated PTPs

The up-regulated expression of the PRLs in tissues that are undergoing dramatic remodeling has been instrumental in their discovery and in propelling investigation into their cellular roles. The prototype PRL, PRL-1 (*PTP4A1*), was identified as an immediate-early gene in regenerating liver (**p**hosphatase of **r**egenerating **l**iver) [1, 2], followed by the identification of two closely related family members PRL-2 (*PTP4A2*) and PRL-3 (*PTP4A3*) [3–5]. Thereafter, PRL-3 was found to be highly overexpressed in colon cancer metastases relative to primary tumors and normal colon epithelia [6]. Numerous studies (reviewed in [7, 8]) have now confirmed the overabundance of PRL-3 in many types of cancer. Its expression is often elevated in association with the manifestation of increasingly malignant features of this disease such as metastasis and angiogenesis, and correlated with poor survival, thus nominating this molecule as a useful biomarker of cancer progression and poor outcome and as a prospective therapeutic target. PRL-1 and PRL-2 have less widespread cancer associations, but are also up-regulated in some tumor types.

The PRLs are members of the protein tyrosine phosphatase (PTP) superfamily, and their structure and phosphoamino acid specificity place them in the subfamily of dual specific phosphatases (DUSPs). Indeed, as shown in Table 1, the PRLs can dephosphorylate phosphotyrosine, -threonine, and -serine sites in proteins as well as, albeit in vitro but akin to the related phosphatase PTEN, a phosphoinositide. The PRLs are small (~22 kDa) proteins with a central PTP catalytic domain and a C-terminal tail region containing a polybasic sequence which precedes the final CAAX motif-unique among PTP superfamily members-that directs their prenylation. CAAX motif-dependent farnesylation of the PRLs is critical for their membrane localization, which together with an intact catalytic site is required for their actions in promoting cell proliferation, migration, invasion, colony and tumor formation, and metastasis. Extensive investigation of the molecular and cellular consequences of increasing or depleting PRL expression, particularly of PRL-3, in varied cell lines has revealed pleiotropic effects. Often these involve the modulation of molecules involved in regulating cell-cell and cell-matrix interactions, cytoskeletal rearrangement, epithelial-to-mesenchymal transition (EMT), and the cell cycle. These are described in previous reviews that provide comprehensive information on PRL structure, cell signaling, and tissue and cancer-specific expression [7, 9, 10]. Potential regulators and/or effectors of the PRLs include interacting proteins and substrates, and those identified to date are shown in Table 1. Still, neither a universal substrate nor a master pathway that coordinates PRL-regulated cellular changes has been identified for any of these phosphatases. Therefore, the functions of the PRLs in normal cells and how they exert their transformative cellular effects remain a mystery and a main focus of investigation.

Here we describe recent progress in PRL research, especially focusing on a growing body of evidence that speaks to a key role for up-regulated PRLs as drivers of altered gene expression. We propose that this action underlies PRL-induced broad-based changes in molecular networks to effect cellular reprogramming and tissue remodeling that promotes the progressive development of malignant cancer phenotypes, or for that matter, liver regeneration following hepatectomy.

Table 1 PRL-interacting proteins and substrates

Interactor	Substrate	PRL	Known or inferred effect of interaction	Ref.
α-tubulin		PRL-1	Localization of PRL-1 to mitotic spindles and centrosomes during mitosis	[59]
ATF-7	In vitro	PRL-1	Dephosphorylation of ATF-7	[66]
p115 RhoGAP		PRL-1	Mitigation of p115 RhoGAP binding and inhibition of MEKK1-promoting ERK1/2 activation	[67]
βGGT II		PRL-2	Unknown	[39, 68]
ARF1 & ARF5		PRL-3	Promotes ARF1 activity and increases integrin α5 recycling	[61]
COPZ1		PRL-3	Unknown	[61]
Ezrin[a]	Thr567	PRL-3	Dephosphorylation of Ezrin at Thr567	[69]
CDH22		PRL-3	Unknown	[17]
FKBP38		PRL-3	Increased degradation of PRL-3	[70]
HDAC4		PRL-3	Protects PRL-3 from proteasomal degradation	[47]
Integrin-α1		PRL-3	May act as a negative regulator of integrin-β1 dephosphorylation by PRL-3	[71]
Integrin-β1	Tyr783	PRL-3	Dephosphorylation of integrin-β1 Tyr783 to promote talin and kindlin interaction	[72]
Keratin 8	Ser73/431	PRL-3	Dephosphorylation of keratin 8 at Ser73/431 to promote intermediate filament reorganization	[73]
Nucleolin	Maybe	PRL-3	Potential dephosphorylation of nucleolin to promote its nuclear localization and nucleoli assembly	[74]
PI(4,5)P_2[b]	In vitro	PRL-3	Unknown	[75]
RAP1		PRL-3	RAP1 localization to the cytosol	[24]
Stathmin	Maybe	PRL-3	Might dephosphorylate stathmin	[26]
TRP32		PRL-3	Reduction and reactivation of oxidized PRL-3 in vitro	[32]

Several proteins and a phosphoinositide interact with a PRL, and some of these interactors are substrates or potential substrates. The known or presumed effect of the interaction is indicated

[a]Ezrin is the only protein that has not been shown to stably interact with a PRL

[b]PI(4,5)P_2 interacts with PRL-3 in silico

Drivers of Altered Gene Expression

There is a huge contrast between the highly consistent findings of altered cell properties, especially those of migration and invasion, and the widely divergent repertoire of molecular changes that are concomitantly induced upon the up- or down-regulation of PRL expression. This repertoire includes the increased activity of key enzymes within cell signaling pathways that are often hyperactivated in cancer cells, in accord with the major role of reversible protein-tyrosine phosphorylation in regulating these pathways and with the identity of the PRLs as PTPs. However, as described below, PRL-enhanced signaling molecule activation has, as yet, not been directly attributable to a specific catalytic target of the PRLs, but is linked to PRL-induced inhibition of the expression of negative regulators. In addition, the perturbed expression of many other molecules is associated with altered PRL levels. The broad spectrum of these changes is being increasingly revealed by recent global "-omic" profiling of PRL-mediated changes in cell proteins, transcripts and miRNAs, and points to a key role for the PRLs as regulators of gene expression.

PRL-Mediated Down-Regulation of Negative Regulators of Signaling

PRL-3 overexpression in numerous cell contexts can induce the activation of SRC, AKT, and ERK1/2; the altered phosphorylation of SRC effectors such as STAT3 or CAS and of the AKT effector GSK3β; and/or altered Rho GTPase activities. These effects are coincident with enhanced proliferation and/or migration, invasion, colony formation, tumor formation, and metastatic- and angiogenesis-promoting effects. PRL-1 overexpression has many of the same effects. The activation of at least some of these key signaling molecules is driven by PRL-3-mediated inhibition of negative regulatory signaling (Fig. 1): for example, by the reduced expression of CSK [11, 12], a kinase that catalyzes inhibitory phosphorylation of SRC; and the reduced expression of PTEN [12, 13], a lipid phosphatase that counterbalances the PI3K-mediated generation of AKT-activating phospholipid. PRL-2 overexpression can also induce PTEN down-regulation [14]. Most recently, the PRL-3-induced activation of EGFR and its downstream signaling in A431 epidermoid and MDA-MB-468 breast cancer cell lines was found to be due to down-regulated expression of PTP1B [15], an EGFR phosphatase that inhibits receptor tyrosine kinase activity. Despite the common action of reducing the expression of inhibitory signaling proteins, the mechanism of PRL-induced down-regulation is different in each instance. PRL-3-dependent reduction of CSK expression results from defective protein synthesis, attributed to slowed translation arising from enhanced phosphorylation and less effective functioning of eIF2α [16]. PRL-2 and -3-mediated PTEN down-regulation is likewise due to posttranscriptional regulation, and, at

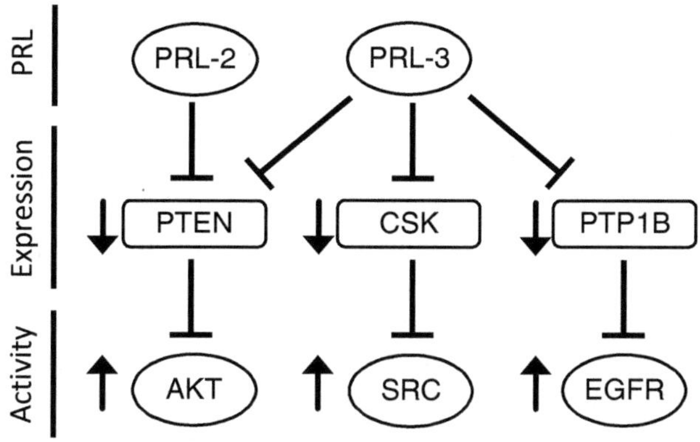

Fig. 1 PRLs downregulate negative regulators to activate signaling. Overexpression of PRL-2 or PRL-3 reduces the expression of some molecules that function as negative regulators of kinases, resulting in kinase activation

least with PRL-2, involves enhanced proteasomal degradation of PTEN [13, 14]. On the other hand, PRL-3-induced reduction of *PTP1B* mRNA indicates that this regulation occurs at the transcriptional level [15].

Regulators of Broad-Based Gene Expression

PRL-1 and PRL-3 overexpression also induce the up- and down-regulated expression of a spectrum of other proteins. Many of these are components of cell adhesive structures (focal adhesions, adherens junctions), such as paxillin, vinculin, integrin α2 and β3, E-cadherin, cadherin-like protein 22, and γ-catenin [13, 17–19]. They also include transcription factors such as c-fos and Snail [13, 18], matrix degrading metalloproteinases (MMPs) and the MMP inhibitor TIMP2 [20, 21], the calcium-activated potassium channel KCNN4 [22], growth factors and cytokines such as VEGF and IL-8 [23, 24], and others. Several lines of evidence suggest these are real changes with valid roles in mediating the actions of the PRLs in/during cancer progression. First, PRL-mediated alterations in protein levels appear to be, for the most part, dependent upon the catalytic activity of the PRLs as they are almost never observed upon expression of inactive PRL C104S mutants. Second, the reciprocal dysregulation of the expression of specific proteins often occurs upon PRL knockdown and overexpression. Third, PRL-3-dependent increases and decreases in specific protein levels in cultured cells can be functionally linked to the concomitant acquisition or abrogation of cancer cell proliferative, migratory, and/or invasive abilities. Finally, the association between PRL-3 overexpression in cultured cells and the consequently altered level of certain proteins is mirrored in corresponding types of tumor tissues, implying that this is a functional linkage that is relevant to the biology of the disease. For example, in accord with the PRL-3-induced down-regulation of integrin α2 in the A2780 ovarian cancer cell line and of PTP1B in the invasive MDA-MB-468 breast cancer cell line, there are significant inverse correlations between *PRL-3* and integrin α2 (*ITGA2*) transcript levels in ovarian cancer samples and between *PRL-3* and *PTP1B* mRNA expression in primary tumors of breast cancer patients that developed brain metastases [15, 18].

Clearly, PRL overexpression (although less investigated with PRL-2 than with PRL-3 and PRL-1) has profound effects on protein expression. Understanding the full extent and repertoire of these changes requires unbiased and broadly directed analyses. Towards this, comparative proteomic analysis has been employed to identify PRL-3-dependent alterations in protein expression. Ten up-regulated and seven downregulated proteins were identified in PRL-3 overexpressing Lovo colon cancer cells compared to vector control cells [25], whereas 24 down-regulated and 15 up-regulated proteins were identified in PRL-3 stable knockdown SW480 colon cancer cells compared to vector control cells [26]. No proteins in common were found between the two studies, although the expression of three different members of the annexin family of calcium- and phospholipid-binding proteins was up-regulated in both studies. The altered expression of several proteins associated with nuclear functions of proteasomal degradation, transcription, and histone demethylation was noted in the PRL-knockdown cells. In a follow-up study, depleting PRL-3 expression in colon cancer cells was confirmed to respectively down- and up-regulate two histone demethylases known as Jumonji domain-containing protein 1B (JMJD1B) and 2B (JMJD2B) and to alter histone H3 lysine 9 (H3K9) methylation [27]. Moreover, nuclear PRL-3 was detectable in colorectal tumors, with highest levels of nuclear PRL-3 detected in advanced stage cancers, and JMJD1B/2B levels were altered in the advanced stage tumors in accord with the observed effects of PRL-3 on their expression in cultured cells. Histone methylation is associated with both transcriptional activation and repression, and represents a novel mechanism by which PRL-3 may regulate gene expression.

In another approach, gene expression profiling of PRL-3-expressing and vector control HCT116 colon cancer cells revealed that nearly 2700 genes exhibited two-fold or greater differences in expression between the two cell types, with about equal numbers of genes being up-regulated and down-regulated [24]. Although the full dataset was not made available, some of the changes that were reported were indicative of altered NFκB, TLR, ERBB, WNT, and JAK-STAT signaling pathways, while others were in the expression of genes encoding players in cell adhesion and cell cycle. The PRL-3-dependent up-regulation of some genes linked to NFκB signaling was verified by qPCR, and NFκB signaling was involved in the PRL-3-induced up-regulation of the PRL-3 interactor RAP1 [24]. Interestingly, RAP1 (repressor activator protein 1) is a multifunctional telomere-binding protein and regulator of telomere-dependent and -independent gene expression [28, 29], suggesting that it could mediate a panoply or subset of the PRL-3-induced changes in gene expression.

A recent and highly informative analysis and alignment of the PRL-1-induced changes in mRNA and protein expression in HEK293 cells [30] revealed concordance between the mRNA and protein data as to the repressive or enhancing effect of PRL-1 on target expression, and found comparable numbers of up-regulated and down-regulated targets. There was an overlapping and significant fold-enrichment of differentially expressed mRNAs and proteins functionally linked to cell proliferation and tumorigenesis, RNA recognition and splicing, and cytoskeletal remodeling. The latter was consistent with numerous PRL-1 induced changes that were detected in the expression of gene products that encode RHO signaling pathway molecules, including down-regulated RhoA expression. Among the many novel targets of PRL-1 that were identified with high confidence and significantly altered expression were three that appeared in both the protein and microarray datasets: up-regulated filamin A (FLNA) and heterogeneous nuclear ribonucleoprotein H2 (HNRNPH2) and down-regulated peroxiredoxin 2 (PRDX2). FLNA is an actin binding and scaffold protein with roles in cytoskeletal remodeling and HNRNPH2 is an RNA-binding protein that regulates alternative splicing. PRDX2 is an antioxidant enzyme that functions in maintaining cellular redox balance, an interesting target in view of the altered ROS signaling in cancer cells and the redox regulation of PRL activity and expression [31, 32]. In addition, the most highly up-regulated transcript produces a secreted extracellular matrix protein (probably explaining its absence from the up-regulated protein dataset) called SPARC or osteonectin that has many roles, including the regulation of cell adhesive interactions, cytoskeletal rearrangement, and the cell cycle, and controversial roles in tumor progression, notably cancer metastasis to the bone [33].

The dataset of PRL-1-regulated targets is a rich resource. It can be mined for links to the PRL-1 literature, as well as, given the similar outcomes of overexpressing PRL-1 or PRL-3 on a variety of cancer-related cell processes, to compare and contrast with knowledge about PRL-3. It will provide valuable comparisons with existing and future PRL-regulated protein, mRNA and also miRNA (particularly specific cancer-type linked miRNA) datasets (see section "MicroRNAs as PRL Effectors"). Even a superficial examination leads to some potentially interesting correlations. The PRL-1 up-regulated target HNRNPH2 (also known as HNRNP H' or DSEF) is also up-regulated in pancreatic adenocarcinoma and in gastric and hepatocellular carcinomas [34], with the latter two cancers being the only types found in an independent survey to have up-regulated *PRL-1* transcript [35]. The expression of several proteins is similarly altered upon PRL-1 and PRL-3 overexpression, such as septin 11 and macrophage-capping protein, whereas others, such as inorganic pyrophosphatase (PPA1) and ATP-dependent RNA helicase DDX3X, are disparately regulated [25, 30]. Stathmin (and stathmin-2) appears in the PRL-1-regulated dataset as a protein(s) with significantly down-regulated expression, whereas another proteomic analysis [26] identified stathmin down-regulation to be an effect of silencing PRL-3 expression in SW480 colon cancer cells. At first glance, this suggests that PRL-1 and PRL-3 have opposing effects on stathmin level. However, this might be reconciled by the surprising finding that PRL-1 overexpression in HEK293 cells significantly down-regulates *PRL-3* mRNA [30]; thus the reduced stathmin could be a consequence of PRL-1-mediated PRL-3 down-regulation.

MicroRNAs as PRL Effectors

Links reported between PRL-3 and specific microRNAs (miRNAs)-small noncoding RNAs that posttranscriptionally inhibit gene expression-suggest a novel mechanism by which the PRLs can control the cellular levels of particular proteins. Zhang et al. (2012) identified miRNAs with increased expression in PRL-3-expressing Lovo colorectal cancer cells compared to control transfected cells [12]. They focused on three miRNAs with known cancer-related associations, miR-17, -19a, and -21, and confirmed that these were significantly up-regulated in several cell lines upon PRL-3 overexpression. Supporting the notion that these miRNAs function as PRL-3 effectors, their introduction into Lovo cells mimicked the effects of overexpressing PRL-3; reducing PTEN expression; and enhancing proliferation and invasion.

How might PRL-3 up-regulate miRNA expression? A CSK-SRC-STAT3 pathway is proposed to control this, based on PRL-3-induced down-regulation of CSK, the consequently reduced inhibitory phosphorylation of SRC at Tyr527 leading to SRC activation, and the SRC-mediated enhanced phosphorylation of STAT3 at Tyr705 [12]. In support of a key role for STAT3, its siRNA-mediated silencing reduced the expression of miR-17, 19a, and -21 in PRL-3-expressing Lovo cells. Additionally, a converse correlation was extended to human malignancy, where elevated PRL-3 expression in colon cancer liver and lymph node metastases was paralleled by enhanced phospho-STAT3 and miR-17, 19a, and -21 expression in these samples over primary tumor and normal colon epithelia [12]. While uniting several reported PRL-3 signaling targets (CSK/SRC, STAT3, PTEN) into a common pathway, the functional connection between PRL-3 and CSK, the most upstream effector identified in this pathway, remains obscure.

Moving Forward with PRL-omics

It appears that the PRLs, certainly PRL-1 and PRL-3, are master regulators of the cellular proteome and transcriptome. Precisely how they effect such regulation is unknown, as although the PRLs are phosphatases by nature, their substrates have remained elusive. Moving forward, the identification of PRL interacting proteins (interactomes), also using unbiased approaches, may reveal unexpected functional partners, especially those with involvement in the broad regulation of gene expression. In this respect the immunoprecipitation of tagged PRL-3 or PRL-2 followed by mass spectrometry identified, with high confidence, 38 candidate PRL-3 interactors and 5 candidate PRL-2 interactors in human cells [36]. The PRL-3 interactor with the top confidence score was GCN1L1, a protein about which little is known, but that is a component of the CDK8 subcomplex within the larger Mediator complex that regulates the expression of all genes [37, 38]. As far as we know, the PRL-3–GCN1L1 interaction has not been independently validated or functionally investigated, but in light of the apparent roles of the PRLs as regulators of gene expression, this may now be an interesting avenue to be explored. Other interactors also warrant further investigation, and overall could provide new insights towards molecular targeting of the PRLs and substrate identification. The development of "PRL-omics" and informatic integration of different datasets should potentiate discovery of the substrate(s) and mechanism(s) whereby the PRLs exert their pervasive effects.

PRL Transgenic and Knockout Animals

Until recently, there has been a dearth of reported studies employing altered PRL expression in animal models. As more groups generate PRL knockout and transgenic animals, and cross these with the many available animal models of cancer, it is expected that new information will be obtained about the physiological and pathophysiological actions of these phosphatases. The following sections highlight findings from the few first such investigations in whole animal systems.

PRLs in Mouse Models of Cancer

Two studies have investigated the consequences of increasing PRL-2 and ablating PRL-3 expression in murine models of breast and colon cancer. Whereas mice expressing an MMTV-directed PRL-2 transgene in mammary epithelium did not develop tumors, in double transgenic mice (MMTV-ERBB2 × MMTV-PRL-2), the enhanced expression of PRL-2 slightly accelerated the development of tumors induced by oncogenic ERBB2 [39]. Half of the double transgenic animals evidenced tumors at 132 days compared to 144 days in the MMTV-ERBB2 mice. This effect was dependent on catalytically intact PRL-2 expression as no reduction in tumor latency occurred in mice expressing ERBB2 together with the inactive PRL-2 C101S mutant.

The converse situation was explored in a mouse model of colitis-associated colon cancer induced by treatment of wild-type and PRL-3 knockout mice with azoxymethane (AOM) and dextran sodium sulfate (DSS) [40]. Both types of mice developed equivalent numbers of similarly sized tumors in the distal colon at 12 weeks after beginning treatment, but by 16 weeks, twice as many tumors had formed in the distal colon of wild-type mice compared to PRL-$3^{-/-}$ animals. Further, the wild-type mice recapitulated findings in human colorectal cancer (CRC), wherein PRL-3 expression was enhanced in tumors relative to normal colon epithelium.

These animal studies indicate that increased expression of PRL-2 may be insufficient to induce breast tumor formation unless a tumorigenic signal is supplied such as activated ERBB2, and that the lack of PRL-3 reduces but does not preclude colon tumor formation when a tumorigenic signal is supplied such as AOM/DSS treatment. Consistent with these in vivo findings, Hardy et al. (2010) also found that heterologous expression of PRL-2 in tumor-forming breast cancer cell lines enhances colony-forming ability, but PRL-2 lacks the ability to promote this property in non-transformed breast epithelial cell lines [39]. Together, these studies indicate that the PRLs act as cellular modifiers that potentiate tumorigenesis and/or tumor survival and progression. The cellular actions reported for the PRLs are diverse, and as yet no consensus substrate or signaling pathway for any PRL has been identified. In the murine cancer models, molecular alterations in the tumors were associated with PRL-2 expression and PRL-3 depletion. ERBB2/PRL-2-expressing breast tumors contained elevated phospho-ERK1/2 and a normal level of phospho-AKT compared with ERBB2- or ERBB2/PRL-2 C101S-expressing tumors, although a global or unbiased approach to identify molecular changes was not undertaken/described [39]. Reverse phase protein array was used to profile and compare colon tumors with up-regulated PRL-3 from wild-type mice and colon tumors that developed in PRL-3-null mice. Despite heterogeneous alterations in protein and phosphoprotein expression profiles, a consistently up-regulated expression of IGF-1Rβ and c-MYC in the PRL-3-null tumors was detected and validated by Western blotting [40]. Further analysis of these and additional in vivo tumor models are promising avenues for understanding the actions of PRLs in the potentiation and progression of cancer.

PRLs in Mouse Growth and Development

New roles of the PRLs in growth regulation have also been revealed by the generation of transgenic and knockout mice. PRL-3$^{-/-}$ mice are viable, although fewer than expected male mice were present at weaning and, by 6 weeks of age, these mice showed a 10 % reduction in body mass [40, 41]. In contrast, both genders of PRL-2-null mice were 20 % smaller at birth than wild type animals and maintained their reduced size through adulthood [14]. The lower birthweight of PRL-2$^{-/-}$ mice seems to be due to defective placental development. PRL-2$^{-/-}$ placental abnormalities, evident by at least E12.5 and pronounced by E16.5, include dramatically smaller decidual and spongiotrophoblast layers, attributed to the reduced proliferation of spongiotrophoblast and glycogen cells that normally express PRL-2 [14]. The PRL-2 mutant placenta also has a somewhat smaller labyrinth area but with normal trophoblasts and fetal vasculature. Nutrient transport, measured by short-term ^{14}C-mannitol transfer from the maternal bloodstream to the embryo, is reduced by about one-third. Overall, in the absence of PRL-2, placental insufficiency results in fetal growth retardation. The basis for the inability of PRL-2$^{-/-}$ animals to postnatally attain normal body mass remains unclear, as no specific organ abnormalities or altered blood metabolites, electrolytes, or growth hormone levels were detected.

In the PRL-2-deficient placenta, AKT phosphorylation and that of its target GSK3β, a regulator of glycogen synthesis, are reduced concomitant with an increased level of PTEN. SRC and ERK1/2 phosphorylation are normal, and PI3K activity is not perturbed. Evidence supports a role for placental PRL-2 in promoting proteasomal degradation of PTEN [14], possibly mediated through vinculin down-regulation, which disturbs the adherens junction-localized PTEN/MAGI2/β-catenin complex, thereby enhancing PTEN susceptibility to degradation. PTEN and vinculin down-regulation also occurred upon PRL-3 overexpression in DLD-1 human colorectal cancer cells. In that study [13], the PI3K inhibitor LY294002 countered the AKT activation and GSK3β inhibition induced upon PRL-3 overexpression, suggesting that PI3K activity was required for PRL-3 signaling effects. The precise mechanisms by which PRL-2 and PRL-3 regulate vinculin down-regulation and PTEN degradation remain to be elucidated. Importantly, the PRL-2 knockout study provides in vivo evidence that these are not only shared, but genuine signaling actions of these PRLs, albeit the distinct phenotype of the PRL-3 knockout mouse suggesting that this is likely operational in distinct cell types or tissues in a PRL-2- or -3-specific manner.

Unexpected Actions of Drosophila *PRL-1 in Normal and Abnormal Growth and Development*

Striking evidence of growth inhibitory and tumor suppressor actions of a PRL has come from a recent study of transgenic *Drosophila* [42]. The protein product of the single PRL gene in *Drosophila*, dPRL-1, shares 58 % amino acid identity with human PRL-1, including an identical P-loop sequence and a divergent "WPD" loop containing only the central Phe and a potential proton donating C-terminal Asp corresponding to those in the conserved human/mouse WPFDD sequence. The catalytic activity of dPRL-1 has not been investigated. Ubiquitous overexpression of dPRL-1 prevented larval growth early in development and resulted in lethality in the absence of evident morphological abnormalities. Tissue-specific expression of dPRL-1 in the posterior wing or dorsal compartments reduced wing size and surface area, while eye-specific expression reduced the size of the eye and head capsule. As in the PRL-2 knockout mouse placenta [14], but in this case involving *overexpression* of a PRL, dPRL-1-induced developmental restrictions were associated with reduced cell proliferation.

In addition to inhibiting growth and development in a normal genetic background, overexpression of dPRL-1 markedly and differentially modified the outcomes of dorsal compartment overexpression of Src and Ras, suppressing Src-induced adult lethality and accelerating Ras-induced pupal lethality [42]. The action of Src in inducing dramatic wing disc overgrowth and developmental disorganization is believed to account for its lethal effects. dPRL-1 appeared to slow cell proliferation to prevent Src-mediated tissue overgrowth, and, because Src phosphorylation was not altered by dPRL-1, to do so by inhibiting Src signaling rather than directly affecting Src catalytic activity.

Prenylation-dependent membrane localization of PRLs is requisite for their ability to promote cancer-like cell properties. The CAAX prenylation motif of dPRL-1 differs from that of all human/mouse PRLs in that it lacks the second cysteine residue (the human PRL CCI/VQ/M tail sequence is CSVQ in *Drosophila*), indicating that it can undergo classical CAAX processing but not alternative posttranslational processing as was recently shown to generate dual prenylated and palmitoylated PRL-3 [43]. Like mammalian PRLs, dPRL-1 is membrane localized, and a novel polarized localization was noted at the apical edge of epithelial cells, together with tissue- and development-dependent cytoplasmic localization [42]. Deletion of the CAAX box motif eliminates the apical occupancy of dPRL-1 but not membrane localization, suggesting that this could be maintained through the well-conserved polybasic region. CAAX box deletion also abolished the growth inhibitory action of dPRL-1. Interestingly, the CAAX prenylation motif of dPRL-1 was dispensable for inhibition of Src-induced growth, indicating that dPRL-1 may utilize distinct mechanisms to inhibit normal and Src-induced cell proliferation and growth. In this context, it is interesting to note that farnesyltransferase activity is essential for cell proliferation during early mouse embryogenesis, but not for adult homeostasis nor for K-RAS or H-RAS-dependent tumor initiation in mice [44]. The latter may reflect alternate compensatory modes of membrane association of H-RAS that maintain its oncogenicity. By analogy, but in keeping with the unexpected growth and tumor inhibitory effects of dPRL-1, prenylation-dependent direction of dPRL-1 may allow it to access asymmetrically localized apical cell targets to inhibit embryonic proliferation whereas prenylation-independent general membrane association is sufficient for dPRL-1 to access and down-regulate the abnormal signaling exerted by Src transgene expression.

Regulation of PRLs

Cellular Levels of the PRLs

PRL expression can undergo dramatic up-regulation, as observed with PRL-3 during cancer progression and for PRL-1 during liver regeneration. A variety of factors have been identified that operate at many levels to control PRL protein expression (Table 2). Some recently discovered regulators and their potential to modulate PRL-3 and PRL-1 levels in specific cancer types are described below.

Several transcription factors can up-regulate PRL-3, although a comprehensive picture of the integrated transcriptional control of PRL-3 expression in normal cells and its alteration in disease settings is lacking. Park et al. (2013) have linked PRL-3 up-regulation to the actions of an oncogene associated with a subset of acute myeloid leukemias (AML) [45]. About one-third of AML patients have a gain-of-function mutation involving internal tandem duplication of the fms-like tyrosine kinase 3 (FLT3-ITD) that represents a driver mutation for leukemia initiation [46]. The FLT3-ITD mutation is associated with higher PRL-3 expression in AML patients, and FLT3-ITD-activated SRC-STAT5 signaling induces *PRL-3* transcription and enhanced PRL-3 protein expression in AML cell lines [45, 47]. Moreover, STAT5 binds to a consensus sequence found in the upstream promoter region of *PRL-3*, suggesting that it is a direct transcriptional regulator of PRL-3 [45].

Table 2 Regulators of PRL expression

Regulator	Type	Mode of regulation	Effect on PRL levels	Target	Ref.
EGR-1	TF	Transcription	Up	PRL-1	[76]
MEF2C (VEGF)	TF	Transcription	Up	PRL-3	[77]
p53	TF	Transcription	Up	PRL-1 & PRL-3	[78, 79]
p73	TF	Transcription	Up	PRL-3	[80]
SMAD3 (TGFβ)	TF	Transcription	Down	PRL-3	[81]
SNAIL	TF	Transcription	Up	PRL-3	[82]
STAT5	TF	Transcription	Down	PRL-3	[45]
mIR-339-5p	miR	Post-transcription	Down	PRL-1	[49]
mIR-495, -551a	miR	Post-transcription	Down	PRL-3	[48]
PCBP1	RNA BP	Translation	Down	PRL-3	[83]
FKBP38	Isom	Protein degradation	Down	PRL-3	[70]
HGF, IGF-1	GF/Cyt	Transcription	Up	PRL-3	[60]
IL-6, -15, -21	GF/Cyt	Transcription	Up	PRL-3	[60]

PRL expression can be regulated by several types of factors, including transcription factors (TF), growth factors and cytokines (GF/Cyt), an RNA binding protein (RNA BP), miRNA (miR) and a peptidyl prolyl *cis/trans* isomerase (Isom). Growth factors known to act upstream of the transcription factor responsible for regulating PRL expression are bracketed in the left column

Acting posttranscriptionally, tumor-specific expression of miRNAs can determine PRL expression. Overall, PRL-3 is elevated and miR-495 and -551a levels are reduced in gastric cancer cell lines and tissues compared with corresponding non-malignant samples [48]. Consistent with the predicted ability of these miRNAs to bind to the 3′UTR of *PRL-3*, their introduction into gastric cancer cells reduced the expression of PRL-3, as well as inhibiting cell migration and invasion. In another study [49], PRL-1 was identified as a target of miR-339-5p. The introduction of miR-339-5p into SW620 colon cancer cells inhibited PRL-1 expression and ERK1/2 phosphorylation and reduced proliferation, migration, invasion, and colony forming ability. It also slowed SW620 cell tumor formation in mice. Consistent with these tumor suppressor-like actions, miR-339-5p expression is down-regulated in colon [49] and breast [50] tumor samples compared to normal tissue. However, in these tumors miR-339-5p may target mRNAs that encode oncoproteins other than PRL-1 as a concomitant up-regulation of PRL-1 was not observed. In fact, miR-339-5p reduction was accompanied by a small decrease in *PRL-1* expression in the tumors among these same colon tissue samples. A survey of small numbers of several tumor types and adjacent normal tissue likewise detected no tumor-specific increase in *PRL-1* transcript level in colon tissues and significantly reduced expression of *PRL-1* transcript in breast carcinoma versus normal samples [35]. Whether PRL-1 up-regulation is associated with particular tumor types and/or their metastases is still unclear. The above survey found increased *PRL-1* transcript in all gastric and hepatocellular carcinomas (HCC) [35], and elevated expression of PRL-1 in HCC has also been reported by Lu et al. [51], indicating that these cancers may be of interest for the further investigation of *PRL-1*-interacting miRNAs, including miR-339-5p.

Dysregulated miRNA expression is being profiled in an ever growing number of cancer types (as for example, those collated in the searchable miR2Disease [http://www.mir2disease.org] and PhenoMiR [http://mips.helmholtz-muenchen.de/phenomir] databases), and both oncogenic and tumor suppressor activities of individual miRNAs, miRNA clusters, and families have been revealed. Not surprisingly, a more complex picture is materializing wherein miRNA regulatory networks are implicated in virtually all aspects of cancer development and progression, including tumor angiogenesis and metastasis [52, 53]. The initial evidence that the PRLs are regulated by and, as described above in section "MicroRNAs as PRL Effectors," are regulators of miRNAs indicates that the future elucidation of such relationships will provide important insights into the mechanisms by which the PRLs, particularly PRL-3, function as mediators and markers of cancer progression. It is also a promising avenue for the development of miRNA-based PRL(-3)-directed therapies.

Regulation of PRL Activity and Function

The cellular effects of the PRLs are widely abolished upon mutation of the conserved Cys residue in the active site or upon deletion or mutation of the CAAX motif, indicating that their functions are dependent upon catalytic activity and prenylation-dependent membrane association. The target(s) towards which their phosphatase activities are directed remain obscure, notwithstanding unique reports of some specific protein substrates (Table 1). Here, we highlight some recent findings that describe the modification of a PRL by phosphorylation, a novel protein interaction that is related to the potential redox regulation of PRL activity, and some new findings pertaining to the subcellular localization, and thus the regulation of the site of action of the PRLs.

Although protein phosphorylation is well documented to control the activity of several PTPs, this modification has not been readily detected for the PRLs, suggesting that it may not be a stable or predominant modification involved in their regulation. Fiordalisi et al. (2013) demonstrate that SRC can mediate the phosphorylation of human PRL-3 on Tyr53, and to a lesser extent, on Tyr126. However, phosphotyrosyl-PRL-3 in cells cannot be detected unless SRC is introduced and/or cells are treated with the PTP inhibitor pervanadate. Mutation of Tyr53 compromised the ability of overexpressed PRL-3 to promote cell migration and invasion, indicating that an intact Tyr is important for these actions of PRL-3 [54]. Whether this is indeed due to Tyr53 phosphorylation or to another undetermined role of this residue, for example in the catalytic activity of PRL-3, is unclear.

The PRLs are subject to redox-regulated disulfide bond formation between the active site Cys and Cys46 in PRL-2 or Cys49 in PRL-1 and -3 [55, 56]. Oxidation of PRL-1 inhibits its phosphatase activity and can be reversed by reducing reagent. Protein redox status is controlled by two major cellular reducing systems, the thioredoxin (TRX) and the glutathione (GSH) systems. Oxidized PRL-1 is detected in vivo and its redox status can be regulated by the GSH system [31]. There is also evidence that the thioredoxin system can regulate reversible oxidation of PRL-1 and -3. The thioredoxin related protein 32 (TRP32) binds to PRL-3 and shows specificity for reducing the PRLs [32]. It thus represents a protein partner and potential cellular (re-)activator of PRLs. Interestingly, about 85 % of cellular TRP32 is associated with the 26 S proteasome, and this interaction is mediated by the same domain in TRP32 that mediates interaction with PRL-3 [57].

The association of the PRLs with intracellular membranes is mediated by CAAX box-directed prenylation (farnesylation) at their C-terminus [58]. Endogenous PRLs have been localized to cell membranes, cytoplasm, and the nucleus, and in some cells undergo cell cycle-dependent shuttling between the cytoplasm and the nucleus [59, 60]. Other subcellular compartments where the PRLs are found include the Golgi region and early and recycling endosomes [58, 61]. The direction of the PRLs to different subcellular locales may be additionally regulated by alternative post-prenylation processing, whereby the dual cysteines in the CAAX (CCAX) motif permit subsequent modification by palmitoylation [43]. PRL-3 can undergo farnesylation and palmitoylation [43], and, as shown with brain Cdc42, this might enrich its localization at particular membrane types, and/or determine its susceptibility to re-localization. A new motif within the PRLs (but absent in dPRL-1), the MXXE sequence, located between the active site and the polybasic region, regulates the interaction of PRL-3 with the Golgi and perinuclear-localized protein ARF1 and the localization of PRL-3 to the Golgi [61]. Thus, while the PRL-3 CAAX motif is required for membrane association, the MXXE sequence mediates a protein–protein interaction that specifies a particular subcellular compartmentalization of PRL-3. ARF1 itself contains a MXXE sequence that is involved in targeting it to the Golgi, likely via binding to the ER-Golgi SNARE, membrin [62]. Whether PRLs can likewise associate directly with membrin or if further binding specificity is conferred by the PRL- and ARF-dissimilar central XX residues (conserved among the vertebrate PRLs as Lys-Tyr) in the motif is unknown. Interestingly, the central tyrosine (Tyr126 in human PRL-3) in the PRL MXXE sequence is a minor site for SRC-mediated PRL-3 phosphorylation under certain conditions [54], raising the possibility that MXXE-dependent protein interactions of the PRLs could be regulated by phosphorylation. Functionally, the interaction of PRL-3 with ARF1 promotes ARF1 activation to accelerate the recycling of integrin α5. This would promote the turnover of integrin-dependent cell adhesions, and is proposed to underlie the PRL-3-induced enhancement of cell migration [61].

Advances Towards Targeting PRL-Expressing Tumors

PRL-3 (and possibly other PRLs) is an attractive target for therapeutic intervention to combat cancer. A number of compounds have been identified as inhibitors of PRL catalytic activity and/or PRL-promoted cell properties that are associated with malignancy. Table 3 shows some different compounds that have been established to target the PRLs or their actions. PRL inhibitors are also discussed in Chapter 10 of this book. Here we highlight studies with implications for the PRL-directed application of therapeutics that are currently in clinical trial or already approved for clinical use, as well as some novel approaches that are being pioneered to target the PRLs in cancer.

Table 3 Inhibitors of PRL signaling

Compound	Molecular target	Effect on PRL signaling	Cellular effect	Ref.
ABT-869[a] (Linifanib)	FLT3-ITD	– Inhibition of FLT3-ITD activation of STAT pathway – Decreased expression of PRL-3	– Increased apoptosis in PRL-3 overexpressing MOLM-14 cells	[47]
Curcumin[a]	Unknown	– p53 independent decreased production of *PRL-3* mRNA in B16BL6 cells in vitro – Decreases SRC/STAT3 activation in B16BL6 cells in vitro	– Decreases B16BL6 tumor growth and invasion	[84]
EGFR inhibitors (Erlotinib and Cetuximab)[b]	EGFR	– Inhibition of EGFR activity, a PRL-3 downstream oncogene	– Inhibition of PRL-induced tumor growth in vivo and colony formation in vitro	[15]
Emodin	PRL-3	– Decreases production of *PRL-3* mRNA in SGC-7901 cells	– Activation of apoptosis in SGC-7901 cells in vitro	[85]
		– Inhibition of PRL-3 phosphatase activity in vitro	– Inhibition of migration in DLD-1 cells overexpressing PRL-3 in vitro	[86]
Ginkgetin and sciadopitysin	PRL-3	– Inhibition of PRL-3 phosphatase activity in vitro		[87]
Halichondramide	Unknown	– Decreases mRNA and protein levels of PRL-3, p53, p85, p110, N-cadherin, MMP-2 and -9 in PC-3 cells in vitro	– Inhibition of cell migration and invasion	[88]
		– Upregulation of E-cadherin in PC-3 cells in vitro	– Decreases MMP-2 and -9 secretion into the extracellular environment	
Pentamidine[b]	PTP1B, MKP-1, PRL-1,-2,-3	– Inhibition of PRL phosphatase activity in vitro	– Growth inhibition of WM9, DU149, HEY, C4-2, SW480 and A549 in culture – Inhibition of WM9 cell tumor formation in nude mice	[89]
Rhodanine based compounds	PRL-3	– Increases phosphorylation of Ezrin and keratin 8	– Inhibition of cell proliferation in GCIY, AZ521, Kato III, SH10, MKN7 and MKN74 cell in vitro	[90]
		– Inhibition of PRL phosphatase activity in vitro	– Increases apoptosis in GCIY, AZ521, SH10, MKN74 and C212 cell in vitro	[91]
			– Inhibition of cell migration and invasion in DLD-1, A375P, MIA PaCa-2, Du 145, MDA-MB-231, AsPc-1 cells in vitro – Decreases expression of EMT markers (Snail and E-cadherin)	[92]
SAHA[b]	HDAC4	– Inhibits HDAC4 interaction with PRL-3 – Increases proteasomal degradation of PRL-3	– Increased apoptosis in PRL-3 overexpressing MOLM-14 cells	[47]
Thienopyridone	PRL-1,-2,-3	– Promotes cleavage of CAS, FAK, PARP, and caspase-8 in HeLa cells	– Inhibition of anchorage-independent RKO and HT-29 cell growth in vitro – Inhibition of HUVEC migration but not proliferation in vitro	[93]

[a]Currently in phase 2 clinical trials
[b]Approved for clinical use

In a recent study by Al-Aidaroos et al. (2013), PRL-3 overexpressing cancer cells were found to become "addicted" to active EGFR signaling and consequently hypersensitive to EGFR-specific inhibitors such as cetuximab, erlotinib, or AG-1478. This relationship is supported by clinical evidence correlating higher PRL-3 expression in colorectal cancer patients with a favorable response and better outcome to treatment with cetuximab, an intriguing twist on the widely reported correlation between elevated PRL-3 expression in human tumors and poor survival. Also, higher PRL-3 expression in non-small-cell lung carcinoma (NSCLC) cell lines correlated with sensitivity to growth inhibition by gefitinib, an EGFR tyrosine kinase inhibitor (TKI) approved for treatment of advanced NSCLC patients. Thus in these and perhaps other types of cancer, high PRL-3 expression may be a predictive marker for sensitivity to anti-EGFR therapies.

Another type of linkage between receptor tyrosine kinase (RTK) signaling and PRL-3 expression influenced the AML cell response to combined treatment with suberoylanilide hydroxamic acid (SAHA; a histone deacetylase [HDAC] inhibitor) and ABT-869 (also known as Linifanib; a FLT3 inhibitor). AML cell lines homozygous or heterozygous for the mutant FLT3-ITD RTK and primary FLT3-ITD-positive leukemia cells from patients exhibited synergistic cell death and inhibition of colony formation responses to combined SAHA/ABT-869 treatment, whereas a synergistic response did not occur in AML cell lines or primary leukemic cells lacking this FLT3 mutation [47]. It was determined that FLT3-ITD signaling up-regulates PRL-3 and promotes PRL-3-mediated expression of the anti-apoptotic protein MCL-1. The synergistic effect of ABT-869 and SAHA on AML cells is due to their independently exerted down-regulation of PRL-3; ABT-869 acting by inhibiting FLT3-ITD signaling and so decreasing PRL-3 expression, and SAHA acting via inhibition of the PRL-3 binding partner HDAC4 to promote proteasomal degradation of PRL-3 [47]. Thus, this combination therapy may be of benefit for the ~30 % of AML patients having the FLT3-ITD mutation.

A number of monoclonal antibody (mAb) treatments are in use for diseases including, but not limited to, rheumatoid arthritis, multiple sclerosis, and various types of cancer. Traditionally, the targets of mAb therapy are proteins with complete or partial extracellular exposure because intracellular proteins were believed to be inaccessible to the antibody. Guo et al. (2008) queried this and found that treating mice with a PRL-1- or PRL-3-specific mAb prevented cancer cells expressing the PRL member that was recognized by the mAb from forming tumors in an in vivo metastasis assay [63]. They also demonstrated that cancer cells grown in culture can uptake antibodies without the need for permeabilization. The anti-tumor mechanism is not known; however, inhibition of metastatic tumor formation was at least partly dependent on host B cells [64]. The therapeutic potential of mAbs is not exclusive to solid tumors; Park et al. (2013) used a PRL-3 antibody to effectively treat a mouse model of leukemia (AML) [45]. Another intervention currently under study is the utilization of DNA vaccination targeted to PRL-3. This method involves producing an immune response to a specific antigen by introducing a modified version of the gene encoding the target antigen. Lv et al. (2013) vaccinated mice with pVAX-Igκ-PRL-3 plasmid (K-P3) and injected the mice with

PRL-3 overexpressing D2F2 breast cancer cells. There was a significant increase in survival of K-P3 vaccinated mice compared to control animals [65]. When the DNA vaccination was used as therapy, rather than for prevention, there was a significant inhibition of tumor growth rate. The DNA vaccination-induced immune response was dependent on T-helper type 1 cells as well as on PRL-3-targeted cytotoxic T lymphocytes. These novel approaches to PRL-associated cancer intervention show promise for future clinical use.

Conclusions

Recent studies reveal a diversity of changes in cell protein, transcript, and even miRNA expression that occurs as a consequence of PRL overexpression, highlighting the challenge of elucidating the molecular mechanisms of action of these phosphatases. Nevertheless, these findings indicate a role for the PRLs as drivers of altered gene expression that reconfigures cells to acquire a progressively malignant phenotype. Transgenic and knockout animal models are providing insight into the roles of the PRLs in development and cancer. They present opportunities to investigate how PRL ablation or overexpression in specific cancer settings can influence tumor biology, disease progression, and therapeutic outcomes, especially as PRL expression in specific oncogenic settings is showing promise in the stratification of therapeutic treatment and effectiveness [66–93].

References

1. Mohn KL, Laz TM, Hsu JC, Melby AE, Bravo R, et al. The immediate-early growth response in regenerating liver and insulin-stimulated H-35 cells: comparison with serum-stimulated 3T3 cells and identification of 41 novel immediate-early genes. Mol Cell Biol. 1991;11:381–90.
2. Diamond RH, Cressman DE, Laz TM, Abrams CS, Taub R. PRL-1, a unique nuclear protein tyrosine phosphatase, affects cell growth. Mol Cell Biol. 1994;14:3752–62.
3. Montagna M, Serova O, Sylla BS, Feunteun J, Lenoir GM. A 100-kb physical and transcriptional map around the EDH17B2 gene: identification of three novel genes and a pseudogene of a human homologue of the rat PRL-1 tyrosine phosphatase. Hum Genet. 1995;96:532–8.
4. Cates CA, Michael RL, Stayrook KR, Harvey KA, Burke YD, et al. Prenylation of oncogenic human PTP(CAAX) protein tyrosine phosphatases. Cancer Lett. 1996;110:49–55.
5. Zeng Q, Hong W, Tan YH. Mouse PRL-2 and PRL-3, two potentially prenylated protein tyrosine phosphatases homologous to PRL-1. Biochem Biophys Res Commun. 1998;244:421–7.
6. Saha S, Bardelli A, Buckhaults P, Velculescu VE, Rago C, et al. A phosphatase associated with metastasis of colorectal cancer. Science. 2001;294:1343–6.
7. Bessette DC, Qiu D, Pallen CJ. PRL PTPs: mediators and markers of cancer progression. Cancer Metastasis Rev. 2008;27:231–52.
8. Guzinska-Ustymowicz K, Pryczynicz A. PRL-3, an emerging marker of carcinogenesis, is strongly associated with poor prognosis. Anticancer Agents Med Chem. 2011;11:99–108.
9. Al-Aidaroos AQ, Zeng Q. PRL-3 phosphatase and cancer metastasis. J Cell Biochem. 2010;111:1087–98.

10. Rios P, Li X, Kohn M. Molecular mechanisms of the PRL phosphatases. FEBS J. 2013; 280:505–24.
11. Liang F, Liang J, Wang WQ, Sun JP, Udho E, et al. PRL3 promotes cell invasion and proliferation by down-regulation of Csk leading to Src activation. J Biol Chem. 2007;282:5413–9.
12. Zhang J, Xiao Z, Lai D, Sun J, He C, et al. miR-21, miR-17 and miR-19a induced by phosphatase of regenerating liver-3 promote the proliferation and metastasis of colon cancer. Br J Cancer. 2012;107:352–9.
13. Wang H, Quah SY, Dong JM, Manser E, Tang JP, et al. PRL-3 down-regulates PTEN expression and signals through PI3K to promote epithelial-mesenchymal transition. Cancer Res. 2007;67:2922–6.
14. Dong Y, Zhang L, Zhang S, Bai Y, Chen H, et al. Phosphatase of regenerating liver 2 (PRL2) is essential for placental development by down-regulating PTEN (Phosphatase and Tensin Homologue Deleted on Chromosome 10) and activating Akt protein. J Biol Chem. 2012; 287:32172–9.
15. Al-Aidaroos AQ, Yuen HF, Guo K, Zhang SD, Chung TH, et al. Metastasis-associated PRL-3 induces EGFR activation and addiction in cancer cells. J Clin Invest. 2013;123:3459–71.
16. Liang F, Luo Y, Dong Y, Walls CD, Liang J, et al. Translational control of C-terminal Src kinase (Csk) expression by PRL3 phosphatase. J Biol Chem. 2008;283:10339–46.
17. Liu Y, Zhou J, Chen J, Gao W, Le Y, et al. PRL-3 promotes epithelial mesenchymal transition by regulating cadherin directly. Cancer Biol Ther. 2009;8:1352–9.
18. Liu H, Al-aidaroos AQ, Wang H, Guo K, Li J, et al. PRL-3 suppresses c-Fos and integrin alpha2 expression in ovarian cancer cells. BMC Cancer. 2013;13:80.
19. Nakashima M, Lazo JS. Phosphatase of regenerating liver-1 promotes cell migration and invasion and regulates filamentous actin dynamics. J Pharmacol Exp Ther. 2010;334:627–33.
20. Peng L, Xing X, Li W, Qu L, Meng L, et al. PRL-3 promotes the motility, invasion, and metastasis of LoVo colon cancer cells through PRL-3-integrin beta1-ERK1/2 and-MMP2 signaling. Mol Cancer. 2009;8:110.
21. Lee SK, Han YM, Yun J, Lee CW, Shin DS, et al. Phosphatase of regenerating liver-3 promotes migration and invasion by upregulating matrix metalloproteinases-7 in human colorectal cancer cells. Int J Cancer. 2012;131:E190–203.
22. Lai W, Chen S, Wu H, Guan Y, Liu L, et al. PRL-3 promotes the proliferation of LoVo cells via the upregulation of KCNN4 channels. Oncol Rep. 2011;26:909–17.
23. Ming J, Jiang Y, Jiang G, Zheng H. Phosphatase of regenerating liver-3 induces angiogenesis by increasing extracellular signal-regulated kinase phosphorylation in endometrial adenocarcinoma. Pathobiology. 2013;81:1–7.
24. Lian S, Meng L, Liu C, Xing X, Song Q, et al. PRL-3 activates NF-kappaB signaling pathway by interacting with RAP1. Biochem Biophys Res Commun. 2013;430:196–201.
25. Chu ZH, Liu L, Zheng CX, Lai W, Li SF, et al. Proteomic analysis identifies translationally controlled tumor protein as a mediator of phosphatase of regenerating liver-3-promoted proliferation, migration and invasion in human colon cancer cells. Chin Med J (Engl). 2011;124: 3778–85.
26. Zheng P, Liu YX, Chen L, Liu XH, Xiao ZQ, et al. Stathmin, a new target of PRL-3 identified by proteomic methods, plays a key role in progression and metastasis of colorectal cancer. J Proteome Res. 2010;9:4897–905.
27. Liu Y, Zheng P, Liu Y, Ji T, Liu X, et al. An epigenetic role for PRL-3 as a regulator of H3K9 methylation in colorectal cancer. Gut. 2013;62:571–81.
28. Zhao Y, McIntosh KB, Rudra D, Schawalder S, Shore D, et al. Fine-structure analysis of ribosomal protein gene transcription. Mol Cell Biol. 2006;26:4853–62.
29. Platt JM, Ryvkin P, Wanat JJ, Donahue G, Ricketts MD, et al. Rap1 relocalization contributes to the chromatin-mediated gene expression profile and pace of cell senescence. Genes Dev. 2013;27:1406–20.
30. Dumaual CM, Steere BA, Walls CD, Wang M, Zhang ZY, et al. Integrated analysis of global mRNA and protein expression data in HEK293 cells overexpressing PRL-1. PLoS One. 2013;8:e72977.

31. Yu L, Kelly U, Ebright JN, Malek G, Saloupis P, et al. Oxidative stress-induced expression and modulation of Phosphatase of Regenerating Liver-1 (PRL-1) in mammalian retina. Biochim Biophys Acta. 2007;1773:1473–82.
32. Ishii T, Funato Y, Miki H. Thioredoxin-related protein 32 (TRP32) specifically reduces oxidized phosphatase of regenerating liver (PRL). J Biol Chem. 2013;288:7263–70.
33. Ribeiro N, Sousa SR, Brekken RA, Monteiro FJ. Role of SPARC in bone remodeling and cancer-related bone metastasis. J Cell Biochem. 2014;115(1):17–26.
34. Honore B, Baandrup U, Vorum H. Heterogeneous nuclear ribonucleoproteins F and H/H′ show differential expression in normal and selected cancer tissues. Exp Cell Res. 2004;294: 199–209.
35. Dumaual CM, Sandusky GE, Soo HW, Werner SR, Crowell PL, et al. Tissue-specific alterations of PRL-1 and PRL-2 expression in cancer. Am J Transl Res. 2012;4:83–101.
36. Ewing RM, Chu P, Elisma F, Li H, Taylor P, et al. Large-scale mapping of human protein-protein interactions by mass spectrometry. Mol Syst Biol. 2007;3:89.
37. Malik S, Roeder RG. Dynamic regulation of pol II transcription by the mammalian Mediator complex. Trends Biochem Sci. 2005;30:256–63.
38. Knuesel MT, Meyer KD, Donner AJ, Espinosa JM, Taatjes DJ. The human CDK8 subcomplex is a histone kinase that requires Med12 for activity and can function independently of mediator. Mol Cell Biol. 2009;29:650–61.
39. Hardy S, Wong NN, Muller WJ, Park M, Tremblay ML. Overexpression of the protein tyrosine phosphatase PRL-2 correlates with breast tumor formation and progression. Cancer Res. 2010;70:8959–67.
40. Zimmerman MW, Homanics GE, Lazo JS. Targeted deletion of the metastasis-associated phosphatase Ptp4a3 (PRL-3) suppresses murine colon cancer. PLoS One. 2013;8:e58300.
41. Yan H, Kong D, Ge X, Gao X, Han X. Generation of conditional knockout alleles for PRL-3. J Biomed Res. 2011;25:438–43.
42. Pagarigan KT, Bunn BW, Goodchild J, Rahe TK, Weis JF, et al. Drosophila PRL-1 is a growth inhibitor that counteracts the function of the Src oncogene. PLoS One. 2013;8:e61084.
43. Nishimura A, Linder ME. Identification of a novel prenyl and palmitoyl modification at the CaaX motif of Cdc42 that regulates RhoGDI binding. Mol Cell Biol. 2013;33:1417–29.
44. Mijimolle N, Velasco J, Dubus P, Guerra C, Weinbaum CA, et al. Protein farnesyltransferase in embryogenesis, adult homeostasis, and tumor development. Cancer Cell. 2005;7:313–24.
45. Park JE, Yuen HF, Zhou JB, Al-Aidaroos AQ, Guo K, et al. Oncogenic roles of PRL-3 in FLT3-ITD induced acute myeloid leukaemia. EMBO Mol Med. 2013;5:1351–66.
46. Leung AY, Man CH, Kwong YL. FLT3 inhibition: a moving and evolving target in acute myeloid leukaemia. Leukemia. 2013;27:260–8.
47. Zhou J, Bi C, Chng WJ, Cheong LL, Liu SC, et al. PRL-3, a metastasis associated tyrosine phosphatase, is involved in FLT3-ITD signaling and implicated in anti-AML therapy. PLoS One. 2011;6:e19798.
48. Li Z, Cao Y, Jie Z, Liu Y, Li Y, et al. miR-495 and miR-551a inhibit the migration and invasion of human gastric cancer cells by directly interacting with PRL-3. Cancer Lett. 2012;323: 41–7.
49. Zhou C, Liu G, Wang L, Lu Y, Yuan L, et al. MiR-339-5p regulates the growth, colony formation and metastasis of colorectal cancer cells by targeting PRL-1. PLoS One. 2013;8:e63142.
50. Wu ZS, Wu Q, Wang CQ, Wang XN, Wang Y, et al. MiR-339-5p inhibits breast cancer cell migration and invasion in vitro and may be a potential biomarker for breast cancer prognosis. BMC Cancer. 2010;10:542.
51. Lu JW, Chang JG, Yeh KT, Chen RM, Tsai JJ, et al. Increased expression of PRL-1 protein correlates with shortened patient survival in human hepatocellular carcinoma. Clin Transl Oncol. 2012;14:287–93.
52. Matejuk A, Collet G, Nadim M, Grillon C, Kieda C. MicroRNAs and tumor vasculature normalization: impact on anti-tumor immune response. Arch Immunol Ther Exp (Warsz). 2013; 61:285–99.
53. Pencheva N, Tavazoie SF. Control of metastatic progression by microRNA regulatory networks. Nat Cell Biol. 2013;15:546–54.

54. Fiordalisi JJ, Dewar BJ, Graves LM, Madigan JP, Cox AD. Src-mediated phosphorylation of the tyrosine phosphatase PRL-3 is required for PRL-3 promotion of Rho activation, motility and invasion. PLoS One. 2013;8:e64309.
55. Kozlov G, Cheng J, Ziomek E, Banville D, Gehring K, et al. Structural insights into molecular function of the metastasis-associated phosphatase PRL-3. J Biol Chem. 2004;279:11882–9.
56. Sun JP, Wang WQ, Yang H, Liu S, Liang F, et al. Structure and biochemical properties of PRL-1, a phosphatase implicated in cell growth, differentiation, and tumor invasion. Biochemistry. 2005;44:12009–21.
57. Andersen KM, Madsen L, Prag S, Johnsen AH, Semple CA, et al. Thioredoxin Txnl1/TRP32 is a redox-active cofactor of the 26 S proteasome. J Biol Chem. 2009;284:15246–54.
58. Zeng Q, Si X, Horstmann H, Xu Y, Hong W, et al. Prenylation-dependent association of protein-tyrosine phosphatases PRL-1, -2, and -3 with the plasma membrane and the early endosome. J Biol Chem. 2000;275:21444–52.
59. Wang J, Kirby CE, Herbst R. The tyrosine phosphatase PRL-1 localizes to the endoplasmic reticulum and the mitotic spindle and is required for normal mitosis. J Biol Chem. 2002;277:46659–68.
60. Fagerli UM, Holt RU, Holien T, Vaatsveen TK, Zhan F, et al. Overexpression and involvement in migration by the metastasis-associated phosphatase PRL-3 in human myeloma cells. Blood. 2008;111:806–15.
61. Krndija D, Munzberg C, Maass U, Hafner M, Adler G, et al. The phosphatase of regenerating liver 3 (PRL-3) promotes cell migration through Arf-activity-dependent stimulation of integrin alpha5 recycling. J Cell Sci. 2012;125:3883–92.
62. Honda A, Al-Awar OS, Hay JC, Donaldson JG. Targeting of Arf-1 to the early Golgi by membrin, an ER-Golgi SNARE. J Cell Biol. 2005;168:1039–51.
63. Guo K, Tang JP, Tan CP, Wang H, Zeng Q. Monoclonal antibodies target intracellular PRL phosphatases to inhibit cancer metastases in mice. Cancer Biol Ther. 2008;7:750–7.
64. Guo K, Li J, Tang JP, Tan CP, Hong CW, et al. Targeting intracellular oncoproteins with antibody therapy or vaccination. Sci Transl Med. 2011;3:99ra85.
65. Lv J, Liu C, Huang H, Meng L, Jiang B, et al. Suppression of breast tumor growth by DNA vaccination against phosphatase of regenerating liver 3. Gene Ther. 2013;20:834–45.
66. Peters CS, Liang X, Li S, Kannan S, Peng Y, et al. ATF-7, a novel bZIP protein, interacts with the PRL-1 protein-tyrosine phosphatase. J Biol Chem. 2001;276:13718–26.
67. Bai Y, Luo Y, Liu S, Zhang L, Shen K, et al. PRL-1 protein promotes ERK1/2 and RhoA protein activation through a non-canonical interaction with the Src homology 3 domain of p115 Rho GTPase-activating protein. J Biol Chem. 2011;286:42316–24.
68. Si X, Zeng Q, Ng CH, Hong W, Pallen CJ. Interaction of farnesylated PRL-2, a protein-tyrosine phosphatase, with the beta-subunit of geranylgeranyltransferase II. J Biol Chem. 2001;276:32875–82.
69. Forte E, Orsatti L, Talamo F, Barbato G, De Francesco R, et al. Ezrin is a specific and direct target of protein tyrosine phosphatase PRL-3. Biochim Biophys Acta. 2008;1783:334–44.
70. Choi MS, Min SH, Jung H, Lee JD, Lee TH, et al. The essential role of FKBP38 in regulating phosphatase of regenerating liver 3 (PRL-3) protein stability. Biochem Biophys Res Commun. 2011;406:305–9.
71. Peng L, Jin G, Wang L, Guo J, Meng L, et al. Identification of integrin alpha1 as an interacting protein of protein tyrosine phosphatase PRL-3. Biochem Biophys Res Commun. 2006;342:179–83.
72. Tian W, Qu L, Meng L, Liu C, Wu J, et al. Phosphatase of regenerating liver-3 directly interacts with integrin beta1 and regulates its phosphorylation at tyrosine 783. BMC Biochem. 2012;13:22.
73. Mizuuchi E, Semba S, Kodama Y, Yokozaki H. Down-modulation of keratin 8 phosphorylation levels by PRL-3 contributes to colorectal carcinoma progression. Int J Cancer. 2009;124:1802–10.
74. Semba S, Mizuuchi E, Yokozaki H. Requirement of phosphatase of regenerating liver-3 for the nucleolar localization of nucleolin during the progression of colorectal carcinoma. Cancer Sci. 2010;101:2254–61.

75. McParland V, Varsano G, Li X, Thornton J, Baby J, et al. The metastasis-promoting phosphatase PRL-3 shows activity toward phosphoinositides. Biochemistry. 2011;50:7579–90.
76. Peng Y, Du K, Ramirez S, Diamond RH, Taub R. Mitogenic up-regulation of the PRL-1 protein-tyrosine phosphatase gene by Egr-1. Egr-1 activation is an early event in liver regeneration. J Biol Chem. 1999;274:4513–20.
77. Xu J, Cao S, Wang L, Xu R, Chen G, et al. VEGF promotes the transcription of the human PRL-3 gene in HUVEC through transcription factor MEF2C. PLoS One. 2011;6:e27165.
78. Basak S, Jacobs SB, Krieg AJ, Pathak N, Zeng Q, et al. The metastasis-associated gene Prl-3 is a p53 target involved in cell-cycle regulation. Mol Cell. 2008;30:303–14.
79. Min SH, Kim DM, Heo YS, Kim YI, Kim HM, et al. New p53 target, phosphatase of regenerating liver 1 (PRL-1) downregulates p53. Oncogene. 2009;28:545–54.
80. Fontemaggi G, Kela I, Amariglio N, Rechavi G, Krishnamurthy J, et al. Identification of direct p73 target genes combining DNA microarray and chromatin immunoprecipitation analyses. J Biol Chem. 2002;277:43359–68.
81. Jiang Y, Liu XQ, Rajput A, Geng L, Ongchin M, et al. Phosphatase PRL-3 is a direct regulatory target of TGFbeta in colon cancer metastasis. Cancer Res. 2011;71:234–44.
82. Zheng P, Meng HM, Gao WZ, Chen L, Liu XH, et al. Snail as a key regulator of PRL-3 gene in colorectal cancer. Cancer Biol Ther. 2011;12:742–9.
83. Wang H, Vardy LA, Tan CP, Loo JM, Guo K, et al. PCBP1 suppresses the translation of metastasis-associated PRL-3 phosphatase. Cancer Cell. 2010;18:52–62.
84. Wang L, Shen Y, Song R, Sun Y, Xu J, et al. An anticancer effect of curcumin mediated by down-regulating phosphatase of regenerating liver-3 expression on highly metastatic melanoma cells. Mol Pharmacol. 2009;76:1238–45.
85. Sun ZH, Bu P. Downregulation of phosphatase of regenerating liver-3 is involved in the inhibition of proliferation and apoptosis induced by emodin in the SGC-7901 human gastric carcinoma cell line. Exp Ther Med. 2012;3:1077–81.
86. Han YM, Lee SK, Jeong DG, Ryu SE, Han DC, et al. Emodin inhibits migration and invasion of DLD-1 (PRL-3) cells via inhibition of PRL-3 phosphatase activity. Bioorg Med Chem Lett. 2012;22:323–6.
87. Choi SK, Oh HM, Lee SK, Jeong DG, Ryu SE, et al. Biflavonoids inhibited phosphatase of regenerating liver-3 (PRL-3). Nat Prod Res. 2006;20:341–6.
88. Shin Y, Kim GD, Jeon JE, Shin J, Lee SK. Antimetastatic effect of halichondramide, a trisoxazole macrolide from the marine sponge Chondrosia corticata, on human prostate cancer cells via modulation of epithelial-to-mesenchymal transition. Mar Drugs. 2013;11:2472–85.
89. Pathak MK, Dhawan D, Lindner DJ, Borden EC, Farver C, et al. Pentamidine is an inhibitor of PRL phosphatases with anticancer activity. Mol Cancer Ther. 2002;1:1255–64.
90. Min G, Lee SK, Kim HN, Han YM, Lee RH, et al. Rhodanine-based PRL-3 inhibitors blocked the migration and invasion of metastatic cancer cells. Bioorg Med Chem Lett. 2013;23:3769–74.
91. Ooki A, Yamashita K, Kikuchi S, Sakuramoto S, Katada N, et al. Therapeutic potential of PRL-3 targeting and clinical significance of PRL-3 genomic amplification in gastric cancer. BMC Cancer. 2011;11:122.
92. Ahn JH, Kim SJ, Park WS, Cho SY, Ha JD, et al. Synthesis and biological evaluation of rhodanine derivatives as PRL-3 inhibitors. Bioorg Med Chem Lett. 2006;16:2996–9.
93. Daouti S, Li WH, Qian H, Huang KS, Holmgren J, et al. A selective phosphatase of regenerating liver phosphatase inhibitor suppresses tumor cell anchorage-independent growth by a novel mechanism involving p130Cas cleavage. Cancer Res. 2008;68:1162–9.

Cdc25 Family Phosphatases in Cancer

Brian Gabrielli and Andrew Burgess

Abstract Loss of normal cell cycle control leading to unregulated proliferation is a hallmark of cancer. In addition, the inactivation of normal responses to stresses such as DNA damage, known as checkpoints controls, also contributes to the initiation and progression of cancer by increasing the level of genomic instability in the cancer genome which is responsible of the ability of cancers to adapt to new environments and conditions as it progresses beyond its normal physiological niche. Dysregulation of range of cell cycle and checkpoint control mechanisms have been identified in cancers. Understanding the defective mechanisms provides an improved understanding of the initiation and progression of disease and also provides an opportunity to either prevent cancer initiation or provide selective targets to treat the existing cancer. Here we will focus on a family of dual specificity Ser/Thr and Tyr phosphatases, the Cdc25 family, that have critical cell cycle and checkpoint control functions, as well as cell cycle independent roles. We will present a review of the current understanding of the normal roles and regulation of this family, the defects in this regulation identified in cancers, and the prospects for using these phosphatases as anticancer targets.

Keywords Cdc25 • Phosphatase • cdk • Cyclin • Auto-amplification loop • Cell cycle • Checkpoint • DNA damage • ATM/ATR • Arsenate reductases • $SCF^{\beta TrCP}$ • Cancer • Oncogene

---

B. Gabrielli (✉)
The University of Queensland Diamantina Institute, Translational Research Institute, Brisbane, QLD, Australia
e-mail: brianG@uq.edu.au

A. Burgess
Garvan Institute of Medical Research, Darlinghurst, NSW, Australia

Faculty of Medicine, St. Vincent's Clinical School, UNSW, Sydney, NSW, Australia
e-mail: a.burgess@garvan.org.au

B.G. Neel, N. Tonks (eds.), *Protein Tyrosine Phosphatases in Cancer*,
DOI 10.1007/978-1-4939-3649-6_11

Cdc25 Family Phosphatases Control Cell Cycle Progression

Cell cycle progression is controlled by a family of protein kinases, the Cyclin dependent kinases (CDKs). In yeast, invertebrates, amphibians and humans, the prototypic member and major driver of entry into mitosis is Cyclin dependent kinase 1 (Cdk1, also known as Cdc2). This was shown to be a component of the protein factor originally defined as "maturation promoting factor" or MPF in amphibian systems. Biochemical purification of this factor from Xenopus eggs revealed this to be a complex of Cdk1 and Cyclin B. The Cyclin subunit was identified in invertebrates based on its periodic accumulation and destruction corresponding to entry and exit from mitosis. Cyclin B/Cdk1 is a major driver of mitosis in all eukaryotes. Subsequently, a number of different Cdk/Cyclin complexes were identified as regulators of progression into and through specific cell cycle phases. Individual Cdks associate with a restricted repertoire of Cyclins at specific stages of the cell cycle, and the activity of discrete Cyclin/Cdks complexes controls progression through the cell cycle phases. Cell cycle progression from G1 into S phase requires the sequential activation of Cyclin D/Cdk4/6, and Cyclin E/Cdk2, although there is a degree of redundancy as Cyclin E/Cdk2 can perform both functions in cells that lack Cyclin D/Cdk4 activity [1]. The Cyclin D and E/Cdk complexes phosphorylate and inactivate Rb, allowing release and full activation of the Rb-bound E2F transcription factor, which is required for the expression of an array of genes essential for progression into and through S phase [2]. Cyclin E/Cdk2 and Cyclin A/Cdk2 also have critical roles in the initiation and maintenance of DNA replication [3–5].

A complex network controls progression through G2 phase and into mitosis. At a basic level, entry into mitosis is controlled through the activation of Cyclin B/Cdk1, which is regulated by factors that directly activate the complex, and down-regulation of inhibitors. Positive regulation of Cyclin B/Cdk1 is by the Cdc25 family of dual specificity phosphatases, and inactivation of negative mitotic regulators such as Wee1. The G2 phase Cyclin A/Cdk2 complex also has a role in regulating the timing of Cyclin B/Cdk1 activation and thus entry into mitosis [6, 7]. Although the mechanism by which this occurs is not fully understood, Cyclin A/Cdk2 has been shown to regulate Cyclin B levels [8], and to negatively regulate Wee1 function [7, 9]. Cdk1/Cyclin A also has a mitotic function although its exact role is poorly understood [4, 10] (Fig. 1).

The Cdk/Cyclin complexes are regulated at multiple levels including cell cycle dependent expression and stability of the regulatory Cyclin subunits, expression of Cdk inhibitory subunits, and phosphorylation of both Cdk and Cyclin subunits. The regulation of the Cdk1/Cyclin B complex is typical of the regulation of a majority of the Cyclin/Cdk complexes. The regulatory Cyclin B subunit is expressed during S and G2 phase where it binds to the stably expressed Cdk1 subunit. This complex is phosphorylated on the Cdk subunit at three sites; the activating Thr161 sites, which is in the T loop, a commonly phosphorylated conserved feature of protein kinases, by the Cdk activating kinase (CAK) [11]; and at the inhibitory Thr14 and Tyr15 sites in the glycine rich loop phosphorylated by Wee1 and Myt1 kinases, that

Fig. 1 Cell cycle progression is driven by the ordered activation of specific Cyclin/Cdk complexes

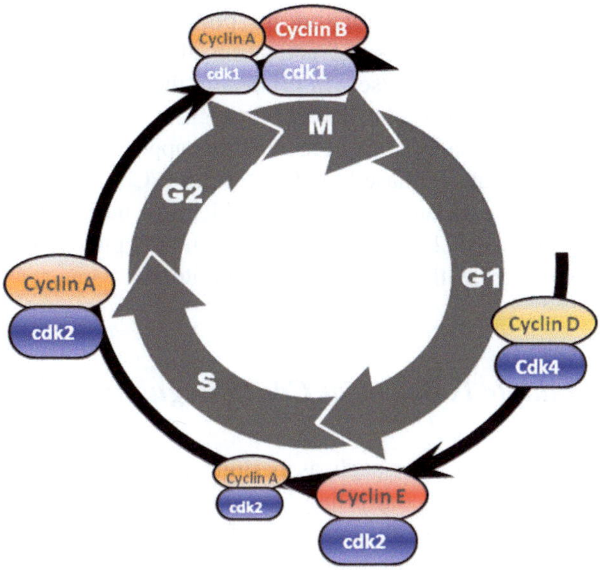

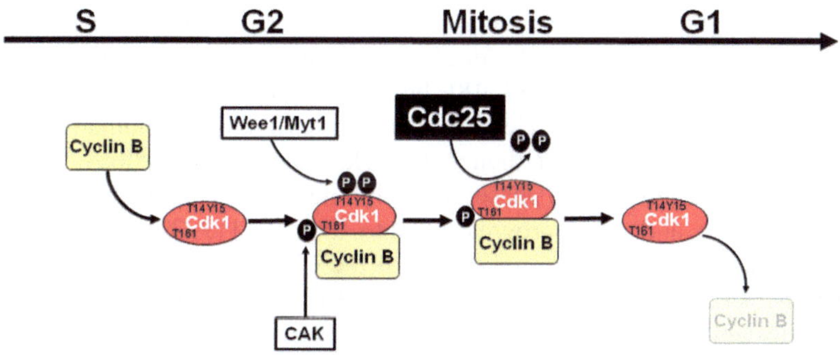

Fig. 2 Cyclin B/Cdk1 activation is typical of many Cyclin/Cdks. Cyclin B is expressed in S and G2 phase where it forms a complex with Cdk1. This is immediately phosphorylated on the activating site Thr161 by CAK, and the inhibitory Thr14 and Tyr15 sites by the Wee1/Myt1 kinases. The latently active kinase is activated at G2/M by the dual specificity Cdc25 phosphatases to drive entry into mitosis. At mitotic exit, the Cyclin B subunit is destroyed, inactivating the kinase

reduces substrate access to the catalytic site [12]. This maintains the complex in a latently active state, requiring only the removal of the inhibitory phosphorylations for activity. These inhibitory modifications are removed by the action of the Cdc25 family of dual specificity phosphatases. Activation of Cdc25 results in activation of Cyclin B/Cdk1 and drives entry into mitosis (Fig. 2). At exit from mitosis, the Cyclin B subunit is destroyed by the anaphase promoting complex/cyclosome (APC/C) inactivating complex and promoting mitotic exit. The role of Cdc25 in controlling mitotic entry is demonstrated by the ability of recombinant human Cdc25C to drive entry into mitosis in Xenopus egg extracts [13].

In addition to regulating the prototypic Cyclin B/Cdk1 complex, this general mechanism controls the activation of other Cyclin/Cdk complexes that regulate other cell cycle phases. In G1 phase, there is evidence that Cyclin D/Cdk4 can be regulated in a similar manner through the phosphorylation of the analogous Thr17 residue in mammalian cells, and this appears to be dephosphorylated prior to progression into S phase [14, 15]. Cyclin E/Cdk2, which regulates progression into S phase, is regulated by an almost identical mechanism to Cyclin B/Cdk1, as is the Cyclin A/Cdk2 complex that has S and G2 phase roles, and Cyclin A/Cdk1, which cooperates with Cyclin B/Cdk1 in mitosis.

Multiple Isoforms of Cdc25 Regulate Cell Cycle Progression

Lower eukaryotes such as fission yeast have a single Cdc25 species, which is responsible for controlling entry into mitosis [16]. In higher eukaryotes, there are three Cdc25 paralogues, Cdc25A, B, and C. All three possess a conserved C-terminal catalytic domain but are highly divergent in their N-terminal regions. One reason for the divergence of the Cdc25 family in higher eukaryotes is the requirement for them to perform more cell cycle roles than solely regulating entry into mitosis. In yeast, the Wee1-Cdc25 dependent regulation of Cdk1 activity is also associated with cell size [17, 18], but there is little evidence for this role in higher eukaryotes. The cell cycle dependent expression of the human Cdc25 isoforms defines to some extent their cell cycle dependent functions. Cdc25A is expressed prior to entry into S phase, where it functions to activate Cyclin E and A/Cdk2 complexes and drive progression into S phase [19–21]. Cdc25A protein appears to persist through the remainder of the cell cycle and has also a role in Cyclin B/Cdk1 activation and entry into mitosis [22]. Cdc25B is expressed in late S phase, destabilized at exit from mitosis [23], and is required for entry into mitosis [23–27]. Cdc25C is expressed and stable throughout the cell cycle, but is only activated in mitosis [28, 29] (Fig. 3). There is also evidence suggesting that Cdc25C plays unique roles in S phase and is required for timely DNA replication [30]. Biochemically, both Cdc25A and Cdc25B are capable of binding and activating Cyclin/Cdk2 as well as Cyclin/Cdk1 complexes, whereas Cdc25C weakly bound Cyclin B/Cdk1 in its unphosphorylated state, but showed a more than 20-fold increase in binding and activity when mitotically phosphorylated [28, 31]. There is a small fraction of Cdc25C activated in normal late G2, most likely by the small centrosomal or cytoplasmic pools of Cyclin B/Cdk1 activated by Cdc25B [23, 32, 33]. The function of this activated Cdc25C appears to ensure rapid activation of the total pool of Cyclin B/Cdk1 during mitosis.

A number of splice variants of each Cdc25 isoform have been identified. There are two Cdc25A (A1, A2) splice variants, and five variants for Cdc25B (1–5) and Cdc25C (1–5) that were shown to encode detectable proteins [34–38]. However, there are several additional predicted splice variants annotated in the NCBI sequence database, suggesting that there may be even more variants of Cdc25 than previously thought.

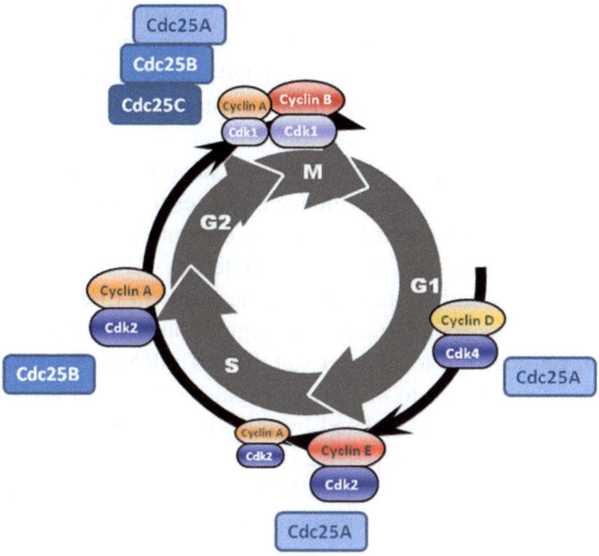

Fig. 3 The cell cycle phase specific Cyclin/Cdks are regulated by specific Cdc25 isoforms

In support of this, several new N-terminal truncated splice variants of Cdc25B have been characterized [39]. Likewise, there is evidence that for additional Cdc25C variants with Cdc25C6 commonly overexpressed in myelodysplastic syndrome [40].

Cdc25A and Cdc25B: Oncogenic Drivers of Cell Cycle Progression?

Regulations of Cdc25 Expression

Cdc25A is the only Cdc25 isoform that is absolutely required for embryonic development in mice, although it can be readily compensated for by Cdc25B, but not Cdc25C, in adult tissue [41–43]. At the transcriptional level, both Cdc25A and Cdc25B, but not Cdc25C, are targets for the proto-oncogene c-myc [44]. Cdc25A has also been shown to be a direct target of E2F-1 [45], while the oncogenic transcription factor Forkhead box protein M1 (FoxM1) can upregulate Cdc25B transcription [46]. Conversely transcriptional repression is a common control mechanism for Cdc25. Most frequently this is achieved by activation of the classical tumor suppressor gene p53. Direct binding of p53 to the promoter region of Cdc25A [47, 48], Cdc25B [49], and Cdc25C [50–53] results in the inhibition of Cdc25 expression. Given that loss of p53 function is a common event in cancer, this lack of p53 mediated repression is likely to contribute to the increased reliance on other mechanisms to inhibit Cdc25 activity in p53-defective cancer cells. Antiproliferative signals such as TGFβ also repress Cdc25A expression, thereby maintaining Cdk4/Cyclin D in an inactive tyrosine phosphorylated state [14]. Cdc25A expression is also upregulated by the HPV type 16 E7 oncoprotein, where it is involved in HPV 16 E7 dependent uncontrolled proliferation [54].

A number of posttranscriptional mechanisms have been identified that regulate Cdc25 isoform expression. Recent evidence indicates that Cdc25A is also regulated at the level of translation by eIF2α [55] and by mutant oncogenic $JAK2^{V617F}$ kinase [56]. Interestingly, Cdc25B may also be regulated by eIF2α, which is phosphorylated in response to UV radiation, causing a decrease in Cdc25B translation [57]. All three Cdc25 isoforms are also targets for regulation via the microRNA (miRNA) pathway, which reduces the levels and stability of mRNA. Specifically, miR-15a and miR-21 have been shown to negatively regulate Cdc25A [58–60], as has miR-449a/b [61]. Cdc25B is downregulated by miR-148a, which binds to the 3'-UTR and suppresses its expression [62]. Cdc25C has been found to be a target of miR-125b [63]. Interestingly, activation of miR-21 by 3,30-Diindolylmethane in human breast cancer cell lines results in a downregulation of Cdc25A, which leads to a subsequent cell cycle arrest and suppression of growth [64]. This suggests that manipulation of the miRNA pathway could provide a novel method of targeting Cdc25 in cancer.

Destabilization and degradation of Cdc25 protein is also a critical mechanism that is aberrantly regulated in cancer cells, leading to hyper-stabilization of the protein and subsequent increase in Cdc25 levels. Both Cdc25A [65, 66] and Cdc25B are degraded by the ubiquitin-proteasome system mediated by the Skp1-Cullin F-box ubiquitin ligase ($SCF^{βTrCP}$) [36, 67, 68], with Cdc25A also degraded during mitotic exit by the APC^{cdh1} [69–71]. The major outcome of Cdc25A destabilization is a block in S phase entry and progression, and destabilization of Cdc25A has also been associated with DNA damage induced G2 phase checkpoint, where it contributes to the cell cycle arrest [72]. The exact mechanisms controlling Cdc25C degradation are still unclear; however, Cdc25C is degraded by the proteasome in response to ATM/ATR checkpoint activation [52], and the translationally controlled tumor protein (TCTP) can also promote the ubiquitin-proteasome mediated degradation of Cdc25C during mitotic progression [73]. Phosphorylations at specific sites in Cdc25 are critical for targeting Cdc25 for proteasomal degradation. For example, Cdc25A requires phosphorylation on Ser79 and Ser82 to create a phosphodegron that is recognized by the $SCF^{βTrCP}$ in response to ATM/ATR checkpoint signaling [65, 74–77]. Several kinases have been implicated in this including NEK11 [78], ERK and p90Rsk [79], Chk1 [80], and GSK-3β [81]. Cdc25B degradation is controlled by stress activated kinases JNK1, JNK2, p38 [68, 82], and MAPK [83], while ERK-mediated phosphorylation on Ser216 of Cdc25C promotes its ubiquitin mediated degradation [84]. Many of these kinases are disrupted in cancer, for example GSK-3β inactivation correlates with Cdc25A overproduction in human cancers [81]. Additionally, $SCF^{βTrCP}$ is often deregulated [85], and this may provide a mechanism for reduced destruction of Cdc25 in some cancers. Alternatively, recent evidence has shown that overexpression of the de-ubiquitylating enzyme Dub3 stabilizes Cdc25A by preventing removal of ubiquitin chains from Cdc25A, thereby preventing its destruction by the proteasome [86]. This study also found that Dub3 overexpression was responsible for increased levels of Cdc25A in some human breast cancers.

Cdc25A expression is upregulated in a range of cancers including breast, non-small cell lung cancer, head and neck cancers, and non-Hodgkins lymphoma [87–90]. The increased expression appears to be through posttranscriptional mechanisms [91], primarily associated with increased stability of the protein [81, 86].

Unlike Cdc25A, Cdc25B is not an essential gene in development, and knockout mice are born without any apparent defects, although they are infertile due to an essential role for Cdc25B in meiosis [92]. Cdc25A and Cdc25B share similar catalytic activities towards Cyclin A, B, and E/Cdks [31], and it is this shared substrate specificity that allows Cdc25A and Cdc25B to complement deletion of the other, whereas Cdc25C is incapable of compensating for loss of these [41]. The inability of Cdc25C to complement deletion of the other Cdc25s is most likely a consequence of its severely limited substrate specificity, essentially it is only capable of activating Cyclin B/Cdk1, and then only after mitotic activation with any efficiency [28].

Cdc25A and Cdc25B, but not Cdc25C, have been demonstrated to possess oncogenic activity in combination with mutant H-Ras or Rb loss [93, 94], and hemizygous deletion of Cdc25A reduces the transformation ability of oncogenes [43]. The contribution of Cdc25A and B to tumorigenesis is likely to be consequence of two functions, both involving their ability to activate essential S and G2/M phase Cyclin/Cdks. The first of these is their ability to promote S phase progression. Overexpression of Cdc25A has been shown to accelerate entry into S phase by activating Cyclin E and A/Cdk2 [21, 95], and is a critical target for oncogene and carcinogen driven transformation [44, 56, 86, 96].

The contribution of Cdc25B appears more complex. Overexpression of Cdc25B drives cells into premature, often faulty mitosis, which frequently leads to cell death [23, 97, 98]. However, it appears that overexpression of different splice variants of Cdc25B has different outcomes. Cdc25B2 overexpression produced less cell death than Cdc25B3 overexpression [38]; Cdc25B2 isoform has been exclusively used in all the transformation studies reported [93, 99]; the Cdc25B2 splice variant is found overexpressed in tumors [89, 100], and is exclusively overexpressed in VCREMS resistant breast cancer cells [101]. It is also mildly increased in human pancreatic cancer, although interestingly the Cdc25B3 isoform showed a much greater upregulation [102].

Biochemically, Cdc25B2 has a stronger relative selectivity for Cdk2 complexes over Cdk1 complexes, the opposite of the more abundant Cdc25B3 variant, which is expressed almost exclusively in G2 and M phase [28, 38]. Additionally, Cdc25B2 is missing a critical mitotic destruction sequence [36], with the consequence that although Cdc25B2 protein level is generally lower than Cdc25B3, it is far more stable throughout mitosis and the cell cycle [38].

Increased expression of specific splice variants of Cdc25A and Cdc25C has also been reported in cancers. Cdc25A2 is overexpressed in breast cancer [101], and a novel Cdc25A isoform (Q110del) is deregulated in non-small cell lung cancer

[103]. An increase in the ratio of Cdc25C5 to Cdc25C1 splice variants is associated with multidrug-resistance in breast cancer cell lines [101]. Interestingly, RNAi mediated knockdown of heterodimeric splicing factor U2AF also altered the ratio of C1 to C5 expression as well as the ratio of Cdc25B2 and B3, providing a mechanism for altering splice variant levels within the cell [104]. Finally, external factors such as exposure to genotoxic agents such as doxorubicin or cisplatin can also alter the expression of Cdc25C splice variants, with the ratio of C1 to C5 decreasing both at the mRNA and protein level in an ATM/ATR dependent manner [105].

Both Cdc25A and Cdc25B have been shown to be critical regulators of centrosome duplication, with increased Cdc25 activity causing aberrant centrosome numbers [106–109]. Excessive centrosome duplication has been strongly linked with incorrect mitotic divisions and subsequent genomic instability [110–113]. This has led to the general consensus that only Cdc25A and Cdc25B are able to promote the transformation process, but none of the isoforms (A, B, or C) can drive oncogenesis by themselves.

Cdc25 Regulation of Cell Cycle Checkpoint Arrest

Equally important are the key roles Cdc25A and Cdc25B play in cell cycle checkpoint responses to stress, particularly DNA damage. The presence of DNA damage triggers activation of apical checkpoint kinases, ATM or ATR dependent on the type of DNA damage detected. ATM is activated in response to double stranded breaks, ATR to single stranded DNA [114]. These in turn activate effector checkpoint kinases, Chk1 and Chk2, which in turn phosphorylate a range of proteins including the Cdc25s, to inhibit Cyclin/Cdk activation and cell cycle progression. Although Chk1 and Chk2 have overlapping substrate specificities, there is little direct evidence for Chk2 regulation of Cdc25 [80]. The consequences of Chk1 phosphorylation are different for each of the Cdc25 family members. In all cases, Chk1-dependent phosphorylation of each Cdc25 isoform produces 14-3-3 binding sites (Fig. 4). 14-3-3 proteins are a highly conserved family of phosphosite specific binding proteins that form homo- and heterodimers, acting as adapters forming intra- and intermolecular bridges to regulate localization and activity [115].

In the case of Cdc25A, 14-3-3 binding to Ser178 and Thr507 regulates substrate access to the catalytic site [116, 117]. This 14-3-3 dependent function appears to be more prominent in regulating the G2/M function of Cdc25A. In addition to this role, it has been proposed that 14-3-3 binding to Cdc25A, possibly through the Chk1-dependent Thr507 site, tethers Chk1, also a 14-3-3 binding protein, to Cdc25A and facilitates phosphorylation of Ser76 which is critical for $SCF^{\beta TrCP}$-dependent destabilization of Cdc25A [118]. The destabilization of Cdc25A in response to DNA damage in G1 and S phase is required for the checkpoint arrest [119]. Cdc25B is phosphorylated at three sites by Chk1, Ser151, Ser230 and Ser323, which act as 14-3-3 binding sites, one end of the dimer binding to the C-terminal site and the other end to the two N-terminal sites in a cooperative

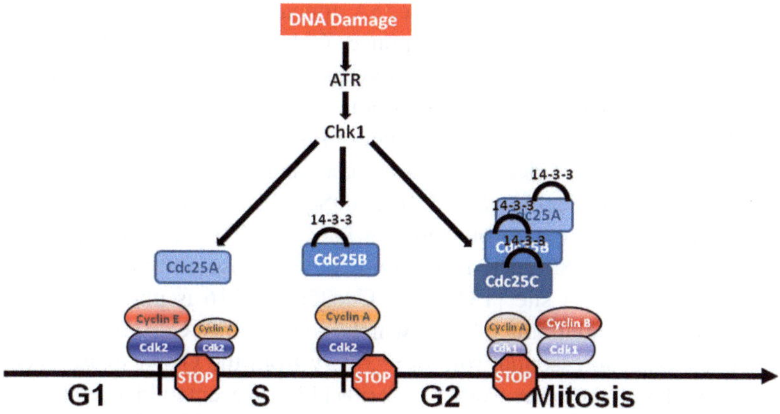

Fig. 4 In response to DNA damage, ATR activates Chk1, which in turn phosphorylates the Cdc25 isoforms, primarily destabilizing Cdc25A in G1/S phase, and promoting 14-3-3 binding to Cdc25B to regulate G2 progression, and 14-3-3 binding in all the Cdc25 isoforms to regulate G2/M progression

manner to form an intramolecular bridge which restricts substrate access to the catalytic site [120, 121]. 14-3-3 binding also regulates the subcellular localization of Cdc25B, by preventing nuclear export [120]. Cdc25B activity is also blocked in checkpoint arrested cells by the stabilized 14-3-3 binding blocking substrate access to the catalytic site inhibiting G2/M phase activation of Cyclin A/Cdk2 and Cyclin B/Cdk1 [24, 122], and overexpression of 14-3-3 binding mutants of Cdc25B readily bypasses the checkpoint arrest [120]. Chk1 phosphorylates Cdc25C on Ser216 which also provides a 14-3-3 binding site that regulates Cdc25C localization and activity [123, 124]. A second 14-3-3 binding site on Cdc25C was recently identified, Ser263, which also regulates Cdc25C localization [125, 126]. The kinase responsible for this phosphorylation is not known but this site is homologous the Ser375 site in Cdc25B, a known Chk1 site [121]. The importance of the Ser216 site in Cdc25C regulation and checkpoint response in general is questionable. Although overexpression of the nonphosphorylatable Ser216Ala mutant does drive cells through a checkpoint arrest [124], this critical site is not conserved in rodents, and knockout of Cdc25C has no discernible phenotype or effect on checkpoint responses [127]. The activation of Cdc25C requires feedback phosphorylation by its primary substrate, Cyclin B/Cdk1 [29, 128]. One of the Cdk1-dependent phosphorylation sites on Cdc25C is the Ser214, phosphorylation of which disrupts 14-3-3 binding to the Ser216 site without dephosphorylation of this residue, effectively removing the inhibitory action of 14-3-3 binding [129]. Thus, Chk1-dependent downregulation of all Cdc25 isoforms contributes to G1/S and G2/M phase checkpoint arrests (Fig. 4). In the G1/S checkpoints, this appears to be primarily through the destabilization of Cdc25A, although 14-3-3 dependent inhibition of substrate accesses to the catalytic sites may ensure the inactivation of the small pool that escapes $SCF^{\beta TrCP}$-dependent degradation. In the G2/M checkpoints, the relative contributions of destabilization and 14-3-3 dependent inhibition of Cdc25A appear reversed. Cdc25B is a major factor in the G2 phase checkpoint response. Chk1 regulation of Cdc25B rather Cd25A regulates mitotic entry [32], and overexpression of wild type Cdc25B, but not Cdc25C, readily overcomes the checkpoint arrest [24, 25].

In addition to Chk1, an increasing number of kinases have been identified that can phosphorylate sites on the Cdc25s that have been identified as Chk1 sites. Cdc25B is phosphorylated on the Ser151, Ser 230, and Ser323 sites by a range of kinases although their exact cell cycle roles are not known [130]. One exception is MK2, which is activated by p38MAPK in response to G2 phase ultraviolet irradiation to phosphorylate the Ser323 site [131–133]. Cdc25C Ser216 is phosphorylated by a wide range of kinases including c-Tak1, which directly binds Cdc25C [134, 135].

Cdc25B also has an essential role in recovery from the G2/M phase checkpoint arrest not shared with the other Cdc25 isoforms [136]. Recent evidence also suggests there are several stable N-terminally truncated isoforms of Cdc25B that have unique and critical roles in checkpoint recovery [39]. Whereas the downregulation of Cdc25B can be Chk1 or MK2 dependent to impose checkpoint arrest in response to a range of DNA damage, recovery from the checkpoint arrest is likely to require more than inactivation of the inhibitory signal. Although inhibition of Chk1 does promote rapid exit from the checkpoint arrest [137–139], it is not clear that inactivation of Chk1 is the primary driver of checkpoint recovery under physiological conditions. In the case of Cdc25B, phosphorylation of the site proximal to the 14-3-3 binding site has been shown to disrupt 14-3-3 binding without affecting the phosphorylation of the 14-3-3 binding site. Ser321 is homologous to Cdc25C Ser214 and is also a target for Cdk1-dependent phosphorylation, but unlike Cdc25C, this site in Cdc25B appears not to contribute significantly to Cdc25B activation [122]. However, phosphorylation of Ser169 does appear to be involved in Cdc25B activation by disrupting binding to the proximal Ser151 14-3-3 binding site [140] (Fig. 4). The kinase responsible for this phosphorylation is still unknown, but deregulation of the Ser169 kinase pathway offers an opportunity to drive premature exit from a checkpoint arrest. Thus, the deregulation of either mechanisms controlling Cdc25A stability such as Dub3 or GSK3β for example, or the Cdc25B Ser169 kinase pathway, could contribute to genomic instability in cancers by weakening the G1/S and G2/M phase checkpoint arrests that protect cells from DNA damage induced mutations.

Non-Cell Cycle Roles for Cdc25s

Roles in Apoptosis

Early studies on Cdc25A identified the proto-oncogene C-Raf (Raf-1) as specific interacting protein [141, 142]. Later studies went on to show that Cdc25A was able to dephosphorylate and thereby regulate the activity of C-Raf [143]. C-Raf has been reported to form a multimeric complex with Akt, and both Akt and C-Raf are components of survival pathways, suggesting that Cdc25A may play a role in regulating apoptosis [144]. The anti-apoptotic functions of Cdc25A appear to be regulated by its subcellular localization, with only cytoplasmic Cdc25A capable of preventing apoptosis [145]. Surprisingly, overexpression of Cdc25A in the nucleus has pro-apoptotic effect, which is mediated by dephosphorylation of the pro-apoptotic transcription factor Forkhead in rhabdomyosarcoma-like 1 (FKHRL1). This dephosphorylation leads to the nuclear retention and stabilization of FKHRL1 [145]. Activation of FKHRL1 can lead to release of cytochrome-c and caspase-dependent apoptosis [146].

There are also anti-apoptotic effects of cytoplasmic Cdc25A which appear to be independent of its phosphatase activity. The C-terminal tail of Cdc25A can bind to apoptosis signal-regulating kinase 1 (Ask1), preventing its homo-oligomerization and thereby increasing resistance to oxidative stress induced cell death [147]. The inhibition of Ask1 function is not dependent on the phosphatase activity of Cdc25A, as specific inhibitors of Cdc25A fail to have an effect [148], while a phosphatase dead double mutant of Cdc25A was as efficient as wild-type Cdc25A at inhibiting Ask1 [147]. The importance of Cdc25A interaction with Ask1 is highlighted by recent work which showed that a single mutation in Cdc25A (Ser88Phe) abolishes Cdc25A binding with Ask1, and subsequently fails to prevent Ask1 mediated apoptosis, leading to early embryonic lethality in mice [149]. Interestingly, this Ser88Phe mutation is also a cancer-predisposing polymorphic variant in humans.

Finally, Cdc25A also has additional pro-apoptotic functions. Activation of caspases results in the cleavage of Cdc25A and subsequent production of a C-terminal fragment, which has higher phosphatase activity than full-length Cdc25A and can activate Cdk1 and Cdk2, leading to amplification of apoptosis signaling [150, 151].

Arsenate Reductase Activity

Arsenic is a ubiquitous environmental element that must be inactivated by cells to avoid arsenic poisoning. A critical process in arsenic detoxification requires the enzymatic reduction of arsenate to arsenite by arsenate reductases. These enzymes have been identified in bacteria [152] and yeast [153], but not in higher eukaryotes and mammals [154], although the human liver is able to efficiently detoxify arsenate [155]. There is mounting evidence that Cdc25 has arsenate reductase activity and could perform this function in human cells. Firstly, Rhodanese domains, which are the critical domain involved in the catalysis of arsenate to arsenite, appear to have significant homology with the C-terminal catalytic region of the Cdc25 phosphatase family, indicating a possible common evolutionary origin [156]. The C-terminal domain of Cdc25 was shown to possess possible Rhodanese activity [157], and has homology to budding yeast arsenate resistance protein Acr2 [157, 158]. Furthermore, the OsACR2 gene, a Cdc25 homologue from rice, has arsenate reductase activity in vitro, and its expression is increased in the roots of rice plants in response to arsenate exposure [159]. While in the plant Holcus lanatus (commonly known as velvet grass), a

Cdc25-like phosphatase is responsible for enhanced arsenate reduction, which may explain the increased tolerance to arsenic in these plants [160]. In fission yeast, Cdc25 mutants have reduced ability to process arsenate, and Cdc25 protein levels increased in response to arsenate exposure [161]. In humans, both Cdc25B and Cdc25C can efficiently catalyze the reduction of inorganic arsenate in vitro [162], indicating that Cdc25 may be the human enzyme responsible for arsenate detoxification. Arsenic is a know carcinogen [163], and arsenic exposure causes a number cell cycle defects including accumulation of cells in S and G2 phase, increased expression of Cyclin A, Cdk1 and Cdk2 [164], mitotic defects [165], and inhibition of Cdc25A gene transcription [166]. Thus the anticancer effects of arsenic exposure could be partially due to the downregulation of Cdc25 in cancers that are reinforced by overexpression of Cdc25.

Steroid Receptor Transactivation

The oncogenic potential of Cdc25 was thought to be limited to its primary role in regulating Cdk activity during the cell cycle. However, an additional non-Cdk role for Cdc25B has been reported. Overexpression of Cdc25B correlated with increased expression of Cyclin D1 (at both the protein and mRNA level), suggested that Cdc25B was either directly or indirectly activating Cyclin D1 transcription [167]. Cyclin D1 expression is regulated by mitogenic and extracellular signaling pathways, often via the estrogen receptor (ER), with ER activation leading to increased Cyclin D1 transcription [168–170]. There is a direct interaction between ER and Cdc25B [167], ER interacting with two regions of Cdc25B, amino acids 81–273 and the catalytic C-term domain (amino acids 352–539). Cdc25B was also able to activate the androgen (AR), glucocorticoid (GR), and progesterone receptors (PR) [167], suggesting that Cdc25B may also play a widespread role in hormone receptor transactivation. The mechanism by which Cdc25B assists ER mediated transcription appears to be via increased recruitment of the histone acetyltransferase enzymes PCAF and CBP [167], which can deacetylate core histones leading to the relaxation of the nucleosome, in turn allowing greater access of transcription factors to the DNA [171, 172].

Cdc25A was found to bind to and dephosphorylate the epidermal growth factor receptor (EGFR) [173], leading to its inactivation. Furthermore, Cdc25A can also bind and dephosphorylate ERK, the downstream kinase of EGFR [174]. Inhibition of Cdc25A with specific inhibitors, such as Cpd5 or NSC95397, leads to hyperactivation of EGFR and the downstream MAPK pathways, leading to decreased transcription, a block in cell cycle progression and increased cell death [173, 175]. Similar to Cdc25B, Cdc25A also appears to have broader hormone receptor regulatory roles, with recent evidence identifying it as a corepressor of Androgen receptor (AR) activity [176].

In terms of oncogenesis, overexpression of Cdc25 could adversely affect hormone receptor signaling, and promote increased transcription, cell cycle progression, and inhibition of cell death, which are all hallmarks of cancer. It is therefore not surprising that overexpression of Cdc25 has been reported in many hormone dependent cancers. For example, Cdc25B and C have been reported to be overexpressed in prostate cancer [177, 178], while both Cdc25A and B have been reported to be overexpressed in breast [87, 93, 94] and in ovarian cancer [179].

Cdc25s as Targets

Given that the major role of Cdc25 is to activate Cdks allowing progression through the cell cycle, specific inhibition of Cdc25 should block cell cycle progression, providing a strong cytostatic effect. The exact point of arrest will depend on which isoform of Cdc25 is inhibited and at what point in the cell cycle the drug is administered. Thus inhibition of Cdc25A and Cdc25B will primarily lead to arrests in G1, S, and G2 phases. Inhibition of Cdc25 activity during mitosis forces cells to prematurely exit mitosis, as Cdc25 is required to maintain Cdk1 activity during mitosis [180].

There are six classes of Cdc25 inhibitors. The first inhibitors of Cdc25, identified in 1994, were the antitumor antibiotics DnacinA1 and B1, which were isolated from a strain of Actinobacteria [181]. This discovery prompted the further development of more potent natural analogues based on Dnacin [182], and together form basis for the Natural class of inhibitors. The second class are based on analogues of Vitamin K and quinones, while lipophilic acids define the third class, electrophilic compounds, phosphate mimics, and peptide analogues corresponding to final three classes [183, 184]. The majority of Cdc25 inhibitors cause the expected cytostatic effects, inducing cell cycle arrest, especially in G1, S, and G2 phases [185–191]. Inhibition during mitosis also causes the expected premature exit in Chinese hamster ovary cells treated with the general Cdc25 inhibitor vanadate [192], while the addition of the specific Cdc25 inhibitor NSC663284 in combination with a Wee1 inhibitor results in collapse of the mitotic state [193].

There are also several surprising effects of Cdc25 inhibition, these include inhibition of microtubule dynamics [194], the induction of apoptosis [195–197], and DNA damage from the production of reactive oxygen species (ROS) [182, 198]. It has been shown that many of the compounds used in commercial small molecule screens are oxidizing agents, and that many of the identified Cdc25 inhibitors are also capable of redox cycling, with cells rapidly producing intracellular reactive oxygen species upon exposure [199]. Therefore, many of the identified Cdc25 inhibitors are likely to be nonspecific.

There has been only limited success of Cdc25 inhibitors to block the growth of xenografted human tumors in nude mice, and therefore the majority of compounds have not entered clinical trials. There have been some exceptions. BN82685, which is effective at reducing tumors derived from pancreatic carcinoma cells [186], IRC-083864 is active against prostatic and pancreatic tumors [195], and is now in Phase II clinical trials [200]. Finally, PM-20 was successful in treating hepatoma tumors [201], and polycystic kidney and liver disease in rodents [202]. Interestingly, tetrasulfides, which occur naturally in vegetables such as garlic and onion and have known anticarcinogenic properties, were recently shown to inhibit Cdc25, which mediated the antiproliferative effects of these compounds [203], suggesting that there may be still significant untapped potential in specific targeting of Cdc25.

There is an obvious need for isoform specific inhibitors. Several recent advances in virtual screen based methods have provided a number of promising novel compounds against the active site of Cdc25A and Cdc25B, and importantly these were able to inhibit the cell cycle progression in human cancer cell lines [204, 205]. Additional isoform specific inhibitors have also been recently identified with two Arylstibonic acids derived compound showing 100-fold selectivity for Cdc25A and Cdc25B compared to Cdc25C [206].

It may also be possible to target Cdc25 via alternative methods. For example, decreasing the protein levels of Cdc25 would provide cytostatic effects, promote increased sensitivity to apoptosis via the Ask-1 pathway, and has reduced hormone receptor signaling. The anti-inflammatory drug mesalazine was shown to decrease Cdc25A protein expression and lead to a block in cell proliferation in colon cancer cells [207]. Hypoxic conditions such as those found in solid tumors have been shown to decrease protein levels of Cdc25A [59], which is mediated by inactivation of the translational regulator eukaryotic initiation factor 2A (eIF2A) [55]. Additional methods could also target the Cdc25A stabilizing protein Dub3 [86], to prevent Cdc25A accumulation, or employ the mi-RNA pathway [60, 62] to block translation of Cdc25 mRNA, thereby reducing protein levels.

In summary, although the initial promise of Cdc25 inhibitors has not yet been realized clinically, there is still significant room for improved targeting of this pathway, and if combined with rational patient selection, this class of compounds could still prove to be an invaluable tool for the treatment of cancer.

References

1. Gray-Bablin J, Zalvide J, Fox MP, Knickerbocker CJ, DeCaprio JA, Keyomarsi K. Cyclin E, a redundant cyclin in breast cancer. Proc Natl Acad Sci U S A. 1996;93(26):15215–20.
2. Sherr CJ. The Pezcoller lecture: cancer cell cycles revisited. Cancer Res. 2000;60(14): 3689–95.
3. Ohtsubo M, Theodoras AM, Schumacher J, Roberts JM, Pagano M. Human cyclin E, a nuclear protein essential for the G1-to-S phase transition. Mol Cell Biol. 1995;15(5):2612–24.
4. Pagano M, Pepperkok R, Verde F, Ansorge W, Draetta G. Cyclin A is required at two points in the human cell cycle. EMBO J. 1992;11(3):961–71.
5. Strausfeld UP, Howell M, Descombes P, Chevalier S, Rempel RE, Adamczewski J, et al. Both cyclin A and cyclin E have S-phase promoting (SPF) activity in Xenopus egg extracts. J Cell Sci. 1996;109(Pt 6):1555–63.
6. De Boer L, Oakes V, Beamish H, Giles N, Stevens F, Somodevilla-Torres M, et al. Cyclin A/cdk2 coordinates centrosomal and nuclear mitotic events. Oncogene. 2008;27(31):4261–8.
7. Fung TK, Ma HT, Poon RY. Specialized roles of the Two mitotic cyclins in somatic cells: cyclin A as an activator of M phase-promoting factor. Mol Biol Cell. 2007;18(5):1861–73.
8. Lindqvist A, Rodriguez-Bravo V, Medema RH. The decision to enter mitosis: feedback and redundancy in the mitotic entry network. J Cell Biol. 2009;185(2):193–202.

9. Li C, Andrake M, Dunbrack R, Enders GH. A bifunctional regulatory element in human somatic Wee1 mediates cyclin A/Cdk2 binding and Crm1-dependent nuclear export. Mol Cell Biol. 2010;30(1):116–30.
10. Rosenblatt J, Gu Y, Morgan DO. Human cyclin-dependent kinase 2 is activated during the S and G2 phases of the cell cycle and associates with cyclin A. Proc Natl Acad Sci U S A. 1992;89(7):2824–8.
11. Jeffrey PD, Russo AA, Polyak K, Gibbs E, Hurwitz J, Massague J, et al. Mechanism of CDK activation revealed by the structure of a cyclinA-CDK2 complex [see comments]. Nature. 1995;376(6538):313–20.
12. Welburn JP, Tucker JA, Johnson T, Lindert L, Morgan M, Willis A, et al. How tyrosine 15 phosphorylation inhibits the activity of cyclin-dependent kinase 2-cyclin A. J Biol Chem. 2007;282(5):3173–81.
13. Gabrielli BG, Lee MS, Walker DH, Piwnica-Worms H, Maller JL. cdc25 regulates the phosphorylation and activity of the *Xenopus* cdk2 protein kinase complex. J Biol Chem. 1992;267(25):18040–6.
14. Iavarone A, Massague J. Repression of the CDK activator Cdc25A and cell-cycle arrest by cytokine TGF-beta in cells lacking the CDK inhibitor p15. Nature. 1997;387(6631):417–22.
15. Terada Y, Tatsuka M, Jinno S, Okayama H. Requirement for tyrosine phosphorylation of Cdk4 in G1 arrest induced by ultraviolet irradiation. Nature. 1995;376(6538):358–62.
16. Russell P, Nurse P. *cdc25*$^{+}$ functions as an inducer in the mitotic control of fission yeast. Cell. 1986;45(11):145–53.
17. Harvey SL, Kellogg DR. Conservation of mechanisms controlling entry into mitosis: budding yeast wee1 delays entry into mitosis and is required for cell size control. Curr Biol. 2003;13(4):264–75.
18. Moseley JB, Mayeux A, Paoletti A, Nurse P. A spatial gradient coordinates cell size and mitotic entry in fission yeast. Nature. 2009;459(7248):857–60.
19. Hoffman I. Giulio Draetta and Eric Karsenti. Activation of the phosphatase activity of human cdc25A by a cdk2-cyclin E dependent phosphorylation at the G1/S transition. EMBO J. 1994;133(18):4302–10.
20. Jinno S, Suto K, Nagata A, Igarashi M, Kanaoka Y, Nojima H, Okayama H. Cdc25A is a novel phosphatase functioning early in the cell cycle. EMBO J. 1994;13(7):1549–56.
21. Sexl V, Diehl JA, Sherr CJ, Ashmun R, Beach D, Roussel MF. A rate limiting function of cdc25A for S phase entry inversely correlates with tyrosine dephosphorylation of Cdk2. Oncogene. 1999;18(3):573–82.
22. Lindqvist A, Kallstrom H, Lundgren A, Barsoum E, Rosenthal CK. Cdc25B cooperates with Cdc25A to induce mitosis but has a unique role in activating cyclin B1-Cdk1 at the centrosome. J Cell Biol. 2005;171(1):35–45.
23. Gabrielli BG, De Souza CP, Tonks ID, Clark JM, Hayward NK, Ellem KA. Cytoplasmic accumulation of cdc25B phosphatase in mitosis triggers centrosomal microtubule nucleation in HeLa cells. J Cell Sci. 1996;109(Pt 5):1081–93.
24. Goldstone S, Pavey S, Forrest A, Sinnamon J, Gabrielli B. Cdc25-dependent activation of cyclin A/cdk2 is blocked in G2 phase arrested cells independently of ATM/ATR. Oncogene. 2001;20(8):921–32.
25. Karlsson C, Katich S, Hagting A, Hoffmann I, Pines J. Cdc25B and Cdc25C differ markedly in their properties as initiators of mitosis. J Cell Biol. 1999;146(3):573–84.
26. Lammer C, Wagerer S, Saffrich R, Mertens D, Ansorge W, Hoffmann I. The cdc25B phosphatase is essential for the G2/M phase transition in human cells. J Cell Sci. 1998;111(Pt 16):2445–53.
27. Lindqvist A, Kallstrom H, Karlsson RC. Characterisation of Cdc25B localisation and nuclear export during the cell cycle and in response to stress. J Cell Sci. 2004;117(Pt 21):4979–90.
28. Gabrielli BG, Clark JM, McCormack AK, Ellem KA. Hyperphosphorylation of the N-terminal domain of Cdc25 regulates activity toward cyclin B1/Cdc2 but not cyclin A/Cdk2. J Biol Chem. 1997;272(45):28607–14.

29. Hoffman I, Paul R. Clarke, M. Jesus Marcote, Eric Karsenti and Giulio Draetta. Phosphorylation and activation of human cdc25C by cdc2-cyclin B and its involvement in the self-amplification of MPF at mitosis. EMBO J. 1993;12(1):53–63.
30. Turowski P, Franckhauser C, Morris MC, Vaglio P, Fernandez A, Lamb NJ. Functional cdc25C dual-specificity phosphatase is required for S-phase entry in human cells. Mol Biol Cell. 2003;14(7):2984–98.
31. Xu X, Burke SP. Roles of active site residues and the NH2-terminal domain in the catalysis and substrate binding of human Cdc25. J Biol Chem. 1996;271(9):5118–24.
32. Loffler H, Rebacz B, Ho AD, Lukas J, Bartek J, Kramer A. Chk1-dependent regulation of Cdc25B functions to coordinate mitotic events. Cell Cycle. 2006;5(21):2543–7.
33. De Souza CP, Ellem KA, Gabrielli BG. Centrosomal and cytoplasmic Cdc2/cyclin B1 activation precedes nuclear mitotic events. Exp Cell Res. 2000;257(1):11–21.
34. Baldin V, Cans C, Superti-Furga G, Ducommun B. Alternative splicing of the human CDC25B tyrosine phosphatase. Possible implications for growth control? Oncogene. 1997;14(20):2485–95.
35. Wegener S, Hampe W, Herrmann D, Schaller HC. Alternative splicing in the regulatory region of the human phosphatases CDC25A and CDC25C. Eur J Cell Biol. 2000;79(11):810–5.
36. Kieffer I, Lorenzo C, Dozier C, Schmitt E, Ducommun B. Differential mitotic degradation of the CDC25B phosphatase variants. Oncogene. 2007;26(57):7847–58.
37. Bureik M, Rief N, Drescher R, Jungbluth A, Montenarh M, Wagner P. An additional transcript of the cdc25C gene from A431 cells encodes a functional protein. Int J Oncol. 2000;17(6):1251–8.
38. Forrest AR, McCormack AK, DeSouza CP, Sinnamon JM, Tonks ID, Hayward NK, et al. Multiple splicing variants of cdc25B regulate G2/M progression. Biochem Biophys Res Commun. 1999;260(2):510–5.
39. Jullien D, Bugler B, Dozier C, Cazales M, Ducommun B. Identification of N-terminally truncated stable nuclear isoforms of CDC25B that are specifically involved in G2/M checkpoint recovery. Cancer Res. 2011;71(5):1968–77.
40. Caudill JS, Porcher JC, Steensma DP. Aberrant pre-mRNA splicing of a highly conserved cell cycle regulator, CDC25C, in myelodysplastic syndromes. Leuk Lymphoma. 2008;49(5):989–93.
41. Lee G, Origanti S, White LS, Sun J, Stappenbeck TS, Piwnica-Worms H. Contributions made by CDC25 phosphatases to proliferation of intestinal epithelial stem and progenitor cells. PLoS One. 2011;6(1), e15561.
42. Lee G, White LS, Hurov KE, Stappenbeck TS, Piwnica-Worms H. Response of small intestinal epithelial cells to acute disruption of cell division through CDC25 deletion. Proc Natl Acad Sci U S A. 2009;106(12):4701–6.
43. Ray D, Terao Y, Nimbalkar D, Hirai H, Osmundson EC, Zou X, et al. Hemizygous disruption of Cdc25A inhibits cellular transformation and mammary tumorigenesis in mice. Cancer Res. 2007;67(14):6605–11.
44. Galaktionov K, Chen X, Beach D. Cdc25 cell-cycle phosphatase as a target of c-myc. Nature. 1996;382(6591):511–7.
45. Vigo E, Muller H, Prosperini E, Hateboer G, Cartwright P, Moroni MC, et al. CDC25A phosphatase is a target of E2F and is required for efficient E2F-induced S phase. Mol Cell Biol. 1999;19(9):6379–95.
46. Wang M, Gartel AL. The suppression of FOXM1 and its targets in breast cancer xenograft tumors by siRNA. Oncotarget. 2011;2(12):1218–26.
47. Rother K, Kirschner R, Sanger K, Bohlig L, Mossner J, Engeland K. p53 downregulates expression of the G1/S cell cycle phosphatase Cdc25A. Oncogene. 2007;26(13):1949–53.
48. Demidova AR, Aau MY, Zhuang L, Yu Q. Dual regulation of Cdc25A by Chk1 and p53-ATF3 in DNA replication checkpoint control. J Biol Chem. 2009;284(7):4132–9. Epub 2008/12/09.

49. Dalvai M, Mondesert O, Bourdon JC, Ducommun B, Dozier C. Cdc25B is negatively regulated by p53 through Sp1 and NF-Y transcription factors. Oncogene. 2011;30(19):2282–8.
50. Badie C, Itzhaki JE, Sullivan MJ, Carpenter AJ, Porter AC. Repression of CDK1 and other genes with CDE and CHR promoter elements during DNA damage-induced G(2)/M arrest in human cells. Mol Cell Biol. 2000;20(7):2358–66.
51. Krause K, Haugwitz U, Wasner M, Wiedmann M, Mossner J, Engeland K. Expression of the cell cycle phosphatase cdc25C is down-regulated by the tumor suppressor protein p53 but not by p73. Biochem Biophys Res Commun. 2001;284(3):743–50.
52. Thanasoula M, Escandell JM, Suwaki N, Tarsounas M. ATM/ATR checkpoint activation downregulates CDC25C to prevent mitotic entry with uncapped telomeres. EMBO J. 2012;31(16):3398–410.
53. St Clair S, Giono L, Varmeh-Ziaie S, Resnick-Silverman L, Liu WJ, Padi A, et al. DNA damage-induced downregulation of Cdc25C is mediated by p53 via two independent mechanisms: one involves direct binding to the cdc25C promoter. Mol Cell. 2004;16(5):725–36.
54. Nguyen DX, Westbrook TF, McCance DJ. Human papillomavirus type 16 E7 maintains elevated levels of the cdc25A tyrosine phosphatase during deregulation of cell cycle arrest. J Virol. 2002;76(2):619–32.
55. Tomko Jr RJ, Lazo JS. Multimodal control of Cdc25A by nitrosative stress. Cancer Res. 2008;68(18):7457–65.
56. Gautier EF, Picard M, Laurent C, Marty C, Villeval JL, Demur C, et al. The cell cycle regulator CDC25A is a target for JAK2V617F oncogene. Blood. 2012;119(5):1190–9.
57. Lemaire M, Ducommun B, Nebreda AR. UV-induced downregulation of the CDC25B protein in human cells. FEBS Lett. 2010;584(6):1199–204.
58. Lee SO, Masyuk T, Splinter P, Banales JM, Masyuk A, Stroope A, et al. MicroRNA15a modulates expression of the cell-cycle regulator Cdc25A and affects hepatic cystogenesis in a rat model of polycystic kidney disease. J Clin Invest. 2008;118(11):3714–24.
59. de Oliveira PE, Zhang L, Wang Z, Lazo JS. Hypoxia-mediated regulation of Cdc25A phosphatase by p21 and miR-21. Cell Cycle. 2009;8(19):3157–64.
60. Wang P, Zou F, Zhang X, Li H, Dulak A, Tomko Jr RJ, et al. microRNA-21 negatively regulates Cdc25A and cell cycle progression in colon cancer cells. Cancer Res. 2009;69(20):8157–65.
61. Yang X, Feng M, Jiang X, Wu Z, Li Z, Aau M, et al. miR-449a and miR-449b are direct transcriptional targets of E2F1 and negatively regulate pRb-E2F1 activity through a feedback loop by targeting CDK6 and CDC25A. Genes Dev. 2009;23(20):2388–93.
62. Liffers ST, Munding JB, Vogt M, Kuhlmann JD, Verdoodt B, Nambiar S, et al. MicroRNA-148a is down-regulated in human pancreatic ductal adenocarcinomas and regulates cell survival by targeting CDC25B. Lab Invest. 2011;91(10):1472–9.
63. Le MT, Shyh-Chang N, Khaw SL, Chin L, Teh C, Tay J, et al. Conserved regulation of p53 network dosage by microRNA-125b occurs through evolving miRNA-target gene pairs. PLoS Genet. 2011;7(9), e1002242.
64. Jin Y. 3,3'-Diindolylmethane inhibits breast cancer cell growth via miR-21-mediated Cdc25A degradation. Mol Cell Biochem. 2011;358(1-2):345–54.
65. Busino L, Donzelli M, Chiesa M, Guardavaccaro D, Ganoth D, Dorrello NV, et al. Degradation of Cdc25A by beta-TrCP during S phase and in response to DNA damage. Nature. 2003;426(6962):87–91.
66. Donzelli M, Busino L, Chiesa M, Ganoth D, Hershko A, Draetta GF. Hierarchical order of phosphorylation events commits Cdc25A to betaTrCP-dependent degradation. Cell Cycle. 2004;3(4):469–71.
67. Kanemori Y, Uto K, Sagata N. Beta-TrCP recognizes a previously undescribed nonphosphorylated destruction motif in Cdc25A and Cdc25B phosphatases. Proc Natl Acad Sci U S A. 2005;102(18):6279–84.
68. Uchida S, Watanabe N, Kudo Y, Yoshioka K, Matsunaga T, Ishizaka Y, et al. SCFbeta(TrCP) mediates stress-activated MAPK-induced Cdc25B degradation. J Cell Sci. 2011;124(Pt 16): 2816–25.

69. Donzelli M, Squatrito M, Ganoth D, Hershko A, Pagano M, Draetta GF. Dual mode of degradation of Cdc25 A phosphatase. EMBO J. 2002;21(18):4875–84.
70. Reber A, Lehner CF, Jacobs HW. Terminal mitoses require negative regulation of Fzr/Cdh1 by Cyclin A, preventing premature degradation of mitotic cyclins and String/Cdc25. Development. 2006;133(16):3201–11.
71. Skaar JR, Pagano M. Cdh1: a master G0/G1 regulator. Nat Cell Biol. 2008;10(7):755–7.
72. Mailand N, Podtelejnikov AV, Groth A, Mann M, Bartek J, Lukas J. Regulation of G(2)/M events by Cdc25A through phosphorylation-dependent modulation of its stability. EMBO J. 2002;21(21):5911–20.
73. Chan TH, Chen L, Liu M, Hu L, Zheng BJ, Poon VK, et al. Translationally controlled tumor protein induces mitotic defects and chromosome missegregation in hepatocellular carcinoma development. Hepatology. 2012;55(2):491–505.
74. Xiao Z, Chen Z, Gunasekera AH, Sowin TJ, Rosenberg SH, Fesik S, et al. Chk1 mediates S and G2 arrests through Cdc25A degradation in response to DNA-damaging agents. J Biol Chem. 2003;278(24):21767–73.
75. Zhao H, Watkins JL, Piwnica-Worms H. Disruption of the checkpoint kinase 1/cell division cycle 25A pathway abrogates ionizing radiation-induced S and G2 checkpoints. Proc Natl Acad Sci U S A. 2002;99(23):14795–800.
76. Sorensen CS, Syljuasen RG, Falck J, Schroeder T, Ronnstrand L, Khanna KK, et al. Chk1 regulates the S phase checkpoint by coupling the physiological turnover and ionizing radiation-induced accelerated proteolysis of Cdc25A. Cancer Cell. 2003;3(3):247–58.
77. Jin J, Shirogane T, Xu L, Nalepa G, Qin J, Elledge SJ, et al. SCFbeta-TRCP links Chk1 signaling to degradation of the Cdc25A protein phosphatase. Genes Dev. 2003;17(24):3062–74.
78. Melixetian M, Klein DK, Sorensen CS, Helin K. NEK11 regulates CDC25A degradation and the IR-induced G2/M checkpoint. Nat Cell Biol. 2009;11(10):1247–53.
79. Isoda M, Kanemori Y, Nakajo N, Uchida S, Yamashita K, Ueno H, et al. The extracellular signal-regulated kinase-mitogen-activated protein kinase pathway phosphorylates and targets Cdc25A for SCF beta-TrCP-dependent degradation for cell cycle arrest. Mol Biol Cell. 2009;20(8):2186–95.
80. Jin J, Ang XL, Ye X, Livingstone M, Harper JW. Differential roles for checkpoint kinases in DNA damage-dependent degradation of the Cdc25A protein phosphatase. J Biol Chem. 2008;283(28):19322–8.
81. Kang T, Wei Y, Honaker Y, Yamaguchi H, Appella E, Hung MC, et al. GSK-3 beta targets Cdc25A for ubiquitin-mediated proteolysis, and GSK-3 beta inactivation correlates with Cdc25A overproduction in human cancers. Cancer Cell. 2008;13(1):36–47.
82. Uchida S, Yoshioka K, Kizu R, Nakagama H, Matsunaga T, Ishizaka Y, et al. Stress-activated mitogen-activated protein kinases c-Jun NH2-terminal kinase and p38 target Cdc25B for degradation. Cancer Res. 2009;69(16):6438–44.
83. Astuti P, Pike T, Widberg C, Payne E, Harding A, Hancock J, et al. MAPK pathway activation delays G2/M progression by destabilizing Cdc25B. J Biol Chem. 2009;284(49):33781–8.
84. Eymin B, Claverie P, Salon C, Brambilla C, Brambilla E, Gazzeri S. p14ARF triggers G2 arrest through ERK-mediated Cdc25C phosphorylation, ubiquitination and proteasomal degradation. Cell Cycle. 2006;5(7):759–65.
85. Frescas D, Pagano M. Deregulated proteolysis by the F-box proteins SKP2 and beta-TrCP: tipping the scales of cancer. Nat Rev Cancer. 2008;8(6):438–49.
86. Pereg Y, Liu BY, O'Rourke KM, Sagolla M, Dey A, Komuves L, et al. Ubiquitin hydrolase Dub3 promotes oncogenic transformation by stabilizing Cdc25A. Nat Cell Biol. 2010;12(4):400–6.
87. Cangi MG, Cukor B, Soung P, Signoretti S, Moreira Jr G, Ranashinge M, et al. Role of the Cdc25A phosphatase in human breast cancer. J Clin Invest. 2000;106(6):753–61.
88. Gasparotto D, Maestro R, Piccinin S, Vukosavljevic T, Barzan L, Sulfaro S, et al. Overexpression of CDC25A and CDC25B in head and neck cancers. Cancer Res. 1997; 57(12):2366–8.

89. Hernandez S, Hernandez L, Bea S, Pinyol M, Nayach I, Bellosillo B, et al. cdc25a and the splicing variant cdc25b2, but not cdc25B1, -B3 or -C, are over-expressed in aggressive human non-Hodgkin's lymphomas. Int J Cancer. 2000;89(2):148–52.
90. Wu W, Fan YH, Kemp BL, Walsh G, Mao L. Overexpression of cdc25A and cdc25B is frequent in primary non-small cell lung cancer but is not associated with overexpression of c-myc. Cancer Res. 1998;58(18):4082–5.
91. Loffler H, Syljuasen RG, Bartkova J, Worm J, Lukas J, Bartek J. Distinct modes of deregulation of the proto-oncogenic Cdc25A phosphatase in human breast cancer cell lines. Oncogene. 2003;22(50):8063–71.
92. Lincoln AJ, Wickramasinghe D, Stein P, Schultz RM, Palko ME, De Miguel MP, et al. Cdc25b phosphatase is required for resumption of meiosis during oocyte maturation. Nat Genet. 2002;30(4):446–9.
93. Galaktionov K, Lee AK, Eckstein J, Draetta G, Meckler J, Loda M, et al. CDC25 phosphatases as potential human oncogenes. Science. 1995;269(5230):1575–7.
94. Ray D, Terao Y, Fuhrken PG, Ma ZQ, DeMayo FJ, Christov K, et al. Deregulated CDC25A expression promotes mammary tumorigenesis with genomic instability. Cancer Res. 2007;67(3):984–91.
95. Blomberg I, Hoffmann I. Ectopic expression of Cdc25A accelerates the G(1)/S transition and leads to premature activation of cyclin E- and cyclin A-dependent kinases. Mol Cell Biol. 1999;19(9):6183–94.
96. Srivastava SK, Bansal P, Oguri T, Lazo JS, Singh SV. Cell division cycle 25B phosphatase is essential for benzo(a)pyrene-7,8-Diol-9,10-epoxide induced neoplastic transformation. Cancer Res. 2007;67(19):9150–7.
97. Varmeh S, Manfredi JJ. Inappropriate activation of cyclin-dependent kinases by the phosphatase Cdc25b results in premature mitotic entry and triggers a p53-dependent checkpoint. J Biol Chem. 2009;284(14):9475–88.
98. Bansal P, Lazo JS. Induction of Cdc25B regulates cell cycle resumption after genotoxic stress. Cancer Res. 2007;67(7):3356–63.
99. Yao Y, Slosberg ED, Wang L, Hibshoosh H, Zhang YJ, Xing WQ, et al. Increased susceptibility to carcinogen-induced mammary tumors in MMTV-Cdc25B transgenic mice. Oncogene. 1999;18(37):5159–66.
100. Hernandez S, Bessa X, Bea S, Hernandez L, Nadal A, Mallofre C, et al. Differential expression of cdc25 cell-cycle-activating phosphatases in human colorectal carcinoma. Lab Invest. 2001;81(4):465–73.
101. Albert H, Santos S, Battaglia E, Brito M, Monteiro C, Bagrel D. Differential expression of CDC25 phosphatases splice variants in human breast cancer cells. Clin Chem Lab Med. 2011;49(10):1707–14.
102. Guo J, Kleeff J, Li J, Ding J, Hammer J, Zhao Y, et al. Expression and functional significance of CDC25B in human pancreatic ductal adenocarcinoma. Oncogene. 2004;23(1):71–81.
103. Younis RH, Cao W, Lin R, Xia R, Liu Z, Edelman MJ, et al. CDC25A(Q110del): a novel cell division cycle 25A isoform aberrantly expressed in non-small cell lung cancer. PLoS One. 2012;7(10), e46464.
104. Pacheco TR, Moita LF, Gomes AQ, Hacohen N, Carmo-Fonseca M. RNA interference knockdown of hU2AF35 impairs cell cycle progression and modulates alternative splicing of Cdc25 transcripts. Mol Biol Cell. 2006;17(10):4187–99.
105. Albert H, Battaglia E, Monteiro C, Bagrel D. Genotoxic stress modulates CDC25C phosphatase alternative splicing in human breast cancer cell lines. Mol Oncol. 2012;6(5): 542–52.
106. Shreeram S, Hee WK, Bulavin DV. Cdc25A serine 123 phosphorylation couples centrosome duplication with DNA replication and regulates tumorigenesis. Mol Cell Biol. 2008;28(24): 7442–50.
107. Boutros R, Ducommun B. Asymmetric localization of the CDC25B phosphatase to the mother centrosome during interphase. Cell Cycle. 2008;7(3):401–6. Epub 2008/02/01.

108. Boutros R, Lobjois V, Ducommun B. CDC25B involvement in the centrosome duplication cycle and in microtubule nucleation. Cancer Res. 2007;67(24):11557–64.
109. Boutros R, Lorenzo C, Mondesert O, Jauneau A, Oakes V, Dozier C, et al. CDC25B associates with a centrin 2-containing complex and is involved in maintaining centrosome integrity. Biol Cell. 2011;103(2):55–68.
110. Duensing A, Duensing S. Centrosomes, polyploidy and cancer. Adv Exp Med Biol. 2010;676:93–103.
111. Kramer A, Maier B, Bartek J. Centrosome clustering and chromosomal (in)stability: a matter of life and death. Mol Oncol. 2011;5(4):324–35.
112. Marthiens V, Piel M, Basto R. Never tear us apart—the importance of centrosome clustering. J Cell Sci. 2012;125(Pt 14):3281–92.
113. Schmidt S, Essmann F, Cirstea IC, Kuck F, Thakur HC, Singh M, et al. The centrosome and mitotic spindle apparatus in cancer and senescence. Cell Cycle. 2010;9(22):4469–73.
114. Medema RH, Macurek L. Checkpoint control and cancer. Oncogene. 2012;31(21): 2601–13.
115. Gardino AK, Yaffe MB. 14-3-3 proteins as signaling integration points for cell cycle control and apoptosis. Semin Cell Dev Biol. 2011;22(7):688–95.
116. Chen MS, Ryan CE, Piwnica-Worms H. Chk1 kinase negatively regulates mitotic function of Cdc25A phosphatase through 14-3-3 binding. Mol Cell Biol. 2003;23(21):7488–97.
117. Uto K, Inoue D, Shimuta K, Nakajo N, Sagata N. Chk1, but not Chk2, inhibits Cdc25 phosphatases by a novel common mechanism. EMBO J. 2004;23(16):3386–96.
118. Kasahara K, Goto H, Enomoto M, Tomono Y, Kiyono T, Inagaki M. 14-3-3gamma mediates Cdc25A proteolysis to block premature mitotic entry after DNA damage. EMBO J. 2010;29(16):2802–12.
119. Ray D, Kiyokawa H. CDC25A levels determine the balance of proliferation and checkpoint response. Cell Cycle. 2007;6(24):3039–42.
120. Giles N, Forrest A, Gabrielli B. 14-3-3 acts as an intramolecular bridge to regulate cdc25B localization and activity. J Biol Chem. 2003;278(31):28580–7.
121. Schmitt E, Boutros R, Froment C, Monsarrat B, Ducommun B, Dozier C. CHK1 phosphorylates CDC25B during the cell cycle in the absence of DNA damage. J Cell Sci. 2006;119(Pt 20):4269–75.
122. Astuti P, Boutros R, Ducommun B, Gabrielli B. Mitotic phosphorylation of cdc25B Ser321 disrupts 14-3-3 binding to the high affinity Ser323 site. J Biol Chem. 2010;285(45):34364–70.
123. Dalal SN, Schweitzer CM, Gan J, DeCaprio JA. Cytoplasmic localization of human cdc25C during interphase requires an intact 14-3-3 binding site. Mol Cell Biol. 1999;19(6):4465–79.
124. Peng CY, Graves PR, Thoma RS, Wu Z, Shaw AS, Piwnica-Worms H. Mitotic and G2 checkpoint control: regulation of 14-3-3 protein binding by phosphorylation of Cdc25C on serine-216. Science. 1997;277(5331):1501–5.
125. Chan PM, Ng YW, Manser E. A robust protocol to map binding sites of the 14-3-3 interactome: Cdc25C requires phosphorylation of both S216 and S263 to bind 14-3-3. Mol Cell Proteomics. 2011;10(3):M110.005157.
126. Esmenjaud-Mailhat C, Lobjois V, Froment C, Golsteyn RM, Monsarrat B, Ducommun B. Phosphorylation of CDC25C at S263 controls its intracellular localisation. FEBS Lett. 2007;581(21):3979–85.
127. Chen MS, Hurov J, White LS, Woodford-Thomas T, Piwnica-Worms H. Absence of apparent phenotype in mice lacking Cdc25C protein phosphatase. Mol Cell Biol. 2001;21(12): 3853–61.
128. Poon RY, Chau MS, Yamashita K, Hunter T. The role of Cdc2 feedback loop control in the DNA damage checkpoint in mammalian cells. Cancer Res. 1997;57(22):5168–78.
129. Bulavin DV, Higashimoto Y, Demidenko ZN, Meek S, Graves P, Phillips C, et al. Dual phosphorylation controls Cdc25 phosphatases and mitotic entry. Nat Cell Biol. 2003;5(6):545–51.
130. Boutros R, Lobjois V, Ducommun B. CDC25 phosphatases in cancer cells: key players? Good targets? Nat Rev Cancer. 2007;7(7):495–507.

131. Bulavin DV, Higashimoto Y, Popoff IJ, Gaarde WA, Basrur V, Potapova O, et al. Initiation of a G2/M checkpoint after ultraviolet radiation requires p38 kinase. Nature. 2001;411(6833): 102–7.
132. Lemaire M, Froment C, Boutros R, Mondesert O, Nebreda AR, Monsarrat B, et al. CDC25B phosphorylation by p38 and MK-2. Cell Cycle. 2006;5(15):1649–53.
133. Manke IA, Nguyen A, Lim D, Stewart MQ, Elia AE, Yaffe MB. MAPKAP kinase-2 is a cell cycle checkpoint kinase that regulates the G2/M transition and S phase progression in response to UV irradiation. Mol Cell. 2005;17(1):37–48.
134. Ogg S, Gabrielli B, Piwnica-Worms H. Purification of a serine kinase that associates with and phosphorylates human cdc25C on Serine 216. J Biol Chem. 1994;269(48):30461–9.
135. Hutchins JR, Clarke PR. Many fingers on the mitotic trigger: post-translational regulation of the Cdc25C phosphatase. Cell Cycle. 2004;3(1):41–5.
136. van Vugt MA, Bras A, Medema RH. Polo-like kinase-1 controls recovery from a G2 DNA damage-induced arrest in mammalian cells. Mol Cell. 2004;15(5):799–811.
137. Dai Y, Grant S. New insights into checkpoint kinase 1 in the DNA damage response signaling network. Clin Cancer Res. 2010;16(2):376–83.
138. Graves PR, Yu L, Schwarz JK, Gales J, Sausville EA, O'Connor PM, et al. The Chk1 protein kinase and the Cdc25C regulatory pathways are targets of the anticancer agent UCN-01. J Biol Chem. 2000;275(8):5600–5.
139. Liu Q, Guntuku S, Cui XS, Matsuoka S, Cortez D, Tamai K, et al. Chk1 is an essential kinase that is regulated by Atr and required for the G(2)/M DNA damage checkpoint. Genes Dev. 2000;14(12):1448–59.
140. Astuti P, Gabrielli B. Phosphorylation of Cdc25B3 Ser169 regulates 14-3-3 binding to Ser151 and Cdc25B activity. Cell Cycle. 2011;10(12):1960–7.
141. Conklin DS, Galaktionov K, Beach D. 14-3-3 proteins associate with cdc25 phosphatases. Proc Natl Acad Sci U S A. 1995;92(17):7892–6.
142. Galaktionov K, Jessus C, Beach D. Raf1 interaction with Cdc25 phosphatase ties mitogenic signal transduction to cell cycle activation. Genes Dev. 1995;9(9):1046–58.
143. Xia K, Lee RS, Narsimhan RP, Mukhopadhyay NK, Neel BG, Roberts TM. Tyrosine phosphorylation of the proto-oncoprotein Raf-1 is regulated by Raf-1 itself and the phosphatase Cdc25A. Mol Cell Biol. 1999;19(7):4819–24.
144. Fuhrmann G, Leisser C, Rosenberger G, Grusch M, Huettenbrenner S, Halama T, et al. Cdc25A phosphatase suppresses apoptosis induced by serum deprivation. Oncogene. 2001;20(33):4542–53.
145. Leisser C, Rosenberger G, Maier S, Fuhrmann G, Grusch M, Strasser S, et al. Subcellular localisation of Cdc25A determines cell fate. Cell Death Differ. 2004;11(1):80–9.
146. Obexer P, Geiger K, Ambros PF, Meister B, Ausserlechner MJ. FKHRL1-mediated expression of Noxa and Bim induces apoptosis via the mitochondria in neuroblastoma cells. Cell Death Differ. 2007;14(3):534–47.
147. Zou X, Tsutsui T, Ray D, Blomquist JF, Ichijo H, Ucker DS, et al. The cell cycle-regulatory CDC25A phosphatase inhibits apoptosis signal-regulating kinase 1. Mol Cell Biol. 2001;21(14):4818–28.
148. Kutuzov MA, Andreeva AV, Voyno-Yasenetskaya TA. Regulation of apoptosis signal-regulating kinase 1 (ASK1) by polyamine levels via protein phosphatase 5. J Biol Chem. 2005;280(27):25388–95.
149. el Bahassi M, Yin M, Robbins SB, Li YQ, Conrady DG, Yuan Z, et al. A human cancer-predisposing polymorphism in Cdc25A is embryonic lethal in the mouse and promotes ASK-1 mediated apoptosis. Cell Div. 2011;6(1):4.
150. Mazars A, Fernandez-Vidal A, Mondesert O, Lorenzo C, Prevost G, Ducommun B, et al. A caspase-dependent cleavage of CDC25A generates an active fragment activating cyclin-dependent kinase 2 during apoptosis. Cell Death Differ. 2009;16(2):208–18.
151. Chou ST, Yen YC, Lee CM, Chen MS. Pro-apoptotic role of Cdc25A: activation of cyclin B1/Cdc2 by the Cdc25A C-terminal domain. J Biol Chem. 2010;285(23):17833–45.

152. Liu J, Rosen BP. Ligand interactions of the ArsC arsenate reductase. J Biol Chem. 1997;272(34):21084–9.
153. Mukhopadhyay R, Rosen BP. Saccharomyces cerevisiae ACR2 gene encodes an arsenate reductase. FEMS Microbiol Lett. 1998;168(1):127–36.
154. Mukhopadhyay R, Rosen BP. Arsenate reductases in prokaryotes and eukaryotes. Environ Health Perspect. 2002;110 Suppl 5:745–8.
155. Radabaugh TR, Aposhian HV. Enzymatic reduction of arsenic compounds in mammalian systems: reduction of arsenate to arsenite by human liver arsenate reductase. Chem Res Toxicol. 2000;13(1):26–30.
156. Bordo D, Bork P. The rhodanese/Cdc25 phosphatase superfamily. Sequence-structure-function relations. EMBO Rep. 2002;3(8):741–6.
157. Hofmann K, Bucher P, Kajava AV. A model of Cdc25 phosphatase catalytic domain and Cdk-interaction surface based on the presence of a rhodanese homology domain. J Mol Biol. 1998;282(1):195–208.
158. Yeo HK, Lee JY. Crystal structure of Saccharomyces cerevisiae Ygr203w, a homolog of single-domain rhodanese and Cdc25 phosphatase catalytic domain. Proteins. 2009;76(2):520–4.
159. Duan GL, Zhou Y, Tong YP, Mukhopadhyay R, Rosen BP, Zhu YG. A CDC25 homologue from rice functions as an arsenate reductase. New Phytol. 2007;174(2):311–21.
160. Bleeker PM, Hakvoort HW, Bliek M, Souer E, Schat H. Enhanced arsenate reduction by a CDC25-like tyrosine phosphatase explains increased phytochelatin accumulation in arsenate-tolerant Holcus lanatus. Plant J. 2006;45(6):917–29.
161. Salgado A, Lopez-Serrano Oliver A, Matia-Gonzalez AM, Sotelo J, Zarco-Fernandez S, Munoz-Olivas R, et al. Response to arsenate treatment in Schizosaccharomyces pombe and the role of its arsenate reductase activity. PLoS One. 2012;7(8), e43208.
162. Bhattacharjee H, Sheng J, Ajees AA, Mukhopadhyay R, Rosen BP. Adventitious arsenate reductase activity of the catalytic domain of the human Cdc25B and Cdc25C phosphatases. Biochemistry. 2010;49(4):802–9.
163. Tapio S, Grosche B. Arsenic in the aetiology of cancer. Mutat Res. 2006;612(3):215–46.
164. Chowdhury R, Chatterjee R, Giri AK, Mandal C, Chaudhuri K. Arsenic-induced cell proliferation is associated with enhanced ROS generation, Erk signaling and CyclinA expression. Toxicol Lett. 2010;198(2):263–71.
165. Yih LH, Wu YC, Hsu NC, Kuo HH. Arsenic trioxide induces abnormal mitotic spindles through a PIP4KIIgamma/Rho pathway. Toxicol Sci. 2012;128(1):115–25.
166. Lehmann GM, McCabe Jr MJ. Arsenite slows S phase progression via inhibition of cdc25A dual specificity phosphatase gene transcription. Toxicol Sci. 2007;99(1):70–8.
167. Ma ZQ, Liu Z, Ngan ES, Tsai SY. Cdc25B functions as a novel coactivator for the steroid receptors. Mol Cell Biol. 2001;21(23):8056–67.
168. Altucci L, Addeo R, Cicatiello L, Dauvois S, Parker MG, Truss M, et al. 17beta-Estradiol induces cyclin D1 gene transcription, p36D1-p34cdk4 complex activation and p105Rb phosphorylation during mitogenic stimulation of G(1)-arrested human breast cancer cells. Oncogene. 1996;12(11):2315–24.
169. Prall OW, Sarcevic B, Musgrove EA, Watts CK, Sutherland RL. Estrogen-induced activation of Cdk4 and Cdk2 during G1-S phase progression is accompanied by increased cyclin D1 expression and decreased cyclin-dependent kinase inhibitor association with cyclin E-Cdk2. J Biol Chem. 1997;272(16):10882–94.
170. Lamb J, Ladha MH, McMahon C, Sutherland RL, Ewen ME. Regulation of the functional interaction between cyclin D1 and the estrogen receptor. Mol Cell Biol. 2000;20(23):8667–75. Epub 2000/11/14.
171. Kuo MH, Allis CD. Roles of histone acetyltransferases and deacetylases in gene regulation. Bioessays. 1998;20(8):615–26.
172. Lee KK, Workman JL. Histone acetyltransferase complexes: one size doesn't fit all. Nat Rev Mol Cell Biol. 2007;8(4):284–95.

173. Wang Z, Wang M, Lazo JS, Carr BI. Identification of epidermal growth factor receptor as a target of Cdc25A protein phosphatase. J Biol Chem. 2002;277(22):19470–5.
174. Wang Z, Zhang B, Wang M, Carr BI. Cdc25A and ERK interaction: EGFR-independent ERK activation by a protein phosphatase Cdc25A inhibitor, compound 5. J Cell Physiol. 2005;204(2):437–44.
175. Melchheier I, von Montfort C, Stuhlmann D, Sies H, Klotz LO. Quinone-induced Cdc25A inhibition causes ERK-dependent connexin phosphorylation. Biochem Biophys Res Commun. 2005;327(4):1016–23.
176. Chiu YT, Han HY, Leung SC, Yuen HF, Chau CW, Guo Z, et al. CDC25A functions as a novel Ar corepressor in prostate cancer cells. J Mol Biol. 2009;385(2):446–56.
177. Ngan ES, Hashimoto Y, Ma ZQ, Tsai MJ, Tsai SY. Overexpression of Cdc25B, an androgen receptor coactivator, in prostate cancer. Oncogene. 2003;22(5):734–9. Epub 2003/02/06.
178. Ozen M, Ittmann M. Increased expression and activity of CDC25C phosphatase and an alternatively spliced variant in prostate cancer. Clin Cancer Res. 2005;11(13):4701–6.
179. Broggini M, Buraggi G, Brenna A, Riva L, Codegoni AM, Torri V, et al. Cell cycle-related phosphatases CDC25A and B expression correlates with survival in ovarian cancer patients. Anticancer Res. 2000;20(6C):4835–40.
180. Lorca T, Bernis C, Vigneron S, Burgess A, Brioudes E, Labbe JC, et al. Constant regulation of both the MPF amplification loop and the Greatwall-PP2A pathway is required for metaphase II arrest and correct entry into the first embryonic cell cycle. J Cell Sci. 2010;123(Pt 13):2281–91.
181. Horiguchi T, Nishi K, Hakoda S, Tanida S, Nagata A, Okayama H. Dnacin A1 and dnacin B1 are antitumor antibiotics that inhibit cdc25B phosphatase activity. Biochem Pharmacol. 1994;48(11):2139–41.
182. Wipf P, Hopkins CR, Phillips EO, Lazo JS. Separation of Cdc25 dual specificity phosphatase inhibition and DNA cleaving activities in a focused library of analogs of the antitumor antibiotic Dnacin. Tetrahedron. 2002;58(32):6367–72.
183. Lazo JS, Wipf P. Is Cdc25 a druggable target? Anticancer Agents Med Chem. 2008;8(8): 837–42.
184. Lavecchia A, Di Giovanni C, Novellino E. CDC25 phosphatase inhibitors: an update. Mini-Rev Med Chem. 2012;12(1):62–73.
185. Wu FYH, Sun TP. Vitamin K-3 induces cell cycle arrest and cell death by inhibiting Cdc25 phosphatase. Eur J Cancer. 1999;35(9):1388–93.
186. Brezak MC, Quaranta M, Contour-Galcera MO, Lavergne O, Mondesert O, Auvray P, et al. Inhibition of human tumor cell growth in vivo by an orally bioavailable inhibitor of CDC25 phosphatases. Mol Cancer Ther. 2005;4(9):1378–87.
187. Tamura K, Southwick EC, Kerns J, Rosi K, Carr BI, Wilcox C, et al. Cdc25 inhibition and cell cycle arrest by a synthetic thioalkyl vitamin K analogue. Cancer Res. 2000;60(5):1317–25.
188. Kar S, Wang MF, Ham SW, Carr BI. H32, a non-quinone sulfone analog of vitamin K3, inhibits human hepatoma cell growth by inhibiting Cdc25 and activating ERK. Cancer Biol Ther. 2006;5(10):1340–7.
189. Brisson M, Foster C, Wipf P, Joo B, Tomko RJ, Nguyen T, et al. Independent mechanistic inhibition of Cdc25 phosphatases by a natural product caulibugulone. Mol Pharmacol. 2007;71(1):184–92.
190. Aoyagi Y, Masuko N, Ohkubo S, Kitade M, Nagai K, Okazaki S, et al. A novel cinnamic acid derivative that inhibits Cdc25 dual-specificity phosphatase activity. Cancer Sci. 2005;96(9):614–9.
191. Brezak MC, Quaranta M, Mondesert O, Galcera MO, Lavergne O, Alby F, et al. A novel synthetic inhibitor of CDC25 phosphatases: BN82002. Cancer Res. 2004;64(9):3320–5.
192. Ajiro K, Yasuda H, Tsuji H. Vanadate triggers the transition from chromosome condensation to decondensation in a mitotic mutant (tsTM13) inactivation of p34cdc2/H1 kinase and dephosphorylation of mitosis-specific histone H3. FEBS J. 1996;241(3):923–30.
193. Potapova TA, Sivakumar S, Flynn JN, Li R, Gorbsky GJ. Mitotic progression becomes irreversible in prometaphase and collapses when Wee1 and Cdc25 are inhibited. Mol Biol Cell. 2011;22(8):1191–206.

194. Cazales M, Boutros R, Brezak MC, Chaumeron S, Prevost G, Ducommun B. Pharmacologic inhibition of CDC25 phosphatases impairs interphase microtubule dynamics and mitotic spindle assembly. Mol Cancer Ther. 2007;6(1):318–25.
195. Brezak MC, Valette A, Quaranta M, Contour-Galcera MO, Jullien D, Lavergne O, et al. IRC-083864, a novel bis quinone inhibitor of CDC25 phosphatases active against human cancer cells. Int J Cancer. 2009;124(6):1449–56.
196. Kar S, Wang M, Carr BI. 2-Methoxyestradiol inhibits hepatocellular carcinoma cell growth by inhibiting Cdc25 and inducing cell cycle arrest and apoptosis. Cancer Chemother Pharmacol. 2008;62(5):831–40.
197. Brault L, Bagrel D. Activity of novel Cdc25 inhibitors and preliminary evaluation of their potentiation of chemotherapeutic drugs in human breast cancer cells. Life Sci. 2008; 82(5-6):315–23.
198. Rajski SR, Williams RM. DNA cross-linking agents as antitumor drugs. Chem Rev. 1998;98(8):2723–96.
199. Brisson M, Nguyen T, Wipf P, Joo B, Day BW, Skoko JS, et al. Redox regulation of Cdc25B by cell-active quinolinediones. Mol Pharmacol. 2005;68(6):1810–20.
200. Lavecchia A, Di Giovanni C, Novellino E. Inhibitors of Cdc25 phosphatases as anticancer agents: a patent review. Expert Opin Ther Pat. 2010;20(3):405–25.
201. Heidenreich E, Eisler H, Steinboeck F. Epistatic participation of REV1 and REV3 in the formation of UV-induced frameshift mutations in cell cycle-arrested yeast cells. Mutat Res. 2006;593(1–2):187–95.
202. Masyuk TV, Radtke BN, Stroope AJ, Banales JM, Masyuk AI, Gradilone SA, et al. Inhibition of Cdc25A suppresses hepato-renal cystogenesis in rodent models of polycystic kidney and liver disease. Gastroenterology. 2012;142(3):622–33. e4.
203. Viry E, Anwar A, Kirsch G, Jacob C, Diederich M, Bagrel D. Antiproliferative effect of natural tetrasulfides in human breast cancer cells is mediated through the inhibition of the cell division cycle 25 phosphatases. Int J Oncol. 2011;38(4):1103–11.
204. Sibille E, Bana E, Chaouni W, Diederich M, Bagrel D, Chaimbault P. Development of a matrix-assisted laser desorption/ionization-mass spectrometry screening test to evidence reversible and irreversible inhibitors of CDC25 phosphatases. Anal Biochem. 2012;430(1): 83–91.
205. Lavecchia A, Di Giovanni C, Pesapane A, Montuori N, Ragno P, Martucci NM, et al. Discovery of new inhibitors of Cdc25B dual specificity phosphatases by structure-based virtual screening. J Med Chem. 2012;55(9):4142–58.
206. Mak LH, Knott J, Scott KA, Scott C, Whyte GF, Ye Y, et al. Arylstibonic acids are potent and isoform-selective inhibitors of Cdc25a and Cdc25b phosphatases. Bioorg Med Chem. 2012;20(14):4371–6.
207. Stolfi C, Fina D, Caruso R, Caprioli F, Fantini MC, Rizzo A, et al. Mesalazine negatively regulates CDC25A protein expression and promotes accumulation of colon cancer cells in S phase. Carcinogenesis. 2008;29(6):1258–66.

Protein-Tyrosine Phosphatases: Linking Metabolism and Cancer

Kendra K. Bence

Abstract Obesity and its associated metabolic complications, including type II diabetes, are increasing at an unprecedented rate in the United States and globally. Recent epidemiological data identify a strong connection between obesity as a predisposing risk factor for many different types of cancer, yet the molecular links between altered metabolism and cancer are not well understood. Protein tyrosine phosphatases (PTPs) are critical cellular enzymes that regulate a multitude of signaling pathways through rapid dephosphorylation of phosphorylated tyrosyl residues, and thus are central to the regulation of both metabolic and oncogenic signaling. In this chapter, the role of PTPs in regulating insulin, IGF-1, and adipokine signaling is discussed in the context of bridging the gap between metabolism and cancer.

Keywords Metabolism • Cancer • Phosphatase • Signaling • Obesity • Diabetes • Insulin • IGF-1 • Leptin • Adipokine

Obesity, Type 2 Diabetes, and Cancer

The prevalence of obesity and related comorbidities is a major global health crisis. In the United States, current estimates suggest that approximately 35 % of adults are classified as obese and 33 % are considered overweight [1]. Obesity often precedes the development of metabolic syndrome, characterized by insulin resistance,

K.K. Bence, Ph.D. (✉)
Cardiovascular and Metabolic Diseases Research Unit, Pfizer, Inc.,
610 Main Street, Cambridge, MA 02139, USA

Department of Animal Biology, School of Veterinary Medicine, University of Pennsylvania,
3800 Spruce Street, Vet 223E, Philadelphia, PA 19104-6046, USA
e-mail: kbence@vet.upenn.edu

B.G. Neel, N. Tonks (eds.), *Protein Tyrosine Phosphatases in Cancer*,
DOI 10.1007/978-1-4939-3649-6_12

dyslipidemia, altered glucose homeostasis, inflammation, and cardiovascular complications. Recent epidemiological data has found that these deleterious features of metabolic syndrome are linked to an increase risk for many types of cancer, including breast, colorectal, esophageal, liver, endometrial, uterine, ovarian, kidney, prostate, pancreatic, gallbladder, and hematologic malignancies [2–6]. In fact, one study suggests that 20 % of cancer deaths can be attributed to obesity [7]. Obesity is generally a state of positive energy balance; that is, more calories are consumed than expended. When this occurs chronically over time, many cellular signaling pathways are disrupted, including the insulin/insulin-like growth factor axis, adipokine signaling, and cytokine signaling [8]. Some of these same pathways are also linked to the anticancer effects of calorie restriction, which induces a state of negative energy balance [9]. Notably, all of these pathways invoke tyrosine phosphorylation cascades that are regulated by the reciprocal actions of protein-tyrosine kinases (PTKs) and protein-tyrosine phosphatases (PTPs). Although the precise mechanisms by which obesity predisposes to or accelerates cancer development are not clear, alterations in the cellular signaling pathways common to obesity and cancer are likely involved.

Similarly, type 2 diabetes is also linked to increased cancer risk. As a comorbidity of obesity, the number of people diagnosed with type 2 diabetes is on the rise. Recent estimates from the Centers for Disease Control (CDC) indicate that over 29 million people in the United States alone (roughly equivalent to 9.3 % of the population) suffer from the disease (http://www.cdc.gov/diabetes/data/statistics/2014StatisticsReport.html). Worldwide, the numbers are staggering: researchers predict that the number of people living with type 2 diabetes will reach 366 million by the year 2030 [10]. Interestingly, type 2 diabetes has been widely noted as an independent risk factor for cancer (aside from altered body weight or adiposity) (reviewed in [11]). This is likely due to disruption of normal endocrine function. Type 2 diabetes is characterized by the development of insulin resistance, or a failure of peripheral tissues such as skeletal muscle, adipose, and liver to respond to the effects of insulin, resulting in chronically elevated levels of insulin (hyperinsulinemia) and an increase in circulating insulin-like growth factor-1 (IGF-1).

Energy balance refers to the ability of an organism to achieve homeostasis of many physiological metabolic processes, including food intake, energy expenditure, thermoregulation, glucose homeostasis, and nutrient metabolism. The ability to manipulate or disrupt "normal" metabolic functions by using rodent models of obesity or calorie restriction has revealed the importance of a multitude of cellular signaling pathways in achieving energy balance, including insulin, IGF-1, and adipokines, such as leptin and adiponectin. Notably, these same signaling pathways are implicated in cancer. The connection of each of these pathways to metabolism and cancer, and the role of PTPs in regulating these pathways, are discussed below. Other chapters within this volume provide an in-depth analysis of individual PTP functions in the context of cancer; in this chapter, I will highlight which of these PTPs may bridge the gap between metabolic and oncogenic signaling.

Insulin/IGF-1 Axis and Regulation by PTPs

The increased prevalence of cancer mortality in patients with obesity and/or type 2 diabetes has been linked to insulin resistance and hyperinsulinemia, placing insulin signaling in the spotlight as a possible cellular connection between cancer and altered metabolic state. Insulin resistance in peripheral tissues such as muscle, liver, and adipose tissue results in a compensatory response from the beta cells of the pancreas to secrete more insulin into the circulation. Elevated insulin levels can lead to elevated insulin-like growth factor levels by increasing hepatic IGF-1 production and by decreasing levels of IGF binding proteins (IGFBPs), which normally sequester bioactive IGF-1 [12]. Notably, many tumors overexpress the insulin receptor and IGF-1R and display elevated activation of the insulin/IGF signaling pathway [12–16]. Conversely, calorie restriction in rodents results in lowering of circulating insulin and IGF levels which is protective against cancer [17].

Insulin is a key regulator of glucose homeostasis and fat metabolism, which promotes the uptake and storage of nutrients in peripheral tissues [18]. Insulin action at the cellular level is mediated by insulin receptors (IRs) on the plasma membranes of target cells [19]. Both the IR and IGF-1R are receptor tyrosine kinases (RTKs) made up of an α- and β-subunit. Ligand binding to the α-subunit results in autophosphorylation of the β-subunit and activation of downstream signaling pathways, including the MAPK, PI3K/AKT, and JAK/STAT pathways. The IR has two isoforms as a result of alternative splicing, IR-A and IR-B. The IR-A isoform is expressed in neonatal and cancer cells and signals primarily to mitogenic pathways, whereas the IR-B isoform is primarily involved in glucose uptake and metabolic signaling [20]. Insulin can bind to IR-A and IR-B, and can also bind to IGF-1R with low affinity. On the other hand, IGF-1 binds to the IGF-1R or hybrid IR/IGF receptors (made up of an α- and β-subunit of each receptor), but binds to the IR with low affinity (reviewed in [12]). Activation of the IR/IGFR leads to transphosphorylation of tyrosine residues in the activation loop, which in turn leads to enhanced ability of the receptor to phosphorylate target proteins, including insulin receptor substrates (IRSs) and SHC. Tyrosyl phosphorylated IRS proteins act as docking sites for SH2 domain-containing proteins, including SHP2 and the p85 regulatory subunit of phosphatidylinositol 3-kinase (PI3K). PI3K becomes activated upon binding to IRS proteins, and its phosphoinositide products promote the activation of downstream targets such as AKT, which helps promote translocation of glucose transporters from intracellular stores to the cell surface and is also important for regulation of cell proliferation, growth, and survival [21]. A second major signaling pathway involves the recruitment of GRB2 to SHC and activation of RAS, which subsequently binds to RAF promoting activation of the ERK mitogen-activated protein kinase (MAPK) phosphorylation cascade [22]. More recent studies also suggest that IGF-1 in particular might also signal through the JAK/STAT signaling pathway (reviewed in [23]).

Although insulin may directly promote cancer risk through its mitogenic effects, it may also indirectly influence cancer risk by promoting increased levels of bioavailable IGF-1 in the circulation. IGF-1 is an important regulator of tissue growth and metabolism; dysregulation of the IGF-1 axis leads to a number of pathologies, including metabolic and growth disorders, as well as cancer. In the presence of elevated insulin, growth hormone receptor (GHR) expression in the liver is elevated and the production and secretion of IGF-1 is increased [11]. Studies using cell lines have demonstrated the requirement of an intact functional IGF-1R for oncogenic transformation and animal studies have shown that altering the levels of IGF-1R expression affects tumor growth [24, 25]. In rodent models, obesity is correlated with elevated IGF-1 levels, whereas caloric restriction reduces IGF-1 levels [9, 26–28]. In humans, studies have found a significant link between higher levels of IGF-1 and cancer (reviewed in [27]).

Because obesity is a pro-inflammatory state, characterized by increased adiposity, elevated levels of cytokines and adipokines, hyperinsulinemia and hyperglycemia, it is difficult to tease apart the individual effects of each of these parameters on cancer risk. LeRoith and colleagues attempted to do just that with their MKR mouse model; these mice express a dominant negative IGF1R specifically within skeletal muscle. The MKR mice are insulin-resistant and glucose intolerant, but are not obese and have normal levels of circulating inflammatory cytokines. Female MKR mice display enhanced growth of mammary tumors, which can be remediated by blocking the IR/IGF-1 pathway or the PI3K/AKT pathway. These findings suggest that the hyperinsulinemia is an important driving force of the accelerated tumor growth [14, 29]. Although the mechanistic link between insulin resistance, hyperinsulinemia, and cancer is largely unclear, this signaling pathway appears to play a prominent role in promoting malignancy and neoplastic transformation. Because the IR and IGF-1R are RTKs and invoke feed-forward tyrosyl phosphorylation cascades, there is a pressing need to understand their regulation by PTPs. Indeed many PTPs have been shown to regulate the insulin/IGF signaling pathway (reviewed in [30] and discussed below).

PTP1B

PTP1B is a member of the classical PTP superfamily of receptor-like and non-transmembrane proteins [31–33], and is an important cellular regulator of insulin and growth factor receptor signaling. Classical PTPs contain a catalytic domain specific for phosphotyrosyl (pTyr) residues, with a central "signature motif" containing an essential cysteine that executes a nucleophilic attack on the phosphate ester of the substrate resulting in dephosphorylation of target pTyr residues [34]. PTP1B is an abundant, ubiquitous cellular enzyme, localized on the endoplasmic reticulum via its hydrophobic C-terminal targeting sequence [35, 36]. Early studies in cultured cells suggested that PTP1B might inhibit insulin signaling by dephosphorylating the IR and possibly IRS-1 [37–46]. Consistent with these studies, two papers described the generation of germline PTP1B-deficiency (*Ptpn1*$^{-/-}$) in mice and showed that in the absence of PTP1B, insulin signaling is enhanced and/or prolonged in muscle and liver, resulting in improved overall insulin sensitivity even in the face of diet-induced obesity [47, 48]. Furthermore, mice treated with PTP1B-specific anti-sense oligos (ASO) displayed improved insulin sensitivity [49–51]. Transgenic overexpression of human *PTPN1* in skeletal muscle of mice results in whole-body insulin resistance [52]; similarly, hepatic re-expression of *PTPN1* in *Ptpn1−/−* mice results in insulin resistance and impaired hepatic insulin-stimulated IR phosphorylation [53]. On the other hand, muscle-specific PTP1B-deficiency results in improved systemic insulin sensitivity, enhanced muscle glucose uptake, and increased insulin signaling [54]. Liver-specific PTP1B-deficiency results in improved glucose tolerance, increased hepatic insulin signaling, and protection against high-fat diet-induced fatty liver and endoplasmic reticulum (ER) stress [55]. Furthermore, there is increased association between the glucose transporter GLUT2 and IR-A in PTP1B-deficient neonatal hepatocytes resulting in increased glucose uptake compared with wildtype controls [56]. Despite the clear role for PTP1B regulation of the IR in liver and muscle, PTP1B does not appear to play a role in regulating IR signaling in white adipose tissue, suggesting that other PTPs play a more predominant role in this tissue [47, 48, 57]. Together, these results along with structural studies [58] conclusively established that PTP1B is a physiologically important regulator of insulin signaling.

PTP1B has also been implicated in the regulation of IGF signaling, although its role in this context is more complex. Co-expression of PTP1B with the β-chain of IGF-1R inhibited IGF-1R kinase activity in yeast, COS cells and fibroblasts, and PTP1B-deficient fibroblasts displayed enhanced IGF-1-mediated survival, growth, and motility [59]. Furthermore, PTP1B has been shown to directly interact with IGF-1 receptors and IR-A/IGF1Rs in cell lines [60, 61]. In addition, the level of hepatic IGF-I mRNA and protein, as well as IGF-IR tyrosine phosphorylation, is elevated in young *Ptpn1−/−* mice [62], consistent with a role for PTP1B as a negative regulator of IGF-1 signaling. However, PTP1B has recently been shown to also have growth promoting effects; PTP1B can dephosphorylate the inhibitory Y529 site in SRC leading to its activation, and has been shown to be a positive mediator of ErbB2-induced breast cancer in mice (reviewed in [63]). Consistent with this notion, IGF-2 induced MCF-7 cell migration is impaired upon siRNA-mediated reduction of PTP1B or its closest homologue TCPTP [64]. Most notably, there are no overt oncogenic phenotypes noted in *Ptpn1−/−* mice suggesting that in vivo, there are compensatory mechanisms for any chronic activation of IGF signaling. Further studies are warranted on the specific effects of PTP1B (and TCPTP) on IGF receptor signaling.

TCPTP

T-cell PTP (TCPTP), encoded by *PTPN2* in humans, is closely related to PTP1B and has also been implicated in the regulation of insulin signaling. Initial studies found that TCPTP could interact with the IR [44], and the IR is hyperphosphorylated in *Ptpn2*-null cells [65, 66]. Interestingly, despite the high degree of similarity in their catalytic domains, phosphorylation of the IR appears to be differentially affected by the absence of these two PTPs:increased phosphorylation of Y972 and Y1162/Y1163 is noted in *Ptpn1*-deficient fibroblasts, whereas increased Y972 phosphorylation and sustained phosphorylation of Y1162/Y1163 are found in *Ptpn2*-deficient fibroblasts [67]. Insulin-induced PI3K/AKT signaling is prolonged in both *Ptpn1*–/– and *Ptpn2*–/– immortalized MEFs, whereas insulin-induced ERK activation is increased only in *Ptpn1*–/– MEFs, furthering the notion that PTP1B and TCPTP are not redundant in insulin signaling within the same cell (reviewed in [68, 69]). It could be that PTP1B regulates the intensity or peak response to insulin, whereas TCPTP controls the temporal response to insulin within the same cell. Indeed both PTPs are transiently oxidized and inactivated by reactive oxygen species (ROS) in response to insulin stimulation [66], which may be a mechanism to enhance and sustain a peak insulin response [68]. In vivo, TCPTP regulates IR signaling and gluconeogenesis in the liver [70], yet in contrast to muscle-specific PTP1B-deficient mice, muscle-specific TCPTP-deficient mice have completely normal insulin signaling and glucose homeostasis [71]. Furthermore, in MIN6 insulinoma beta cells, lack of PTP1B promoted ER stress-induced cell death whereas TCPTP-deficiency was protective [72]. Notably, TCPTP exists as two distinct isoforms which may have different functions within the cell: a nuclear 45 kDa form (TC45) and an endoplasmic reticulum-associated 48 kDa form (TC48) [68]. Combined, these studies further highlight the distinct physiological roles of PTP1B and TCPTP in different tissues.

SHP2

The SH2 domain-containing PTP SHP2 plays an essential role in RTK signaling through its activation of the downstream ERK pathway. In response to insulin activation of the IR, SHP2 is recruited to a complex with IRS1/2, GAB1, and SHPS1. Whether SHP2 acts as a negative or positive mediator of insulin signaling likely depends on the cell type examined. Studies using 32D myeloid cells found that SHP2 attenuates phosphorylation of IRS-1, thus acting as a negative regulator of insulin's metabolic effects [73], and overexpression of SHP2 in 3T3L1 cells inhibits insulin signaling and glycogen synthesis [74]. On the other hand, expressing catalytically inactive mutants of SHP2 in CHO-IR cells attenuates insulin-induced RAS activation [75]. Studies in vivo are also not entirely consistent. Hemizygous *Ptpn11*$^{+/-}$ mice have normal metabolic responses to insulin, but transgenic expression of a dominant negative version of SHP2 in skeletal muscle leads to insulin resistance [76, 77]. Tissue-specific deletion of *Ptpn11* in skeletal muscle or pancreas also leads to insulin resistance and impaired glucose tolerance [78, 79]. By contrast, deletion of *Ptpn11* in the liver results in improved insulin signaling and enhanced glucose tolerance [80], and deletion of *Ptpn11* in adipose tissue has no effect on systemic insulin sensitivity or glucose homeostasis [81]. These studies highlight the complex role SHP2 plays in regulating insulin signaling in various cellular contexts.

SHP2 regulation of IGF signaling may also be cell type dependent. For example, in vascular smooth muscle cells, IGF-1 stimulates SHPS-1 phosphorylation and recruitment of SHP2 to the receptor resulting in IGF-1R dephosphorylation [82, 83]. In C2C12 myoblasts, the GAB1–SHP2 complex inhibits IGF-1-dependent myogenesis through ERK1/2 [84]. In fibroblasts and hematopoietic cells, on the other hand, SHP2 is found in complex with IRS proteins and acts as a positive mediator of IGF-1 signaling to ERK in response to IGF-1 (although IGF-1 does not induce tyrosyl phosphorylation of SHP2 in these cells as other growth factors do, including FGF and PDGF) [85]. Overall, it appears that, depending on the cell/tissue type or context, SHP2 can act to promote or inhibit IR/IGF-1 signaling, perhaps by acting on different substrates or in differing subcellular locations. Given the important role of SHP2 in cancer and metabolism, it will be important to more fully understand how SHP2 functions in these different cellular contexts.

LAR

The transmembrane PTP LAR has been shown in cell lines to be a negative regulator of insulin signaling and can associate directly with the IR [86, 87]. Transgenic overexpression of human *PTPRF*in muscle of mice is consistent with a physiological role for LAR as a negative regulator of insulin signaling [88] and reviewed in [89]. In addition, LAR-knockdown improved insulin-stimulated glucose uptake in cultured myotubes treated with palmitate [90], and upregulation of LAR in response to TNF-α impairs insulin-mediated tyrosine phosphorylation of FAK in liver cells [91]. A mouse model of *Ptprf*-deficiency shows some evidence of insulin sensitivity but overall deficits in glucose homeostasis [92], making it unclear what role LAR plays in insulin signaling in vivo. Notably, obese, insulin-resistant patients show high levels of LAR in muscle and liver [93], suggesting a probable link between altered metabolism and levels of LAR in humans.

Receptor PTPs

RPTPsigma, encoded by the *Ptprs* gene, is a member of the LAR-RPTP family that is 85 % identical to the intracellular domain of LAR at the amino acid level. It is more abundantly expressed than LAR in adipose and muscle, suggesting it may also be a regulator of insulin sensitivity [94, 95]. Chagnon et al. analyzed *Ptprs−/−* mice and found that they have reduced plasma glucose and insulin levels in the fasted state and are more sensitive to insulin compared to controls in an insulin tolerance test. Interestingly, however, tyrosine phosphorylation of the IR was not altered in the absence of RPTPsigma, suggesting an indirect effect on insulin signaling. Indeed, these mice displayed GH/IGF-1 deficiency and the improved insulin sensitivity was normalized by low dose GH injection [96].

The receptor PTPs, PTPalpha (*Ptpra*) and PTPepsilon (*Ptpre*), have also been shown to interact with the IR and act as negative regulators of IR tyrosine kinase activity in cell lines [97–100]. In primary rat hepatocytes, expression of PTPepsilon resulted in dephosphorylation of the IR at Y972, Y1158, and Y1162/Y1163 and suppression of insulin-induced glycogen synthesis [101]. Furthermore, overexpression of a cytosolic form of PTPepsilon (cytPTPepsilon) in L6 muscle cells impairs insulin action, whereas insulin signaling and insulin-stimulated glucose uptake were enhanced in the absence of cytPTPepsilon [102]. *Ptpre–/–* mice have defects in myelination and increased phosphorylation and activation of voltage-gated potassium channels [103]. In addition, *Ptpre–/–* mice display improved glucose homeostasis and female *Ptpre–/–* mice are hypersensitive to leptin and are protected from high-fat diet-induced weight gain [104]. It will be interesting to determine how PTPepsilon regulates these important metabolic pathways, and whether there might be a connection between the regulation of hormone pathways and ion channels.

In isolated rat adipose cells, expression of wildtype PTPalpha led to a decrease in insulin-stimulated GLUT4 translocation to the plasma membrane [105]. However, *Ptpra–/–* mice have normal glucose and insulin levels and normal performance in insulin or glucose tolerance tests, indicating that PTPalpha might not play a major role in mediating the physiological action of insulin [106]. Interestingly, IGF-1 was recently shown to potently induce PTPalpha phosphorylation at Y789 via a Src-dependent mechanism. PTPalpha interacts directly with the IGF-1R and is required for IGF-1-induced cell migration (although PTPalpha does not appear to be an IGF-1R phosphatase) [107]. Overall, the roles of these receptor PTPs are less well studied in the physiological context of insulin and IGF-1 signaling in vivo, although it should be noted that many receptor PTPs appear to have links to both cancer and metabolism [108].

Finally, although *PTPRC*/CD45 has been shown in some contexts to function as a negative regulator of insulin and IGF-1 signaling, its expression is restricted to cells of hematopoietic origin, making it unlikely to be an important regulator of insulin signaling in vivo [30, 109–112].

SHP-1

The Src-homology 2-domain-containing protein tyrosine phosphatase SHP-1 (*Ptpn6*) is involved in the regulation of multiple signaling pathways and has been proposed as a candidate tumor suppressor gene in lymphoma and leukemia (but is overexpressed in some non-lymphocytic cell lines) [113, 114]. Viable motheaten (me^{v}) mice expressing catalytically impaired SHP-1 display increased systemic insulin sensitivity via increased insulin signaling to AKT in liver and muscle [115]. Consistent with this study, hepatocyte-specific *Ptpn6–/–* mice are insulin hypersensitive and protected against high-fat diet-induced hepatic insulin resistance [116]. Furthermore, expression of a DN-SHP-1 in L6 myocytes increases insulin-induced AKT activation, GLUT4 activity, and glucose uptake [117].

PTP-MEG2

PTP-MEG2 is a non-receptor PTP with a putative lipid-binding domain at the N-terminus; indeed, it has been shown to bind to PIP3, leading to its activation at the plasma membrane [118]. Although its many cellular functions are still being revealed, it has been shown to dephosphorylate the IR and antagonize insulin signaling in hepatocytes [119]. Consistent with this study, treatment of diet-induced obese mice with a PTP-MEG2 inhibitor resulted in increased insulin sensitivity and improved glucose homeostasis [120]. PTP-MEG2 has been recently shown to regulate EGFR and ErbB2 phosphorylation, suggesting a possible role in breast cancer cells [121].

DUSPs

The dual-specificity phosphatases, or DUSPs, are less well-conserved than classical PTPs but retain the conserved catalytic cysteine signature motif. The structure of the active site allows DUSPs to dephosphorylate phosphoserine/phosphothreonine residues in addition to phosphotyrosine (reviewed in [122, 123]). The mitogen-activated protein kinase (MAPK) phosphatase family is the most well characterized family of enzymes, which act as critical regulators of MAPK signaling [124, 125]. MKP-1 is capable of interacting with and dephosphorylating ERK1/2, but displays greater preference for the stress-responsive MAPKs JNK and p38 MAPK. Levels of the gene encoding MKP-1, *Dusp1*, are elevated in states of obesity and type II diabetes, and recent genetic studies revealed that *Dusp1−/−* mice are resistant to age-related and high-fat diet-induced obesity due to increased energy expenditure and transcriptional effects in various metabolically active tissues ([126] and reviewed in [125]). Importantly, elevated expression of *DUSP-1* has been reported in some cancers, including melanoma, advanced stage breast cancers, pancreatic cancer and non-small cell lung cancer; however, *DUSP-1* expression is lost or reduced in during cancer progression in colon, bladder, and prostate tumors (reviewed in [127]). Thus, targeting this particular phosphatase for either cancer or metabolic reasons could be challenging and requires further study.

PTEN

The tumor suppressor phosphatase and tensin homologue (PTEN) is frequently disrupted in cancer, and is distinct from the classical PTPs in that it is a dual-specificity phosphatase capable of removing phosphates from both protein and lipid substrates [128, 129]. In the context of cancer and metabolism, the primary targets of PTEN are lipid second messengers. Downstream of insulin and IGF-1 signaling, activated PI3K converts PIP2 to PIP3, counterbalanced by the actions of PTEN, which converts PIP3 back to PIP2. PIP3 activates PDK1 leading to the activation of downstream AKT, which increases cell growth and proliferation, decreases apoptosis, and promotes carcinogenesis via many mechanisms (reviewed in [130, 131] and in additional chapters in this volume).

PRLs

Members of the phosphatase of regenerating liver (PRL) family of PTPs, including PRL-1, PRL-2, and PRL-3, contain a unique prenylation motif at the C-terminus, and are known to be highly expressed in many human cancers [132]. Although little is known about a potential metabolic role for PRLs, there are some intriguing links emerging in the literature connecting PRLs to signaling pathways common to metabolic and oncogenic signaling. For example, DLD-1 colon carcinoma cells overexpressing PRL-3 display elevated AKT activity likely due to downregulation of PTEN expression [133]. Mice lacking PRL3 (*Ptp4a3–/–* mice) display lower fasted blood glucose levels and improved glucose tolerance during an intraperitoneal glucose tolerance test, despite having similar body weights compared to their wildtype controls [134]. Furthermore, colon tumors from *Ptp4a3–/–* mice have elevated levels of IGF-1R, suggesting potential enhancement of IGF signaling in the absence of PRL-3 [135]. Colon tumors isolated from *Ptp4a3*-null mice also have reduced density of tumor microvessels and attenuated VEGF-mediated vascular permeability, suggesting PRL-3 could be an important regulator of endothelial cell function and VEGF-mediated angiogenesis [136]. More studies are needed to elucidate the possible role of the PRL family of PTPs in linking metabolic and oncogenic signaling.

LMW-PTPs

Low molecular weight PTPs (LMW-PTPs), also known as ACP1, are a group of 18 kDa proteins with broad tissue distribution that have been implicated in the regulation of several RTKs [137]. LMW-PTP is a positive mediator of tumor onset and proliferation, and increased expression of LMW-PTPs has been noted in various tumor types, including breast, colon, lung, and neuroblastoma [138, 139]. In vitro, LMW-PTP can act as a negative regulator of insulin-mediated mitogenic and metabolic signaling [140]. Further, an in vivo study found that reducing LMW-PTP levels in liver and fat by using ASO resulted in enhanced IR signaling and improvements in insulin and glucose tolerance in diet-induced and leptin-deficient obese mice [141]. Importantly, LMW-PTP levels in humans are associated with hyperglycemia and dyslipidemia [142, 143], highlighting yet another PTP that may play a cross-functional role in metabolism and cancer in humans.

Obesity, Adipokines, and Cancer: Regulation by PTPs

As outlined in the introduction, obesity is a major global health crisis and is a contributing factor to many different types of cancer. Obesity is a low-grade inflammatory state resulting in over production of adipokines, including adiponectin and leptin, as well as inflammatory molecules such as IL-6, TNFα, monocyte chemoattractant protein (MCP-1), and plasminogen activator inhibitor-1 (PAI-1) [28, 144]. An increase in inflammatory cells in adipose tissue of the obese could promote systemic inflammation, leading to a pro-tumorigenic environment. The regulation of inflammatory cytokine signaling by PTPs in this context is likely to be critically important and has been reviewed elsewhere [145–151]. In this chapter, the role of PTPs in linking oncogenic and metabolic leptin signaling will be highlighted.

Adipokines are emerging as key molecular mediators linking obesity and increased cancer risk. The most well studied of these is the adipocyte-secreted hormone, leptin, which is positively correlated with adipose tissue mass and is a critical regulator of feeding, energy balance, reproduction, inflammation, hematopoiesis, and proliferation [152]. The major role of leptin is in regulation of energy balance. As such, humans and rodents deficient in leptin or its receptor are morbidly obese, hyperphagic, infertile, and diabetic [153]. Leptin is a 16 kDa adipokine produced largely in white adipose tissue, but it can also be produced in the placenta, brown adipose tissue, intestinal cells, ovaries, bone marrow, brain, pituitary, liver, and myocytes [154, 155]. There is also evidence that leptin is expressed in human mammary epithelial cells [156, 157], and overexpressed in some human endometrial and colorectal cancers [158, 159]. Importantly, despite the strong correlation between body mass index (BMI)/waist-hip ratio (WHR) and cancer risk, attempts to correlate *circulating* levels of leptin with cancer incidence or progression have been conflicting [155]. It may be that circulating serum leptin levels are less important than local leptin production and action within the tumor microenvironment [160].

Leptin has been associated with many types of cancer, and promotes growth, migration, and invasiveness of certain cancer cells, likely through activation of the PI3K, ERK, and STAT3 pathways [154, 155, 161, 162]. Leptin can also promote angiogenesis [163, 164]. The long form of the leptin receptor (LRb) is a member of the class I cytokine receptor family; the intracellular domain of the receptor does not have intrinsic tyrosine kinase activity, and thus transmits signals via association with the cytoplasmic Janus tyrosine kinase 2 (JAK2) [165]. Ligand binding causes autophosphorylation and activation of JAK2, which in turn phosphorylates critical tyrosine residues on the intracellular domain of the LRb including Y^{985}, Y^{1077}, and Y^{1138} (reviewed in [166–171]). Signals stemming from activated JAK2 (independent of LRb phosphorylation) lead to activation of the IRS-PI3K pathway, and phosphorylated tyrosine residues on LRb provide docking sites for downstream signaling molecules. For example, phosphorylation of Y^{1077} leads to recruitment and activation of the transcription factor, Signal Transducer and Activator of Transcription 5 (STAT5). Phosphorylation of Y^{1138} recruits STAT3 (Signal Transducer and Activator of Transcription 3), resulting in tyrosine phosphorylation and nuclear translocation of pSTAT3 to mediate gene transcription. One target of activated STAT3 is the suppressor of cytokine signaling SOCS3, a feedback inhibitor of leptin signaling. Inhibition of leptin signaling by SOCS3 occurs by SOCS3 binding to phosphorylated Y^{985} (and to a lesser extent by binding to JAK2). In addition to providing a docking site for SOCS3, phosphorylation of Y^{985} leads to GRB2 and SHP2 recruitment and activation of the ERK pathway. Leptin also appears to influence other important cellular signaling pathways, including mammalian target of Rapamycin (mTOR) and AMP-dependent protein kinase (AMPK), although precisely how this occurs and how/whether this influences cancer risk requires further investigation (summarized in [155, 171]).

The leptin signaling pathway relies largely on tyrosine phosphorylation of key signaling intermediates; thus, opposing regulation of these sites by PTPs is critically important. To date, PTP1B, TCPTP, RPTPε, and SHP2 have been implicated in the direct regulation of leptin signaling, whereas PTEN might also influence leptin signaling through proximal negative regulation of the PI3K pathway (reviewed in [172, 173]). PTP1B has been shown in vitro and in vivo to be an important negative regulator of leptin signaling through direct dephosphorylation of the active site tyrosines of JAK2 (Y^{1007}/Y^{1008}) [174–177]. Similar to whole-body *Ptpn1−/−* mice, PTP1B-deficiency restricted to leptin receptor-expressing cells (*LepRb-Ptpn1−/−* mice) also results in a lean phenotype and improved leptin sensitivity, demonstrating the importance of PTP1B as a physiological "brake" on this pathway in vivo [178]. Levels of the PTP1B protein are elevated in response to high-fat feeding and inflammatory cytokines [179, 180], which might contribute to the molecular events underlying the development of leptin resistance in the obese state. Similar to PTP1B, RPTPε can also act as a negative regulator of leptin signaling through dephosphorylation of JAK2; *Ptpre−/−* mice are resistant to high-fat diet-induced weight gain and are leptin hypersensitive [104]. Interestingly, the PTP most closely related to PTP1B, TCPTP also was recently shown to be induced by high-fat feeding and exogenous leptin. Loh and colleagues describe a role for TCPTP as a negative regulator of leptin signaling through its ability to dephosphorylate STAT3. Brain-specific *Ptpn2−/−* mice are lean and leptin hypersensitive. Importantly, compound deletion of *Ptpn1* and *Ptpn2* in the brain results in additive effects on leptin sensitivity, indicating that these two PTPs can act sequentially on distinct targets within the same signaling pathway [181].

In contrast to PTP1B, TCPTP, and RPTPε, the tyrosine phosphatase SHP2 acts as a positive mediator of leptin signaling in vivo. In response to leptin, tyrosine phosphorylation of LRb-Y^{985} recruits and binds SHP2, which competes for binding with Socs3 and mediates activation of the GRB2-ERK pathway [182]. Although in vitro studies found that SHP2 may also act to suppress JAK2 phosphorylation, deficiency of SHP2 in post-mitotic forebrain neurons in mice (CaSKO mice) reveals a positive role for SHP2 in promoting leptin signaling [183]. CaSKO mice, as well as POMC-*Ptpn11−/−* mice, are obese and leptin resistant [184, 185]. Consistent with these studies, transgenic expression of a dominant-active mutant, SHP2-D61A, in forebrain neurons confers sex-specific resistance to high-fat diet-induced obesity [186]. Notably, this study identifies a physical interaction between SHP2 and the estrogen receptor alpha which is required for the synergistic activation of ERK by leptin and estrogen. It will be interesting to pursue the role of this phosphatase in linking leptin and estrogen signaling in the context of obesity and estrogen receptor-positive breast cancer.

In summary, several PTPs have now been shown to be important regulators of signaling through the leptin-JAK/STAT axis. Given that constitutive activation of the JAK/STAT pathway is implicated in neoplastic transformation and abnormal cell proliferation in many malignancies [150], the role of PTPs in the interplay between metabolic and oncogenic signaling and the effects on individual downstream molecules will be essential to interrogate.

ROS, Insulin Resistance, and PTPs

An over production of cellular ROS in obese and diabetic patients might cause increased DNA damage and underlie the increased risk of cancer in this population. Interestingly, states of nutrient deprivation, such as hypoxia and starvation, as well as states of nutrient excess, such as obesity and oncogenic activation, can lead to cellular oxidative stress [187]. Chronic hyperglycemia and obesity-associated inflammation are correlated with increased mitochondrial ROS production and a state of chronic oxidative stress [188]. Excessive ROS production, in turn, can activate serine/threonine kinases (including IKKβ, JNK, and PKCs) that phosphorylate critical serine residues on IRS proteins and decrease activity of downstream signaling pathways, effectively promoting insulin resistance [189]. Paradoxically, however, transient small bursts of ROS in response to a variety of cytokines, growth factors, and hormones (including insulin) can enhance the response to these ligands through transient, reversible inhibition of PTP activity and other signaling enzymes within the cell [122, 190, 191]. A classic example of this is the inactivation of PTP1B in response to insulin: insulin triggers the transient generation of H_2O_2 within the cell, leading to rapid oxidation of the catalytic cysteine of PTP1B, and promotion of IR activity [66, 192, 193]. In a similar manner, transient oxidation of the active site cysteine in PTEN results in enhanced insulin-induced PI3K activity [194–196], which may contribute to long-term enhancements in insulin sensitivity in vivo [197]. SHP2 and TCPTP (and likely other PTPs) also undergo oxidation in response to various growth factors [66, 198]. Thus, PTPs clearly play an integral role in fine tuning the cellular response to ROS generation, under both normal and pathological conditions. It will be important in future studies to evaluate the role of PTPs in the cellular response to transient vs. sustained ROS, and in the context of varying energy substrate availability.

The free radical nitric oxide (NO) is an important intracellular signaling molecule, yet at the same time can also be a potent mediator of cellular damage. Given the pleiotropic role of this molecule within the cell, it is not surprising that its role in carcinogenesis and tumor progression is equally conflicting (summarized in [199]). NO reacts with superoxide to form the more potent oxidant peroxynitrite ($ONOO^-$). In vivo, excessive NO and peroxynitrite generation are associated with many pathological conditions including obesity, diabetes, and cancer [188, 200]. Both insulin and leptin have been shown to induce NO production within certain cell types [201, 202], and endogenous NO plays an integral role in insulin responsiveness. Many PTPs have been shown to be inhibited by NO-induced *S*-nitrosylation of the active site cysteine residue [203–208]. For example, NO can enhance insulin signaling at least in part by NO-induced *S*-nitrosylation and inactivation of PTP1B,

SHP2, and SHP1 (but not TCPTP) [209]. Thus, the redox environment (oxidizing vs. reducing) within the cell is a critical factor in determining how PTPs may function to regulate various signaling pathways.

Oxidative stress can be reduced by controlling elevated blood glucose levels and by caloric restriction [210]. Interestingly, aging in rats is associated with impaired responsiveness to insulin and an increase in the activity of PTP1B in liver and skeletal muscle, which can be reversed by long-term caloric restriction [211]. Further, chronic inflammation associated with obesity and aging can induce hepatic insulin resistance and increase risk for some types of liver cancer; notably, acute exercise can reverse the elevation in hepatic PTP1B levels associated with aging in rats and improve insulin sensitivity [212]. Thus, the energy status of an organism plays an integral role in setting the oxidative environment and regulating the activity of PTPs. Understanding how PTPs are regulated by ROS during aging and under conditions of metabolic stress may shed light on the molecular link between ROS and cancer risk.

Summary: Implications for Targeting PTPs at the Intersection of Cancer and Metabolism

Given their biological roles as tumor suppressors or oncogenes, many PTPs have emerged as attractive drug targets for treating cancer [213–218], and have more recently been recognized as desirable targets in the treatment of obesity and diabetes [219, 220]. Due to the intersection and complexity of metabolic and oncogenic signaling pathways, it is not surprising that many medications used to treat type 2 diabetes can influence risk factors associated with cancer, including insulin and IGF-1 levels, hormone signaling, and inflammation. For example, in the context of overt type 2 diabetes, exogenous insulin is a common treatment to control hyperglycemia. However, some studies have found an increased rate of cancer mortality in type 2 diabetic patients treated with insulin (or insulin analogues) or drugs that stimulate insulin secretion (sulphonylureas), raising caution about targeting PTPs that may enhance signaling through these pathways (reviewed in [11, 221]).

Given the prominent role for IGF-1 signaling in cancer, it is not surprising that many therapies for cancer treatment have been developed that target IGF-1 and its receptor. One such class of compounds is tyrosine kinase inhibitors which compete for the ATP binding site in the kinase domain of the RTK [222]. As this site is conserved between the IGF-1R and IR, however, these compounds inhibit both IGF-1 and insulin signaling, and raise the potential problem that although blocking IGF-1/IR signaling may be anti-oncogenic, chronically blocking the IR may lead to hyperglycemia. Indeed, this side effect has been reported in animal studies and early clinical trials in humans (reviewed in [27]). If PTPs can be identified that have differential access to/preference for dephosphorylating IGF-1R vs. IR (or hybrid receptors) in certain types of cancer cells, perhaps this would be a novel way to selectively modulate either signaling pathway if that is deemed desirable in a particular context.

The widely used anti-diabetic biguanide compound, metformin, and the insulin-sensitizing thiazolidinedione (TZD) rosiglitazone, have been proposed to have anti-cancer activity in many model systems, potentially through targeting intracellular signaling molecules [221, 223–225]. The effects of these various anti-diabetic compounds on the activity/levels of specific PTPs are largely unknown, although a study in humans did not find a correlation with metformin treatment of Scottish male subjects and PTP1B mRNA levels in adipose tissue [226]. Another study found that metformin reduces PTP1B activity and improves response to IFNα in insulin-resistant obese mice [227]. Treatment of diabetic rats with rosiglitazone reduces levels of PTP1B in skeletal muscle, suggesting this may be a possible mechanism for improved insulin sensitivity [228]. However, rosiglitazone treatment improves insulin sensitivity further in muscle-specific PTP1B-deficient mice, suggesting that TZDs may act additively with PTP1B inhibition and likely work through a PTP1B-independent mechanism [54].

In conclusion, tremendous strides have been made to enhance our understanding of how metabolic and oncogenic signaling might be linked. PTPs clearly play a central role in cancer and metabolism, and future work in understanding how PTPs are themselves regulated, as well as determining the cellular localization and substrate specificity of PTPs under different cellular conditions, should move the field even further along. Of course many other signaling molecules and pathways not mentioned in depth in this chapter are implicated in both metabolism and cancer, including PI3K/AKT, MAPK, c-MYC, p53, and AMPK/LKB1/mTOR. Components of these signaling cascades also invoke tyrosyl phosphorylation and are likely targets of regulation by PTPs. In addition, a key glycolytic regulator linking glucose metabolism to cancer cell metabolism, pyruvate kinase M2 (PKM2), is regulated by tyrosine phosphorylation and was recently shown to be a substrate of PTP1B in adipocytes [229]. This intriguing finding may shed light on how a PTP functions at the intersection of glucose and cancer cell metabolism. Finally, as we move toward the age of personalized medicine, it will be critical to keep in mind how various therapeutic compounds (including those that target PTPs) might differentially affect a lean cancer patient vs. an obese and/or diabetic cancer patient.

References

1. Flegal KM, Carroll MD, Kit BK, Ogden CL. Prevalence of obesity and trends in the distribution of body mass index among US adults, 1999-2010. JAMA. 2012;307(5):491–7. doi:10.1001/jama.2012.39.
2. Calle EE, Rodriguez C, Walker-Thurmond K, Thun MJ. Overweight, obesity, and mortality from cancer in a prospectively studied cohort of U.S. adults. N Engl J Med. 2003;348(17):1625–38. doi:10.1056/NEJMoa021423.
3. Chen J. Multiple signal pathways in obesity-associated cancer. Obes Rev. 2011;12(12):1063–70. doi:10.1111/j.1467-789X.2011.00917.x.
4. Hursting SD. Minireview: the year in obesity and cancer. Mol Endocrinol. 2012;26(12):1961–6. doi:10.1210/me.2012-1283.

5. Calle EE, Kaaks R. Overweight, obesity and cancer: epidemiological evidence and proposed mechanisms. Nat Rev Cancer. 2004;4(8):579–91. doi:10.1038/nrc1408.
6. Hursting SD, Digiovanni J, Dannenberg AJ, Azrad M, Leroith D, Demark-Wahnefried W, Kakarala M, Brodie A, Berger NA. Obesity, energy balance, and cancer: new opportunities for prevention. Cancer Prev Res. 2012;5(11):1260–72. doi:10.1158/1940-6207.CAPR-12-0140.
7. Wolin KY, Carson K, Colditz GA. Obesity and cancer. Oncologist. 2010;15(6):556–65. doi:10.1634/theoncologist.2009-0285.
8. Hursting SD, Berger NA. Energy balance, host-related factors, and cancer progression. J Clin Oncol. 2010;28(26):4058–65. doi:10.1200/JCO.2010.27.9935.
9. Hursting SD, Smith SM, Lashinger LM, Harvey AE, Perkins SN. Calories and carcinogenesis: lessons learned from 30 years of calorie restriction research. Carcinogenesis. 2010;31(1):83–9. doi:10.1093/carcin/bgp280.
10. Wild S, Roglic G, Green A, Sicree R, King H. Global prevalence of diabetes: estimates for the year 2000 and projections for 2030. Diabetes Care. 2004;27(5):1047–53.
11. Cohen DH, LeRoith D. Obesity, type 2 diabetes, and cancer: the insulin and IGF connection. Endocr Relat Cancer. 2012;19(5):F27–45. doi:10.1530/ERC-11-0374.
12. Gallagher EJ, LeRoith D. Minireview: IGF, insulin, and cancer. Endocrinology. 2011;152(7):2546–51. doi:10.1210/en.2011-0231.
13. Papa V, Pezzino V, Costantino A, Belfiore A, Giuffrida D, Frittitta L, Vannelli GB, Brand R, Goldfine ID, Vigneri R. Elevated insulin receptor content in human breast cancer. J Clin Invest. 1990;86(5):1503–10. doi:10.1172/JCI114868.
14. Novosyadlyy R, Lann DE, Vijayakumar A, Rowzee A, Lazzarino DA, Fierz Y, Carboni JM, Gottardis MM, Pennisi PA, Molinolo AA, Kurshan N, Mejia W, Santopietro S, Yakar S, Wood TL, LeRoith D. Insulin-mediated acceleration of breast cancer development and progression in a nonobese model of type 2 diabetes. Cancer Res. 2010;70(2):741–51. doi:10.1158/0008-5472.CAN-09-2141.
15. Weroha SJ, Haluska P. The insulin-like growth factor system in cancer. Endocrinol Metab Clin North Am. 2012;41(2):335–50. doi:10.1016/j.ecl.2012.04.014. vi.
16. Shaw LM. The insulin receptor substrate (IRS) proteins: at the intersection of metabolism and cancer. Cell Cycle. 2011;10(11):1750–6.
17. Lee C, Longo VD. Fasting vs dietary restriction in cellular protection and cancer treatment: from model organisms to patients. Oncogene. 2011;30(30):3305–16. doi:10.1038/onc.2011.91.
18. Holman GD, Kasuga M. From receptor to transporter: insulin signalling to glucose transport. Diabetologia. 1997;40(9):991–1003.
19. Lizcano JM, Alessi DR. The insulin signalling pathway. Curr Biol. 2002;12(7):R236–8.
20. Singh SK, Brito C, Tan QW, De Leon M, De Leon D. Differential expression and signaling activation of insulin receptor isoforms A and B: a link between breast cancer and diabetes. Growth Factors. 2011;29(6):278–89. doi:10.3109/08977194.2011.616200.
21. Cheng Z, Tseng Y, White MF. Insulin signaling meets mitochondria in metabolism. Trends Endocrinol Metab. 2010;21(10):589–98. doi:10.1016/j.tem.2010.06.005.
22. Ceresa BP, Pessin JE. Insulin regulation of the Ras activation/inactivation cycle. Mol Cell Biochem. 1998;182(1–2):23–9.
23. Himpe E, Kooijman R. Insulin-like growth factor-I receptor signal transduction and the Janus Kinase/Signal Transducer and Activator of Transcription (JAK-STAT) pathway. Biofactors. 2009;35(1):76–81. doi:10.1002/biof.20.
24. Jones RA, Campbell CI, Wood GA, Petrik JJ, Moorehead RA. Reversibility and recurrence of IGF-IR-induced mammary tumors. Oncogene. 2009;28(21):2152–62. doi:10.1038/Onc.2009.79.
25. Sell C, Dumenil G, Deveaud C, Miura M, Coppola D, Deangelis T, Rubin R, Efstratiadis A, Baserga R. Effect of a null mutation of the insulin-like growth-factor-I receptor gene on growth and transformation of mouse embryo fibroblasts. Mol Cell Biol. 1994;14(6):3604–12.

26. Hursting SD, Lavigne JA, Berrigan D, Perkins SN, Barrett JC. Calorie restriction, aging, and cancer prevention: mechanisms of action and a applicability to humans. Annu Rev Med. 2003;54:131–52. doi:10.1146/annurev.med.54.101601.15216.
27. Gallagher EJ, Fierz Y, Ferguson RD, LeRoith D. The pathway from diabetes and obesity to cancer, on the route to targeted therapy. Endocr Pract. 2010;16(5):864–73. doi:10.4158/Ep10098.Ra.
28. Arcidiacono B, Iiritano S, Nocera A, Possidente K, Nevolo MT, Ventura V, Foti D, Chiefari E, Brunetti A. Insulin resistance and cancer risk: an overview of the pathogenetic mechanisms. Exp Diabetes Res. 2012;2012:789174. doi:10.1155/2012/789174.
29. Gallagher EJ, Fierz Y, Vijayakumar A, Haddad N, Yakar S, LeRoith D. Inhibiting PI3K reduces mammary tumor growth and induces hyperglycemia in a mouse model of insulin resistance and hyperinsulinemia. Oncogene. 2012;31(27):3213–22. doi:10.1038/onc.2011.495.
30. Asante-Appiah E, Kennedy BP. Protein tyrosine phosphatases: the quest for negative regulators of insulin action. Am J Physiol Endocrinol Metab. 2003;284(4):E663–70. doi:10.1152/ajpendo.00462.2002.
31. Tonks NK, Neel BG. Combinatorial control of the specificity of protein tyrosine phosphatases. Curr Opin Cell Biol. 2001;13(2):182–95.
32. Andersen JN, Mortensen OH, Peters GH, Drake PG, Iversen LF, Olsen OH, Jansen PG, Andersen HS, Tonks NK, Moller NP. Structural and evolutionary relationships among protein tyrosine phosphatase domains. Mol Cell Biol. 2001;21:7117–36.
33. Alonso A, Sasin J, Bottini N, Friedberg I, Osterman A, Godzik A, Hunter T, Dixon J, Mustelin T. Protein tyrosine phosphatases in the human genome. Cell. 2004;117:699–711.
34. Denu JE, Stuckey JA, Saper MA, Dixon JE. Form and function in protein dephosphorylation. Cell. 1996;87:361–4.
35. Frangioni JV, Beahm PH, Shifrin V, Jost CA, Neel BG. The nontransmembrane tyrosine phosphatase PTP-1B localizes to the endoplasmic reticulum via its 35 amino acid C-terminal sequence. Cell. 1992;68(3):545–60.
36. Woodford-Thomas TA, Rhodes JD, Dixon JE. Expression of a protein tyrosine phosphatase in normal and v-src-transformed mouse 3T3 fibroblasts. J Cell Biol. 1992;117(2):401–14.
37. Kenner KA, Anyanwu E, Olefsky JM, Kusari J. Protein-tyrosine-phosphatase 1B is a negative regulator of insulin- and insulin-like growth-factor-I-stimulated signaling. J Biol Chem. 1996;271:9810–6.
38. Byon JC, Kenner KA, Kusari AB, Kusari J. Regulation of growth factor-induced signaling by protein-tyrosine- phosphatases. Proc Soc Exp Biol Med. 1997;216(1):1–20.
39. Byon JC, Kusari AB, Kusari J. Protein-tyrosine phosphatase-1B acts as a negative regulator of insulin signal transduction. Mol Cell Biochem. 1998;182:101–8.
40. Ahmad F, Li PM, Meyerovitch J, Goldstein BJ. Osmotic loading of neutralizing antibodies demonstrates a role for protein-tyrosine phosphatase 1B in negative regulation of the insulin action pathway. J Biol Chem. 1995;270(35):20503–8.
41. Goldstein BJ, Bittner-Kowalczyk A, White MF, Harbeck M. Tyrosine dephosphorylation and deactivation of insulin receptor substrate-1 by protein-tyrosine phosphatase 1B. Possible facilitation by the formation of a ternary complex with the Grb2 adaptor protein. J Biol Chem. 2000;275(6):4283–9.
42. Cicirelli MF, Tonks NK, Diltz CD, Weiel JE, Fischer EH, Krebs EG. Microinjection of a protein-tyrosine-phosphatase inhibits insulin action in Xenopus oocytes. Proc Natl Acad Sci U S A. 1990;87(14):5514–8.
43. Bandyopadhyay D, Kusari A, Kenner KA, Liu F, Chernoff J, Gustafson TA, Kusari J. Protein-tyrosine phosphatase 1B complexes with the insulin receptor in vivo and is tyrosine-phosphorylated in the presence of insulin. J Biol Chem. 1997;272(3):1639–45.
44. Walchli S, Curchod ML, Gobert RP, Arkinstall S, Hooft van Huijsduijnen R. Identification of tyrosine phosphatases that dephosphorylate the insulin receptor. A brute force approach based on "substrate-trapping" mutants. J Biol Chem. 2000;275(13):9792–6.

45. Seely BL, Staubs PA, Reichart DR, Berhanu P, Milarski KL, Saltiel AR, Kusari J, Olefsky JM. Protein tyrosine phosphatase 1B interacts with the activated insulin receptor. Diabetes. 1996;45(10):1379–85.
46. Flint AJ, Tiganis T, Barford D, Tonks NK. Development of "substrate-trapping" mutants to identify physiological substrates of protein tyrosine phosphatases. Proc Natl Acad Sci U S A. 1997;94(5):1680–5.
47. Elchebly M, Payette P, Michaliszyn E, Cromlish W, Collins S, Loy AL, Normandin D, Cheng A, Himms-Hagen J, Chan CC, Ramachandran C, Gresser MJ, Tremblay ML, Kennedy BP. Increased insulin sensitivity and obesity resistance in mice lacking the protein tyrosine phosphatase-1B gene. Science. 1999;283(5407):1544–8.
48. Klaman LD, Boss O, Peroni OD, Kim JK, Martino JL, Zabolotny JM, Moghal N, Lubkin M, Kim YB, Sharpe AH, Stricker-Krongrad A, Shulman GI, Neel BG, Kahn BB. Increased energy expenditure, decreased adiposity, and tissue-specific insulin sensitivity in protein-tyrosine phosphatase 1B-deficient mice. Mol Cell Biol. 2000;20(15):5479–89.
49. Waring JF, Ciurlionis R, Clampit JE, Morgan S, Gum RJ, Jolly RA, Kroeger P, Frost L, Trevillyan J, Zinker BA, Jirousek M, Ulrich RG, Rondinone CM. PTP1B antisense-treated mice show regulation of genes involved in lipogenesis in liver and fat. Mol Cell Endocrinol. 2003;203(1–2):155–68.
50. Gum RJ, Gaede LL, Heindel MA, Waring JF, Trevillyan JM, Zinker BA, Stark ME, Wilcox D, Jirousek MR, Rondinone CM, Ulrich RG. Antisense protein tyrosine phosphatase 1B reverses activation of p38 mitogen-activated protein kinase in liver of ob/ob mice. Mol Endocrinol. 2003;17(6):1131–43. doi:10.1210/me.2002-0288.
51. Zinker BA, Rondinone CM, Trevillyan JM, Gum RJ, Clampit JE, Waring JF, Xie N, Wilcox D, Jacobson P, Frost L, Kroeger PE, Reilly RM, Koterski S, Opgenorth TJ, Ulrich RG, Crosby S, Butler M, Murray SF, McKay RA, Bhanot S, Monia BP, Jirousek MR. PTP1B antisense oligonucleotide lowers PTP1B protein, normalizes blood glucose, and improves insulin sensitivity in diabetic mice. Proc Natl Acad Sci U S A. 2002;99(17):11357–62. doi:10.1073/pnas.142298199.
52. Zabolotny JM, Haj FG, Kim YB, Kim HJ, Shulman GI, Kim JK, Neel BG, Kahn BB. Transgenic overexpression of protein-tyrosine phosphatase 1B in muscle causes insulin resistance, but overexpression with leukocyte antigen-related phosphatase does not additively impair insulin action. J Biol Chem. 2004;279(23):24844–51. doi:10.1074/jbc.M310688200.
53. Haj FG, Zabolotny JM, Kim YB, Kahn BB, Neel BG. Liver-specific protein-tyrosine phosphatase 1B (PTP1B) re-expression alters glucose homeostasis of PTP1B-/-mice. J Biol Chem. 2005;280(15):15038–46. doi:10.1074/jbc.M413240200.
54. Delibegovic M, Bence KK, Mody N, Hong EG, Ko HJ, Kim JK, Kahn BB, Neel BG. Improved glucose homeostasis in mice with muscle-specific deletion of protein-tyrosine phosphatase 1B. Mol Cell Biol. 2007;27(21):7727–34. doi:10.1128/MCB.00959-07.
55. Delibegovic M, Zimmer D, Kauffman C, Rak K, Hong EG, Cho YR, Kim JK, Kahn BB, Neel BG, Bence KK. Liver-specific deletion of protein-tyrosine phosphatase 1B (PTP1B) improves metabolic syndrome and attenuates diet-induced endoplasmic reticulum stress. Diabetes. 2009;58(3):590–9. doi:10.2337/db08-0913.
56. Gonzalez-Rodriguez A, Nevado C, Escriva F, Sesti G, Rondinone CM, Benito M, Valverde AM. PTP1B deficiency increases glucose uptake in neonatal hepatocytes: involvement of IRA/GLUT2 complexes. Am J Physiol Gastrointest Liver Physiol. 2008;295(2):G338–47. doi:10.1152/ajpgi.00514.2007.
57. Owen C, Czopek A, Agouni A, Grant L, Judson R, Lees EK, McIlroy GD, Goransson O, Welch A, Bence KK, Kahn BB, Neel BG, Mody N, Delibegovic M. Adipocyte-specific protein tyrosine phosphatase 1B deletion increases lipogenesis, adipocyte cell size and is a minor regulator of glucose homeostasis. PLoS One. 2012;7(2):e32700. doi:10.1371/journal.pone.0032700.
58. Salmeen A, Andersen JN, Myers MP, Tonks NK, Barford D. Molecular basis for the dephosphorylation of the activation segment of the insulin receptor by protein tyrosine phosphatase 1B. Mol Cell. 2000;6:1401–12.

59. Buckley DA, Cheng A, Kiely PA, Tremblay ML, O'Connor R. Regulation of insulin-like growth factor type I (IGF-I) receptor kinase activity by protein tyrosine phosphatase 1B (PTP-1B) and enhanced IGF-I-mediated suppression of apoptosis and motility in PTP-1B-deficient fibroblasts. Mol Cell Biol. 2002;22(7):1998–2010.
60. Blanquart C, Boute N, Lacasa D, Issad T. Monitoring the activation state of the insulin-like growth factor-1 receptor and its interaction with protein tyrosine phosphatase 1B using bioluminescence resonance energy transfer. Mol Pharmacol. 2005;68(3):885–94. doi:10.1124/mol.105.013151.
61. Blanquart C, Gonzalez-Yanes C, Issad T. Monitoring the activation state of insulin/insulin-like growth factor-1 hybrid receptors using bioluminescence resonance energy transfer. Mol Pharmacol. 2006;70(5):1802–11. doi:10.1124/mol.106.026989.
62. Escriva F, Gonzalez-Rodriguez A, Fernandez-Millan E, Rondinone CM, Alvarez C, Valverde AM. PTP1B deficiency enhances liver growth during suckling by increasing the expression of insulin-like growth factor-I. J Cell Physiol. 2010;225(1):214–22. doi:10.1002/jcp.22246.
63. Yip SC, Saha S, Chernoff J. PTP1B: a double agent in metabolism and oncogenesis. Trends Biochem Sci. 2010;35(8):442–9. doi:10.1016/j.tibs.2010.03.004.
64. Blanquart C, Karouri SE, Issad T. Protein tyrosine phosphatase-1B and T-cell protein tyrosine phosphatase regulate IGF-2-induced MCF-7 cell migration. Biochem Biophys Res Commun. 2010;392(1):83–8. doi:10.1016/j.bbrc.2009.12.176.
65. Galic S, Klingler-Hoffmann M, Fodero-Tavoletti MT, Puryer MA, Meng TC, Tonks NK, Tiganis T. Regulation of insulin receptor signaling by the protein tyrosine phosphatase TCPTP. Mol Cell Biol. 2003;23(6):2096–108.
66. Meng TC, Buckley DA, Galic S, Tiganis T, Tonks NK. Regulation of insulin signaling through reversible oxidation of the protein-tyrosine phosphatases TC45 and PTP1B. J Biol Chem. 2004;279(36):37716–25. doi:10.1074/jbc.M404606200.
67. Galic S, Hauser C, Kahn BB, Haj FG, Neel BG, Tonks NK, Tiganis T. Coordinated regulation of insulin signaling by the protein tyrosine phosphatases PTP1B and TCPTP. Mol Cell Biol. 2005;25(2):819–29. doi:10.1128/MCB.25.2.819-829.2005.
68. Tiganis T. PTP1B and TCPTP—nonredundant phosphatases in insulin signaling and glucose homeostasis. FEBS J. 2013;280(2):445–58. doi:10.1111/j.1742-4658.2012.08563.x.
69. Dube N, Tremblay ML. Involvement of the small protein tyrosine phosphatases TC-PTP and PTP1B in signal transduction and diseases: from diabetes, obesity to cell cycle, and cancer. Biochim Biophys Acta. 2005;1754(1–2):108–17. doi:10.1016/j.bbapap.2005.07.030.
70. Fukushima A, Loh K, Galic S, Fam B, Shields B, Wiede F, Tremblay ML, Watt MJ, Andrikopoulos S, Tiganis T. T-cell protein tyrosine phosphatase attenuatcs STAT3 and insulin signaling in the liver to regulate gluconeogenesis. Diabetes. 2010;59(8):1906–14. doi:10.2337/db09-1365.
71. Loh K, Merry TL, Galic S, Wu BJ, Watt MJ, Zhang S, Zhang ZY, Neel BG, Tiganis T. T cell protein tyrosine phosphatase (TCPTP) deficiency in muscle does not alter insulin signalling and glucose homeostasis in mice. Diabetologia. 2012;55(2):468–78. doi:10.1007/s00125-011-2386-z.
72. Bettaieb A, Liu S, Xi Y, Nagata N, Matsuo K, Matsuo I, Chahed S, Bakke J, Keilhack H, Tiganis T, Haj FG. Differential regulation of endoplasmic reticulum stress by protein tyrosine phosphatase 1B and T cell protein tyrosine phosphatase. J Biol Chem. 2011;286(11):9225–35. doi:10.1074/jbc.M110.186148.
73. Myers Jr MG, Mendez R, Shi P, Pierce JH, Rhoads R, White MF. The COOH-terminal tyrosine phosphorylation sites on IRS-1 bind SHP-2 and negatively regulate insulin signaling. J Biol Chem. 1998;273(41):26908–14.
74. Ouwens DM, van der Zon GC, Maassen JA. Modulation of insulin-stimulated glycogen synthesis by Src homology phosphatase 2. Mol Cell Endocrinol. 2001;175(1–2):131–40.
75. Noguchi T, Matozaki T, Horita K, Fujioka Y, Kasuga M. Role of SH-PTP2, a protein-tyrosine phosphatase with Src homology 2 domains, in insulin-stimulated Ras activation. Mol Cell Biol. 1994;14(10):6674–82.

76. Arrandale JM, Gore-Willse A, Rocks S, Ren JM, Zhu J, Davis A, Livingston JN, Rabin DU. Insulin signaling in mice expressing reduced levels of Syp. J Biol Chem. 1996;271(35):21353–8.
77. Maegawa H, Hasegawa M, Sugai S, Obata T, Ugi S, Morino K, Egawa K, Fujita T, Sakamoto T, Nishio Y, Kojima H, Haneda M, Yasuda H, Kikkawa R, Kashiwagi A. Expression of a dominant negative SHP-2 in transgenic mice induces insulin resistance. J Biol Chem. 1999;274(42):30236–43.
78. Zhang SS, Hao E, Yu J, Liu W, Wang J, Levine F, Feng GS. Coordinated regulation by Shp2 tyrosine phosphatase of signaling events controlling insulin biosynthesis in pancreatic beta-cells. Proc Natl Acad Sci U S A. 2009;106(18):7531–6. doi:10.1073/pnas.0811715106.
79. Princen F, Bard E, Sheikh F, Zhang SS, Wang J, Zago WM, Wu D, Trelles RD, Bailly-Maitre B, Kahn CR, Chen Y, Reed JC, Tong GG, Mercola M, Chen J, Feng GS. Deletion of Shp2 tyrosine phosphatase in muscle leads to dilated cardiomyopathy, insulin resistance, and premature death. Mol Cell Biol. 2009;29(2):378–88. doi:10.1128/MCB.01661-08.
80. Matsuo K, Delibegovic M, Matsuo I, Nagata N, Liu S, Bettaieb A, Xi Y, Araki K, Yang W, Kahn BB, Neel BG, Haj FG. Altered glucose homeostasis in mice with liver-specific deletion of Src homology phosphatase 2. J Biol Chem. 2010;285(51):39750–8. doi:10.1074/jbc.M110.153734.
81. Bettaieb A, Matsuo K, Matsuo I, Nagata N, Chahed S, Liu S, Haj FG. Adipose-specific deletion of Src homology phosphatase 2 does not significantly alter systemic glucose homeostasis. Metabolism. 2011;60(8):1193–201. doi:10.1016/j.metabol.2011.01.004.
82. Maile LA, Clemmons DR. The alphaVbeta3 integrin regulates insulin-like growth factor I (IGF-I) receptor phosphorylation by altering the rate of recruitment of the Src-homology 2-containing phosphotyrosine phosphatase-2 to the activated IGF-I receptor. Endocrinology. 2002;143(11):4259–64.
83. Maile LA, Clemmons DR. Regulation of insulin-like growth factor I receptor dephosphorylation by SHPS-1 and the tyrosine phosphatase SHP-2. J Biol Chem. 2002;277(11):8955–60. doi:10.1074/jbc.M109258200.
84. Koyama T, Nakaoka Y, Fujio Y, Hirota H, Nishida K, Sugiyama S, Okamoto K, Yamauchi-Takihara K, Yoshimura M, Mochizuki S, Hori M, Hirano T, Mochizuki N. Interaction of scaffolding adaptor protein Gab1 with tyrosine phosphatase SHP2 negatively regulates IGF-I-dependent myogenic differentiation via the ERK1/2 signaling pathway. J Biol Chem. 2008;283(35):24234–44. doi:10.1074/jbc.M803907200.
85. Araki T, Nawa H, Neel BG. Tyrosyl phosphorylation of Shp2 is required for normal ERK activation in response to some, but not all, growth factors. J Biol Chem. 2003;278(43):41677–84. doi:10.1074/jbc.M306461200.
86. Kulas DT, Zhang WR, Goldstein BJ, Furlanetto RW, Mooney RA. Insulin receptor signaling is augmented by antisense inhibition of the protein tyrosine phosphatase LAR. J Biol Chem. 1995;270(6):2435–8.
87. Ahmad F, Goldstein BJ. Functional association between the insulin receptor and the transmembrane protein-tyrosine phosphatase LAR in intact cells. J Biol Chem. 1997;272(1):448–57.
88. Zabolotny JM, Kim YB, Peroni OD, Kim JK, Pani MA, Boss O, Klaman LD, Kamatkar S, Shulman GI, Kahn BB, Neel BG. Overexpression of the LAR (leukocyte antigen-related) protein-tyrosine phosphatase in muscle causes insulin resistance. Proc Natl Acad Sci U S A. 2001;98(9):5187–92. doi:10.1073/pnas.071050398.
89. Chagnon MJ, Uetani N, Tremblay ML. Functional significance of the LAR receptor protein tyrosine phosphatase family in development and diseases. Biochem Cell Biol. 2004;82(6):664–75. doi:10.1139/o04-120.
90. Gorgani-Firuzjaee S, Bakhtiyari S, Golestani A, Meshkani R. Leukocyte antigen-related inhibition attenuates palmitate-induced insulin resistance in muscle cells. J Endocrinol. 2012;215(1):71–7. doi:10.1530/JOE-12-0160.
91. Cheung AT, Wang J, Ree D, Kolls JK, Bryer-Ash M. Tumor necrosis factor-alpha induces hepatic insulin resistance in obese Zucker (fa/fa) rats via interaction of leukocyte antigen-related tyrosine phosphatase with focal adhesion kinase. Diabetes. 2000;49(5):810–9.

92. Ren JM, Li PM, Zhang WR, Sweet LJ, Cline G, Shulman GI, Livingston JN, Goldstein BJ. Transgenic mice deficient in the LAR protein-tyrosine phosphatase exhibit profound defects in glucose homeostasis. Diabetes. 1998;47(3):493–7.
93. Ahmad F, Azevedo JL, Cortright R, Dohm GL, Goldstein BJ. Alterations in skeletal muscle protein-tyrosine phosphatase activity and expression in insulin-resistant human obesity and diabetes. J Clin Invest. 1997;100(2):449–58. doi:10.1172/JCI119552.
94. Norris K, Norris F, Kono DH, Vestergaard H, Pedersen O, Theofilopoulos AN, Moller NP. Expression of protein-tyrosine phosphatases in the major insulin target tissues. FEBS Lett. 1997;415(3):243–8.
95. Pulido R, Serra-Pages C, Tang M, Streuli M. The LAR/PTP delta/PTP sigma subfamily of transmembrane protein-tyrosine-phosphatases: multiple human LAR, PTP delta, and PTP sigma isoforms are expressed in a tissue-specific manner and associate with the LAR-interacting protein LIP.1. Proc Natl Acad Sci U S A. 1995;92(25):11686–90.
96. Chagnon MJ, Elchebly M, Uetani N, Dombrowski L, Cheng A, Mooney RA, Marette A, Tremblay ML. Altered glucose homeostasis in mice lacking the receptor protein tyrosine phosphatase sigma. Can J Physiol Pharmacol. 2006;84(7):755–63. doi:10.1139/y06-020.
97. Moller NP, Moller KB, Lammers R, Kharitonenkov A, Hoppe E, Wiberg FC, Sures I, Ullrich A. Selective down-regulation of the insulin receptor signal by protein-tyrosine phosphatases alpha and epsilon. J Biol Chem. 1995;270(39):23126–31.
98. Lacasa D, Boute N, Issad T. Interaction of the insulin receptor with the receptor-like protein tyrosine phosphatases PTPalpha and PTPepsilon in living cells. Mol Pharmacol. 2005;67(4):1206–13. doi:10.1124/mol.104.009514.
99. Lammers R, Moller NP, Ullrich A. The transmembrane protein tyrosine phosphatase alpha dephosphorylates the insulin receptor in intact cells. FEBS Lett. 1997;404(1):37–40.
100. Andersen JN, Elson A, Lammers R, Romer J, Clausen JT, Moller KB, Moller NP. Comparative study of protein tyrosine phosphatase-epsilon isoforms: membrane localization confers specificity in cellular signalling. Biochem J. 2001;354(Pt 3):581–90.
101. Nakagawa Y, Aoki N, Aoyama K, Shimizu H, Shimano H, Yamada N, Miyazaki H. Receptor-type protein tyrosine phosphatase epsilon (PTPepsilonM) is a negative regulator of insulin signaling in primary hepatocytes and liver. Zoolog Sci. 2005;22(2):169–75.
102. Aga-Mizrachi S, Brutman-Barazani T, Jacob AI, Bak A, Elson A, Sampson SR. Cytosolic protein tyrosine phosphatase-epsilon is a negative regulator of insulin signaling in skeletal muscle. Endocrinology. 2008;149(2):605–14. doi:10.1210/en.2007-0908.
103. Peretz A, Gil-Henn H, Sobko A, Shinder V, Attali B, Elson A. Hypomyelination and increased activity of voltage-gated K(+) channels in mice lacking protein tyrosine phosphatase epsilon. EMBO J. 2000;19(15):4036–45. doi:10.1093/emboj/19.15.4036.
104. Rousso-Noori L, Knobler H, Levy-Apter E, Kuperman Y, Neufeld-Cohen A, Keshet Y, Akepati VR, Klinghoffer RA, Chen A, Elson A. Protein tyrosine phosphatase epsilon affects body weight by downregulating leptin signaling in a phosphorylation-dependent manner. Cell Metab. 2011;13(5):562–72. doi:10.1016/j.cmet.2011.02.017.
105. Cong LN, Chen H, Li Y, Lin CH, Sap J, Quon MJ. Overexpression of protein tyrosine phosphatase-alpha (PTP-alpha) but not PTP-kappa inhibits translocation of GLUT4 in rat adipose cells. Biochem Biophys Res Commun. 1999;255(2):200–7. doi:10.1006/bbrc.1999.0183.
106. Le HT, Ponniah S, Pallen CJ. Insulin signaling and glucose homeostasis in mice lacking protein tyrosine phosphatase alpha. Biochem Biophys Res Commun. 2004;314(2):321–9.
107. Chen SC, Khanna RS, Bessette DC, Samayawardhena LA, Pallen CJ. Protein tyrosine phosphatase-alpha complexes with the IGF-I receptor and undergoes IGF-I-stimulated tyrosine phosphorylation that mediates cell migration. Am J Physiol Cell Physiol. 2009;297(1):C133–9. doi:10.1152/ajpcell.00110.2009.
108. Lamprianou S, Harroch S. Receptor protein tyrosine phosphatase from stem cells to mature glial cells of the central nervous system. J Mol Neurosci. 2006;29(3):241–55. doi:10.1385/JMN:29:3:241.

109. Kulas DT, Freund GG, Mooney RA. The transmembrane protein-tyrosine phosphatase CD45 is associated with decreased insulin receptor signaling. J Biol Chem. 1996;271(2):755–60.
110. Descamps G, Pellat-Deceunynck C, Szpak Y, Bataille R, Robillard N, Amiot M. The magnitude of Akt/phosphatidylinositol 3′-kinase proliferating signaling is related to CD45 expression in human myeloma cells. J Immunol. 2004;173(8):4953–9.
111. Ramachandran C, Aebersold R, Tonks NK, Pot DA. Sequential dephosphorylation of a multiply phosphorylated insulin receptor peptide by protein tyrosine phosphatases. Biochemistry. 1992;31(17):4232–8.
112. Tonks NK, Diltz CD, Fischer EH. CD45, an integral membrane protein tyrosine phosphatase. Characterization of enzyme activity. J Biol Chem. 1990;265(18):10674–80.
113. Wu C, Sun M, Liu L, Zhou GW. The function of the protein tyrosine phosphatase SHP-1 in cancer. Gene. 2003;306:1–12.
114. Neel BG, Gu H, Pao L. The 'Shp'ing news: SH2 domain-containing tyrosine phosphatases in cell signaling. Trends Biochem Sci. 2003;28(6):284–93. doi:10.1016/S0968-0004(03)00091-4.
115. Dubois MJ, Bergeron S, Kim HJ, Dombrowski L, Perreault M, Fournes B, Faure R, Olivier M, Beauchemin N, Shulman GI, Siminovitch KA, Kim JK, Marette A. The SHP-1 protein tyrosine phosphatase negatively modulates glucose homeostasis. Nat Med. 2006;12(5):549–56. doi:10.1038/nm1397.
116. Xu E, Charbonneau A, Rolland Y, Bellmann K, Pao L, Siminovitch KA, Neel BG, Beauchemin N, Marette A. Hepatocyte-specific Ptpn6 deletion protects from obesity-linked hepatic insulin resistance. Diabetes. 2012;61(8):1949–58. doi:10.2337/db11-1502.
117. Bergeron S, Dubois MJ, Bellmann K, Schwab M, Larochelle N, Nalbantoglu J, Marette A. Inhibition of the protein tyrosine phosphatase SHP-1 increases glucose uptake in skeletal muscle cells by augmenting insulin receptor signaling and GLUT4 expression. Endocrinology. 2011;152(12):4581–8. doi:10.1210/en.2011-1268.
118. Huynh H, Wang X, Li W, Bottini N, Williams S, Nika K, Ishihara H, Godzik A, Mustelin T. Homotypic secretory vesicle fusion induced by the protein tyrosine phosphatase MEG2 depends on polyphosphoinositides in T cells. J Immunol. 2003;171(12):6661–71.
119. Cho CY, Koo SH, Wang Y, Callaway S, Hedrick S, Mak PA, Orth AP, Peters EC, Saez E, Montminy M, Schultz PG, Chanda SK. Identification of the tyrosine phosphatase PTP-MEG2 as an antagonist of hepatic insulin signaling. Cell Metab. 2006;3(5):367–78. doi:10.1016/j.cmet.2006.03.006.
120. Zhang S, Liu S, Tao R, Wei D, Chen L, Shen W, Yu ZH, Wang L, Jones DR, Dong XC, Zhang ZY. A highly selective and potent PTP-MEG2 inhibitor with therapeutic potential for type 2 diabetes. J Am Chem Soc. 2012;134(43):18116–24. doi:10.1021/ja308212y.
121. Yuan T, Wang Y, Zhao ZJ, Gu H. Protein-tyrosine phosphatase PTPN9 negatively regulates ErbB2 and epidermal growth factor receptor signaling in breast cancer cells. J Biol Chem. 2010;285(20):14861–70. doi:10.1074/jbc.M109.099879.
122. Tonks NK. Protein tyrosine phosphatases: from genes, to function, to disease. Nat Rev Mol Cell Biol. 2006;7(11):833–46. doi:10.1038/nrm2039.
123. Patterson KI, Brummer T, O'Brien PM, Daly RJ. Dual-specificity phosphatases: critical regulators with diverse cellular targets. Biochem J. 2009;418(3):475–89.
124. Caunt CJ, Keyse SM. Dual-specificity MAP kinase phosphatases (MKPs): shaping the outcome of MAP kinase signalling. FEBS J. 2013;280(2):489–504. doi:10.1111/j.1742-4658.2012.08716.x.
125. Roth Flach RJ, Bennett AM. Mitogen-activated protein kinase phosphatase-1—a potential therapeutic target in metabolic disease. Expert Opin Ther Targets. 2010;14(12):1323–32. doi:10.1517/14728222.2010.528395.
126. Wu JJ, Roth RJ, Anderson EJ, Hong EG, Lee MK, Choi CS, Neufer PD, Shulman GI, Kim JK, Bennett AM. Mice lacking MAP kinase phosphatase-1 have enhanced MAP kinase activity and resistance to diet-induced obesity. Cell Metab. 2006;4(1):61–73. doi:10.1016/j.cmet.2006.05.010.
127. Wancket LM, Frazier WJ, Liu Y. Mitogen-activated protein kinase phosphatase (MKP)-1 in immunology, physiology, and disease. Life Sci. 2012;90(7–8):237–48. doi:10.1016/j.lfs.2011.11.017.

128. Song MS, Salmena L, Pandolfi PP. The functions and regulation of the PTEN tumour suppressor. Nat Rev Mol Cell Biol. 2012;13(5):283–96. doi:10.1038/nrm3330.
129. Maehama T, Taylor GS, Dixon JE. PTEN and myotubularin: novel phosphoinositide phosphatases. Annu Rev Biochem. 2001;70:247–79. doi:10.1146/annurev.biochem.70.1.247.
130. Carnero A. The PKB/AKT pathway in cancer. Curr Pharm Des. 2010;16(1):34–44.
131. Santos CR, Schulze A. Lipid metabolism in cancer. FEBS J. 2012;279(15):2610–23. doi:10.1111/j.1742-4658.2012.08644.x.
132. Bessette DC, Wong PC, Pallen CJ. PRL-3: a metastasis-associated phosphatase in search of a function. Cells Tissues Organs. 2007;185(1–3):232–6. doi:10.1159/000101324.
133. Wang H, Quah SY, Dong JM, Manser E, Tang JP, Zeng Q. PRL-3 down-regulates PTEN expression and signals through PI3K to promote epithelial-mesenchymal transition. Cancer Res. 2007;67(7):2922–6. doi:10.1158/0008-5472.CAN-06-3598.
134. Yan H, Kong D, Ge X, Gao X, Han X. Generation of conditional knockout alleles for PRL-3. J Biomed Res. 2011;25(6):438–43. doi:10.1016/S1674-8301(11)60058-4.
135. Zimmerman MW, Homanics GE, Lazo JS. Targeted deletion of the metastasis-associated phosphatase Ptp4a3 (PRL-3) suppresses murine colon cancer. PLoS One. 2013;8(3):e58300. doi:10.1371/journal.pone.0058300.
136. Zimmerman MW, McQueeney KE, Isenberg JS, Pitt BR, Wasserloos KA, Homanics GE, Lazo JS. PTP4A3 phosphatase promotes VEGF signaling and enables endothelial cell motility. J Biol Chem. 2014. doi:10.1074/jbc.M113.480038.
137. Raugei G, Ramponi G, Chiarugi P. Low molecular weight protein tyrosine phosphatases: small, but smart. Cell Mol Life Sci. 2002;59(6):941–9.
138. Chiarugi P, Taddei ML, Schiavone N, Papucci L, Giannoni E, Fiaschi T, Capaccioli S, Raugei G, Ramponi G. LMW-PTP is a positive regulator of tumor onset and growth. Oncogene. 2004;23(22):3905–14. doi:10.1038/sj.onc.1207508.
139. Malentacchi F, Marzocchini R, Gelmini S, Orlando C, Serio M, Ramponi G, Raugei G. Up-regulated expression of low molecular weight protein tyrosine phosphatases in different human cancers. Biochem Biophys Res Commun. 2005;334(3):875–83. doi:10.1016/j.bbrc.2005.06.176.
140. Chiarugi P, Cirri P, Marra F, Raugei G, Camici G, Manao G, Ramponi G. LMW-PTP is a negative regulator of insulin-mediated mitotic and metabolic signalling. Biochem Biophys Res Commun. 1997;238(2):676–82. doi:10.1006/bbrc.1997.7355.
141. Pandey SK, Yu XX, Watts LM, Michael MD, Sloop KW, Rivard AR, Leedom TA, Manchem VP, Samadzadeh L, McKay RA, Monia BP, Bhanot S. Reduction of low molecular weight protein-tyrosine phosphatase expression improves hyperglycemia and insulin sensitivity in obese mice. J Biol Chem. 2007;282(19):14291–9. doi:10.1074/jbc.M609626200.
142. Bottini N, MacMurray J, Peters W, Rostamkhani M, Comings DE. Association of the acid phosphatase (ACP1) gene with triglyceride levels in obese women. Mol Genet Metab. 2002;77(3):226–9.
143. Iannaccone U, Bergamaschi A, Magrini A, Marino G, Bottini N, Lucarelli P, Bottini E, Gloria-Bottini F. Serum glucose concentration and ACP1 genotype in healthy adult subjects. Metabolism. 2005;54(7):891–4. doi:10.1016/j.metabol.2005.02.002.
144. Giovannucci E, Harlan DM, Archer MC, Bergenstal RM, Gapstur SM, Habel LA, Pollak M, Regensteiner JG, Yee D. Diabetes and cancer: a consensus report. Diabetes Care. 2010;33(7):1674–85. doi:10.2337/Dc10-0666.
145. Jung KJ, Lee EK, Yu BP, Chung HY. Significance of protein tyrosine kinase/protein tyrosine phosphatase balance in the regulation of NF-kappaB signaling in the inflammatory process and aging. Free Radic Biol Med. 2009;47(7):983–91. doi:10.1016/j.freeradbiomed.2009.07.009.
146. Stuible M, Doody KM, Tremblay ML. PTP1B and TC-PTP: regulators of transformation and tumorigenesis. Cancer Metastasis Rev. 2008;27(2):215–30. doi:10.1007/s10555-008-9115-1.
147. Mustelin T, Vang T, Bottini N. Protein tyrosine phosphatases and the immune response. Nat Rev Immunol. 2005;5(1):43–57. doi:10.1038/nri1530.
148. Xu D, Qu CK. Protein tyrosine phosphatases in the JAK/STAT pathway. Front Biosci. 2008;13:4925–32.

149. Tonks NK. Redox redux: revisiting PTPs and the control of cell signaling. Cell. 2005;121(5):667–70. doi:10.1016/j.cell.2005.05.016.
150. Valentino L, Pierre J. JAK/STAT signal transduction: regulators and implication in hematological malignancies. Biochem Pharmacol. 2006;71(6):713–21. doi:10.1016/j.bcp.2005.12.017.
151. Pike KA, Tremblay ML. Regulating naive and memory CD8 T cell homeostasis—a role for protein tyrosine phosphatases. FEBS J. 2013;280(2):432–44. doi:10.1111/j.1742-4658.2012.08587.x.
152. Ahima RS, Flier JS. Leptin. Annu Rev Physiol. 2000;62:413–37. doi:10.1146/annurev.physiol.62.1.413.
153. Huang L, Li C. Leptin: a multifunctional hormone. Cell Res. 2000;10(2):81–92. doi:10.1038/sj.cr.7290038.
154. Paz-Filho G, Lim EL, Wong ML, Licinio J. Associations between adipokines and obesity-related cancer. Front Biosci. 2011;16:1634–50.
155. Garofalo C, Surmacz E. Leptin and cancer. J Cell Physiol. 2006;207(1):12–22. doi:10.1002/jcp.20472.
156. Smith-Kirwin SM, O'Connor DM, De Johnston J, Lancey ED, Hassink SG, Funanage VL. Leptin expression in human mammary epithelial cells and breast milk. J Clin Endocrinol Metab. 1998;83(5):1810–3.
157. Koda M, Sulkowska M, Kanczuga-Koda L, Jarzabek K, Sulkowski S. Expression of leptin and its receptor in female breast cancer in relation with selected apoptotic markers. Folia Histochem Cytobiol. 2007;45 Suppl 1:S187–91.
158. Koda M, Sulkowska M, Kanczuga-Koda L, Surmacz E, Sulkowski S. Overexpression of the obesity hormone leptin in human colorectal cancer. J Clin Pathol. 2007;60(8):902–6. doi:10.1136/jcp.2006.041004.
159. Koda M, Sulkowska M, Wincewicz A, Kanczuga-Koda L, Musiatowicz B, Szymanska M, Sulkowski S. Expression of leptin, leptin receptor, and hypoxia-inducible factor 1 alpha in human endometrial cancer. Ann N Y Acad Sci. 2007;1095:90–8. doi:10.1196/annals.1397.013.
160. Ando S, Catalano S. The multifactorial role of leptin in driving the breast cancer microenvironment. Nat Rev Endocrinol. 2012;8(5):263–75. doi:10.1038/nrendo.2011.184.
161. Howard JM, Pidgeon GP, Reynolds JV. Leptin and gastro-intestinal malignancies. Obes Rev. 2010;11(12):863–74. doi:10.1111/j.1467-789X.2010.00718.x.
162. Gao J, Tian J, Lv Y, Shi F, Kong F, Shi H, Zhao L. Leptin induces functional activation of cyclooxygenase-2 through JAK2/STAT3, MAPK/ERK, and PI3K/AKT pathways in human endometrial cancer cells. Cancer Sci. 2009;100(3):389–95. doi:10.1111/j.1349-7006.2008.01053.x.
163. Bouloumie A, Drexler HC, Lafontan M, Busse R. Leptin, the product of Ob gene, promotes angiogenesis. Circ Res. 1998;83(10):1059–66.
164. Sierra-Honigmann MR, Nath AK, Murakami C, Garcia-Cardena G, Papapetropoulos A, Sessa WC, Madge LA, Schechner JS, Schwabb MB, Polverini PJ, Flores-Riveros JR. Biological action of leptin as an angiogenic factor. Science. 1998;281(5383):1683–6.
165. Ghilardi N, Skoda RC. The leptin receptor activates janus kinase 2 and signals for proliferation in a factor-dependent cell line. Mol Endocrinol. 1997;11(4):393–9.
166. Sweeney G. Leptin signalling. Cell Signal. 2002;14(8):655–63.
167. Bjorbaek C, Kahn BB. Leptin signaling in the central nervous system and the periphery. Recent Prog Horm Res. 2004;59:305–31.
168. Ahima RS, Osei SY. Leptin signaling. Physiol Behav. 2004;81(2):223–41.
169. Myers MG, Cowley MA, Munzberg H. Mechanisms of leptin action and leptin resistance. Annu Rev Physiol. 2008;70:537–56.
170. Leshan RL, Bjornholm M, Munzberg H, Myers Jr MG. Leptin receptor signaling and action in the central nervous system. Obesity (Silver Spring). 2006;14 Suppl 5:208S–12.
171. Villanueva EC, Myers Jr MG. Leptin receptor signaling and the regulation of mammalian physiology. Int J Obes (Lond). 2008;32 Suppl 7:S8–12. doi:10.1038/ijo.2008.232.
172. St-Pierre J, Tremblay ML. Modulation of leptin resistance by protein tyrosine phosphatases. Cell Metab. 2012;15(3):292–7. doi:10.1016/j.cmet.2012.02.004.
173. Tsou RC, Bence KK. Central regulation of metabolism by protein tyrosine phosphatases. Front Neurosci. 2012;6:192. doi:10.3389/fnins.2012.00192.

174. Cheng A, Uetani N, Simoncic PD, Chaubey VP, Lee-Loy A, McGlade CJ, Kennedy BP, Tremblay ML. Attenuation of leptin action and regulation of obesity by protein tyrosine phosphatase 1B. Dev Cell. 2002;2(4):497–503.
175. Zabolotny JM, Bence-Hanulec KK, Stricker-Krongrad A, Haj F, Wang Y, Minokoshi Y, Kim YB, Elmquist JK, Tartaglia LA, Kahn BB, Neel BG. PTP1B regulates leptin signal transduction in vivo. Dev Cell. 2002;2(4):489–95.
176. Myers MP, Andersen JN, Cheng A, Tremblay ML, Horvath CM, Parisien JP, Salmeen A, Barford D, Tonks NK. TYK2 and JAK2 are substrates of protein-tyrosine phosphatase 1B. J Biol Chem. 2001;276(51):47771–4. doi:10.1074/jbc.C100583200.
177. Kaszubska W, Falls HD, Schaefer VG, Haasch D, Frost L, Hessler P, Kroeger PE, White DW, Jirousek MR, Trevillyan JM. Protein tyrosine phosphatase 1B negatively regulates leptin signaling in a hypothalamic cell line. Mol Cell Endocrinol. 2002;195(1–2):109–18.
178. Tsou RC, Zimmer DJ, De Jonghe BC, Bence KK. Deficiency of PTP1B in leptin receptor-expressing neurons leads to decreased body weight and adiposity in mice. Endocrinology. 2012;153(9):4227–37. doi:10.1210/en.2012-1548.
179. White CL, Whittington A, Barnes MJ, Wang Z, Bray GA, Morrison CD. HF diets increase hypothalamic PTP1B and induce leptin resistance through both leptin-dependent and -independent mechanisms. Am J Physiol Endocrinol Metab. 2009;296(2):E291–9. doi:10.1152/ajpendo.90513.2008.
180. Zabolotny JM, Kim YB, Welsh LA, Kershaw EE, Neel BG, Kahn BB. Protein-tyrosine phosphatase 1B expression is induced by inflammation in vivo. J Biol Chem. 2008;283(21):14230–41. doi:10.1074/jbc.M800061200.
181. Loh K, Fukushima A, Zhang X, Galic S, Briggs D, Enriori PJ, Simonds S, Wiede F, Reichenbach A, Hauser C, Sims NA, Bence KK, Zhang S, Zhang ZY, Kahn BB, Neel BG, Andrews ZB, Cowley MA, Tiganis T. Elevated hypothalamic TCPTP in obesity contributes to cellular leptin resistance. Cell Metab. 2011;14(5):684–99. doi:10.1016/j.cmet.2011.09.011.
182. Bjorbaek C, Buchholz RM, Davis SM, Bates SH, Pierroz DD, Gu H, Neel BG, Myers Jr MG, Flier JS. Divergent roles of SHP-2 in ERK activation by leptin receptors. J Biol Chem. 2001;276(7):4747–55.
183. Feng GS. Shp2 as a therapeutic target for leptin resistance and obesity. Expert Opin Ther Targets. 2006;10(1):135–42. doi:10.1517/14728222.10.1.135.
184. Banno R, Zimmer D, De Jonghe BC, Atienza M, Rak K, Yang W, Bence KK. PTP1B and SHP2 in POMC neurons reciprocally regulate energy balance in mice. J Clin Invest. 2010;120(3):720–34. doi:10.1172/JCI39620.
185. Zhang EE, Chapeau E, Hagihara K, Feng GS. Neuronal Shp2 tyrosine phosphatase controls energy balance and metabolism. Proc Natl Acad Sci U S A. 2004;101(45):16064–9. doi:10.1073/pnas.0405041101.
186. He Z, Zhang SS, Meng Q, Li S, Zhu HH, Raquil MA, Alderson N, Zhang H, Wu J, Rui L, Cai D, Feng GS. Shp2 controls female body weight and energy balance by integrating leptin and estrogen signals. Mol Cell Biol. 2012;32(10):1867–78. doi:10.1128/MCB.06712-11.
187. Wellen KE, Thompson CB. Cellular metabolic stress: considering how cells respond to nutrient excess. Mol Cell. 2010;40(2):323–32. doi:10.1016/j.molcel.2010.10.004.
188. Furukawa S, Fujita T, Shimabukuro M, Iwaki M, Yamada Y, Nakajima Y, Nakayama O, Makishima M, Matsuda M, Shimomura I. Increased oxidative stress in obesity and its impact on metabolic syndrome. J Clin Invest. 2004;114(12):1752–61. doi:10.1172/JCI21625.
189. Kim JA, Wei Y, Sowers JR. Role of mitochondrial dysfunction in insulin resistance. Circ Res. 2008;102(4):401–14. doi:10.1161/CIRCRESAHA.107.165472.
190. Rhee SG. Cell signaling. H_2O_2, a necessary evil for cell signaling. Science. 2006;312(5782):1882–3. doi:10.1126/science.1130481.
191. Ostman A, Frijhoff J, Sandin A, Bohmer FD. Regulation of protein tyrosine phosphatases by reversible oxidation. J Biochem. 2011;150(4):345–56. doi:10.1093/jb/mvr104.
192. Mahadev K, Motoshima H, Wu X, Ruddy JM, Arnold RS, Cheng G, Lambeth JD, Goldstein BJ. The NAD(P)H oxidase homolog Nox4 modulates insulin-stimulated generation of H_2O_2 and plays an integral role in insulin signal transduction. Mol Cell Biol. 2004;24(5):1844–54.

193. Goldstein BJ, Mahadev K, Wu X. Redox paradox: insulin action is facilitated by insulin-stimulated reactive oxygen species with multiple potential signaling targets. Diabetes. 2005;54(2):311–21.
194. Kwon J, Lee SR, Yang KS, Ahn Y, Kim YJ, Stadtman ER, Rhee SG. Reversible oxidation and inactivation of the tumor suppressor PTEN in cells stimulated with peptide growth factors. Proc Natl Acad Sci U S A. 2004;101(47):16419–24. doi:10.1073/pnas.0407396101.
195. Lee SR, Yang KS, Kwon J, Lee C, Jeong W, Rhee SG. Reversible inactivation of the tumor suppressor PTEN by H_2O_2. J Biol Chem. 2002;277(23):20336–42. doi:10.1074/jbc.M111899200.
196. Seo JH, Ahn Y, Lee SR, Yeol Yeo C, Chung Hur K. The major target of the endogenously generated reactive oxygen species in response to insulin stimulation is phosphatase and tensin homolog and not phosphoinositide-3 kinase (PI-3 kinase) in the PI-3 kinase/Akt pathway. Mol Biol Cell. 2005;16(1):348–57. doi:10.1091/mbc.E04-05-0369.
197. Loh K, Deng H, Fukushima A, Cai X, Boivin B, Galic S, Bruce C, Shields BJ, Skiba B, Ooms LM, Stepto N, Wu B, Mitchell CA, Tonks NK, Watt MJ, Febbraio MA, Crack PJ, Andrikopoulos S, Tiganis T. Reactive oxygen species enhance insulin sensitivity. Cell Metab. 2009;10(4):260–72. doi:10.1016/j.cmet.2009.08.009.
198. Meng TC, Fukada T, Tonks NK. Reversible oxidation and inactivation of protein tyrosine phosphatases in vivo. Mol Cell. 2002;9(2):387–99.
199. Burke AJ, Sullivan FJ, Giles FJ, Glynn SA. The yin and yang of nitric oxide in cancer progression. Carcinogenesis. 2013;34(3):503–12. doi:10.1093/carcin/bgt034.
200. Pacher P, Beckman JS, Liaudet L. Nitric oxide and peroxynitrite in health and disease. Physiol Rev. 2007;87(1):315–424. doi:10.1152/physrev.00029.2006.
201. Joffin N, Niang F, Forest C, Jaubert AM. Is there NO help for leptin? Biochimie. 2012;94(10):2104–10. doi:10.1016/j.biochi.2012.06.017.
202. Zeng G, Quon MJ. Insulin-stimulated production of nitric oxide is inhibited by wortmannin. Direct measurement in vascular endothelial cells. J Clin Invest. 1996;98(4):894–8. doi:10.1172/JCI118871.
203. Barrett DM, Black SM, Todor H, Schmidt-Ullrich RK, Dawson KS, Mikkelsen RB. Inhibition of protein-tyrosine phosphatases by mild oxidative stresses is dependent on S-nitrosylation. J Biol Chem. 2005;280(15):14453–61. doi:10.1074/jbc.M411523200.
204. Takakura K, Beckman JS, MacMillan-Crow LA, Crow JP. Rapid and irreversible inactivation of protein tyrosine phosphatases PTP1B, CD45, and LAR by peroxynitrite. Arch Biochem Biophys. 1999;369(2):197–207. doi:10.1006/abbi.1999.1374.
205. Curcio MF, Batista WL, Linares E, Nascimento FD, Moraes MS, Borges RE, Sap J, Stern A, Monteiro HP. Regulatory effects of nitric oxide on Src kinase, FAK, p130Cas, and receptor protein tyrosine phosphatase alpha (PTP-alpha): a role for the cellular redox environment. Antioxid Redox Signal. 2010;13(2):109–25. doi:10.1089/ars.2009.2534.
206. Ashraf QM, Haider SH, Katsetos CD, Delivoria-Papadopoulos M, Mishra O. Nitric oxide-mediated alterations of protein tyrosine phosphatase activity and expression during hypoxia in the cerebral cortex of newborn piglets. Neurosci Lett. 2004;362(2):108–12. doi:10.1016/j.neulet.2004.02.069.
207. Chen YY, Huang YF, Khoo KH, Meng TC. Mass spectrometry-based analyses for identifying and characterizing S-nitrosylation of protein tyrosine phosphatases. Methods. 2007;42(3):243–9. doi:10.1016/j.ymeth.2007.03.002.
208. Forrester MT, Foster MW, Stamler JS. Assessment and application of the biotin switch technique for examining protein S-nitrosylation under conditions of pharmacologically induced oxidative stress. J Biol Chem. 2007;282(19):13977–83. doi:10.1074/jbc.M609684200.
209. Hsu MF, Meng TC. Enhancement of insulin responsiveness by nitric oxide-mediated inactivation of protein-tyrosine phosphatases. J Biol Chem. 2010;285(11):7919–28. doi:10.1074/jbc.M109.057513.
210. Rains JL, Jain SK. Oxidative stress, insulin signaling, and diabetes. Free Radic Biol Med. 2011;50(5):567–75. doi:10.1016/j.freeradbiomed.2010.12.006.

211. Zhu M, de Cabo R, Anson RM, Ingram DK, Lane MA. Caloric restriction modulates insulin receptor signaling in liver and skeletal muscle of rat. Nutrition. 2005;21(3):378–88. doi:10.1016/j.nut.2004.06.030.
212. de Moura LP, Souza Pauli LS, Cintra DE, de Souza CT, da Silva AS, Marinho R, de Melo MA, Ropelle ER, Pauli JR. Acute exercise decreases PTP-1B protein level and improves insulin signaling in the liver of old rats. Immun Ageing. 2013;10(1):8. doi:10.1186/1742-4933-10-8.
213. Easty D, Gallagher W, Bennett DC. Protein tyrosine phosphatases, new targets for cancer therapy. Curr Cancer Drug Targets. 2006;6(6):519–32.
214. Hardy S, Julien SG, Tremblay ML. Impact of oncogenic protein tyrosine phosphatases in cancer. Anticancer Agents Med Chem. 2012;12(1):4–18.
215. Julien SG, Dube N, Hardy S, Tremblay ML. Inside the human cancer tyrosine phosphatome. Nat Rev Cancer. 2011;11(1):35–49. doi:10.1038/nrc2980.
216. Ostman A, Hellberg C, Bohmer FD. Protein-tyrosine phosphatases and cancer. Nat Rev Cancer. 2006;6(4):307–20. doi:10.1038/nrc1837.
217. Jiang ZX, Zhang ZY. Targeting PTPs with small molecule inhibitors in cancer treatment. Cancer Metastasis Rev. 2008;27(2):263–72. doi:10.1007/s10555-008-9113-3.
218. Scott LM, Lawrence HR, Sebti SM, Lawrence NJ, Wu J. Targeting protein tyrosine phosphatases for anticancer drug discovery. Curr Pharm Des. 2010;16(16):1843–62.
219. Kasibhatla B, Wos J, Peters KG. Targeting protein tyrosine phosphatase to enhance insulin action for the potential treatment of diabetes. Curr Opin Investig Drugs. 2007;8(10):805–13.
220. Tautz L, Pellecchia M, Mustelin T. Targeting the PTPome in human disease. Expert Opin Ther Targets. 2006;10(1):157–77. doi:10.1517/14728222.10.1.157.
221. Khandekar MJ, Cohen P, Spiegelman BM. Molecular mechanisms of cancer development in obesity. Nat Rev Cancer. 2011;11(12):886–95. doi:10.1038/nrc3174.
222. Clemmons DR. Modifying IGF1 activity: an approach to treat endocrine disorders, atherosclerosis and cancer. Nat Rev Drug Discov. 2007;6(10):821–33. doi:10.1038/nrd2359.
223. Pollak MN. Investigating metformin for cancer prevention and treatment: the end of the beginning. Cancer Discov. 2012;2(9):778–90. doi:10.1158/2159-8290.CD-12-0263.
224. Shaw RJ, Lamia KA, Vasquez D, Koo SH, Bardeesy N, Depinho RA, Montminy M, Cantley LC. The kinase LKB1 mediates glucose homeostasis in liver and therapeutic effects of metformin. Science. 2005;310(5754):1642–6. doi:10.1126/science.1120781.
225. Kalender A, Selvaraj A, Kim SY, Gulati P, Brule S, Viollet B, Kemp BE, Bardeesy N, Dennis P, Schlager JJ, Marette A, Kozma SC, Thomas G. Metformin, independent of AMPK, inhibits mTORC1 in a rag GTPase-dependent manner. Cell Metab. 2010;11(5):390–401. doi:10.1016/j.cmet.2010.03.014.
226. Hoggard N, Agouni A, Mody N, Delibegovic M. Serum levels of RBP4 and adipose tissue levels of PTP1B are increased in obese men resident in northeast Scotland without associated changes in ER stress response genes. Int J Gen Med. 2012;5:403–11. doi:10.2147/IJGM.S25879.
227. Garcia-Ruiz I, Solis-Munoz P, Gomez-Izquierdo E, Munoz-Yague MT, Valverde AM, Solis-Herruzo JA. Protein-tyrosine phosphatases are involved in interferon resistance associated with insulin resistance in HepG2 cells and obese mice. J Biol Chem. 2012;287(23):19564–73. doi:10.1074/jbc.M112.342709.
228. Wu Y, Ouyang JP, Wu K, Wang SS, Wen CY, Xia ZY. Rosiglitazone ameliorates abnormal expression and activity of protein tyrosine phosphatase 1B in the skeletal muscle of fat-fed, streptozotocin-treated diabetic rats. Br J Pharmacol. 2005;146(2):234–43. doi:10.1038/sj.bjp.0706306.
229. Bettaieb A, Bakke J, Nagata N, Matsuo K, Xi Y, Liu S, Aboubechara D, Melhem R, Stanhope K, Cummings B, Graham J, Bremer A, Zhang S, Lyssiotis CA, Zhang ZY, Cantley LC, Havel PJ, Haj FG. Protein-tyrosine phosphatase 1B regulates pyruvate kinase M2 tyrosine phosphorylation. J Biol Chem. 2013;288(24):17360–71. doi:10.1074/jbc.M112.441469.

Current Status of PTP-Based Therapeutics

Rongjun He and Zhong-Yin Zhang

Abstract Protein tyrosine phosphorylation regulates virtually all aspects of cellular function. Proper levels of protein tyrosine phosphorylation are controlled by protein-tyrosine kinases (PTKs) and protein-tyrosine phosphatases (PTPs). Deregulation of tyrosine phosphorylation-mediated signaling is a major cause of cancer. Indeed, PTKs represent an important class of anti-cancer drug targets with more than 20 drugs already approved by the FDA. Several PTPs are also implicated in tumor onset, progression and metastasis, and have attracted attention as potential therapeutic targets for cancer treatment. The most recognized members of the PTP family include PTP1B, SHP2, CDC25, and PRLs. PTP1B plays a positive role in breast cancer progression; SHP2 germ line and somatic gain-of-function mutations cause Noonan Syndrome, juvenile myelomonocytic leukemia, acute myeloid leukemia, and solid tumors; CDC25 regulates cell cycle progression and is overexpressed in a number of human cancers; and the PRLs are overexpressed in cancer and have been linked to tumor metastasis. Hence, inhibitors directed against these PTPs can serve as potential novel anti-cancer agents. This chapter provides an overview of the mechanisms of these oncogenic PTPs in promoting cancer, the approaches for inhibitor discovery, as well as the current status of drug development targeting these phosphatases.

Keywords Protein-tyrosine phosphatase • PTP1B • SHP2 • CDC25 • PRL • Inhibitor design • Potency and specificity • Drug development • Therapeutics • Cancer

---

R. He • Z.-Y. Zhang (✉)
Department of Biochemistry and Molecular Biology, Indiana University School of Medicine, 635 Barnhill Drive, Indianapolis, IN 46202, USA
e-mail: zyzhang@iu.edu

B.G. Neel, N. Tonks (eds.), *Protein Tyrosine Phosphatases in Cancer*,
DOI 10.1007/978-1-4939-3649-6_13

Introduction

Protein tyrosine phosphorylation is a key strategy for the regulation of cellular processes, such as growth, differentiation, metabolism, migration, survival, and apoptosis [1]. Inappropriate protein tyrosine phosphorylation is a hallmark of many human pathologic conditions, including cancer, diabetes, neurological, and infectious diseases. Protein tyrosine kinases (PTKs), which catalyze tyrosine phosphorylation, represent the single most important class of anti-cancer drug targets with more than 20 small molecule inhibitors and biomolecules targeting the PTKs already approved by the FDA and many currently under clinical development [2]. Given the reversible nature of protein tyrosine phosphorylation, it is not surprising that dysregulation of the activity of protein tyrosine phosphatases (PTPs) has been linked to many human diseases, including type 2 diabetes. Consequently, PTPs have been recognized as next-generation drug targets for cancer, as well as other diseases [3–6].

The PTP superfamily is composed of 107 members, 99 of which are class I cysteine-based PTPs sharing a conserved sequence (H/V)C(X)$_5$R(S/T), namely the PTP signature motif [3, 4]. This class is further divided into two groups: classical tyrosine-specific PTPs (38 members), and tyrosine- and serine/threonine dual-specific phosphatases (DSPs) (61 members). Classical tyrosine-specific PTPs comprise 17 non-receptor or intracellular PTPs and 21 receptor-based or transmembrane tyrosine phosphatases.

Many PTPs have been shown to play a role in oncogenic events, such as transformation, tumor progression and metastasis, rendering them potential drug targets for various cancers [7–11]. The chemistry and biology of these PTPs has been described in detail in other chapters of this book. Here, we will focus on the most established oncogenic PTPs, including PTP1B, SHP2, CDC25, and phosphatases of regenerating liver (PRLs). We will begin with a brief introduction of their involvement in cancer, followed by discussion of the current status of small molecule therapeutics targeting these PTPs.

PTP1B

PTP1B, encoded by *PTPN1*, is the prototypic mammalian PTP, first purified from human placenta [12–15]. PTP1B is a classic non-receptor PTP enzyme, is ubiquitously expressed, and consists of two domains: a conserved N-terminal catalytic domain and a non-catalytic C-terminal domain harboring a hydrophobic tail responsible for localizing PTP1B to the cytoplasmic face of endoplasmic reticulum [16, 17].

PTP1B has been studied extensively since its initial characterization. Early studies indicated that PTP1B functions as a key negative regulator of insulin- and leptin-mediated pathways, which control body weight, glucose homeostasis, and energy expenditure [18]. PTP1B regulates insulin signaling by directly dephosphorylating the phosphorylated insulin receptor (IR) and insulin receptor substrates (IRS), hence blocking the activation of downstream events, while it modulates leptin signaling by dephosphorylating JAK2 and possibly STAT3. Inhibition of PTP1B by antibody or small molecule inhibitors increases insulin-stimulated phosphorylation of IR and IRS and activates STAT3 [19, 20]. PTP1B knockout mice exhibit lower blood glucose and insulin levels, enhanced insulin sensitivity, reduced body fat ratio, and leptin hypersensitivity [21–23]. These data suggest that inhibition of PTP1B could be an effective strategy to address both type 2 diabetes and obesity.

Interestingly, recent evidence suggests that PTP1B might also play an important role in promoting cancer [24]. For example, the activity of c-SRC, one the earliest proto-oncogene, whose activity is elevated in many types of human cancers, is regulated by phosphorylation at two distinct tyrosine residues. Autophosphorylation of Tyr416 in the kinase domain is important for SRC activity. By contrast, phosphorylation of Tyr527 in the C-terminal tail by CSK inactivates SRC due to an intramolecular pTyr-SH2 interaction. PTP1B is capable of activating c-SRC by dephosphorylating its inhibitory pTyr527 in breast cancer cells [25]. PTP1B has also been reported to dephosphorylate p62Dok, a scaffold protein involved in binding and activating p120Ras GAP, resulting in activation of the oncogenic RAS-ERK cascade [26, 27]. In a study of 29 human breast cancer samples, elevated PTP1B expression was recorded in 72 %, in comparison to normal mammary epithelium samples [28]. *ERBB2* is a frequently amplified gene in breast cancer. Interestingly, it was found that 90 % tumors overexpressing ERBB2 also had elevated PTP1B expression, and PTP1B overexpression was present in many stages of tumor development. These data suggest that PTP1B might participate in mammary tumorigenesis. In order to define the role of PTP1B in breast cancer, NDL2 mice, which express an activated mutant *ErbB2* with a deletion in the extra cellular segment under control of the mouse mammary tumor virus (MMTV) promoter, were crossed with *Ptpn1* knockout mice [27]. *Ptp1b* deletion led, on average, to an 85-day delay in tumor onset and decreased lung metastasis. Independently, *MMTV-NeuNT* mice, which express an activated *ErbB2* with a point mutation in the transmembrane region (V664Q), were crossed to *Ptpn1* knockout mice [29]. In this model, *Ptp1b* deletion resulted in an even more pronounced delay in tumor development. These data suggest that PTP1B is functionally linked to ERBB2 in promoting breast cancer tumorigenesis, and that PTP1B inhibition could be a novel strategy for breast cancer treatment.

Understandably, the pursuit of PTP1B inhibitors with therapeutic potential not only for type 2 diabetes and obesity but also for cancer has been a central topic in medicinal chemistry [30]. A major challenge in this area has been the development of compounds with exquisite specificity to PTP1B, due to the highly conserved PTP active site structure. In particular, PTP1B and its close homologue TC-PTP share >75 % sequence similarity in their catalytic domains. TC-PTP is required for hematopoiesis and immune function, and its absence in mice causes premature death 3–5 weeks after birth [31] with elevated levels of inflammatory mediators such as NO and TNF-α [32]. Hence, selectivity for PTP1B over TC-PTP should be an important consideration for PTP1B inhibitor development.

An effective strategy to address the specificity issue is to design bidentate ligands targeting both the active site and a unique secondary binding pocket in the vicinity of the active site [33]. Initial efforts using this strategy generated a number of highly potent and specific PTP1B inhibitors. Compound **1** (Fig. 1) represents the first

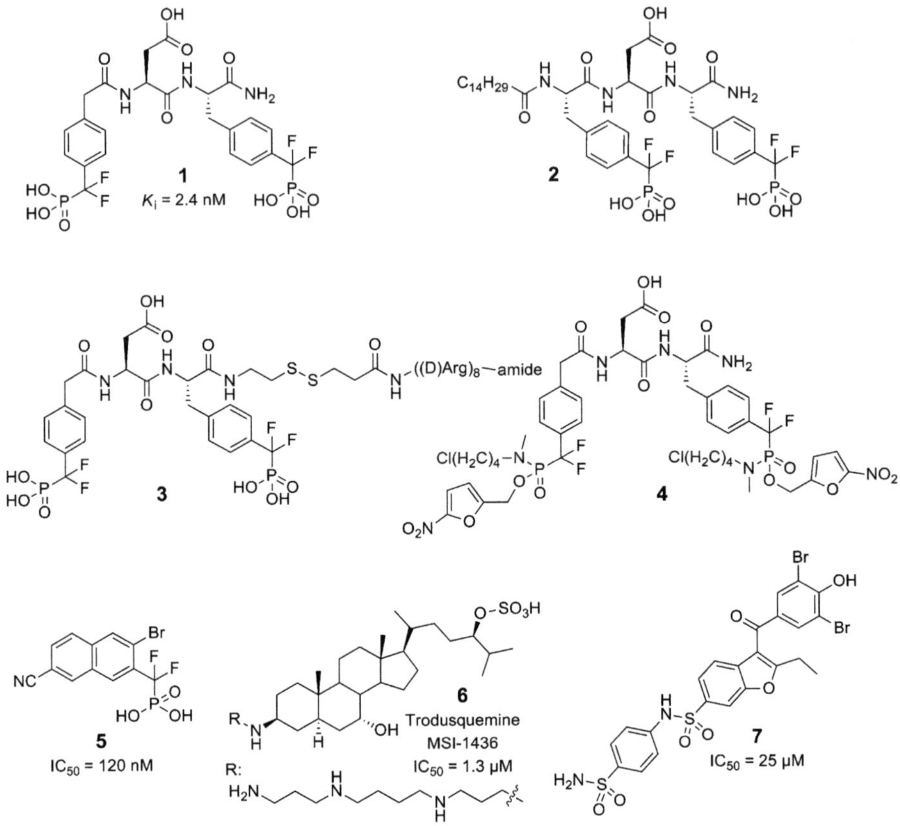

Fig. 1 Structures and activities of PTP1B inhibitors

successful example, with an IC_{50} of 2.4 nM and a specificity of at least ten-fold over TC-PTP and a large panel of PTPs [34]. Engagement of both the active site and a unique secondary pocket by compound **1** was validated by biochemical and structural studies [35]. Unfortunately, this compound has five negative charges and thus is not cell-permeable under physiological conditions. To improve permeability, derivatives were prepared with a fatty acid chain (compound **2**) [36], a cell-permeable poly Arg peptide (compound **3**) [37], or a nitrofuryl protecting group on the phosphonic acid (compound **4**) [38]. These derivatives exhibited in vivo activity in sensitizing insulin and leptin signaling pathways, suggesting their potential for the treatment of type 2 diabetes and obesity. Compound **5** is a difluorophosphonate-based PTP1B inhibitor developed by Merck Frosst [39], with an IC_{50} value of 120 nM for both PTP1B and TC-PTP. Compound **5** exhibited a dose-dependent anti-diabetic effect in diet-induced diabetic mice. Importantly, this compound also displayed anti-cancer effects: similar to the effects of *Ptpn1* deletion, administration of compound **5** to NDL2 *Ptpn1*$^{+/+}$ mice also delayed tumor formation by 29 days, compared with mice treated with vehicle [27]. It is noteworthy that this experiment only demonstrated compound **5**'s effects on tumor formation. It would be interesting to test whether it has effect on tumor maintenance. Nevertheless, these results indicate that small molecule inhibitors of PTP1B might have therapeutic potentials for breast cancer. Trodusquemine (compound **6**, MSI-1436) is a PTP1B inhibitor developed by Genaera for the treatment of type 2 diabetes and obesity [40–42]. It inhibits PTP1B in a reversible, allosteric, noncompetitive, and highly selective manner. In HepG2 cells and hypothalamic tissue, Trodusquemine significantly enhanced tyrosine phosphorylation of IR and STAT3. In obese animal models, Trodusquemine decreased appetite, caused weight loss, and normalized fasting blood glucose, cholesterol, and triglyceride levels. Trodusquemine was well tolerated in a phase 1 clinical trial; however, phase 2 clinical trials with type 2 diabetes and obesity patients have not been conducted.

Targeting allosteric sites of PTP1B has also been demonstrated to be a good strategy for the development of inhibitors with high specificity [43]. For example, compound **7** binds to a novel allosteric site 20 Å away from PTP1B catalytic pocket, which is formed by helices α3 and α6 and is not well conserved among PTPs. Compound **7** inhibits PTP1B by preventing the closure of the WPD loop, an essential process for PTP1B catalytic activity. It has an IC_{50} of 25 μM, and not surprisingly, high selectivity over TC-PTP (IC_{50} at 129 μM) and LAR (IC_{50} at >500 μM). Importantly, compound **7** exhibited excellent cellular activity in enhancing insulin signaling pathway. Unlike active site-directed inhibitors, allosteric inhibitors are not required to interact with the PTP catalytic pocket; hence, they have been suggested to possess more favorable pharmacological properties.

In addition to small molecules, ISIS 113715 is an antisense oligonucleotide (ASO) targeting PTP1B [44–46]. It is a 20-mer ASO with a sequence of "GCTCCTTCCACTGATCCTGC," complementary to *PTPN1* mRNA, where a ten-deoxynucleotide gap region flanks on its 3′ and 5′ ends with five 2′-*O*-(2-methoxyethyl) nucleotides, with all cytosine nucleotides being 5-methylcytosines and all linkages being phosphorothioates. In diabetic animal models, ISIS 113715 increased insulin sensitivity, reduced body glucose level and weight gain, indicating its anti-diabetic effects. ISIS 113715 has progressed into phase 2 clinical trials with positive results; however, further reports on its progress are not available.

There are more than 100 reported PTP1B inhibitors [47]; however, almost all of them have been developed for the treatment of type 2 diabetes and obesity, and their applications as anti-cancer agents have been largely unexplored. Further studies of PTP1B inhibition may offer new avenues for the treatment of diabetes and obesity as well as cancer.

SHP2

The SRC homology 2 (SH2)-domain containing protein tyrosine phosphatase-2 (SHP2), encoded by *PTPN11*, plays a positive role in signal transduction downstream of growth factor and cytokine receptors to regulate proliferation, differentiation, motility, and apoptosis [48–52]. Biochemical and genetic evidence place SHP2 upstream of RAS, an essential component of the signaling pathway that underlies growth factor/cytokine-induced cell proliferation and survival, and SHP2 activity is required for full activation of the RAS-extracellular signal-regulated kinase (ERK) cascade. A classic non-receptor PTP, SHP2 has two SH2 domains at its N-terminus (N-SH2, C-SH2) and a catalytic (PTP) domain followed by C-terminal tyrosine phosphorylation sites at Y542 and Y580 and a proline-rich region [53]. In the basal state, the N-SH2 domain interacts with the PTP domain to block the access of substrates, and hence is auto-inhibited. Upon growth factor or cytokine stimulation, the N-SH2 domain binds tyrosine-phosphorylated proteins, such as receptor-tyrosine kinases, cytokine receptors, or scaffold proteins, to release the PTP domain, enabling catalysis. Alternatively, Y542 or Y580, phosphorylated by PTKs or other kinase, can also bind to N-SH2 domain, thereby disrupting auto-inhibition and activating SHP2.

Not surprisingly, SHP2 activation mechanisms have been subverted in human disease. For example, *PTPN11* mutations are the most common cause of Noonan Syndrome (NS) and juvenile myelomonocytic leukemia (JMML) and have also been reported in acute myelogenous leukemia (AML) and several types of solid tumors [48, 50, 54–59, 60, 61]. Germ-line mutations in *PTPN11* cause approximately 50 % of NS cases, and somatic mutations of *PTPN11* are present in ~35 % of patients with sporadic JMML. In addition, *PTPN11* somatic mutations have been reported in ~3 % of neuroblastoma cases, making it one of the most common mutated genes in this disease [54, 62]. Most of the mutations in these disorders affect residues in the interface between the N-SH2 and PTP domains, which are involved in basal auto-inhibition [60]. The PTP activity of these disease-causing SHP2 mutants is greater than that of wild-type SHP2, indicating that these substitutions disrupt the auto-inhibitory interface between the N-SH2 and catalytic domains to favor the open, active conformation, lowering the threshold for growth factor or cytokine stimulation. In addition to its critical role in the activation of the RAS-ERK signaling cascade, SHP2 has also been reported to regulate the PI3K/AKT pathway, which is frequently enhanced in various cancers, although the role of SHP2 in PI3K/AKT regulation is ligand- and cell-type dependent [63, 64]. Furthermore, SHP2 has been indicated in JAK/STAT, JNK, and NF-κB signaling, supporting its broad activity in cellular environment [60]. Very recently, SHP2 was shown to promote maintenance and progression of HER2-positive and triple negative breast cancers [65]. Deletion of SHP2 by shRNAmiRs significantly blocked HER2/3-evoked invasion in MCF10A human breast epithelial cells in a three-dimensional cell culture model. In xenograft models, treatment with SHP2 shRNAmiRs eradicated breast tumor-initiating cells, and blocked the growth and metastasis of established breast tumors. Mechanistically, SHP2 promotes breast cancer by activating the ERK pathway and stemness-associated transcription factors, such as cMYC and zinc finger E-box binding homeobox 1 (ZEB1). Taken together, biochemical, structural, cellular, animal, and clinical studies indicate that SHP2 is a *bona fide* PTP proto-oncogene.

In view of the importance of SHP2 in various leukemias and solid tumors, SHP2 represents an exciting therapeutic target. Moreover, given the obligatory requirement of SHP2 function in multiple oncogenic receptor tyrosine kinase pathways, inhibition of SHP2 might also prove effective for cancers with activated receptor PTKs, which respond poorly to kinase inhibitor monotherapy. Significant efforts have been devoted into the development of SHP2 inhibitors. However, it is still a difficult endeavor due to the aforementioned challenges in acquiring inhibitor potency, specificity, and cell permeability. In particular, SHP1 is a close homologue of SHP2 and belongs to the same subgroup of non-receptor-based PTPs. As SHP1 is primarily involved in downregulation of cytokine signaling and inhibitory receptor signaling, and its cellular function in cancer is largely unknown, it is advisable to avoid inhibiting SHP1 while targeting SHP2. High throughput screening of an NCI compound library resulted in the identification of NSC-87877 (Fig. 2, compound **8**) as one of the earliest reported SHP2 inhibitors, with an IC_{50} value of 0.32 μM [66]. Although NSC-87877 is not selective over SHP1, it is 5- to 475-fold selective over PTP1B, HePTP, DEP1, CD45, and LAR. Applied to cells, NSC-87877 blocked EGF-stimulated SHP2, RAS, and ERK activation, but not GAB1 tyrosine phosphorylation or GAB1-SHP2 association. From the same NCI library, oxindole derivatives were also discovered to be SHP2 inhibitors [67]. Optimization of the oxindole scaffold yielded SPI-112 (compound **9**) with an IC_{50} of 1 μM against SHP2 and a selectivity of 14-fold over SHP1 and PTP1B. Given that its specificity for SHP2 over SHP1 is the best reported to date, it will be important to obtain SHP2 ligand binding information for this compound, which could provide invaluable insights for selective SHP2 inhibitor design. Although compound **9** did not display cellular activity, its methyl ester analogue SPI-112Me (compound **10**) was able to

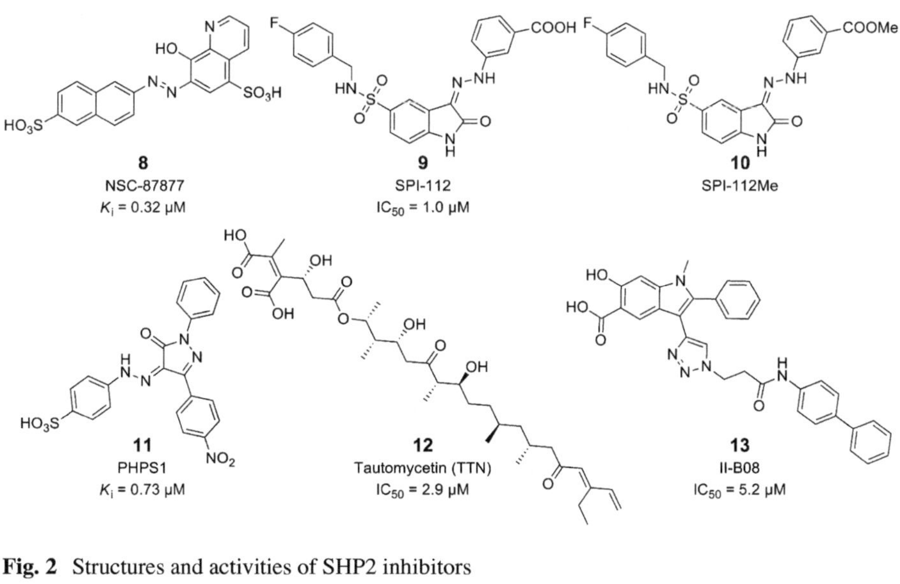

Fig. 2 Structures and activities of SHP2 inhibitors

block EGF-stimulated SHP2 activity, paxillin dephosphorylation, ERK1/2 activation, and cell migration [68]. SPI-112Me also inhibits SHP2 activity, ERK1/2 activation, and cell survival in SHP2/E76K mutant-transformed TF-1 myeloid cells and enhances IFN-γ-stimulated STAT1 tyrosine phosphorylation, ISRE-luciferase reporter activity, and p21 expression, while inhibiting proliferation of HT-29 cells. In silico screening of 2.7 million compounds identified PHPS1 (compound **11**) as a potent SHP2 inhibitor (K_i=0.73 μM) [69]. This compound displayed a selectivity of two-fold over EC-PTP, eight-fold over PTP1B, 15-fold over SHP1, and >20-fold over PTPH1, STEP, PTPN7, PTPRK, GLEPP1, LAR2, and mPTPA. PHPS1 exhibited excellent cellular activity in blocking HGF/SF-stimulated phosphorylation of ERK1/2 and dephosphorylation of paxillin, a downstream target of SHP2-dependent signaling. PHPS1 also inhibited ERK1/2 activation in cells with SHP2/E76K mutation, which is associated with leukemia, and blocked the anchorage-independent growth of a number of human cancer cell lines.

Tautomycetin (TTN, compound **12**) is a polyketide natural product used as an immunosuppressing agent in organ transplantation [70, 71]; however, it was recently found to be an SHP2 inhibitor with an IC_{50} of 2.9 μM [72]. With regard to its specificity, it has IC_{50} values of 14.6 μM for SHP1, 20 μM for LYP, 41.2 μM for PTP1B, and greater than 50 μM for HePTP, PTPα, CD45, VHR, and CDC14. TTN was able to block SHP2-mediated cellular signaling, including T cell receptor-mediated tyrosine phosphorylation and ERK1/2 activation in Jurkat T cells. TTN also blocked hematopoietic progenitor hyper-proliferation and monocytic differentiation induced by the gain-of-function SHP2/E76K mutant. Inhibition of SHP2 was validated by a crystal structure of the SHP2 in complex with a TTN analogue, TTN-D1, which showed that TTN binds to the SHP2 active site in a manner similar to a peptide substrate. In addition to its immunosuppressive activity, TTN also inhibits the growth of colorectal cancer cells [73], which might be due to SHP2 inhibition. Optimization of TTN might afford more potent and specific SHP2 inhibitors with enhanced anti-cancer and immunosuppressive activities.

II-B08 (compound **13**) is an indole salicylic acid-based SHP2 inhibitor identified by click chemistry, with a K_i of 5.2 μM [74]. II-B08 exhibited three-fold selectivity over SHP1, and several-fold selectivity over a panel of mammalian PTPs, including the cytosolic PTPs PTP1B, LYP, HePTP, and FAP1, the receptor-like PTPs, CD45, LAR, and PTPα, the dual specificity phosphatases VHR, and CDC14, and the low molecular weight PTP. II-B08 displayed excellent cellular activity, blocking growth factor-stimulated ERK1/2 activation in HEK293 cells as well as SHP2/E76K mutant-induced hematopoietic progenitor hyper-proliferation in response to GM-CSF. The importance of this compound as an SHP2 inhibitor is underscored by the fact that its activity in inhibiting SHP2 has been validated by several independent research groups. For instance, in bone marrow-derived mast cells (BMMCs), wherein SHP2 signaling downstream of KIT is essential for cell survival and homeostasis, SHP2 deletion or treatment with II-B08 resulted in survival defects upon removal of cytokines [75]. SHP2 has also been reported to be essential for ligand-independent growth and survival in oncogenic KIT/D814V mutant cells. Treatment of KIT/D814V mutant cells with II-B08 reversed these processes and caused a dose-dependent increase in apoptosis by disrupting formation of p85α, SHP2, and GAB2 complex, a required component for PI3K/AKT activation [76]. Recent studies have also demonstrated that SHP2 activity is associated with FLT3-ITD-induced hematopoietic progenitor hyper-proliferation and malignant disease [74, 77]. SHP2 knockdown reduced hyper-proliferation in FLT3-ITD-expressing cells, and similar results were obtained with treatment of SHP2 inhibitor II-B08 [77]. To provide additional evidence that the effect of II-B08 on intracellular signaling is SHP2 dependent, the effect of II-B08 on EGF-mediated ERK1/2 activation was evaluated after the level of SHP2 was downregulated by siRNA [78]. Indeed, the ability of II-B08 to inhibit EGF-mediated ERK1/2 activation was blunted when SHP2 was knocked down with siRNA. As a negative control, II-B05, a structurally related but inactive compound, had no effect on ERK1/2 activation. To further establish that SHP2 inhibitors specifically block SHP2-dependent signaling, the effect of II-B08 on ERK1/2 activation was determined after the cells were stimulated by phorbol 12-myristate 13-acetate (PMA) [78]. The PMA-induced ERK1/2 activation has been shown to be independent of SHP2 and instead depends on activation of protein kinase C and Raf in an Ras-independent manner. Thus, one would predict that SHP2 inhibitors would not inhibit PMA-induced ERK1/2 phosphorylation. As predicted, II-B08 did not inhibit PMA-induced ERK1/2 phosphorylation [78]. These independent studies suggest that II-B08 is an authentic SHP2 inhibitor.

Given the efficacious activity of II-B08 in a number of cell-based systems, II-B08 was also evaluated in animal models of cancer. In a mouse model of KIT/D814V-induced mast cell leukemia, II-B08 treatment was found to prolong mouse survival and reduce splenomegaly and hepatomegaly [76]. Interestingly, II-B08 synergizes with PI3K inhibitor (LY294002) in repressing KIT/D814V-induced mast cell leukemia. More recent results show that II-B08 also exhibited significant anti-tumor activity in non-small cell lung cancer xenograft mice carrying activating EGFR mutations that are resistant to existing kinase inhibitor treatment [78]. Taken together, the results represent the first demonstration that targeting an oncogenic PTP may be an effective cancer treatment. The results also suggest that II-B08 could be used as a pharmacological tool to study SHP2's functions in vivo, as well as a lead compound for the development of anti-cancer therapeutics. X-ray crystallographic analysis of II-B08-bound SHP2 structure reveals that the salicylic acid core occupies the PTP active site, while the distal biphenyl ring makes hydrophobic contacts with a region highly divergent among the PTPs. The atomic-level information on SHP2•II-B08 furnishes a solid foundation for the design of more potent and selective SHP2-based small molecule therapeutics.

CDC25

The CDC25 phosphatases are dual-specificity phosphatases first identified in yeast as proteins influencing cell division [79–82]. There are three isoforms in humans, CDC25A, CDC25B, and CDC25C, encoded by genes on three different chromosomes [83]. CDC25A, CDC25B, and CDC25C are further divided into 2, 5, and 5 variants, respectively, by differential splicing. However, the PTP domain is located at the C-terminus, which is conserved among all isoforms. The N-terminal regulatory domain varies in length, shares low sequence homology, and contains sites for phosphorylation, ubiquitylation, proline isomerization, and 14-3-3 sequestration [84].

All CDC25 isoforms share a common mode of action, dephosphorylating the two negative regulatory residues in cyclin-dependent kinases (CDKs), which are required for cell cycle progression [85]. These two residues (Thr14 and Tyr15 in CDK1) are located in the ATP binding loop of CDKs [86]. CDC25A is expressed in early G1-phase and is mainly responsible for the activation of cyclin A- and cyclin E-CDK2 complexes during the G1-S transition [80, 87]. CDC25B and CDC25C are responsible for the transition into mitosis [80, 88]. CDC25B partially activates the cyclin B-CDK1 complex at the centrosome during the transition from G2 to M, after which CDC25C in the nucleus fully activates the cyclin B-CDK1 complex at the onset of mitosis [89].

As cell cycle deregulation is a shared characteristic of all cancer cells, it is not surprising that CDC25 family members have been associated with various human cancers. For example, CDC25A and CDC25B, in cooperation with oncogenic RAS or loss of RB1, exhibits oncogenic properties, promoting growth of normal mouse embryonic fibroblasts in soft agar and nude mice [90]. *Cdc25B* transgenic mice show enhanced proliferation of mammary epithelial cells, precocious alveolar hyperplasia [91], and increased susceptibility to carcinogen (9,10-dimethyl-1,2-benzanthracene, DMBA)-induced tumors [92]. CDC25A and CDC25B overexpression has also been reported in a number of human malignancies, such as breast, colon, gastric, lung, colorectal, pancreatic, and head and neck carcinoma, neuroblastoma, and non-Hodgkin's lymphoma [80]. Current data suggest that CDC25 overexpression occurs at different stages of tumorigenesis and, in cooperation with other oncogenes, causes cell cycle deregulation and genome instability. Therefore, the CDC25 family also constitutes a class of attractive targets for cancer drug discovery.

Early efforts in CDC25 inhibitor discovery focused on the exploration of natural products. Dysidiolide (Fig. 3, compound **14**), isolated from the Caribbean sponge *Dysideaetheria*, inhibits CDC25A with an IC_{50} of 9.4 μM [93]. Dysidiolide also inhibits the growth of human A549 lung carcinoma cells and the mouse P388 leukemia cells, with IC_{50} values of 4.7 and 1.5 μM, respectively. Further structure activity relationship analysis yielded several diastereomers, e.g., **15**, which differs at carbon centers 4 and 6 and processes similar inhibitory activity against CDC25A and cell growth, indicating that the stereo-configuration of dysidiolides do not influence activity [94]. Solid-phase synthesis of dysidiolide analogues yielded compound **16**, capable of inhibiting CDC25C with an IC_{50} of 0.8 μM, as well as good activity in non-small cell lung, colon, breast, and ovarian cancer cells [95]. It is reasonable to predict that the γ-hydroxybutenolide moiety in these compounds functions as a phosphate mimetic, which is consistent with a recent report that RE44 (compound **17**) is a CDC25A and B inhibitor with IC_{50} of 13.5 μM and 4.26 μM, respectively [96]. RE44 also inhibits VHR with an IC_{50} of 24.9 μM, but it does not inhibit PTP1B, MKP1, and MKP3 at 100 μM. Importantly, this compound inhibits

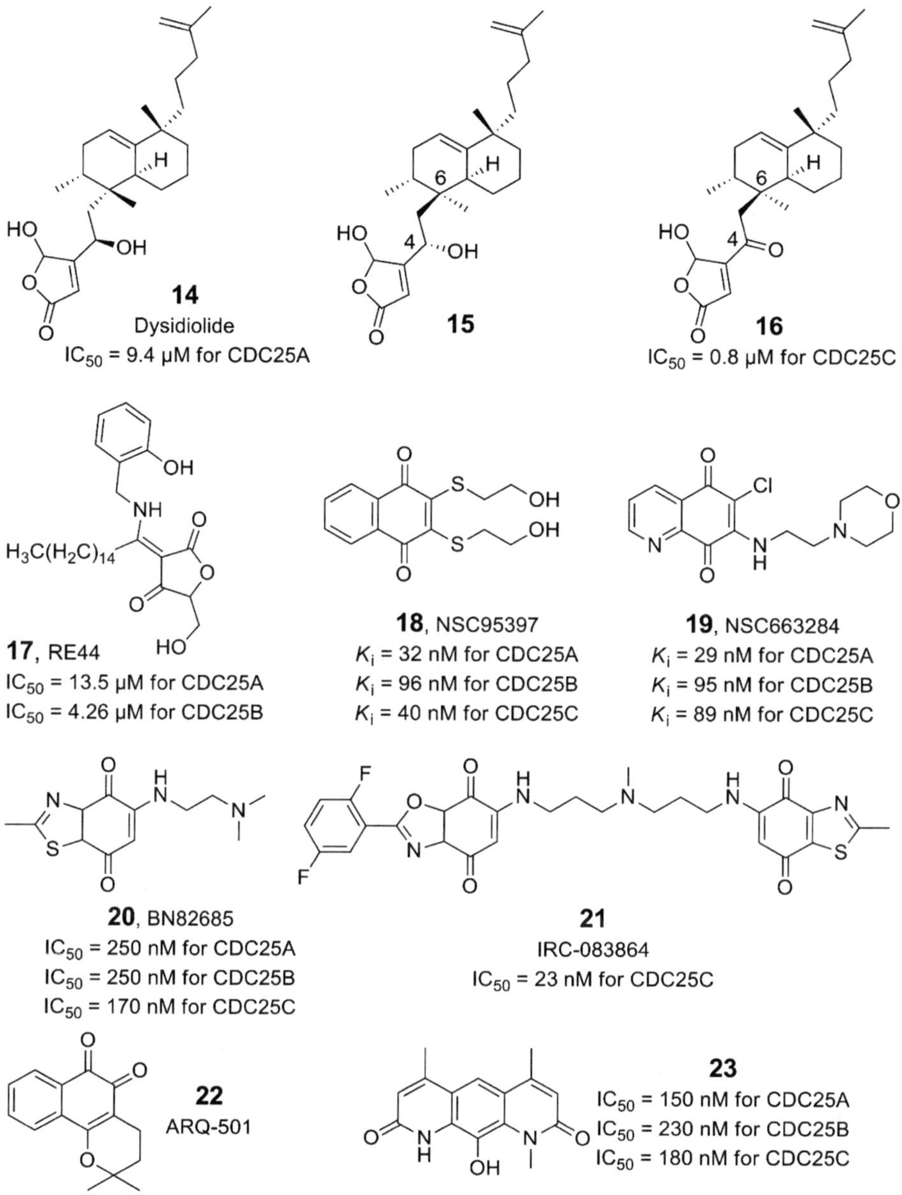

Fig. 3 Structures and activities of CDC25 inhibitors

the proliferation of HL60 cells with an IC_{50} of 2.44 μM, induces cell cycle arrest at the G2-M transition and blocks CDK1 dephosphorylation in tsFT210 cells, indicating excellent cellular activity against CDC25A and B. However, its long aliphatic carbon chain might cause promiscuous binding to proteins, an obvious concern in inhibitor development.

Quinone derivatives related to vitamin K3 constitute another major class of CDC25 inhibitors. NSC95397 (compound **18**) [97] and NSC663284 (compound **19**) [98] were identified from NCI's diversity sets as CDC25 inhibitors. Compound **18** inhibited CDC25A, B and C with K_i values of 32 nM, 96 nM, and 40 nM, respectively. Similarly, compound **19** inhibited CDC25A, B, and C with K_i values of 29 nM, 95 nM, and 89 nM, respectively. The observed inhibitory potencies toward the CDC25 phosphatases were nearly two orders of magnitude higher than those for VHR and PTP1B. Compound **18** also blocked the G_2-M phase transition and attenuated growth of human and murine carcinoma cells.

BN82685 and IRC-083864 (compounds **20** and **21**) were developed by IPSEN research laboratory as CDC25 inhibitors. BN82685 is a heterocyclic quinone that inhibits CDC25A, B, and C with IC_{50} values of 250 nM, 250 nM and 170 nM, respectively [99]. BN82685 also blocked the proliferation of a large panel of human cancer cell lines, including DU-145, Mia PaCa-2, and A2058. Importantly, BN82685, when administrated orally, inhibited tumor growth of Mia PaCa-2 xenografts. A combination of BN82685 and paclitaxel at low concentrations inhibited the proliferation of colon cancer cells [100], suggesting that CDC25 inhibitors could be used alone or together with agents targeting microtubules to treat cancer. IRC-083864 is a bis-quinonoid compound inhibiting CDC25C with an IC_{50} value of 23 nM [101]. This compound was found to inhibit LNCaP and DU-145 cell proliferation with IC_{50} values of 23 nM and 20 nM, respectively. IRC-083864, when taken orally, also exhibited significant anti-tumor effect on LNCaP xenografts. ARQ-501 (compound **22**), developed by ArQule, has undergone phase 1 and phase 2 clinical trials as a single agent treatment for leiomysarcoma, head and neck cancer [102]. Unfortunately, information on the status of development of BN82685, IRC-083864, and ARQ-501 is not available. Compound **23** and related compounds were patented by InterMedDiscovery GmbH as potent CDC25 inhibitors [103]. Compound **23** exhibited IC_{50} values of 150 nM, 230 nM, and 180 nM against CDC25A, CDC25B, and CDC25C, respectively. Compound **23** showed anti-proliferative effects on the HCT116, DU145, A549, and H460 cancer cell lines, with IC_{50}s in the 100–300 nM range. Despite the impressive biochemical and cell-based efficacies exhibited by the quinone-based CDC25 inhibitors, quinone compounds are known to produce reactive oxygen species and also undergo Michael addition reaction with reactive thiol groups, which may inactivate various redox-sensitive enzymes and cause toxicity in vivo [104]. Hence, it is not clear whether their in vivo effects are really specific for CDC25, nor whether they can be developed into *bona fide* drugs. Identification of non-quinone inhibitors should be an important focus for future efforts targeting the CDC25 phosphatases as cancer therapeutics.

PRLs

PRLs are a subfamily of DSPs with three members, PRL-1, PRL-2, and PRL-3 [105, 106]. As its name implies, PRL-1 was first identified in regenerating rat liver [107, 108]. PRL-1 and PRL-2 are expressed in most tissues, whereas PRL-3 is primarily present in heart and skeletal muscle. PRLs are relatively small proteins (about 20 kDa) and have a high degree of sequence homology between them (87 % for PRL-1 and PRL-2; 79 % for PRL-1 and PRL-3; and 76 % for PRL-2 and PRL-3 in human), but low sequence identity with other dual-specific phosphatases (<30 %) [106]. All PRLs have a single catalytic domain with the typical PTP signature motif CX_5R, followed by a C-terminal polybasic region and a CAAX sequence for prenylation, which is responsible for plasma or intracellular membrane localization [109, 110].

Ectopic expression of PRLs in various cell lines results in increased cell proliferation and invasion, although the precise biochemical mechanisms remain complex and controversial. Biochemical studies suggest the PRL-1 and PRL-3 may promote these processes by activating c-SRC [111, 112]. PRL-1 expression in HCT116 or HeLa cells augments AKT-mediated phosphorylation/activation of MDM2, a p53 ubiquitin ligase, leading to p53 downregulation [113]. In addition, PRL-3 activates the PI3K/AKT pathway and promotes cell survival under growth factor-deprivation conditions, which can lead to p53 downregulation by stabilizing MDM2 and increasing transcription of PIRH2, another p53 ubiquitin ligase [114, 115]. PRLs can also positively regulate the expression and activation of MMPs (matrix metalloproteinase), which are extracellular secreted proteins important for tumor metastasis. PRL-1 upregulates MMP2 and MMP9 through c-SRC and ERK1/2 activation [116], whereas PRL-3 increases the activity and expression of MMP2 and MMP7, but not MMP9, via integrin β1, ERK1/2, and PI3K/AKT pathways [117]. Finally, PRLs have been reported to regulate the RHO family of small G-proteins [118–121] to promote cell adhesion and migration, although the mechanism of PRLs coupling to RHO is unclear.

PRL-1 overexpression has been documented in lung and pancreatic cancer cell lines with increased cell motility and invasive properties, which can be reduced by *PRL-1* knockdown [111, 116, 122, 123]. Specifically, the PRL-1 expression level was significantly increased in lymph node metastases of adenocarcinomas [120, 121]. PRL-2 was also found to be overexpressed in many cancer cell lines and tumor samples, including pancreatic, breast, and lung malignancies, and the level of PRL-2 expression was associated with tumor progression [123–125]. In addition, PRL-2 mRNA levels are elevated and associated with poor prognosis in pediatric AML [126]. The most studied member of the PRL family, PRL-3, is overexpressed in various cancers, including colon, breast, gastric, oral, cervix, and ovarian carcinomas, as well as hematological malignancies, such as multiple myeloma and AML [105, 106]. In colon cancer, PRL-3 overexpression is associated with poor prognosis and the development of liver metastasis [127]. Stable expression of *PRL-3* enhances cell motility and invasiveness, and induces tumor formation with metastasis in mice [128], whereas *PRL-3* knockdown (by siRNA) abrogates cell motility and metastasis [129, 130].

Hence, PRL-3, together with PRL-1 and PRL-2, are a class of oncogenes that contribute to tumor formation and/or metastasis, and small molecules targeting PRLs could represent novel agents for anti-cancer treatments. The earliest example of a PRLs inhibitor was the finding that the anti-leishmaniasis drug pentamidine (Fig. 4, compound **24**) inhibits all three PRLs at 1–10 μg/mL (2.9–29 μM) [131]. Pentamidine also blocked the growth of several human cancer cell lines, such as melanoma, prostate carcinoma, ovarian carcinoma, colon carcinoma, and lung

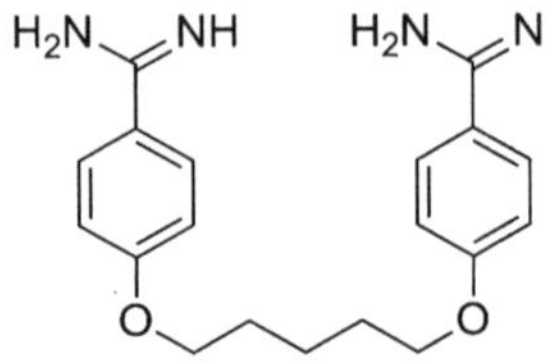

24
Pentamidine
complete inhibition for PRLs at 29 μM

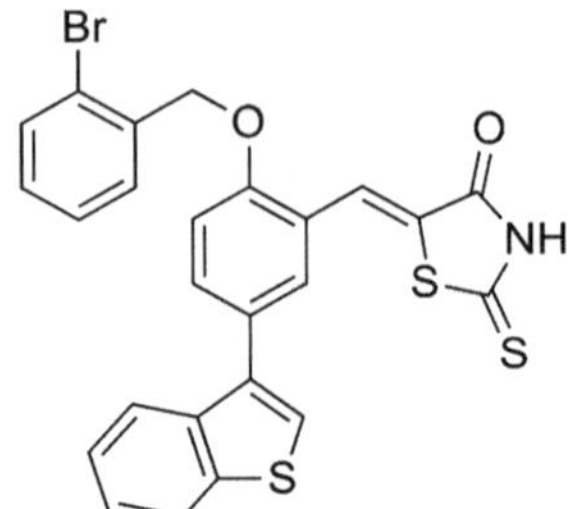

25
IC_{50} = 0.9 μM for PRL-3

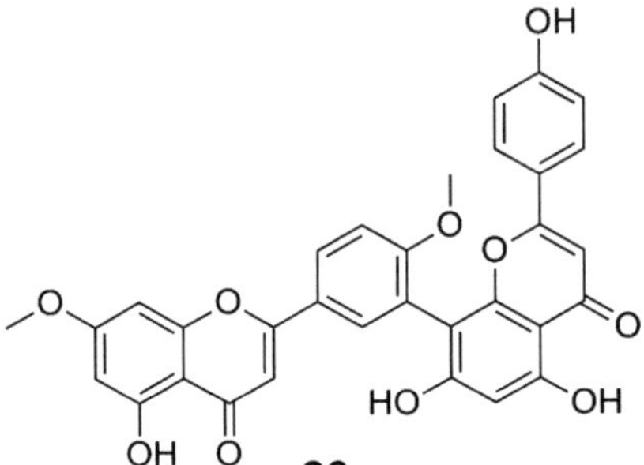

26
IC_{50} = 25.8 μM for PRL-3

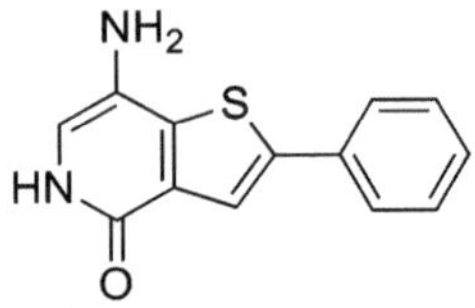

27, Thienopridone
IC_{50} = 0.173 μM for PRL-1
IC_{50} = 0.277 μM for PRL-2
IC_{50} = 0.128 μM for PRL-3

Fig. 4 Structures and activities of PRLs inhibitors

carcinoma, in a dose-dependent manner. At the highest tolerable dose (6–10 mg/kg), pentamidine inhibited the growth of melanoma tumor in nude mice induced by WM9 cells. It was reported that the tissue concentration could reach 1.6–34 μg/g in rats at 24 h after injection of pentamidine at 4 mg/kg [132], thus pentamidine at highest tolerable dose should also inhibit other PTPs such as PTP1B (complete inhibition at 1 μg/mL). Moreover, pentamidine is a known DNA minor groove binder and has been shown to disrupt hERG protein processing [133]; its anti-tumor effects may not be purely due to PRLs inhibition. Rhodanine derivatives have been reported as PRL-3 inhibitors, the most active one being compound **25**, with an IC_{50} of 0.9 μM [134]. Treatment of B16F10 murine metastatic melanoma cells with compound **25** (2–60 μM) reduced their invasive properties. However, the specificity profile of compound **25** is not available, thus there is no evidence that it only inhibits PRL-3 in the family of PTPs. Biflavonoids isolated from the young branches of *Taxuscuspidate* are active against PRL-3. To this end, ginkgetin (compound **26**) inhibits PRL-3 in vitro with an IC_{50} of 25.8 μM [135].

More recently, thienopyridone (compound **27**) was reported to inhibit all three PRLs with IC_{50}~0.2 μM [136]. Thienopyridone is at least 100-fold less active against a panel of PTPs, including PTP1B, TC-PTP, SHP2, CD45, PTPβ, and LAR, marking it as the first truly selective inhibitor against the PRL family.

Thienopyridone also blocked anchorage-independent growth of RKO and HT-29 colorectal cancer cells in soft agar, with EC_{50} values of 3.29 μM and 3.05 μM, respectively. The mechanism of action of thienopyridone reportedly involves the proteolytic cleavage of p130CAS and loss of cell-matrix interaction induced apoptosis, which is similar to siRNA-mediated PRL knockdown effects. p130CAS is required for SRC-induced oncogenic transformation and cell growth [137], and SRC has been reported to be regulated by PRL-1 and PRL-3 [112, 116], it is worth exploring further to have a better understanding of the exact signaling networks mediated by PRLs, SRC, and p130CAS. This study suggests that PRLs are attractive targets for cancer, and small molecule inhibitors are valuable for therapeutic intervention. Nevertheless, it is obvious that very limited success has been achieved for PRL inhibitor development, in contrast to the extensive study of their biological functions. Future work in design and synthesis of potent, selective, and cell permeable PRL inhibitors with diverse chemical structures would be highly desirable.

Another unique feature of PRL family is that they may form trimmers. PRL-1 trimer formation was reported by two groups and was found to be essential for PRL1-mediated cell growth and migration [109, 138, 139]. Trimerization could be a novel regulatory mechanism for PRLs, and in which case, blocking trimerization with small molecules could be a viable but distinct approach for PRL inhibition.

Conclusion

Deregulation of protein-tyrosine phosphorylation levels is a major cause of human cancers. Targeting PTKs is a proven strategy for cancer treatment, and accounts for a significant amount of drug discovery activities in pharmaceutical industry [2]. As the other class of tyrosine phosphorylation regulators, PTPs have also been strongly associated with human cancer. Table 1 summarizes the association of PTPs with cancer and the status of PTP-based therapeutics. PTP1B, a key negative regulator of insulin pathway, plays a positive role in breast cancer onset, development, and metastasis. *PTPN11* mutations lead to loss of SHP2 auto-inhibition and cause Noonan Syndrome, JMML, AML, and solid tumors. CDC25 family members play critical roles in cell cycle control, and their overexpression has been detected in many tumor samples. PRLs are linked to many cancers with metastatic properties, and their expression levels correlate with tumor invasiveness. The oncogenic properties of these PTPs have been validated in cells and animals by both RNA interfering and small molecule inhibitor approaches. One of the most striking results is that inhibiting SHP2 by shRNAmiRs in a xenograft model completely eradiated breast tumor-initiating cells, and significantly reduced the size of established breast tumors. Small molecules targeting these PTPs thus confer broad opportunities for cancer treatments. Although various PTP1B inhibitors are available, their specificity and bioavailability remain lacking. Moreover, in contrast to the extensive studies of their anti-diabetic effects, the anti-cancer properties of PTP1B inhibitors have not been widely pursued.

Table 1 The association of PTPs with cancer and the status of PTP-based therapeutics

PTP	Association with cancer	Therapeutic status and comments	Key references
PTP1B	PTP1B is capable of activating c-SRC by dephosphorylating its inhibitory Tyr527 in breast cancer cells; PTP1B promotes oncogenic RAS-ERK pathway by dephosphorylating p62Dok and activating p120Ras GAP; PTP1B promotes breast cancer tumorigenesis and metastasis in mice	Most PTP1B inhibitors are developed for type 2 diabetes and obesity, their use for cancer treatment is scarce; Compound **5** has been demonstrated to delay breast tumor onset	[25–27, 29, 30, 39]
SHP2	SHP2 activity is required for the activation of RAS-ERK signaling pathway; SHP2 gain-of-function mutations have been identified in Noonan Syndrome (NS), juvenile myelomonocytic leukemia (JMML), acute myelogenous leukemia (AML), and several types of solid tumors	SHP2 shRNAmiRs blocks breast cancer growth and metastasis; small molecule inhibitors have been tested in various cell and animal models, but none of them have optimal potency and specificity for drug development	[49, 50, 54, 60, 62, 65, 66, 74, 76]
CDC25	CDC25 dephosphorylates negative regulatory residues in cyclin-dependent kinases (CDKs) thus promoting cell cycle progressions CDC25 overexpression have been found in a number of human cancers, such as breast, colon, gastric, lung, colorectal, and pancreatic	Quinones have been extensively studied as CDC25 inhibitors, some have progressed into clinical trials; however, quinones may inactivate CDC25 via their redox properties; non-quinone-based compounds should be the focus for drug development	[80, 91, 97–99, 101, 103]
PRL	PRLs activate c-SRC, ERK, PI3K/AKT, and RHO family of small G-proteins; PRL-1 and PRL-3 downregulates p53 via MDM2 stabilization; PRLs overexpression is found in various cancers, such as colon, breast, gastric, oral, and cervix	Very limited success has been obtained for PRLs inhibitor development; thienopyridone is the only selective PRL inhibitor to date, which blocks growth of colorectal cancer cells in soft agar	[105, 106, 114, 115, 120, 121, 127, 128, 130, 136]

Overall, efforts to develop inhibitors against SHP2, CDC25, and the PRLs have met only limited success. SHP2 inhibitors with excellent potency, specificity, and diverse chemical structures are urgently required for target validation. Non-quinone-based CDC25 inhibitors with potential for preclinical and clinical examinations should be in high demand. Significant research effort should also be devoted into the development of PRLs inhibitors. In addition, agents disrupting PRLs trimer formation might be a novel strategy to block PRLs functions *in vivo*. Success in obtaining small molecule inhibitors targeting the oncogenic PTPs will require challenges such as potency, specificity, and cell permeability to be overcome, and will likely need multidisciplinary approaches, such as high throughput screening, medicinal chemistry, enzymology, structure-based design, and computer modeling [140–142]. Translation of small molecular PTP inhibitors into novel anti-cancer therapeutic drugs could encounter even greater challenges associated with pharmacokinetics, efficacy, toxicity, and authority regulations. The current decade will be a critical time for such endeavors to afford the first PTP-based anti-cancer agents.

Acknowledgments This work was supported in part by National Institutes of Health Grant CA152194.

References

1. Hunter T. Cell. 1999;80:225–36.
2. Cohen P, Alessi DR. ACS Chem Biol. 2013;8:96–104.
3. Alonso A, Sasin J, Bottini N, et al. Cell. 2004;117:699–711.
4. Tonks NK. Nat Rev Mol Cell Biol. 2006;7:833–46.
5. Tonks NK, Neel BG. Curr Opin Cell Biol. 2001;13:182–95.
6. Zhang ZY. Curr Opin Chem Biol. 2001;5:416–23.
7. Hardy S, Tremblay ML. Curr Oncol. 2008;15:5–8.
8. Jiang ZX, Zhang ZY. Cancer Metastasis Rev. 2008;27:263–72.
9. Julien SG, Dubé N, Hardy S, et al. Nat Rev Cancer. 2011;11:35–49.
10. McCain DF, Zhang ZY. Tyrosine phosphatases in cancer: targets for therapeutic intervention. In: Joaquin A, Denis A, editors. Topics in current genetics, vol. 5. Berlin: Springer; 2004. p. 359–74.
11. Ostman A, Hellberg C, Böhmer FD. Nat Rev Cancer. 2006;6:307–20.
12. Brown-Shimer S, Johnson KA, Lawrence JB, et al. Proc Natl Acad Sci U S A. 1990;87:5148–52.
13. Chernoff J, Schievella AR, Jost CA, et al. Proc Natl Acad Sci U S A. 1990;87:2735–9.
14. Tonks NK, Diltz CD, Fischer EH. J Biol Chem. 1988;263:6722–30.
15. Tonks NK, Diltz CD, Fischer EH. J Biol Chem. 1988;263:6731–7.
16. Frangioni JV, Beahm PH, Shifrin V, et al. Cell. 1992;68:545–60.
17. Woodford-Thomas TA, Rhodes JD, Dixon JE. J Cell Biol. 1992;117:401–14.
18. Zhang ZY, Lee SY. Expert Opin Investig Drugs. 2003;12:223–33.
19. Ahmad F, Li PM, Meyerovitch J, et al. J Biol Chem. 1995;270:20503–8.
20. Zhang S, Zhang ZY. Drug Discov Today. 2007;12:373–81.
21. Elchebly M, Payette P, Michaliszyn E, et al. Science. 1999;283:1544–8.
22. Klaman LD, Boss O, Peroni OD, et al. Mol Cell Biol. 2000;20:5479–89.
23. Zabolotny JM, Bence-Hanulec KK, Stricker-Krongrad A, et al. Dev Cell. 2002;2:489–95.
24. Lessard L, Stuible M, Tremblay ML. Biochim Biophys Acta. 2009;1804:613–9.
25. Bjorge JD, Pang A, Fujita DJ. J Biol Chem. 2000;275:41439–46.
26. Dubé N, Cheng A, Tremblay ML. Proc Natl Acad Sci U S A. 2004;101:1834–9.
27. Julien SG, Dubé N, Read M, et al. Nat Genet. 2007;39:338–46.
28. Wiener JR, Kerns BJ, Harvey EL, et al. J Natl Cancer Inst. 1994;86:372–8.
29. Bentires-Alj M, Neel BG. Cancer Res. 2007;67:2420–4.
30. Combs AP. J Med Chem. 2010;53:2333–44.
31. You-Ten KE, Muise ES, Itié A, et al. J Exp Med. 1997;186:683–93.
32. Heinonen KM, Nestel FP, Newell EW, et al. Blood. 2004;103:3457–64.
33. Puius YA, Zhao Y, Sullivan M, et al. Proc Natl Acad Sci U S A. 1997;94:13420–5.
34. Shen K, Keng YF, Wu L, et al. J Biol Chem. 2001;276:47311–9.
35. Sun JP, Fedorov AA, Lee SY, et al. J Biol Chem. 2003;278:12406–14.
36. Xie L, Lee SY, Andersen JN, et al. Biochemistry. 2003;42:12792–804.
37. Lee SY, Liang F, Guo XL, et al. Angew Chem Int Ed Engl. 2005;44:4242–4.

38. Boutselis IG, Yu X, Zhang ZY, et al. J Med Chem. 2007;50:856–64.
39. Han Y, Belley M, Bayly CI, et al. Bioorg Med Chem Lett. 2008;18:3200–5.
40. Lantz KA, Hart SG, Planey SL, et al. Obesity. 2010;18:1516–23.
41. Mclane M, Ruiz-White, I, Wolfe, H, et al. PCT Int Appl WO 2009/032321 (2009).
42. Shrestha S, Bhattarai BR, Cho H, et al. Bioorg Med Chem Lett. 2007;17:2728–30.
43. Wiesmann C, Barr KJ, Kung J, et al. Nat Struct Mol Biol. 2004;11:730–7.
44. Bhanot S, Monia BP, Geary RS, et al. PCT Int Appl WO 2006/044531 (2006).
45. Geary RS, Bradley JD, Watanabe T, et al. Clin Pharmacokinet. 2006;45:789–801.
46. Rondinone CM, Trevillyan JM, Clampit J, et al. Diabetes. 2002;51:2405–11.
47. He R, Zeng LF, He Y, et al. Recent advances in PTP1B inhibitor development for the treatment of type-2 diabetes and obesity. In: Jones RM, editor. New therapeutic strategies for type 2 diabetes: small molecule approaches. London: RSC; 2012.
48. Chan RJ, Feng GS. Blood. 2007;109:862–7.
49. Grossmann KS, Rosário M, Birchmeier C, et al. Adv Cancer Res. 2010;106:53–89.
50. Mohi MG, Neel BG. Curr Opin Genet Dev. 2007;7:23–30.
51. Neel BG, Gu H, Pao L. Trends Biochem Sci. 2003;28:284–93.
52. Pao LI, Badour K, Siminovitch KA, et al. Annu Rev Immunol. 2007;25:473–523.
53. Freeman Jr RM, Plutzky J, Neel BG. Proc Natl Acad Sci U S A. 1992;89:11239–43.
54. Bentires-Alj M, Paez JG, David FS, et al. Cancer Res. 2004;64:8816–20.
55. Kratz CP, Niemeyer CM, Castleberry RP, et al. Blood. 2005;106:2183–5.
56. Loh ML, Reynolds MG, Vattikuti S, et al. Leukemia. 2004;18:1831–4.
57. Loh ML, Vattikuti S, Schubbert S, et al. Blood. 2004;103:2325–31.
58. Tartaglia M, Martinelli S, Cazzaniga G, et al. Blood. 2004;104:307–13.
59. Tartaglia M, Niemeyer CM, Fragale A. Nat Genet. 2003;34:148–50.
60. Chan G, Kalaitzidis D, Neel BG. Cancer Metastasis Rev. 2008;27:179–92.
61. Miyamoto D, Miyamoto M, Takahashi A, et al. Oncogene. 2008;27:3508–15.
62. Pugh TJ, Morozova O, Attiyeh EF, et al. Nat Genet. 2013;45:279–84.
63. Mattoon DR, et al. BMC Biol. 2004;2:24.
64. Zhang SQ, Tsiaras WG, Araki T, et al. Mol Cell Biol. 2002;22:4062–72.
65. Aceto N, Sausgruber N, Brinkhaus H, et al. Nat Med. 2012;18:529–37.
66. Chen L, Sung SS, Yip ML, et al. Mol Pharmacol. 2006;70:562–70.
67. Lawrence HR, Pireddu R, Chen L, et al. J Med Chem. 2008;51:4948–56.
68. Chen L, Pernazza D, Scott LM, et al. Biochem Pharmacol. 2010;80:801–10.
69. Hellmuth K, Grosskopf S, Lum CT, et al. Proc Natl Acad Sci U S A. 2008;105:7275–80.
70. Cheng XC, Kihara T, Ying X, et al. J Antibiot (Tokyo). 1989;42:141–4.
71. Shim JH, Lee HK, Chang EJ, et al. Proc Natl Acad Sci U S A. 2002;99:10617–22.
72. Liu S, Yu Z, Yu X, et al. Chem Biol. 2011;18:101–10.
73. Lee JH, Lee JS, Kim SE, et al. Mol Cancer Ther. 2006;5:3222–31.
74. Zhang X, He Y, Liu S, et al. J Med Chem. 2010;53:2482–93.
75. Sharma N, Kumar V, Everingham S, et al. Mol Cell Biol. 2012;32:2653–63.
76. Mali RS, Ma P, Zeng LF, et al. Blood. 2012;120:2669–78.
77. Nabinger SC, Li XJ, Ramdas B, et al. Leukemia. 2013;27:398–408.
78. Xu J, Zeng LF, Shen W, et al. Biochem Biophys Res Commun. 2013;439:586–90.
79. Boutros R, Dozier C, Ducommun B. Curr Opin Cell Biol. 2006;18:185–91.
80. Boutros R, Lobjois V, Ducommun B. Nat Rev Cancer. 2007;7:495–507.
81. Rudolph J. Biochemistry. 2007;46:3595–604.
82. Russell P, Nurse P. Cell. 1986;45:145–53.
83. Lyon MA, Ducruet AP, Wipf P, et al. Nat Rev Drug Discov. 2002;1:961–76.
84. Kristjánsdóttir K, Rudolph J. Chem Biol. 2004;11:1043–51.
85. Karlsson-Rosenthal C, Millar JB. Trends Cell Biol. 2006;16:285–92.
86. Malumbres M, Barbacid M. Trends Biochem Sci. 2005;30:630–41.
87. Blomberg I, Hoffmann I. Mol Cell Biol. 1999;19:6183–94.
88. Millar JB, Blevitt J, Gerace L. Proc Natl Acad Sci U S A. 1991;88:10500–4.
89. Gabrielli BG, Clarck JM, McCormack A, et al. J Biol Chem. 1997;272:28607–14.

90. Galaktionov K, Lee AK, Eckstein J, et al. Science. 1995;269:1575–7.
91. Ma ZQ, Chua SS, DeMayo FJ, et al. Oncogene. 1999;18:4564–76.
92. Yao Y, Slosberg ED, Wang L, et al. Oncogene. 1999;18:5159–66.
93. McCarthy PJ, Kelly-Borges M, Lobkovsky E, et al. J Am Chem Soc. 1996;118:8759–60.
94. Takahashi M, Dodo K, Sugimoto Y, et al. Bioorg Med Chem Lett. 2000;10:2571–4.
95. Brohm D, Metzger S, Bhargava A, et al. Angew Chem Int Ed Engl. 2002;41:307–11.
96. Tsuchiya A, Hirai G, Koyama Y, et al. ACS Med Chem Lett. 2012;3:294–8.
97. Lazo JS, Nemoto K, Pestell KE, et al. Mol Pharmacol. 2002;61:720–8.
98. Lazo JS, Aslan DC, Southwick EC, et al. J Med Chem. 2001;44:4042–9.
99. Brezak MC, Quaranta M, Contour-Galcera MO, et al. Mol Cancer Ther. 2005;4:1378–87.
100. Cazales M, Boutros R, Brezak MC, et al. Mol Cancer Ther. 2007;6:318–25.
101. Galcera-Contour MO, Bigg D, Prevost G, et al. PCT Int Appl WO 2006/067311 (2006).
102. Siddik ZH, editor. Checkpoint controls and targets in cancer therapy. Totowa: Humana Press; 2009. p. 265.
103. Lavecchia A, Di Giovanni C, Novellino E. Mini Rev Med Chem. 2012;12:62–73.
104. Wang X, Thomas B, Sachdeva R, et al. Proc Natl Acad Sci U S A. 2006;103:3604–9.
105. Bessette DC, Qiu D, Pallen CJ. Cancer Metastasis Rev. 2008;27:231–52.
106. Rios P, Li X, Köhn M. FEBS J. 2013;280:505–24.
107. Diamond RH, Cressman DE, Laz TM, et al. Mol Cell Biol. 1994;14:3752–62.
108. Mohn KL, Laz TM, Hsu JC, et al. Mol Cell Biol. 1991;11:381–90.
109. Sun JP, Wang WQ, Yang H, et al. Biochemistry. 2005;44:12009–21.
110. Zeng Q, Si X, Horstmann H, et al. J Biol Chem. 2000;275:21444–52.
111. Achiwa H, Lazo JS. Cancer Res. 2007;67:643–50.
112. Liang F, Liang J, Wang WQ, et al. J Biol Chem. 2007;282:5413–9.
113. Min SH, Kim DM, Heo YS, et al. Oncogene. 2009;28:545–54.
114. Jiang Y, Liu XQ, Rajput A, et al. Cancer Res. 2011;71:234–44.
115. Min SH, Kim DM, Heo YS, et al. Life Sci. 2010;86:66–72.
116. Luo Y, Liang F, Zhang ZY. Biochemistry. 2009;48:1838–46.
117. Peng L, Xing X, Li W, et al. Mol Cancer. 2009;8:110.
118. Bai Y, Luo Y, Liu S, et al. J Biol Chem. 2011;286:42316–24.
119. Fiordalisi JJ, Keller PJ, Cox AD. Cancer Res. 2006;66:3153–61.
120. Wang H, Quah SY, Dong JM, et al. Cancer Res. 2007;67:2922–6.
121. Wang Y, Li ZF, He J, et al. Int J Colorectal Dis. 2007;22:1179–84.
122. Nakashima M, Lazo JS. J Pharmacol Exp Ther. 2010;334:627–33.
123. Stephens B, Han H, Hostetter G. Mol Cancer Ther. 2008;7:202–10.
124. Hardy S, Wong NN, Muller WJ, et al. Cancer Res. 2010;70:8959–67.
125. Wang Y, Lazo JS. Oncogene. 2012;31:818–27.
126. Yagi T, Morimoto A, Eguchi M, et al. Blood. 2003;102:1849–56.
127. Peng L, Ning J, Meng L, et al. J Cancer Res Clin Oncol. 2004;130:521–6.
128. Zeng Q, Dong JM, Guo K, et al. Cancer Res. 2003;63:2716–22.
129. Kato H, Semba S, Miskad UA, et al. Clin Cancer Res. 2004;10:7318–28.
130. Rouleau C, Roy A, St Martin T, et al. Mol Cancer Ther. 2006;5:219–29.
131. Pathak MK, Dhawan D, Lindner DJ, et al. Mol Cancer Ther. 2002;1:1255–64.
132. Farinotti R, Pascaud MB, Girard PM, et al. J Infect Dis. 1989;160:507–12.
133. Stephens BJ, Han H, Gokhale V, et al. Mol Cancer Ther. 2005;4:1653–61.
134. Ahn JH, Kim SJ, Park WS, et al. Bioorg Med Chem Lett. 2006;16:2996–9.
135. Choi SK, Oh HM, Lee SK, et al. Nat Prod Res. 2006;20:341–6.
136. Daouti S, Li WH, Qian H, et al. Cancer Res. 2008;68:1162–9.
137. Defilippi P, Di Stefano P, Cabodi S. Trends Cell Biol. 2006;16:257–63.
138. Jeong DG, Kim SJ, Kim JH, et al. J Mol Biol. 2005;345:401–13.
139. Sun JP, Luo Y, Yu X, et al. J Biol Chem. 2007;282:29043–51.
140. Bialy L, Waldmann H. Angew Chem Int Ed Engl. 2005;44:3814–39.
141. Hooft van Huijsduijnen R, Bombrun A, Swinnen D. Drug Discov Today. 2002;7:1013–9.
142. Vintonyak VV, Antonchick AP, Rauh D, et al. Curr Opin Chem Biol. 2009;13:272–83.

Index

B.G. Neel, N. Tonks (eds.), *Protein Tyrosine Phosphatases in Cancer*,
DOI 10.1007/978-1-4939-3649-6

R

MIX
Papier aus verantwortungsvollen Quellen
Paper from responsible sources
FSC® C105338

If you have any concerns about our products,
you can contact us on
ProductSafety@springernature.com

In case Publisher is established outside the EU,
the EU authorized representative is:
Springer Nature Customer Service Center GmbH
Europaplatz 3, 69115 Heidelberg, Germany

Printed by Libri Plureos GmbH
in Hamburg, Germany